Maple by Example

Third Edition

Maple by Example

Third Edition

Martha L. Abell and James P. Braselton

ELSEVIER
ACADEMIC
PRESS

Amsterdam Boston Heidelberg London New York Oxford
Paris San Diego San Francisco Singapore Sydney Tokyo

Senior Acquisition Editor	*Barbara Holland*
Project Manager	*Brandy Lilly*
Associate Editor	*Tom Singer*
Marketing Manager	*Linda Beattie*
Cover Design	*Eric DeCicco*
Composition	*Cepha*
Cover Printer	*Phoenix Color*
Interior Printer	*Maple Vail Book Manufacturing Group*

Elsevier Academic Press
30 Corporate Drive, Suite 400, Burlington, MA 01803, USA
525 B Street, Suite 1900, San Diego, California 92101-4495, USA
84 Theobald's Road, London WC1X 8RR, UK

This book is printed on acid-free paper. ♾

Library of Congress Cataloging-in-Publication Data
Application submitted

British Library Cataloguing in Publication Data
A catalogue record for this book is available from the British Library

ISBN: 0-12-088526-3

For all information on all Elsevier Academic Press Publications
visit our Web site at www.books.elsevier.com

PRINTED IN THE UNITED STATES OF AMERICA
05 06 07 08 09 10 9 8 7 6 5 4 3 2 1

Contents

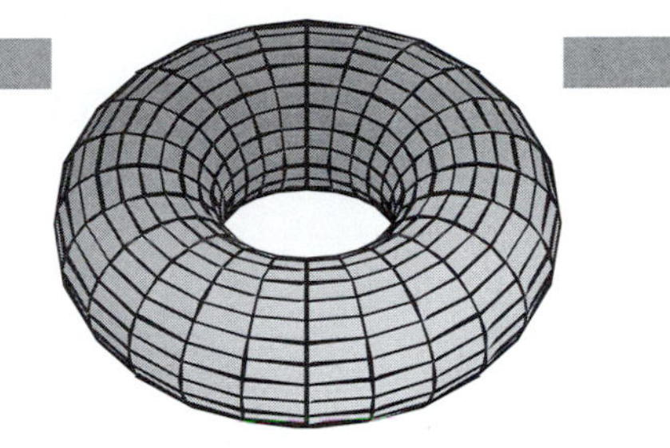

Preface

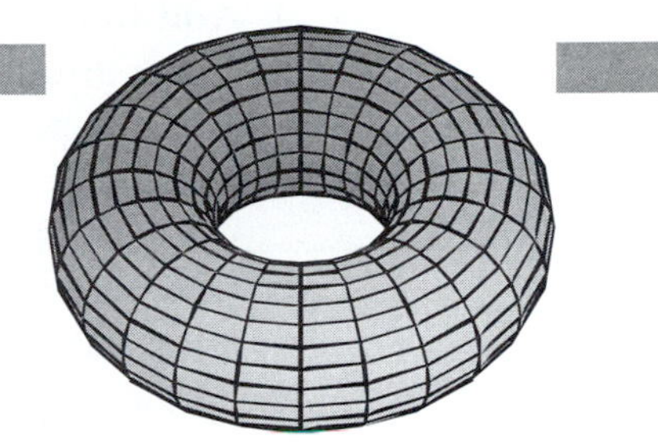

Maple by Example bridges the gap that exists between the very elementary handbooks available on Maple and those reference books written for the advanced Maple users. *Maple by Example* is an appropriate reference for all users of Maple and, in particular, for beginning users like students, instructors, engineers, business people, and other professionals first learning to use Maple. *Maple by Example* introduces the very basic commands and includes typical examples of applications of these commands. In addition, the text also includes commands useful in areas such as calculus, linear algebra, business mathematics, ordinary and partial differential equations, and graphics. In all cases, however, examples follow the introduction of new commands. Readers from the most elementary to advanced levels will find that the range of topics covered addresses their needs.

Taking advantage of Version 9 of Maple, *Maple by Example*, Third Edition, introduces the fundamental concepts of Maple to solve typical problems of interest to students, instructors, and scientists. Other features to help make *Maple by Example*, Third Edition, as easy to use and as useful as possible include the following.

1. **Version 9 Compatibility.** All examples illustrated in *Maple by Example*, Third Edition, were completed using Version 9 of Maple. Although most computations can continue to be carried out with earlier versions of Maple, like Versions 5–8, we have taken advantage of the new features in Version 9 as much as possible.
2. **Applications.** New applications, many of which are documented by references, from a variety of fields, especially biology, physics, and engineering, are included throughout the text.
3. **Detailed Table of Contents.** The table of contents includes all chapter, section, and subsection headings. Along with the comprehensive index, we hope that users will be able to locate information quickly and easily.

4. **Additional Examples.** We have considerably expanded the topics in Chapters 1 through 6. The results should be more useful to instructors, students, business people, engineers, and other professionals using Maple on a variety of platforms. In addition, several sections have been added to help make locating information easier for the user.
5. **Comprehensive Index.** In the index, mathematical examples and applications are listed by topic, or name, as well as commands along with frequently used options: particular mathematical examples as well as examples illustrating how to use frequently used commands are easy to locate. In addition, commands in the index are cross-referenced with frequently used options. Functions available in the various packages are cross-referenced both by package and alphabetically.
6. **Included CD.** All Maple code that appears in *Maple by Example*, Third Edition, is included on the CD packaged with the text.

We began *Maple by Example* in 1991 and the first edition was published in 1992. Back then, we were on top of the world using Macintosh IIcx's with 8 megs of RAM and 40 meg hard drives. We tried to choose examples that we thought would be relevant to beginning users – typically in the context of mathematics encountered in the undergraduate curriculum. Those examples could also be carried out by Maple in a timely manner on a computer as powerful as a Macintosh IIcx.

Now, we are on top of the world with Power Macintosh G4's with 768 megs of RAM and 50 gig hard drives, which will almost certainly be obsolete by the time you are reading this. The examples presented in *Maple by Example* continue to be the ones that we think are most similar to the problems encountered by beginning users and are presented in the context of someone familiar with mathematics typically encountered by undergraduates. However, for this third edition of *Maple by Example* we have taken the opportunity to expand on several of our favorite examples because the machines now have the speed and power to explore them in greater detail.

Other improvements to the third edition include:

1. Throughout the text, we have attempted to eliminate redundant examples and added several interesting ones. The following changes are especially worth noting.

   (a) In Chapter 2, we have increased the number of parametric and polar plots in two and three dimensions. For a sample, see Examples 2.3.8, 2.3.9, 2.3.10, 2.3.11, 2.3.17, and 2.3.18.

   (b) In Chapter 3, Calculus, we have added examples dealing with parametric and polar coordinates to every section. Examples 3.2.9, 3.3.9, and 3.3.10 are new examples worth noting.

(c) Chapter 4, Introduction to Lists and Tables, contains several new examples illustrating various techniques of how to quickly create plots of bifurcation diagrams, Julia sets, and the Mandelbrot set. See Examples 4.1.7, 4.2.5, 4.2.7, 4.4.6, 4.4.7, 4.4.8, 4.4.9, 4.4.10, 4.4.11, 4.4.12, and 4.4.13.

(d) Several examples illustrating how to determine graphically if a surface is nonorientable have been added to Chapter 5, Matrices and Vectors. See especially Examples 5.5.8 and 5.5.9.

(e) Chapter 6, Differential Equations, has been completely reorganized. More basic – and more difficult – examples have been added throughout.

2. We have included references that we find particularly interesting in the **Bibliography**, even if they are not specific Maple-related texts. A comprehensive list of Maple-related publications can be found at the Maple website.

`http://www.maplesoft.com/publications/`

Finally, we must express our appreciation to those who assisted in this project. We would like to express appreciation to our editors, Tom Singer and Barbara Holland, and our production editor, Brandy Lilly, at Academic Press for providing a pleasant environment in which to work. In addition, Frances Morgan, our project manager at Keyword Typesetting Services, deserves thanks for making the production process run smoothly. Finally, we thank those close to us, especially Imogene Abell, Lori Braselton, Ada Braselton, and Mattie Braselton for enduring with us the pressures of meeting a deadline and for graciously accepting our demanding work schedules. We certainly could not have completed this task without their care and understanding.

Martha Abell (email: `martha@georgiasouthern.edu`)
James Braselton (email: `jbraselton@georgiasouthern.edu`)
Statesboro, Georgia
June, 2004

1 Getting Started

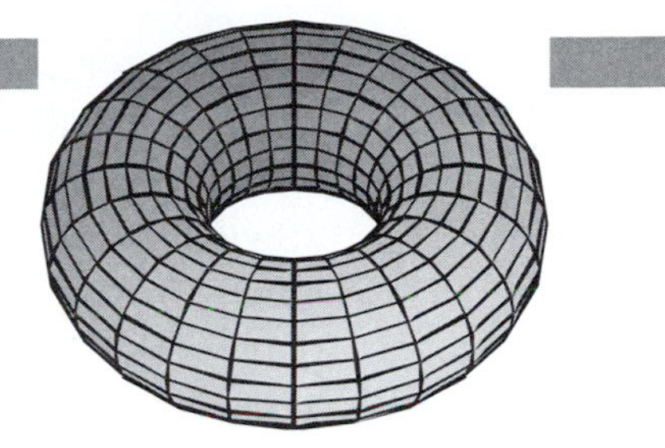

1.1 Introduction to Maple

Maple, first released in 1981 by Waterloo Maple, Inc.,

`http://www.maplesoft.com/`,

is a system for doing mathematics on a computer. Maple combines symbolic manipulation, numerical mathematics, outstanding graphics, and a sophisticated programming language. Because of its versatility, Maple has established itself as the computer algebra system of choice for many computer users including commercial and government scientists and engineers, mathematics, science, and engineering teachers and researchers, and students enrolled in mathematics, science, and engineering courses. However, due to its special nature and sophistication, beginning users need to be aware of the special syntax required to make Maple perform in the way intended. You will find that calculations and sequences of calculations most frequently used by beginning users are discussed in detail along with many typical examples. In addition, the comprehensive index not only lists a variety of topics but also cross-references commands with frequently used options. *Maple by Example* serves as a valuable tool and reference to the beginning user of Maple as well as to the more sophisticated user, with specialized needs.

For information, including purchasing information, about Maple contact:

Corporate Headquarters:
Maplesoft
615 Kumpf Drive, Waterloo
Ontario, Canada N2V 1K8
telephone: 519-747-2373
fax: 519-747-5284

email: info@maplesoft.com
web: http://www.maplesoft.com

Europe:
Maplesoft Europe GmbH
Grienbachstrasse 11
CH-6300 Zug
Switzerland
telephone: +41-(0)41-763.33.11
fax: +41-(0)41-763.33.15
email: info-europe@maplesoft.com

A Note Regarding Different Versions of Maple

With the release of Version 9 of Maple, many new functions and features have been added to Maple. We encourage users of earlier versions of Maple to update to Version 9 as soon as they can. All examples in *Maple by Example*, Third Edition, were completed with Version 9. In most cases, the same results will be obtained if you are using earlier versions of Maple, although the appearance of your results will almost certainly differ from that presented here. Occasionally, however, particular features of Version 9 are used and in those cases, of course, these features are not available in earlier versions. If you are using an earlier or later version of Maple, your results may not appear in a form identical to those found in this book: some commands found in Version 9 are not available in earlier versions of Maple; in later versions some commands will certainly be changed, new commands added, and obsolete commands removed.

On-line help for upgrading older versions of Maple and installing new versions of Maple is available at the Maple website:

http://www.maplesoft.com/.

1.1.1 Getting Started with Maple

We begin by introducing the essentials of Maple. The examples presented are taken from algebra, trigonometry, and calculus topics that you are familiar with to assist you in becoming acquainted with the Maple computer algebra system.

We assume that Maple has been correctly installed on the computer you are using. If you need to install Maple on your computer, please refer to the documentation that came with the Maple software package.

Start Maple on your computer system. Using Windows or Macintosh mouse or keyboard commands, activate the Maple program by selecting the Maple icon or an existing Maple document (or worksheet), and then clicking or double-clicking on the icon.

Maple worksheets are platform-independent and can be exchanged by users of different platforms. Even the appearance of Maple worksheets looks the same across platforms. To illustrate, we have included screenshots for both Windows and Macintosh versions of Maple throughout *Maple by Example*.

If you start Maple by selecting the Maple icon, a blank untitled worksheet is opened, as illustrated in the following screenshot.

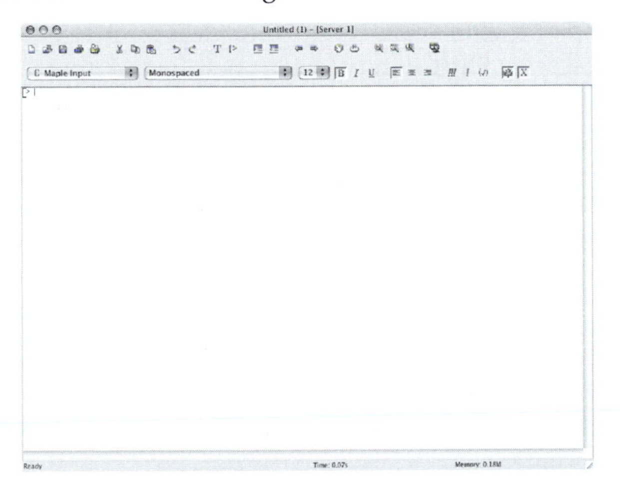

When you start typing, your typing appears to the right of the prompt.

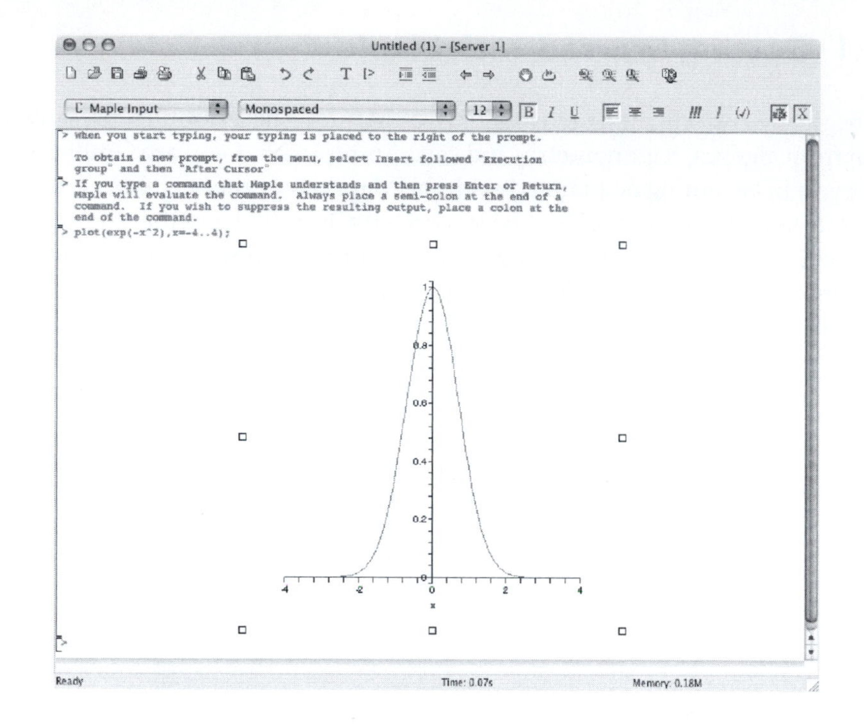

If you forget to include a semicolon (or colon) at the end of a command, Maple will remind you that you have forgotten it but try to evaluate the command anyway.

With some operating systems, **Enter** evaluates commands and **Return** yields a new line.

Once Maple has been started, computations can be carried out immediately. Maple commands are typed to the right of the prompt. End a command by placing a semicolon at the end and then evaluate the command by pressing **Enter**. If you wish to suppress the resulting output, place a colon at the end of the command instead of a semicolon. Note that pressing **Enter** or **Return** evaluates commands and pressing **Shift-Return** yields a new line. Output is displayed below input. We illustrate some of the typical steps involved in working with Maple in the calculations that follow. In each case, we type the command, end the command with a semicolon, and press **Enter**. Maple evaluates the command, displays the result, and inserts a prompt after the result. For example, typing `evalf(Pi,25);` and then pressing the **Enter** key

```
> evalf(Pi,25);
```

$$3.141592653589793238462643$$

returns a 25-digit approximation of π.

The next calculation can then be typed and entered in the same manner as the first. For example, entering

```
> plot(sin(x),2*cos(2*x),x=0..3*Pi);
```

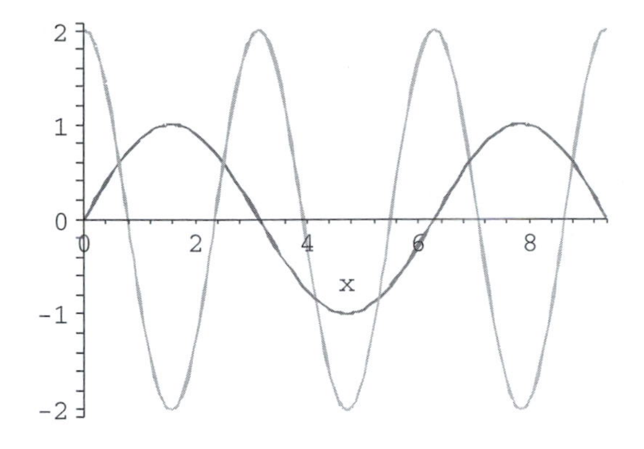

Figure 1-1 A two-dimensional plot

Figure 1-2 A three-dimensional plot

graphs the functions $y = \sin x$ and $y = 2\cos 2x$ and on the interval $[0, 3\pi]$ shown in Figure 1-1. Similarly, entering

```
> plot3d(sin(x+cos(y)),x=0..4*Pi,y=0..4*Pi);
```

graphs the function $z = \sin(x + \cos y)$ for $0 \le x \le 4\pi$ and $0 \le y \le 4\pi$ shown in Figure 1-2.

Similarly,

```
> solve(x^3-2*x+1=0);
```

$$1, -1/2 + 1/2\sqrt{5}, -1/2 - 1/2\sqrt{5}$$

solves the equation $x^3 - 2x + 1 = 0$ for x.

You can control how input and output are displayed by following the Maple menu from **Maple** to **Preferences**.

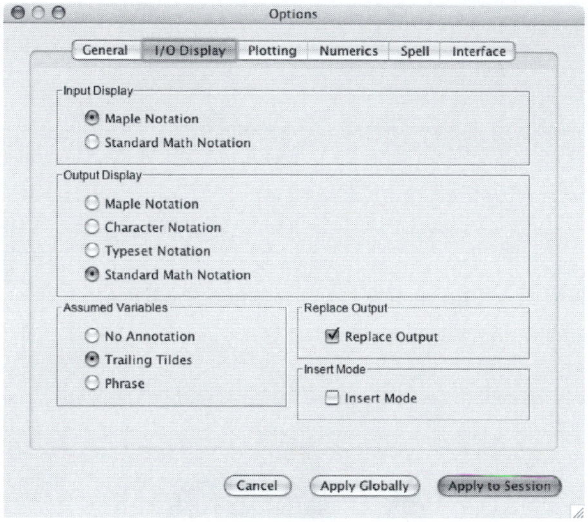

Options

General | I/O Display | Plotting | Numerics | Spell | Interface

Input Display
- (•) Maple Notation
- ( ) Standard Math Notation

Output Display
- ( ) Maple Notation
- ( ) Character Notation
- ( ) Typeset Notation
- (•) Standard Math Notation

Assumed Variables
- ( ) No Annotation
- (•) Trailing Tildes
- ( ) Phrase

Replace Output
- [x] Replace Output

Insert Mode
- [ ] Insert Mode

Cancel | Apply Globally | Apply to Session

In the following screenshot, we illustrate the appearance of output for each of the four output options.

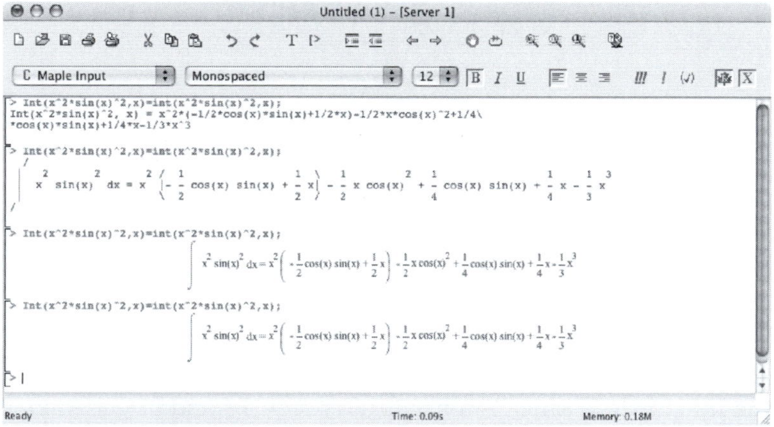

Untitled (1) - [Server 1]

C Maple Input | Monospaced | 12

```
> Int(x^2*sin(x)^2,x)=int(x^2*sin(x)^2,x);
Int(x^2*sin(x)^2, x) = x^2*(-1/2*cos(x)*sin(x)+1/2*x)-1/2*x*cos(x)^2+1/4\
*cos(x)*sin(x)+1/4*x-1/3*x^3

> Int(x^2*sin(x)^2,x)=int(x^2*sin(x)^2,x);
   /
  |   2       2         2 /  1                  1  \    1           2   1                   1     1  3
  |  x  sin(x)  dx = x   |- - cos(x) sin(x) + - x| - - x cos(x)  + - cos(x) sin(x) + - x - - x
  |                       \  2                  2  /    2               4                   4     3
 /

> Int(x^2*sin(x)^2,x)=int(x^2*sin(x)^2,x);
```

$$\int x^2 \sin(x)^2\,dx = x^2\left(-\frac{1}{2}\cos(x)\sin(x)+\frac{1}{2}x\right)-\frac{1}{2}x\cos(x)^2+\frac{1}{4}\cos(x)\sin(x)+\frac{1}{4}x-\frac{1}{3}x^3$$

```
> Int(x^2*sin(x)^2,x)=int(x^2*sin(x)^2,x);
```

$$\int x^2 \sin(x)^2\,dx = x^2\left(-\frac{1}{2}\cos(x)\sin(x)+\frac{1}{2}x\right)-\frac{1}{2}x\cos(x)^2+\frac{1}{4}\cos(x)\sin(x)+\frac{1}{4}x-\frac{1}{3}x^3$$

```
> |
```

Ready   Time: 0.09s   Memory: 0.18M

Maple sessions are terminated by selecting **Quit** from the **File** menu, or by using a keyboard shortcut, like **command-Q**, as with other applications. They can be saved by referring to **Save** from the **File** menu.

Maple allows you to save worksheets (as well as combinations of cells) in a variety of formats, in addition to the standard Maple format.

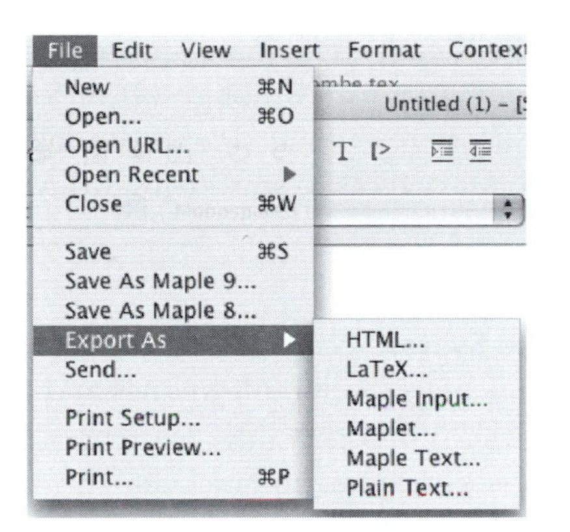

Remark. Input and text regions in worksheets can be edited. Editing input can create a worksheet in which the mathematical output does not make sense in the sequence it appears. It is also possible to simply go into a worksheet and alter input without doing any recalculation. To insert command prompts, go to the menu and select **Insert** followed by **Execution Group**.

Text ⌘T
Standard Math ⌘R
Maple Input ⌘M
Standard Math Input ⌘G
Execution Group
Paragraph
Plot
Spreadsheet...
Section
Subsection
Hyperlink...
Image...
Sketch
Page Break

You may then choose to insert an execution group before or after the cursor.

However, this can create misleading worksheets. Hence, common sense and caution should be used when editing the input regions of worksheets. Recalculating all commands in the worksheet will clarify any confusion.

Preview

In order for the Maple user to take full advantage of this powerful software, an understanding of its syntax is imperative. The goal of *Maple by Example* is to introduce the reader to the Maple commands and sequences of commands most frequently used by beginning users. Although all of the rules of Maple syntax are far too numerous to list here, knowledge of the following five rules equips the beginner with the necessary tools to start using the Maple program with little trouble.

Five Basic Rules of Maple Syntax

1. The arguments of *all* functions (both built-in ones and ones that you define) are given in parentheses `(...)`. Brackets `[...]` are used for grouping operations: vectors, matrices, and lists are given in brackets.
2. A semicolon (`;`) or colon (`:`) must be included at the end of each command. Maple does not display the result when a colon is included at the end of a command. Never name a user-defined object with the same name as that of a built-in Maple object.
3. Multiplication is represented by an asterisk, `*`. Enter `2*x*y` to evaluate $2xy$ *not* `2xy`.
4. Powers are denoted by a `^`. Enter `(8*x^3)^(1/3)` to evaluate $(8x^3)^{1/3} = 8^{1/3}(x^3)^{1/3} = 2x$ instead of `8*x^1/3`, which returns `8x/3`.
5. Maple follows the order of operations *exactly*. Thus, entering `(1+x)^1/x` returns $\frac{(1+x)^1}{x}$ while `(1+x)^(1/x)` returns $(1+x)^{1/x}$. Similarly, entering `x^3*x` returns $x^3 \cdot x = x^4$ while entering `x^(3*x)` returns x^{3x}.

   Remark. If you get no response or an incorrect response, you may have entered or executed the command incorrectly. In some cases, the amount of memory allocated to Maple can cause a crash. Like people, Maple is not perfect and errors can occur.

1.2 Loading Packages

Although Maple contains many built-in functions, some other functions are contained in **packages** that must be loaded separately. A tremendous number of additional commands are available in various packages that are shipped with each version of Maple. Experienced users can create their own packages; other packages are available from user groups and Maplesoft, which electronically distributes Maple-related products. Also see

`http://www.mapleapps.com/`

Enter `index[packages]` at the prompt to see a list of the standard packages.

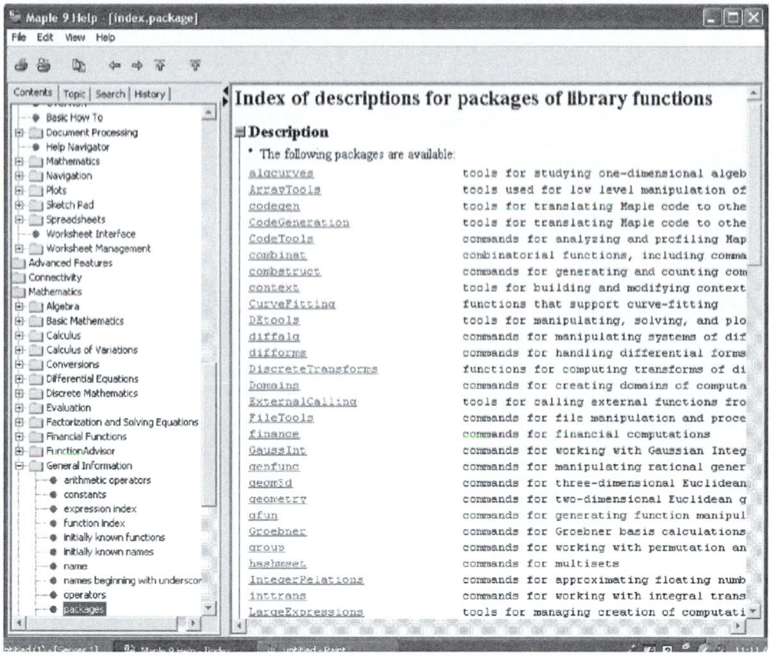

Information regarding the packages in each category is obtained by clicking on the package name from the **Help Browser**'s menu.

Commands that are contained in packages can be entered in their *long form* or, after the particular package has been loaded, in their *short form*. For example, the `display` command, which allows us to show multiple graphics together, is contained in the `plots` package. The long form of this command is

```
plots[display](arguments).
```

On the other hand, after the `plots` package has been loaded, you can use the short form:

```
display(arguments).
```

Much work is done by trial and error so our convention throughout *Maple by Example* is to load a package when we need it rather than repeatedly re-enter commands in their long form.

Packages are loaded by entering the command

```
with(packagename).
```

For example, to load the `plots` and `plottools` packages,

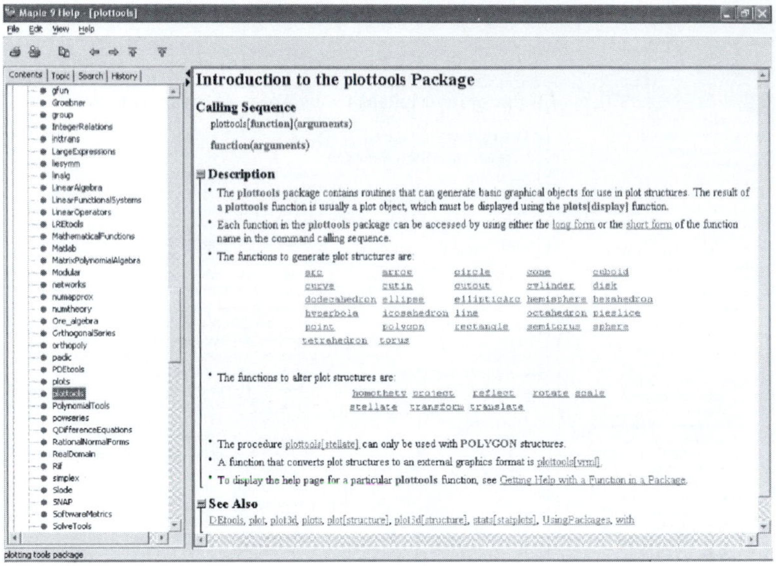

we enter

```
> with(plots):
> with(plottools);
```

[*arc*, *arrow*, *circle*, *cone*, *cuboid*, *curve*, *cutin*, *cutout*, *cylinder*, *disk*, *dodecahedron*, *ellipse*, *ellipticArc*, *hemisphere*, *hexahedron*, *homothety*, *hyperbola*, *icosahedron*, *line*, *octahedron*, *pieslice*, *point*, *polygon*, *project*, *rectangle*, *reflect*, *rotate*, *scale*, *semitorus*, *sphere*, *stellate*, *tetrahedron*, *torus*, *transform*, *translate*, *vrml*]

In this case, the commands contained in the `plottools` package are displayed because we have included a semicolon at the end of the command; the commands contained in the `plots` package are not displayed because we have included a colon at the end of the command. After the `plottools` package has been loaded, entering

```
> display(torus(1,0.5,grid=[30,30]),
> scaling=constrained);
```

generates the graph of a torus shown in Figure 1-3. Note that `torus` is contained in the `plottools` package and `display` is contained in the `plots` package. Next, we generate an icosahedron and a sphere and display the two side-by-side in Figure 1-4.

Figure 1-3 A torus created with `torus`

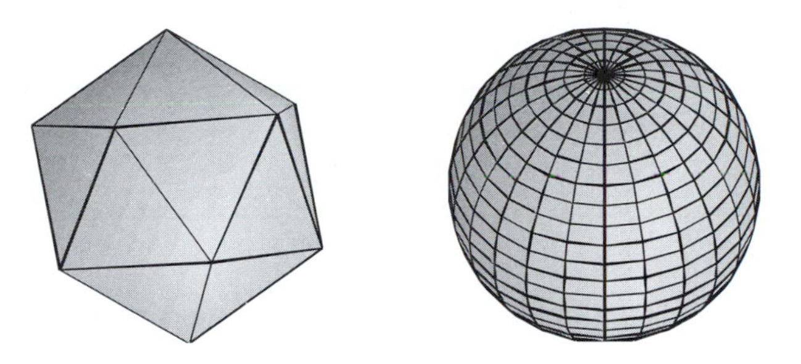

Figure 1-4 An icosahedron and a sphere

```
> display(sphere(grid=[30,30]),scaling=constrained);

> display(icosahedron(1,0.5,grid=[30,30]),
> scaling=constrained);
```

The `plottools` package contains definitions of familiar three-dimensional shapes. In addition, it contains tools that allow us to perform transformations like rotations and translations on three-dimensional graphics.

In *Maple by Example,* we use the `plots`, `linalg`, and `LinearAlgebra` packages frequently. We will make occasional use of the `DEtools`, `finance`, and `PDEtools` packages, as well.

1.3 Getting Help from Maple

Becoming competent with Maple can take a serious investment of time. Hopefully, messages that result from syntax errors will be viewed lightheartedly.

Ideally, instead of becoming frustrated, beginning Maple users will find it challenging and fun to locate the source of errors. Frequently, Maple's error messages indicate where the error(s) has (have) occurred. In this process, it is natural that you will become more proficient with Maple. In addition to Maple's extensive help facilities, which are described next, a tremendous amount of information is available for all Maple users at the Maplesoft website.

`http://www.maplesoft.com/`

One way to obtain information about commands and functions, including user-defined functions, is the command `?`. `?object` gives a basic description and syntax information of the Maple object `object`.

EXAMPLE 1.3.1: Use `?` to obtain information about the command `plot`.

SOLUTION: `?plot` uses basic information about the `plot` function.

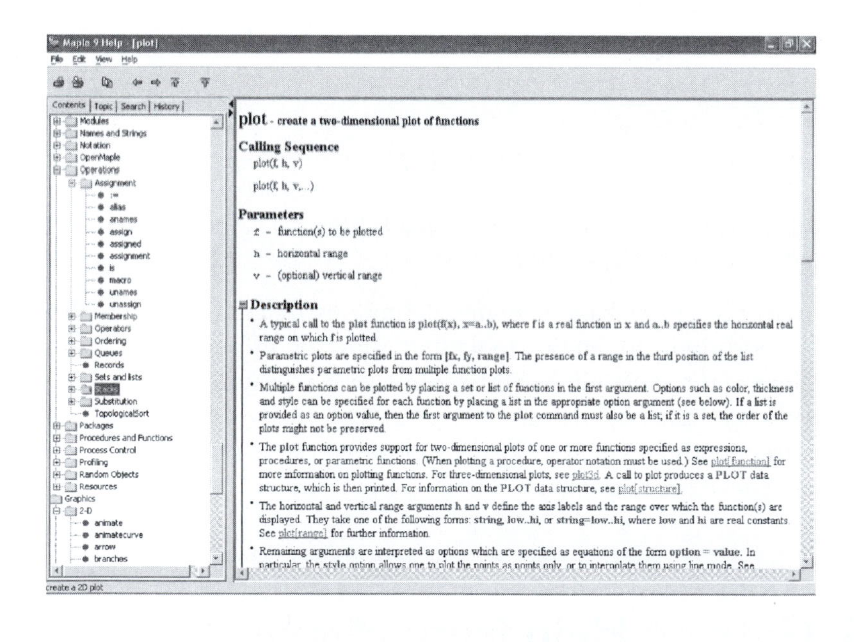

■

For packages, Maple's help facility provides links to package commands. For example, entering `?plots` returns the main help page for the `plots` package.

Maple 9 Help - [plots]

Introduction to the plots package

Calling Sequence

plots[function](arguments)

function(arguments)

Description

- Each function in the plots package can be accessed by using either the long form or the short form of the function name in the command calling sequence.
- The functions available are:

animate, animate3d, animatecurve, arrow, changecoords, complexplot, complexplot3d, conformal, conformal3d, contourplot, contourplot3d, coordplot, coordplot3d, densityplot, display, display3d, fieldplot, fieldplot3d, gradplot, gradplot3d, graphplot3d, implicitplot, implicitplot3d, inequal, interactive, listcontplot, listcontplot3d, listdensityplot, listplot, listplot3d, loglogplot, logplot, matrixplot, odeplot, pareto, plotcompare, pointplot, pointplot3d, polarplot, polygonplot, polygonplot3d, polyhedraplot, polyhedra_supported, replot, rootlocus, semilogplot, setoptions, setoptions3d, spacecurve, sparsematrixplot, surfdata, textplot, textplot3d, tubeplot

- To display the help page for a particular plots function, see Getting Help with a Function in a Package.

See Also

DEtools, plot, plot3d, plottools, stats[statplots], UsingPackages, with

The main page contains links to all commands contained in the package. Thus, clicking on `display` gives us Maple's help page for the `display` command, which is contained in the `plots` package.

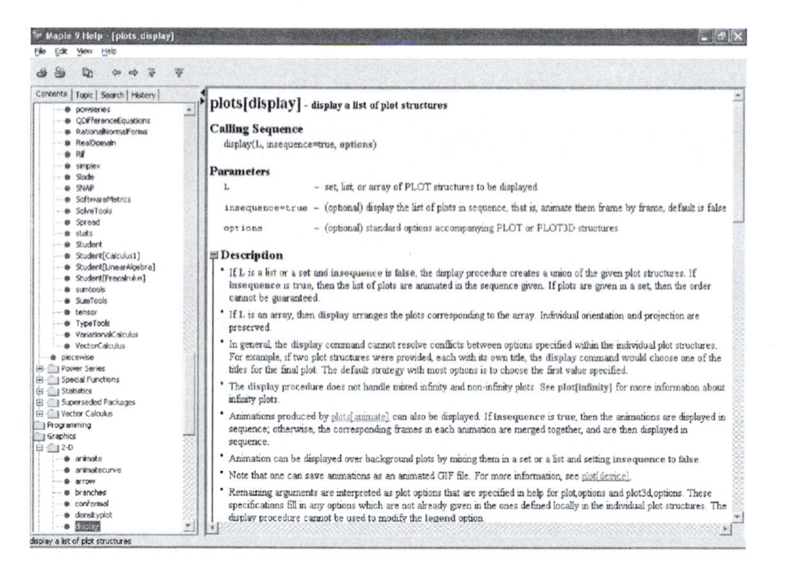

Maple Help

Additional help features are accessed from the Maple menu under **Help**. For basic information about Maple, go to the menu and select **Help**. If you are a beginning Maple user, you might choose to select **New Users** followed by **Quick Tour**

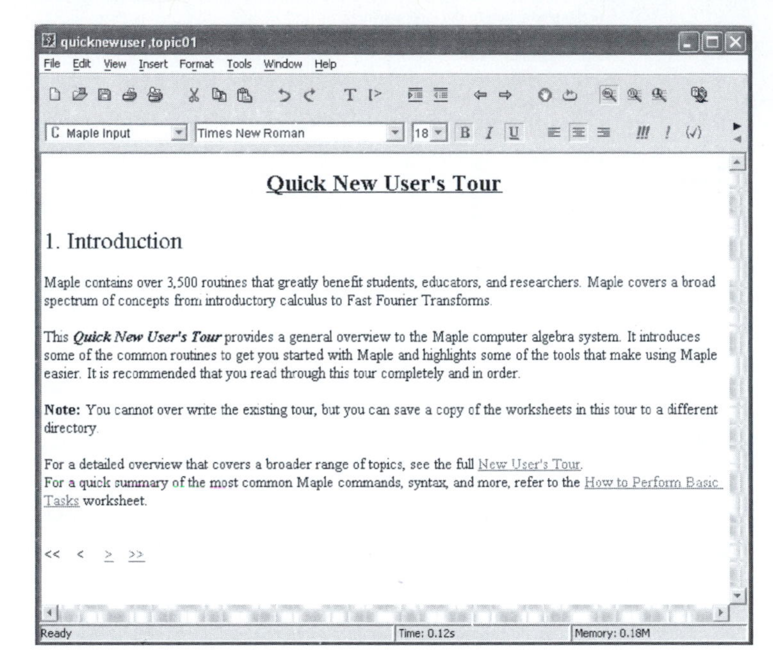

or you might select **Using Help** or **Basic How To**

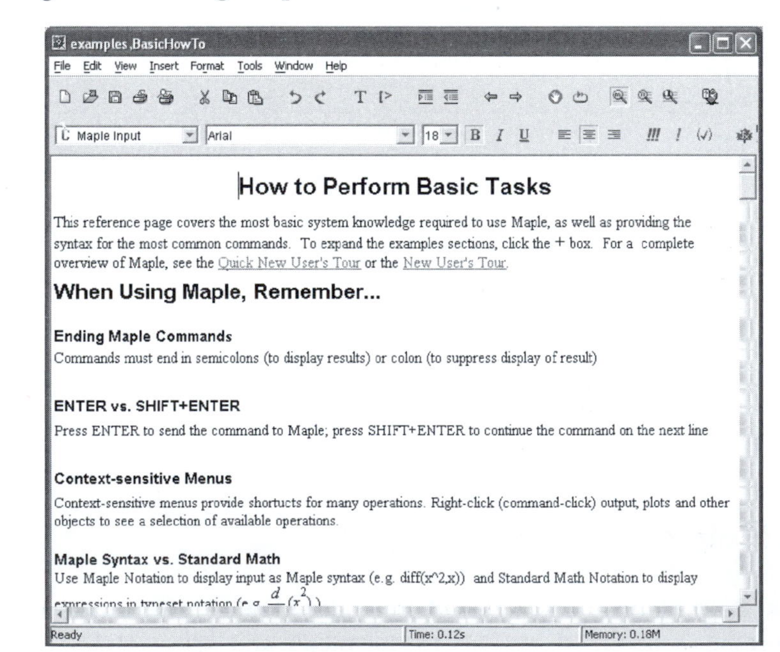

The Maple Menu

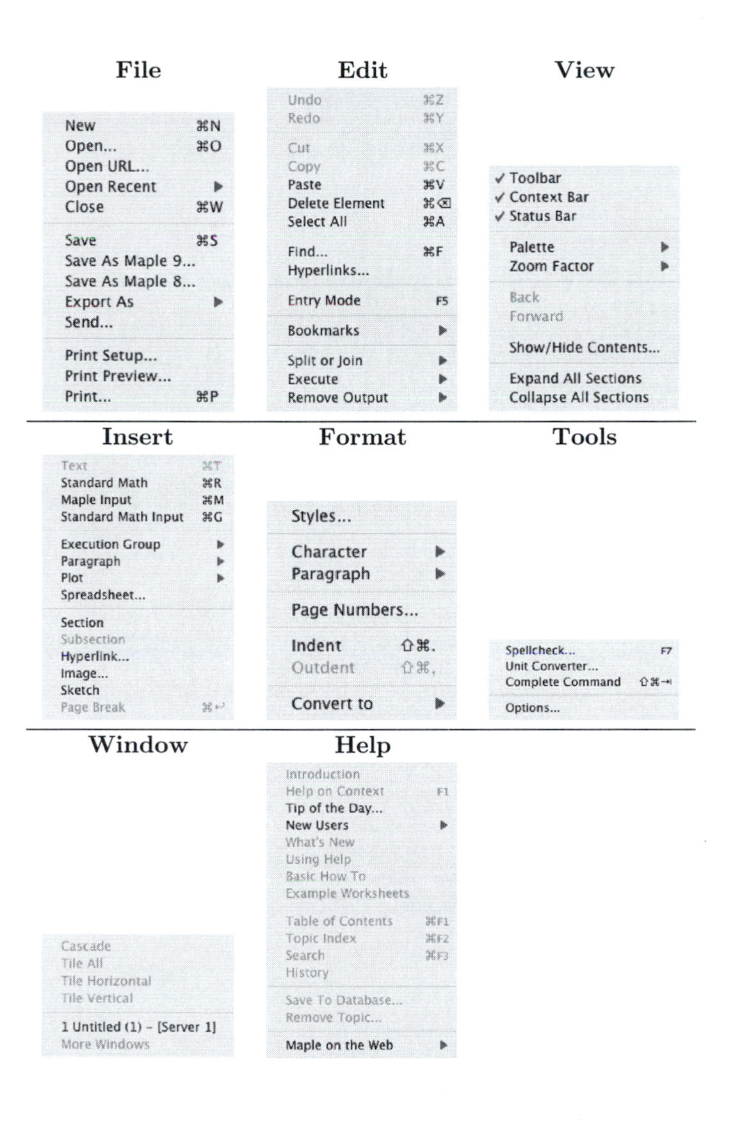

Many features of Maple worksheets can be controlled from the Maple menu. Because worksheets are platform-independent, you can format an entire document on one platform and then deliver it to an individual using a different platform and they will see the same worksheet that you do.

Within a worksheet, you can incorporate text, Maple input and output, and graphics as well as organize your work into sections, subsections, and so on.

Many features of a worksheet can be controlled from the Maple menu.

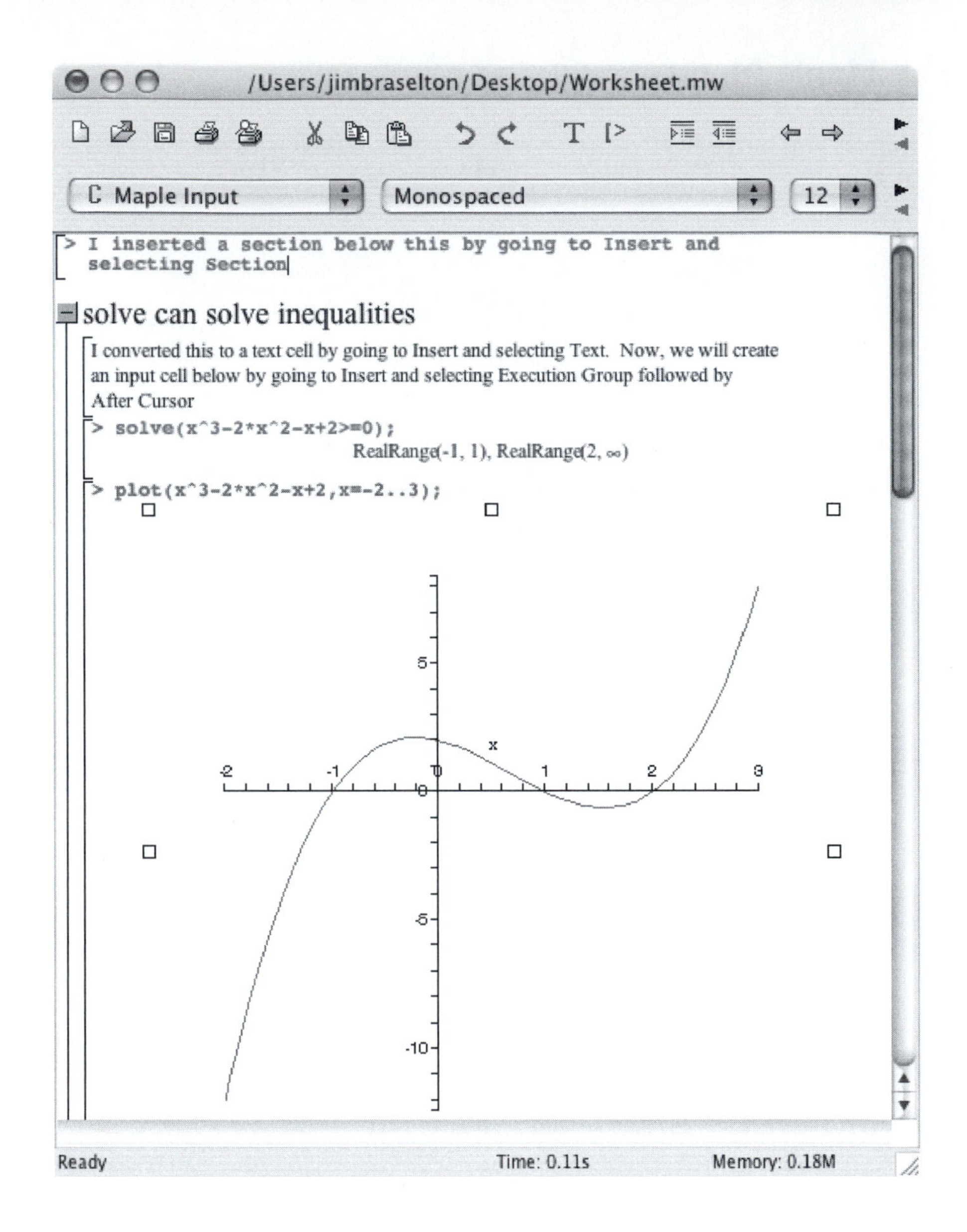

In the worksheet shown, we have inserted a section, text, Maple input, and Maple output using the formatting options available from the Maple menu.

Subsections (and sub-subsections) are inserted within a section (or subsection) by selecting **Insert** followed by **Subsection**

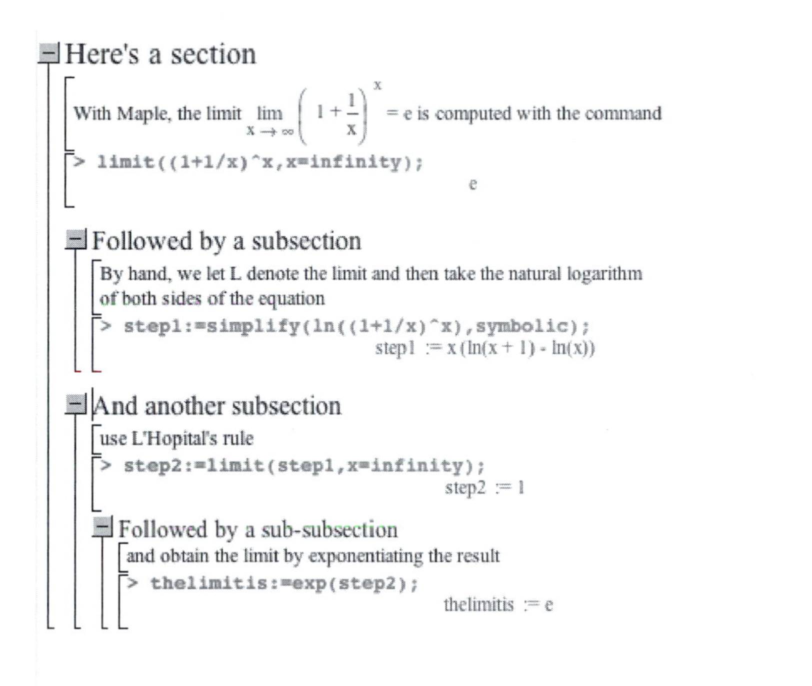

The $+/-$ toggle switch at the top of each group opens and closes the group. When the group is closed, its contents are not seen. Open the group by pressing on the $+$ icon.

Here's a section

With Maple, the limit $\lim_{x \to \infty} \left(1+\frac{1}{x}\right)^x = e$ is computed with the command

```
> limit((1+1/x)^x,x=infinity);
```

$$e$$

⊞ Followed by a subsection

⊞ And another subsection

Basic Operations on Numbers, Expressions, and Functions

2

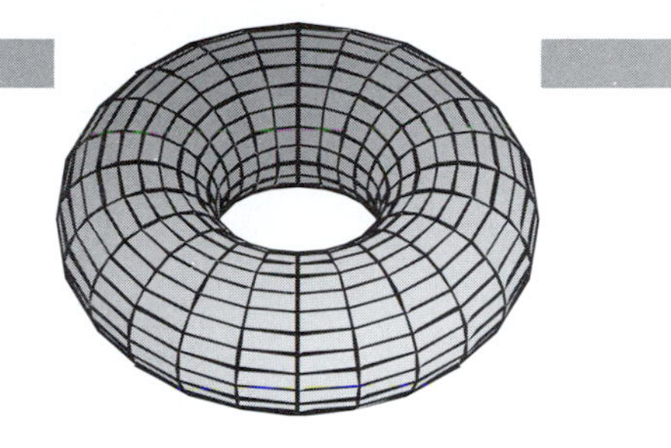

Chapter 2 introduces the essential commands of Maple. Basic operations on numbers, expressions, and functions are introduced and discussed.

2.1 Numerical Calculations and Built-In Functions

2.1.1 Numerical Calculations

The basic arithmetic operations (addition, subtraction, multiplication, division, and exponentiation) are performed in the natural way with Maple. Whenever possible, Maple gives an exact answer and reduces fractions.

1. Maple follows the standard order of operations exactly.
2. "a plus b," $a+b$, is entered as `a+b`;
3. "a minus b," $a-b$, is entered as `a-b`;
4. "a times b," ab, is entered as `a*b`;
5. "a divided by b," a/b, is entered as `a/b`. Executing the command `a/b` results in a fraction reduced to lowest terms; and
6. "a raised to the bth power," a^b, is entered as `a^b`.

When entering commands, be sure to follow the order of operations exactly and pay particular attention to nesting symbols (parentheses), multiplication operators (like `*` and the noncommutative multiplication operator, `&*`), and the exponentiation symbol (`^`).

EXAMPLE 2.1.1: Calculate (a) $121 + 542$; (b) $3231 - 9876$; (c) $(-23)(76)$; (d) $(22341)(832748)(387281)$; (e) $\dfrac{467}{31}$; and (f) $\dfrac{12315}{35}$.

SOLUTION: These calculations are carried out in the following screenshot. In (f), Maple simplifies the quotient because the numerator and denominator have a common factor of 5. In each case, the input is typed, a semicolon is placed at the end of the command, and then evaluated by pressing **Enter**.

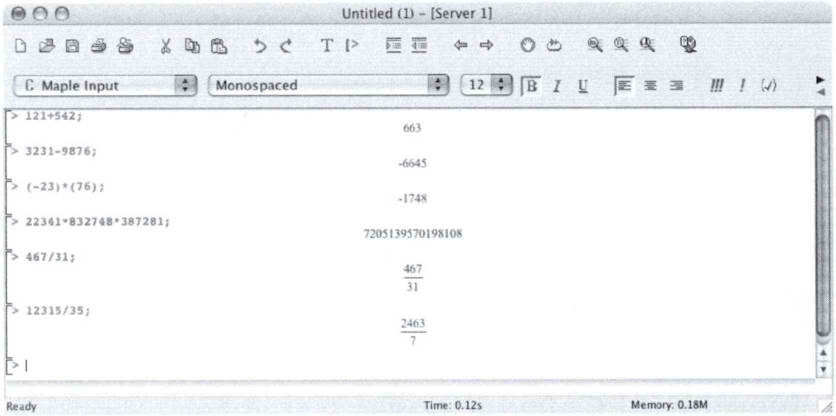

```
Untitled (1) - [Server 1]
> 121+542;
                                663
> 3231-9876;
                               -6645
> (-23)*(76);
                               -1748
> 22341*832748*387281;
                         7205139570198108
> 467/31;
                                467
                                ---
                                31
> 12315/35;
                               2463
                               ----
                                 7
> |
Ready                     Time: 0.12s            Memory: 0.18M
```

■

The term $a^{n/m} = \sqrt[m]{a^n} = \left(\sqrt[m]{a}\right)^n$ is entered as `a^(n/m)`. For $n/m = 1/2$, the command `sqrt(a)` can be used instead. Usually, the result is returned in unevaluated form but `evalf` can be used to obtain numerical approximations to virtually any degree of accuracy. With `evalf(expr,n)`, Maple yields a numerical approximation of `expr` to n digits of precision, if possible. At other times, `simplify` can be used to produce the expected results.

Remark. If the expression b in a^b contains more than one symbol, be sure that the exponent is included in parentheses. Entering `a^n/m` computes $a^n/m = \frac{1}{m}a^n$ while entering `a^(n/m)` computes $a^{n/m}$.

EXAMPLE 2.1.2: Compute (a) $\sqrt{27}$ and (b) $\sqrt[3]{8^2} = 8^{2/3}$.

SOLUTION: (a) Maple automatically simplifies $\sqrt{27} = 3\sqrt{3}$.

```
> sqrt(27);
```

$$3\sqrt{3}$$

We use `evalf` to obtain an approximation of $\sqrt{27}$.

`evalf(number)` returns a numerical approximation of `number`.

```
> evalf(sqrt(27));
```

$$5.196152424$$

(b) Maple does not automatically simplify $8^{2/3}$ so we use `simplify`. Generally,

```
simplify(expression)
```

performs routine simplification on `expression`.

```
> 8^(2/3);
```

$$8^{2/3}$$

```
> simplify(8^(2/3));
```

$$4$$

■

When computing odd roots of negative numbers, Maple's results are surprising to the novice. Namely, Maple returns a complex number. We will see that this has important consequences when graphing certain functions.

EXAMPLE 2.1.3: Calculate (a) $\frac{1}{3}\left(-\frac{27}{64}\right)^2$ and (b) $\left(-\frac{27}{64}\right)^{2/3}$.

SOLUTION: (a) Because Maple follows the order of operations, `(-27/64)^2/3` first computes $(-27/64)^2$ and then divides the result by 3.

```
> (-27/64)^2/3;
```

$$\frac{243}{4096}$$

(b) On the other hand, `(-27/64)^(2/3)` raises $-27/64$ to the $2/3$ power. Maple does not automatically simplify $\left(-\frac{27}{64}\right)^{2/3}$.

```
> (-27/64)^(2/3);
```

$$\frac{1}{64}(-27)^{2/3}\sqrt[3]{64}$$

However, when we use `simplify`, Maple returns the principal root of $\left(-\frac{27}{64}\right)^{2/3}$.

```
> simplify((-27/64)^(2/3));
```

$$\frac{9}{64}\left(1+i\sqrt{3}\right)^2$$

To obtain the result

$$\left(-\frac{27}{64}\right)^{2/3}=\left(\sqrt[3]{\frac{-27}{64}}\right)^2=\left(-\frac{3}{4}\right)^2=\frac{9}{16},$$

which would be expected by most algebra and calculus students, we use the `surd` function:

$$\text{surd}(\text{x},\text{n})=\begin{cases}x^{1/n}, & x\geq 0\\ -(-x)^{1/n}, & x<0\end{cases}.$$

Then,

```
> surd((-27/64),3);
```

$$-3/4$$

```
> surd((-27/64),3)^2;
```

$$\frac{9}{16}$$

returns the result $9/16$.

■

2.1.2 Built-In Constants

Maple has built-in definitions of many commonly used constants. In particular, $e\approx 2.71828$ is denoted by `exp(1)`, $\pi\approx 3.14159$ is denoted by `Pi`, and $i=\sqrt{-1}$ is denoted by `I`. Usually, Maple performs complex arithmetic automatically.

Other built-in constants include ∞, denoted by `infinity`, Euler's constant, $\gamma \approx 0.577216$, denoted by `gamma`, and Catalan's constant, approximately 0.915966, denoted by `Catalan`.

EXAMPLE 2.1.4: Entering

```
> evalf(exp(1),50);
```

$$2.7182818284590452353602874713526624977572470937000$$

returns a 50-digit approximation of e. Entering

```
> evalf(Pi,25);
```

$$3.141592653589793238462643$$

returns a 25-digit approximation of π. Entering

```
> (3+I)/(4-I);
```

$$\frac{11}{17} + \frac{7}{17} i$$

performs the division $(3 + i)/(4 - i)$ and writes the result in standard form.

2.1.3 Built-In Functions

Maple contains numerous mathematical functions.

Functions frequently encountered by beginning users include the exponential function, `exp(x)`; the natural logarithm, `ln(x)`; the absolute value function, `abs(x)`; the square root function, `sqrt(x)`; the trigonometric functions `sin(x)`, `cos(x)`, `tan(x)`, `sec(x)`, `csc(x)`, and `cot(x)`; the inverse trigonometric functions `arcsin(x)`, `arccos(x)`, `arctan(x)`, `arcsec(x)`, `arccsc(x)`, and `arccot(x)`; the hyperbolic trigonometric functions `sinh(x)`, `cosh(x)`, and `tanh(x)`; and their inverses `arcsinh(x)`, `arccosh(x)`, and `arctanh(x)`. Generally, Maple tries to return an exact value unless otherwise specified with `evalf`.

Several examples of the natural logarithm and the exponential functions are given next. Maple often recognizes the properties associated with these functions and simplifies expressions accordingly.

EXAMPLE 2.1.5: Entering

```
> evalf(exp(-5));
```

$$0.006737946999$$

returns an approximation of $e^{-5} = 1/e^5$. Entering

`evalf(number)` returns an approximation of `number`.
`exp(x)` computes e^x. Enter `exp(1)` to compute $e \approx 2.718$.
`ln(x)` computes $\ln x$. $\ln x$ and e^x are inverse functions ($\ln e^x = x$ and $e^{\ln x} = x$) and Maple uses these properties when simplifying expressions involving these functions.

```
> ln(exp(3));
```

$$3$$

computes $\ln e^3 = 3$. Entering

```
> exp(ln(4));
```

$$4$$

computes $e^{\ln 4} = 4$. Entering

```
> abs(-Pi);
```

$$\pi$$

computes $|-\pi| = \pi$. Entering

`abs(x)` returns the absolute value of x, $|x|$.

```
> abs((3+2*I)/(2-9*I));
```

$$\frac{1}{85}\sqrt{1105}$$

computes $|(3+2i)/(2-9i)|$. Entering

```
> sin(Pi/12);
```

$$\sin\left(1/12\,\pi\right)$$

returns $\sin(\pi/12)$ because it does not know a formula for the explicit value of $\sin(\pi/12)$. Although Maple cannot compute the exact value of $\tan 1000$, entering

```
> evalf(tan(1000));
```

$$1.470324156$$

returns an approximation of $\tan 1000$. Similarly, entering

```
> evalf(arcsin(1/3));
```

$$0.3398369094$$

returns an approximation of $\sin^{-1}(1/3)$ and entering

```
> (evalf@arccos)(2/3);
```

$$0.8410686705$$

returns an approximation of $\cos^{-1}(2/3)$, where we have used the composition operator, `@` to compose `evalf` and `arccos`: `(f@g)(x)`$=f\left(g(x)\right)$.

Maple is able to apply many identities that relate the trigonometric and exponential functions.

1. `simplify(expression,trig)` applies the circular identities to `expression`.
2. `combine(expression,trig)` applies the product to sum identities to `expression`.
3. `expand(expression)` expands `expression`; for trigonometric functions it applies the angle sum and difference identities.
4. `convert(expression,form)` tries to convert `expression` to the indicated form. For trigonometric functions, `form` is typically `sincos` (converts to sines and cosines), `exp` (converts to exponentials), or `tan` (converts to tangents).

EXAMPLE 2.1.6: Maple does not automatically apply the identity $\sin^2 x + \cos^2 x = 1$.

```
> cos(x)^2+sin(x)^2;
```

$$(\sin(x))^2 + (\cos(x))^2$$

To apply the identity, we use `simplify`. Note that in this case there is no need to include the `trig` option.

```
> simplify(cos(x)^2+sin(x)^2);
```

$$1$$

Use `expand` to multiply expressions or to rewrite trigonometric functions. In this case, entering

```
> expand(cos(3*x));
```

$$4\,(\cos(x))^3 - 3\cos(x)$$

writes $\cos 3x$ in terms of trigonometric functions with argument x. We use the `combine` function to convert products to sums.

```
> combine(sin(3*x)*cos(4*x));
```

$$1/2 \sin(7x) - 1/2 \sin(x)$$

We use `simplify` to write

```
> simplify(sin(3*x)*cos(4*x));
```

$$\left(-1 + 32 (\cos(x))^6 - 40 (\cos(x))^4 + 12 (\cos(x))^2\right) \sin(x)$$

in terms of trigonometric functions with argument x. We use `convert` with the `trig` option to convert exponential expressions to trigonometric expressions.

```
> convert(1/2*(exp(x)+exp(-x)),trig);
```

$$\cosh(x)$$

Similarly, we use `convert` with the `exp` option to convert trigonometric expressions to exponential expressions.

```
> convert(sin(x),exp);
```

$$-1/2\, i \left(e^{ix} - \left(e^{ix}\right)^{-1}\right)$$

Usually, you can use `expand` to apply elementary identities.

```
> expand(cos(2*x));
```

$$2 (\cos(x))^2 - 1$$

A Word of Caution

Remember that there are certain ambiguities in traditional mathematical notation. For example, the expression $\sin^2(\pi/6)$ is usually interpreted to mean "compute $\sin(\pi/6)$ and square the result." That is, $\sin^2(\pi/6) = [\sin(\pi/6)]^2$. The symbol sin is not being squared; the number $\sin(\pi/6)$ *is* squared. With Maple, we must be especially careful and follow the standard order of operations exactly.

2.2 Expressions and Functions: Elementary Algebra

2.2.1 Basic Algebraic Operations on Expressions

Expressions involving unknowns are entered in the same way as numbers. Maple performs standard algebraic operations on mathematical expressions. For example, the commands

1. `factor(expression)` factors `expression`;
2. `expand(expression)` multiplies `expression`;
3. `simplify(expression)` performs basic algebraic manipulations on `expression` and returns the simplest form it finds.

For basic information about any of these commands (or any other) enter `?command` as we do here for `factor`.

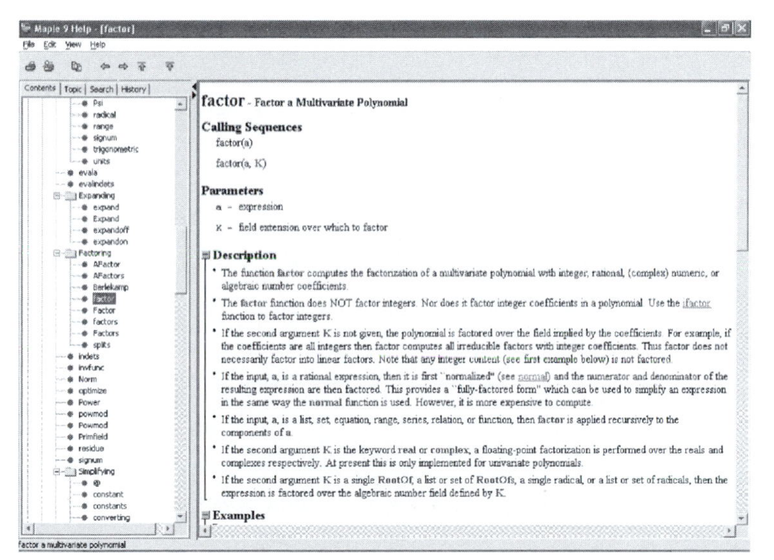

When entering expressions, be sure to include an asterisk, `*`, between variables to denote multiplication.

EXAMPLE 2.2.1: (a) Factor the polynomial $12x^2 + 27xy - 84y^2$. (b) Expand the expression $(x + y)^2(3x - y)^3$. (c) Write the sum $\dfrac{2}{x^2} - \dfrac{x^2}{2}$ as a single fraction.

SOLUTION: The result obtained with factor indicates that $12x^2 + 27xy - 84y^2 = 3(4x - 7y)(x + 4y)$. When typing the command, be sure to include an asterisk, *, between the x and y terms to denote multiplication. xy represents an expression while x*y denotes x multiplied by y.

```
> factor(12*x^2+27*x*y-84*y^2);
```

$$3\left(x + 4y\right)\left(4x - 7y\right)$$

We use expand to compute the product $(x+y)^2(3x-y)^3$ and simplify to express $\frac{2}{x^2} - \frac{2}{x^2}$ as a single fraction.

```
> expand((x+y)^2*(3*x-y)^3);
```

$$27x^5 + 27x^4y - 18x^3y^2 - 10x^2y^3 + 7xy^4 - y^5$$

```
> simplify(2/x^2-x^2/2);
```

$$-1/2\,\frac{-4 + x^4}{x^2}$$

■

factor(x^2-3) returns $x^2 - 3$.

To factor an expression like $x^2 - 3 = x^2 - (\sqrt{3})^2 = (x - \sqrt{3})(x + \sqrt{3})$, use factor and specify the extension, which in this case is $\sqrt{3}$.

```
> factor(x^2-3,sqrt(3));
```

$$(x + \sqrt{3})(x - \sqrt{3})$$

Similarly, use factor and indicate the extension I to factor expressions like $x^2 + 1 = x^2 - i^2 = (x + i)(x - i)$.

```
> factor(x^2+1,I);
```

$$(x - i)(x + i)$$

Maple does not automatically simplify $\sqrt{x^2}$ to the expression x

```
> simplify(sqrt(x^2));
```

$$csgn(x)x$$

because without restrictions on x, $\sqrt{x^2} = |x|$. The commands radsimp (expression) and simplify(expression,symbolic) simplify expression assuming that all variables are positive.

```
> simplify(sqrt(x^2),symbolic);
```

$$x$$

```
> radsimp(sqrt(x^2));
```

$$x$$

Thus, entering

```
> simplify(sqrt(a^2*b^4));
```

$$\sqrt{a^2b^4}$$

returns $\sqrt{a^2b^4}$ but entering

```
> simplify(sqrt(a^2*b^4),symbolic);
```

$$ab^2$$

returns ab^2. If x is truly positive (or negative), you can instruct Maple to assume that x is positive with the `assume` function. In this case, Maple uses a tilde, ~, to indicate that assumptions have been made about the variable.

```
> assume(x>0):
> sqrt(x^2);
```

$$x$$

When multiplying two expressions *always* include an asterisk, `*`, between the expressions being multiplied.

1. `cat*dog` means "variable `cat` times variable `dog`."
2. But, `catdog` is interpreted as a variable `catdog`.

The command `convert(expression,parfrac,variable)` computes the partial fraction decomposition of `expression` in terms of the variable `variable`. `normal(expression)` factors the numerator and denominator of `expression` then reduces `expression` to lowest terms. For a rational expression, `simplify(expression)` does the same.

EXAMPLE 2.2.2: (a) Determine the partial fraction decomposition of $\dfrac{1}{(x-3)(x-1)}$. (b) Simplify $\dfrac{x^2-1}{x^2-2x+1}$.

SOLUTION: `convert` with the `parfrac` option is used to see that

$$\frac{1}{(x-3)(x-1)} = \frac{1}{2(x-3)} - \frac{1}{2(x-1)}.$$

Then, normal is used to find that

$$\frac{x^2-1}{x^2-2x+1}=\frac{(x-1)(x+1)}{(x-1)^2}=\frac{x+1}{x-1}.$$

In this calculation, we have assumed that $x \neq 1$.

```
> convert(1/((x-3)*(x-1)),parfrac,x);
```

$$1/2\,(x-3)^{-1}-1/2\,(x-1)^{-1}$$

```
> normal((x^2-1)/(x^2-2*x+1));
```

$$\frac{x+1}{x-1}$$

■

In addition, Maple has several built-in functions for manipulating parts of fractions:

1. `numer(fraction)` yields the numerator of a `fraction`.
2. `denom(fraction)` yields the denominator of a `fraction`.

EXAMPLE 2.2.3: Given $\frac{x^3+2x^2-x-2}{x^3+x^2-4x-4}$, (a) factor both the numerator and denominator; (b) reduce $\frac{x^3+2x^2-x-2}{x^3+x^2-4x-4}$ to lowest terms; and (c) find the partial fraction decomposition of $\frac{x^3+2x^2-x-2}{x^3+x^2-4x-4}$.

SOLUTION: The numerator of $\frac{x^3+2x^2-x-2}{x^3+x^2-4x-4}$ is extracted with numer. We then use factor to factor the result of executing the numer command.

```
> numer((x^3+2*x^2-x-2)/(x^3+x^2-4*x-4));
```

$$x^3+2x^2-x-2$$

```
> factor(x^3+2*x^2-x-2);
```

$$(x-1)\,(x+2)\,(x+1)$$

Similarly, we use denom to extract the denominator of the fraction. Again, factor is used to factor the denominator of the fraction.

```
> denom((x^3+2*x^2-x-2)/(x^3+x^2-4*x-4));
```

$$x^3 + x^2 - 4x - 4$$

```
> factor(x^3+x^2-4*x-4);
```

$$(x-2)(x+2)(x+1)$$

`normal` is used to reduce the fraction to lowest terms.

```
> normal((x^3+2*x^2-x-2)/(x^3+x^2-4*x-4));
```

$$\frac{x-1}{x-2}$$

Finally, `convert` with the `parfrac` option is used to find its partial fraction decomposition.

```
> convert((x^3+2*x^2-x-2)/(x^3+x^2-4*x-4),parfrac,x);
```

$$1 + (x-2)^{-1}$$

■

2.2.2 Naming and Evaluating Expressions

In Maple, objects can be named. Naming objects is convenient: we can avoid typing the same mathematical expression repeatedly (as we did in Example 2.2.3) and named expressions can be referenced throughout a notebook or Maple session. Every Maple object can be named – expressions, functions, graphics and so on can be named with Maple. Objects are named by using a colon followed by a single equals sign (:=).

Expressions are easily evaluated using `subs`. For example, entering the command

```
subs(x=3,x^2)
```

returns the value of the expression x^2 if $x = 3$. Note, however, this does not assign the symbol x the value 3: entering `x:=3` assigns x the value 3. `eval(expression)` evaluates `expression` immediately. `evalf(expression)` attempts to numerically evaluate `expression`.

EXAMPLE 2.2.4: Evaluate $\dfrac{x^3 + 2x^2 - x - 2}{x^3 + x^2 - 4x - 4}$ if $x = 4$, $x = -3$, and $x = 2$.

SOLUTION: To avoid retyping $\dfrac{x^3+2x^2-x-2}{x^3+x^2-4x-4}$, we define f to be $\dfrac{x^3+2x^2-x-2}{x^3+x^2-4x-4}$.

Of course, you can simply copy and paste this expression if you want neither to name it nor to retype it.

```
> f:=(x^3+2*x^2-x-2)/(x^3+x^2-4*x-4);
```

$$f := \frac{x^3+2x^2-x-2}{x^3+x^2-4x-4}$$

subs is used to evaluate f if $x = 4$ and then if $x = -3$.

If you include a colon (:) at the end of the command, the resulting output is suppressed.

```
> subs(x=4,f);
```

$$3/2$$

```
> subs(x=-3,f);
```

$$4/5$$

The eval command is closely related to the subs command. Entering

```
> eval(f,x=1/2);
```

$$1/3$$

evaluates f if $x = 1/2$.

When we try to replace each x in f by -2, we see that the result is undefined: division by 0 is always undefined.

```
> eval(f,x=-2);
Error, numeric exception: division by zero
```

However, when we use simplify to first simplify and then use subs to evaluate,

```
> g:=simplify(f);
```

$$g := \frac{x-1}{x-2}$$

```
> subs(x=-2,g);
```

$$3/4$$

we see that the result is $3/4$. The result indicates that $\lim_{x\to-2} \frac{x^3+2x^2-x-2}{x^3+x^2-4x-4} = \frac{3}{4}$. We confirm this result with limit.

```
> limit(g,x=-2);
```

$$3/4$$

Generally, use `limit(f(x),x=a)` to compute $\lim_{x\to a} f(x)$. The `limit` function is discussed in more detail in Chapter 3.

■

Two Words of Caution

Be aware that Maple *does not* remember anything defined in a previous Maple session. That is, if you define certain symbols during a Maple session, quit the Maple session, and then continue later, the previous symbols must be redefined to be used. When you assign a name to an object that is similar to a previously defined or built-in function, Maple issues an error message.

2.2.3 Defining and Evaluating Functions

It is important to remember that functions, expressions, and graphics can be named anything that is not the name of a built-in Maple function or command. Because definitions of functions and names of objects are frequently modified, we introduce the command clear command: `expression:='expression'` clears all definitions of `expression`, if any. You can see if a particular symbol has a definition by entering `?symbol`.

If you wish to clear many symbols, you may find it easier to enter `restart`, which clears Maple's internal memory.

In Maple, an elementary function of a single variable, $y = f(x) = expression\ in\ x$, is typically defined using the form

```
f:=x->expression in x
```

EXAMPLE 2.2.5: Entering

```
> f:=x->x/(x^2+1);
```

$$f := x \mapsto \frac{x}{x^2+1}$$

defines and computes $f(x) = x/\left(x^2+1\right)$. Entering

```
> f(3);
```

$$3/10$$

computes $f(3) = 3/\left(3^2+1\right) = 3/10$. Entering

```
> f(a);
```

$$\frac{a}{a^2+1}$$

computes $f(a) = a/\left(a^2+1\right)$. Entering

```
> f(3+h);
```

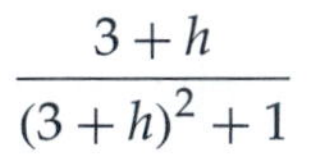

computes $f(3+h) = (3+h)/\left((3+h)^2+1\right)$. Entering

```
> n1:=simplify((f(3+h)-f(3))/h);
```

$$n1 := -1/10\,\frac{8+3h}{10+6h+h^2}$$

computes and simplifies $\dfrac{f(3+h)-f(3)}{h}$ and names the result n1. Entering

```
> subs(h=0,n1);
```

$$-\frac{2}{25}$$

evaluates n1 if $h=0$. Entering

```
> n2:=simplify((f(a+h)-f(a))/h);
```

$$n2 := -\frac{a^2-1+ah}{\left(a^2+2\,ah+h^2+1\right)\left(a^2+1\right)}$$

computes and simplifies $\dfrac{f(a+h)-f(a)}{h}$ and names the result n2. Entering

```
> subs(h=0,n2);
```

$$-\frac{a^2-1}{\left(a^2+1\right)^2}$$

evaluates n2 if $h=0$.

Often, you will need to evaluate a function for the values in a **list**,

$$\texttt{list} = [a_1, a_2, a_3, \ldots, a_n].$$

Once $f(x)$ has been defined, `map(f,list)` returns the list

$$[f(a_1), f(a_2), f(a_3), \ldots, f(a_n)]$$

Also,

The `seq` function will be discussed in more detail as needed as well as in Chapters 4 and 5.

1. `[seq(f(n),n=n1..n2)]` returns the list

$$[f(n_1), f(n_1+1), f(n_1+2), \ldots, f(n_2)]$$

2. `[seq([n,f(n)],n=n1..n2)]` returns the list of ordered pairs

$$\{(n_1, f(n_1)), (n_1+1, f(n_1+1)), (n_1+2, f(n_1+2)), \ldots, (n_2, f(n_2))\}$$

3. `[seq(f(n),n=nvals)]` returns the list consisting of $f(n)$ evaluated for each n in the list `nvals`.

EXAMPLE 2.2.6: Entering

```
> h:='h':
> h:=t->(1+t)^(1/t):
> h(1);
```

$$2$$

defines $h(t) = (1+t)^{1/t}$ and then computes $h(1) = 2$. Because division by 0 is always undefined, $h(0)$ is undefined.

```
> h(0);

Error, (in h) numeric exception: division by zero
```

However, $h(t)$ is defined for all $t > 0$. In the following, we use `rand` together with `seq` to generate 6 random numbers "close" to 0 and name the resulting list `t1`. Because we are using `rand`, your results will almost certainly differ from those here.

`rand()` returns a random 12 digit integer.

```
> t1:=[seq(evalf(rand()*10^(-n)),n=12..17)];
```

$$t1 := [0.4293926737, 0.05254285110, 0.002726006090, 0.0002197600994,$$
$$0.00006759829338, 0.000008454735095]$$

We then use map to compute $h(t)$ for each of the values in the list t1.

```
> map(h,t1);
```

$$[2.297882921, 2.650144108, 2.714585947, 2.717981974,$$
$$2.718178163, 2.718355508]$$

From the result, we suspect that $\lim_{t \to 0^+} h(t) = e$.

Remember to always include arguments of functions in parentheses.

Defining functions as procedures using proc offers more flexibility, especially for more complicated functions. For a simple function like $y = f(x) =$ *formula in terms of the variable x*,

```
f:=proc(x) formula in terms of the variable x
```

defines $y = f(x)$ as a procedure.

Remark. Remember that pressing **Enter** or **Return** evaluates commands while pressing **Shift-Return** and **Shift-Enter** give new lines so that you can continue typing Maple input.

Including a colon at the end of a command suppresses the resulting output.

EXAMPLE 2.2.7: Entering

```
> f:='f':
> f:=proc(n)
> f(n-1)+f(n-2)
> end proc:
> f(0):=1:
> f(1):=1:
```

defines the recursively defined function defined by $f(0) = 1$, $f(1) = 1$, and $f(n) = f(n-1)+f(n-2)$. For example, $f(2) = f(1)+f(0) = 1+1 = 2$; $f(3) = f(2) + f(1) = 2 + 1 = 3$. We use seq to create a list of ordered pairs $(n, f(n))$ for $n = 0, 1, \ldots, 10$.

```
> seq([n,f(n)],n=0..10);
```

$$[0,1], [1,1], [2,2], [3,3], [4,5], [5,8], [6,13], [7,21], [8,34], [9,55], [10,89]$$

In this case, the same result is obtained with

```
> f:=n->f(n-1)+f(n-2):
> f(0):=1:
```

```
> f(1):=1:
> seq([n,f(n)],n=0..10);
```

[0,1], [1,1], [2,2], [3,3], [4,5], [5,8], [6,13], [7,21], [8,34], [9,55], [10,89]

but `proc` offers more flexibility, especially when dealing with more complicated functions.

To define piecewise-defined functions, we usually use `proc` or `piecewise`. A basic piecewise-defined function like $f(t) = \begin{cases} g(t), & t \leq a \\ h(t), & t > a \end{cases}$ is defined using `piecewise` with

```
f:=t->piecewise(t<=a,g(t),t>a,h(t))
```

For more complicated functions, the pattern follows: condition followed by formula.

Remember that **Shift-Return** and **Shift-Enter** give a new line; **Return** and **Enter** evaluate Maple commands.

EXAMPLE 2.2.8: With

```
> f:=t->piecewise(t >0, sin(1/t), t <=0,-t):
> f(-1);
```

$$1$$

we have defined the piecewise-defined function

$$f(t) = \begin{cases} \sin\frac{1}{t}, & t > 0 \\ -t, & t \leq 0 \end{cases}.$$

We can now evaluate $f(t)$ for any real number t.

```
> f(1/(10*Pi));
```

$$0$$

```
> f(0);
```

$$0$$

However, $f(a)$ returns unevaluated because Maple does not know if $a \leq 0$ or if $a > 0$.

```
> f(a);
```

$$PIECEWISE\left([\sin\left(a^{-1}\right), 0 < a], [-a, a \leq 0]\right)$$

However, if you make specific assumptions about a with `assume`, Maple can evaluate. In this case, we instruct Maple to assume that $a \leq 0$. Maple is then able to evaluate $f(a)$.

```
> assume(a<=0);
> f(a);
```

$$-a$$

Virtually the same results are obtained by defining f as a procedure with `proc`.

```
> f:=proc(t)
> if t>0 then sin(1/t) else -t fi
> end proc:
> f(0);
> f(evalf(1/(10*Pi)));
```

$$0$$

$$0.000000004102067615$$

```
> f(a);
```

```
Error, (in f) cannot determine if this expression is
   true or false: -a < 0
```

Recursively defined functions are handled in the same way. The following example shows how to define a periodic function with `proc`.

End procedures with `end` or `end proc`. End an `if` statement with `fi`.

EXAMPLE 2.2.9: Entering

```
> g:='g':
> g:=proc(x)
> if x>=0 and x<1 then x
> elif x>=1 and x<2 then 1
> elif x>=2 and x<3 then 3-x
> elif x>=3 then g(x-3) fi
> end:
```

defines the recursively defined function $g(x)$. For $0 \leq x < 3$, $g(x)$ is defined by

$$g(x) = \begin{cases} x, & 0 \leq x < 1 \\ 1, & 1 \leq x < 2 \\ 3 - x, & 2 \leq x < 3 \end{cases}.$$

In the procedure, `elif` represents "else-if," which lets us avoid repeated nestings of `if...fi`. For $x \geq 3$, $g(x) = g(x - 3)$. We use `seq` to create a list of ordered pairs $(x, g(x))$ for 25 equally spaced values of x between 0 and 6.

```
> xvals:=seq(6*i/24,i=0..24):
> seq([x,g(x)],x=xvals);
```

$$[0,0],\ [1/4,1/4],\ [1/2,1/2],\ [3/4,3/4],\ [1,1],\ [5/4,1],$$
$$[3/2,1],\ [7/4,1],\ [2,1],\ [9/4,3/4],\ [5/2,1/2],$$
$$[11/4,1/4],\ [3,0],\ \left[\frac{13}{4},1/4\right],\ [7/2,1/2],\ \left[\frac{15}{4},3/4\right],$$
$$[4,1],\ \left[\frac{17}{4},1\right],\ [9/2,1],\ \left[\frac{19}{4},1\right],\ [5,1],$$
$$\left[\frac{21}{4},3/4\right],\ [11/2,1/2],\ \left[\frac{23}{4},1/4\right],\ [6,0]$$

Be especially careful when plotting piecewise-defined and recursively defined functions. For the function $g(x)$ defined here, Maple cannot compute $g(x)$ unless the value of x is known. In the next section, we see that for the standard plot command,

```
plot(f(x),x=a..b),
```

Maple evaluates $f(x)$ first and then the domain, which is impossible for a function like $g(x)$. In this case, the x-values need to be first and then $g(x)$. To delay the evaluation of $g(x)$ enclose $g(x)$ in single quotation marks, '. Thus,

```
> plot('g(x)',x=0..12);
```

gives us the plot of $g(x)$ shown in Figure 2-1.

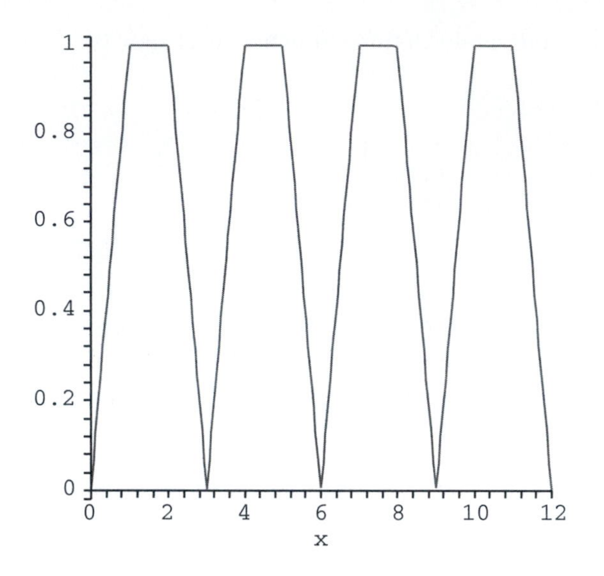

Figure 2-1 Plot of a recursively defined function

We will discuss additional ways to define, manipulate, and evaluate functions as needed. However, Maple's extensive programming language allows a great deal of flexibility in defining functions, many of which are beyond the scope of this text.

2.3 Graphing Functions, Expressions, and Equations

One of the best features of Maple is its graphics capabilities. In this section, we discuss methods of graphing functions, expressions, and equations, and several of the options available to help graph functions.

2.3.1 Functions of a Single Variable

The command

```
plot(f(x),x=a..b)
```

graphs the function $y = f(x)$ on the interval $[a, b]$. Maple returns detailed information regarding the `plot` command with `?plot`.

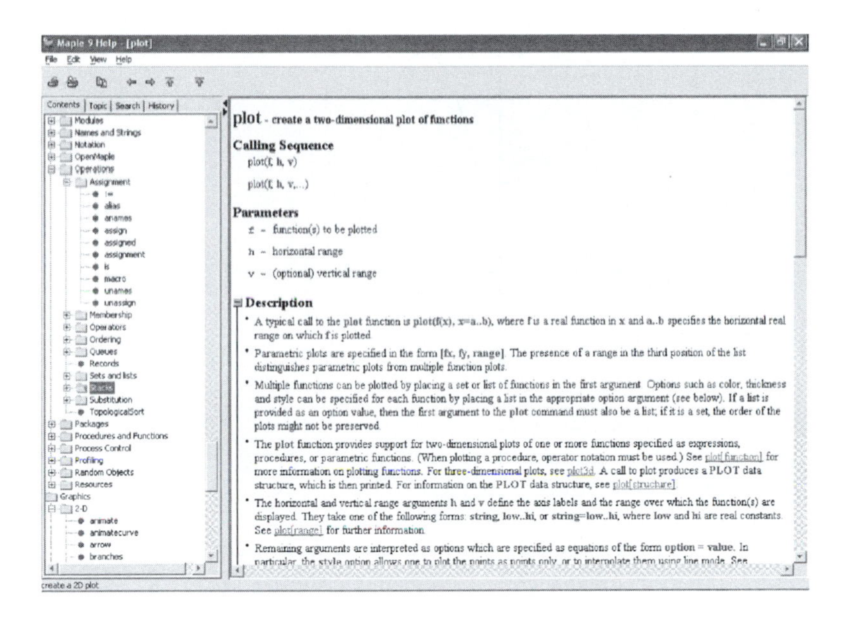

Remember that every Maple object can be assigned a name, including graphics. `display(p1,p2,...pn)` displays the graphics `p1, p2, ..., pn` together. The `display` command is contained in the `plots` package so be sure to load the `plots` package before using the `display` command by entering `with(plots)` unless you choose to use the long form of the command, `plots[display](p1,p2,...,pn)`.

EXAMPLE 2.3.1: Graph $y = \sin x$ for $-\pi \le x \le 2\pi$. $y = \cos x$, and $y = \tan x$.

SOLUTION: Entering

```
> plot(sin(x),x=-Pi..2*Pi);
```

graphs $y = \sin x$ for $-\pi \le x \le 2\pi$. The plot is shown in Figure 2-2.
■

Use delayed evaluation by enclosing the function in single quotation marks, ', to plot functions that are defined using `proc`.

EXAMPLE 2.3.2: Graph $s(t)$ for $0 \le t \le 5$ where $s(t) = 1$ for $0 \le t < 1$ and $s(t) = 1 + s(t-1)$ for $t \ge 1$.

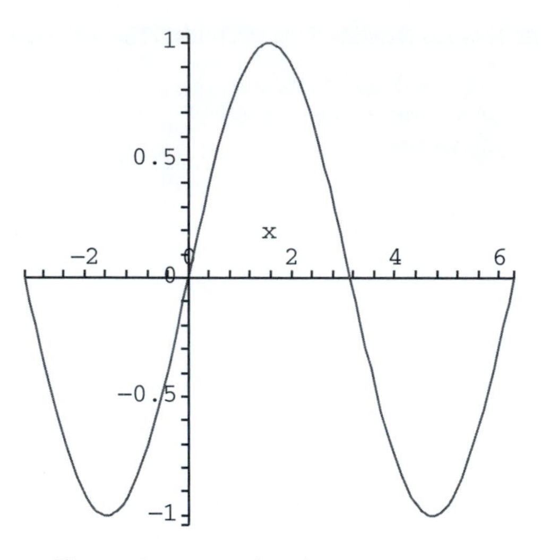

Figure 2-2 $y = \sin x$ for $-\pi \le x \le 2\pi$

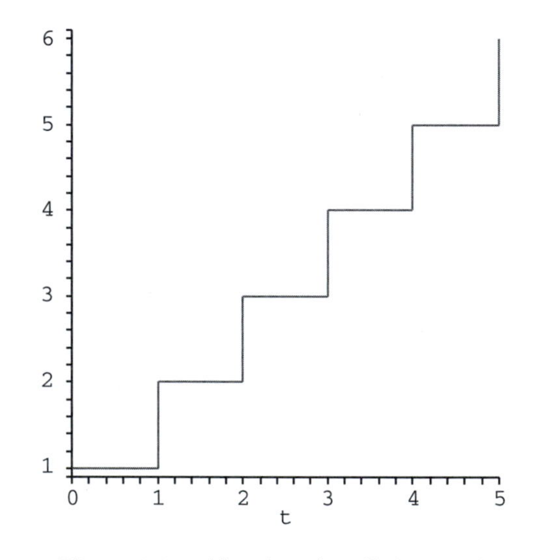

Figure 2-3 $s(t) = 1 + s(t-1), 0 \le t \le 5$

SOLUTION: After defining $s(t)$ with `proc`,

```
> s:=proc(t)
> if t>=0 and t<1 then 1
> else 1+s(t-1) fi
> end proc:
```

we use `plot` to graph $s(t)$ for $0 \le t \le 5$ in Figure 2-3.

```
> plot('s(t)',t=0..5,scaling=constrained);
```

Of course, Figure 2-3 is not completely precise: vertical lines are never the graphs of functions. In this case, discontinuities occur at $t = 1, 2, 3, 4$, and 5. If we were to redraw the figure by hand, we would erase the vertical line segments, and then for emphasis place open dots at $(1,1)$, $(2,2)$, $(3,3)$, $(4,4)$, and $(5,5)$ and then filled dots at $(1,2)$, $(2,3)$, $(3,4)$, $(4,5)$, and $(5,6)$.

■

Entering `?plot[options]` lists all `plot` options and their default values.

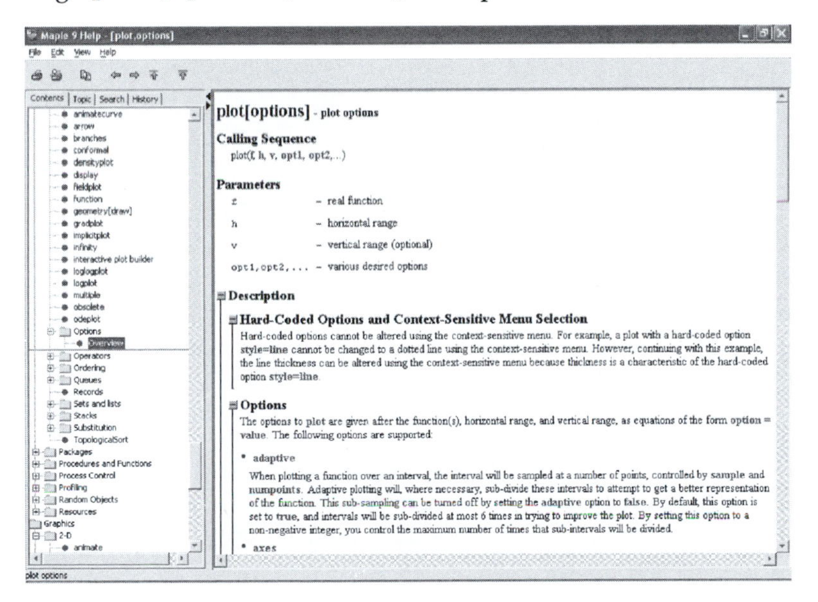

The options most frequently used by beginning users include `color`, `coords`, `symbol`, `thickness`, `view`, `linestyle`, and `scaling`, which are illustrated in the following examples.

EXAMPLE 2.3.3: Graph $y = \sin x$, $y = \cos x$, and $y = \tan x$ together with their inverse functions.

SOLUTION: In `p1`, `p2`, and `p3`, we use `plot` to graph $y = \sin^{-1} x$ and $y = x$, respectively. None of the plots are displayed because we included a colon at the end of each command. `p1`, `p2`, and `p3` are displayed together with `display` in Figure 2-4. The plot is shown to scale because we included the option `scaling=constrained`; the graph of $y = \sin x$ is in black (because we used the option `color=black` in `p1`), $y = \sin^{-1} x$ is in gray (because we used the option `color=gray` in `p3`),

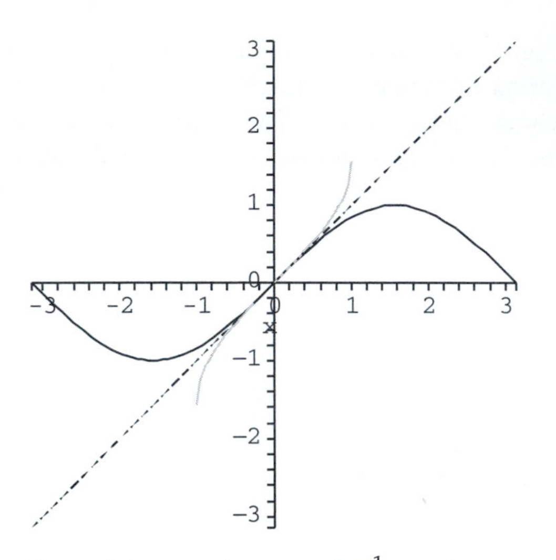

Figure 2-4 $y = \sin x$, $y = \sin^{-1} x$, and $y = x$

and $y = x$ is dashed (because we used the option `linestyle=DASH` in `p2`). Generally, including the option `view=[a..b,c..d]` instructs Maple that the horizontal axis displayed should correspond to the interval $[a, b]$ and that the vertical axis displayed should correspond to the interval $[c, d]$.

```
> p1:=plot(sin(x),x=-Pi..Pi,color=black):
> p2:=plot(x,x=-Pi..Pi,linestyle=DASH,color=black):
> p3:=plot(arcsin(x),x=-1..1,color=gray):
```

```
> with(plots):
> display(p1,p2,p3,view=[-Pi..Pi,-Pi..Pi],
   scaling=constrained);
```

The command `plot([f1(x),f2(x),...,fn(x)],x=a..b)` plots $f_1(x)$, $f_2(x)$, ..., $f_n(x)$ together for $a \leq x \leq b$. `color` and `linestyle` options are incorporated with `color=[color1,color2,..., colorn]` and `linestyle=[style1,style2,...,stylen]`.

In the following, we use `plot` to graph $y = \cos x$, $y = \cos^{-1} x$, and $y = x$ together. We show the plot in Figure 2-5. The plot is shown to scale; the graph of $y = \cos x$ is in black, $y = \cos^{-1} x$ is in gray, and $y = x$ is in light gray.

For two-dimensional plots, you can specify the `linestyle` to be `SOLID`, `DOT`, `DASH`, or `DASHDOT`. These options are case-sensitive so be sure to use all caps if you change from the default, `SOLID`.

```
> plot([cos(x),arccos(x),x],x=-Pi..Pi,
> color=[COLOR(RGB,0,0,0),COLOR(RGB,.25,.25,.25),
   COLOR(RGB,.75,.75,.75)],
> view=[-Pi..Pi,-Pi..Pi],scaling=constrained);
```

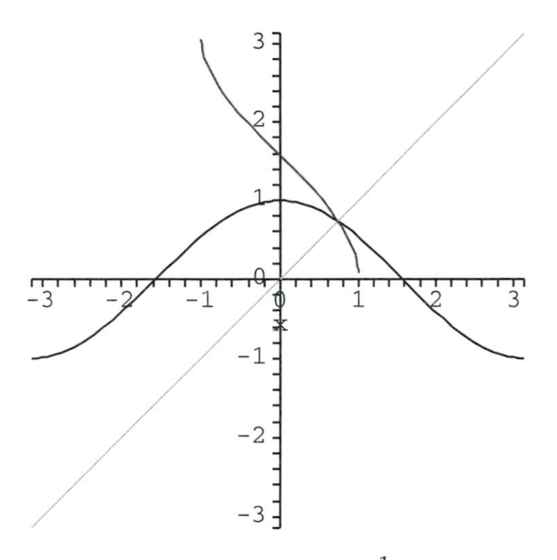

Figure 2-5 $y = \cos x, y = \cos^{-1} x$, and $y = x$

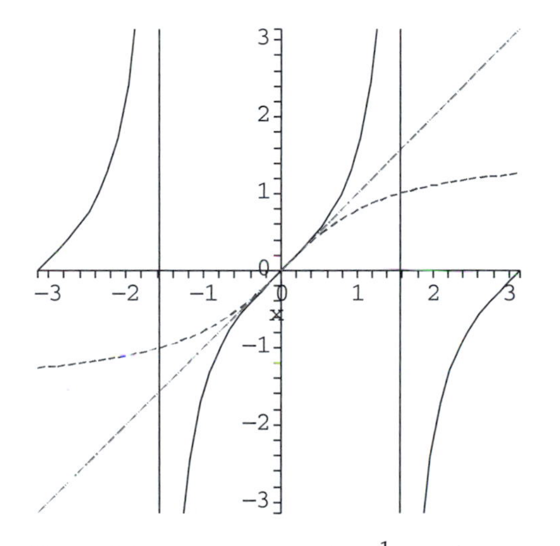

Figure 2-6 $y = \tan x, y = \tan^{-1} x$, and $y = x$

We use the same idea to graph $y = \tan x$, $y = \tan^{-1} x$, and $y = x$ and incorporate the `linestyle` option in Figure 2-6.

```
> plot([tan(x),arctan(x),x],x=-Pi..Pi,
> color=[COLOR(RGB,0,0,0),COLOR(RGB,.25,.25,.25),
   COLOR(RGB,.5,.5,.5)],
> view=[-Pi..Pi,-Pi..Pi],linestyle=[SOLID,DASH,DOT],
   scaling=constrained);
```

■

The previous example illustrates the graphical relationship between a function and its inverse.

EXAMPLE 2.3.4 (Inverse Functions): $f(x)$ and $g(x)$ are **inverse functions** if

$$f(g(x)) = g(f(x)) = x.$$

If $f(x)$ and $g(x)$ are inverse functions, their graphs are symmetric about the line $y = x$.

The `@` symbol is Maple's composition operator. The command

```
(f1@f2@f3...@fn)(x)
```

computes the composition

$$\left(f_1 \circ f_2 \circ \cdots f_n\right)(x) = f_1\left(f_2\left(\cdots\left(f_n(x)\right)\right)\right).$$

For two functions $f(x)$ and $g(x)$, it is usually easiest to compute the composition $f(g(x))$ with `f(g(x))` or `(f@g)(x)`.

Show that

$$f(x) = \frac{-1-2x}{-4+x} \qquad \text{and} \qquad g(x) = \frac{4x-1}{x+2}$$

are inverse functions.

$f(x)$ and $g(x)$ are not returned because a colon is included at the end of each command.

SOLUTION: After defining $f(x)$ and $g(x)$,

```
> f:=x->(-1-2*x)/(-4+x):
> g:=x->(4*x-1)/(x+2):
```

we compute and simplify the compositions $f(g(x))$ and $g(f(x))$. Because both results are x, $f(x)$ and $g(x)$ are inverse functions.

```
> simplify(f(g(x)));
```

$$x$$

```
> simplify((f@g)(x));
```

$$x$$

```
> simplify(g(f(x)));
```

$$x$$

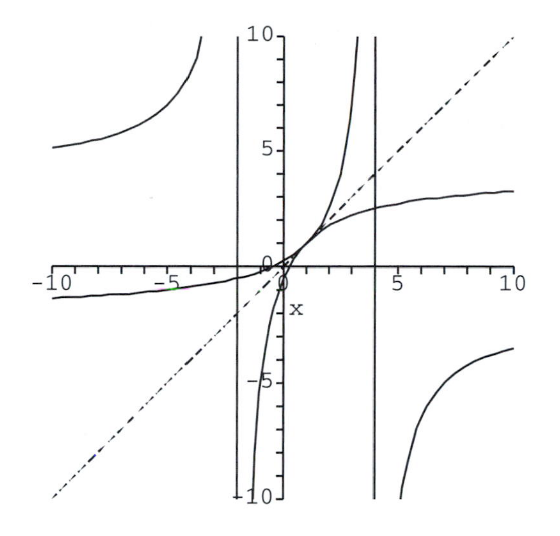

Figure 2-7 $f(x)$ in black, $g(x)$ in gray, and $y = x$ dashed

```
> simplify((g@f)(x));
```

$$x$$

To see that the graphs of $f(x)$ and $g(x)$ are symmetric about the line $y = x$, we use `plot` to graph $f(x)$, $g(x)$, and $y = x$ together in Figure 2-7, illustrating the use of the `color` and `linestyle` options.

```
> plot([f(x),g(x),x],x=-10..10,
> color=[COLOR(RGB,0,0,0),COLOR(RGB,.25,.25,.25),
   COLOR(RGB,0,0,0)],
> linestyle=[SOLID,SOLID,DASH],view=[-10..10,-10..10],
> scaling=constrained);
```

In the plot, observe that the graphs of $f(x)$ and $g(x)$ are symmetric about the line $y = x$. The plot also illustrates that the domain and range of a function and its inverse are interchanged: $f(x)$ has domain $(-\infty, 4) \cup (4, \infty)$ and range $(-\infty, -2) \cup (-2, \infty)$; $g(x)$ has domain $(-\infty, -2) \cup (-2, \infty)$ and range $(-\infty, 4) \cup (4, \infty)$.
■

For repeated compositions of a function with itself, use the repeated composition operator, `@@`: `(f@@n)(x)` computes the composition

$$\underbrace{(f \circ f \circ f \circ \cdots f)}_{n \text{ times}}(x) = \underbrace{(f\,(f\,(f\,(f \cdots))))}_{n \text{ times}}(x) = f^n(x).$$

EXAMPLE 2.3.5: Graph $f(x)$, $f^{10}(x)$, $f^{20}(x)$, $f^{30}(x)$, $f^{40}(x)$, and $f^{50}(x)$ if $f(x) = \sin x$ for $0 \leq x \leq 2\pi$.

SOLUTION: After defining $f(x) = \sin x$, we graph $f(x)$ in `p1` with `plot`

```
> with(plots):
> f:=x->sin(x):
> p1:=plot(f(x),x=0..2*Pi,color=black):
```

and then illustrate the use of the repeated composition operator, `@@`, by computing $f^5(x)$.

```
> (f@@5)(x);
```

$$\sin(\sin(\sin(\sin(\sin(x)))))$$

Next, we use `seq` together with `@@` to create the list of functions

$$\left\{f^{10}(x), f^{20}(x), f^{30}(x), f^{40}(x), f^{50}(x)\right\}.$$

Because the resulting output is rather long, we include a colon at the end of the `seq` command to suppress the resulting output.

```
> toplot:=[seq((f@@(10*n))(x),n=1..5)]:
```

In `grays`, we compute a list of `COLOR(RGB,i,i,i)` for five equally spaced values of i between 0.2 and 0.8. We then graph the functions in `toplot` on the interval $[0, 2\pi]$ with `plot`. The graphs are shaded according to `grays` and named `p2`.

Finally, we use `display` to display `p1` and `p2` together in Figure 2-8.

```
> grays:=[seq(COLOR(RGB,.2+.6*i/4,.2+.6*i/4,.2+.6*i/4),
   i=0..4)]:
```

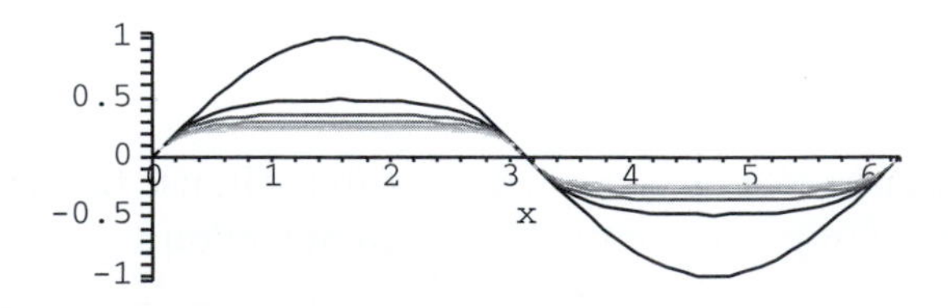

Figure 2-8 $f(x)$ in black; the graphs of $f^{10}(x)$, $f^{20}(x)$, $f^{30}(x)$, $f^{40}(x)$, and $f^{50}(x)$ are successively lighter – the graph of $f^{50}(x)$ is the lightest

```
> p2:=plot(toplot,x=0..2*Pi,color=grays):
> display(p1,p2,scaling=constrained);
```

In the plot, we see that repeatedly composing sine with itself has a flattening effect on $y = \sin x$.

■

Usually, Maple's `plot` command selects an appropriate vertical axis for the displayed graphic. If it does not make a wise choice, use the `view` option (Figure 2-9). Including `view=[a..b,c..d]` in your `plot` or `display` command instructs Maple that the horizontal axis displayed should correspond to the interval $[a, b]$ and that the vertical axis displayed should correspond to the interval $[c, d]$. Include the option `scaling=constrained` if you wish your plot to be displayed to scale.

EXAMPLE 2.3.6: Graph $y = \dfrac{\sqrt{9 - x^2}}{x^2 - 4}$.

SOLUTION: We use `plot` to generate the basic graph of y shown in Figure 2-10(a). The asymptotes result in a plot that we do not expect.

Maple's error messages do not always mean that you have made a mistake entering a command.

```
> g:=x->sqrt(4-x^2)/(x^ 2-1);
```

$$g := x \mapsto \frac{\sqrt{4 - x^2}}{x^2 - 1}$$

```
> plot(g(x),x=-10..10);
```

Observe that the domain of y is $[-3, -2) \cup (-2, 2) \cup (2, 3]$: the values of x where the denominator is *not* equal to zero and where the radicand

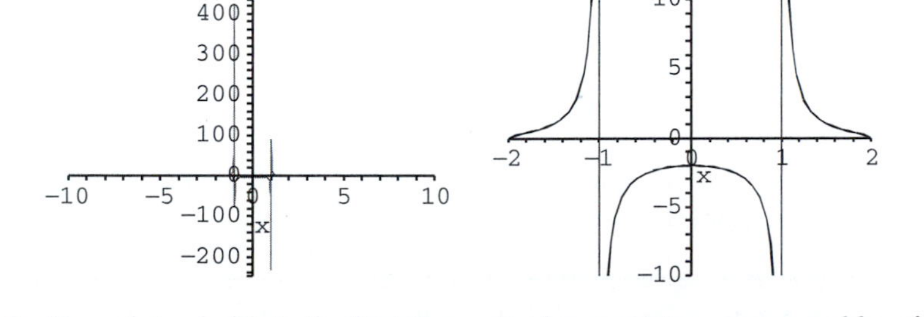

Figure 2-9 Two plots of $g(x)$. In the first, the vertical asymptotes cause a problem for Maple and it does not select a vertical range that we desire. In the second, we use the `view` option to specify the vertical range displayed resulting in a more interesting plot

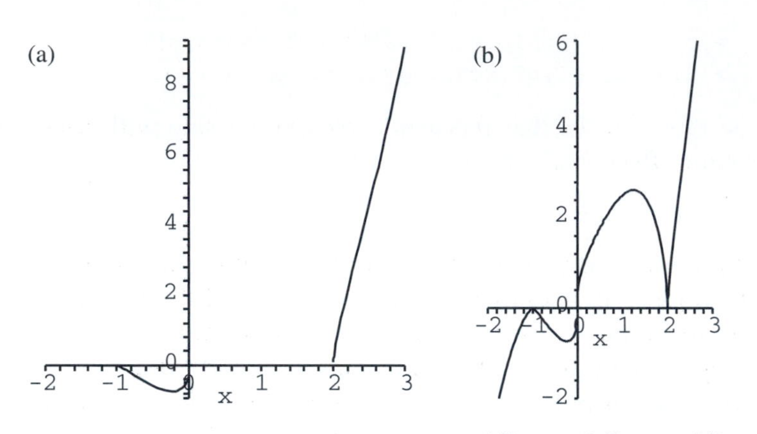

Figure 2-10 (a) and (b) Two plots of $y = x^{1/3}(x-2)^{2/3}(x+1)^{4/3}$

of the numerator is greater than or equal to zero. We determine these values with `solve`. The `solve` command is discussed in more detail in the next section.

```
> solve(x^2-1=0,x);
```

$$1, -1$$

```
> solve(4-x^2>=0,x);
```

$$RealRange\,(-2, 2)$$

A better graph of y is obtained by plotting y for $-3 \le x \le 3$ and shown in Figure 2-10(b). We then use the `view` option to specify that the displayed horizontal axis corresponds to $-2 \le x \le 2$ and that the displayed vertical axis corresponds to $-10 \le y \le 10$.

```
> plot(g(x),x=-2..2,view=[-2..2,-10..10],color=BLACK);
```

■

When graphing functions involving odd roots, Maple's results may be surprising to the beginner. The key is to use the `surd` function when defining the function to be graphed.

EXAMPLE 2.3.7: Graph $y = x^{1/3}(x-2)^{2/3}(x+1)^{4/3}$.

SOLUTION: Entering

```
> f:=x->x^(1/3)*(x-2)^(2/3)*(x+1)^(4/3):
> plot(f(x),x=-2..3,color=black);
```

does not produce the graph we expect (see Figure 2-10(a)) because many of us consider $y = x^{1/3}(x-2)^{2/3}(x+1)^{4/3}$ to be a real-valued function with domain $(-\infty, \infty)$.

Generally, Maple does return a real number when computing the odd root of a negative number. For example, $x^3 = -1$ has three solutions.

solve is discussed in more detail in the next section. evalf(number) returns an approximation of number.

```
> s1:=solve(x^3+1=0);
```

$$s1 := -1, 1/2 + 1/2\,i\sqrt{3}, 1/2 - 1/2\,i\sqrt{3}$$

```
> evalf(s1);
```

$$-1.0, 0.5000000000 + 0.8660254040\,i, 0.5000000000 - 0.8660254040\,i$$

When computing an odd root of a negative number, Maple has many choices (as illustrated above) and chooses a root with positive imaginary part – the result is not a real number.

```
> evalf((-1)^(1/3));
```

$$0.5000000001 + 0.8660254037\,i$$

To obtain real values when computing odd roots of negative numbers, use surd: if x is negative, surd(x,n) returns $-(-x)^{1/n}$. Thus,

```
> plot(surd(x,3)*surd((x-2),3)^2*surd((x+1),3)^4,x=-2..3,
> view=[-2..3,-2..6],color=black,numpoints=200,
   scaling=constrained);
```

produces the expected graph (see Figure 2-10(b)).
■

2.3.2 Parametric and Polar Plots in Two Dimensions

To graph the parametric equations $x = x(t)$, $y = y(t)$, $a \le t \le b$, use

```
plot([x(t),y(t),t=a..b])
```

and to graph the polar function $r = r(\theta)$, $\alpha \leq \theta \leq \beta$, use `plot` with the `coords=polar` option

```
plot(f(theta),theta=alpha..beta,coords=polar)
```

or use `polarplot`

```
polarplot(r(theta),theta=alpha..beta)
```

The `polarplot` function is contained in the `plots` package, so load this by entering `with(plots)` before using the `polarplot` function *or* enter it in its long form: `plots[polarplot](...)`.

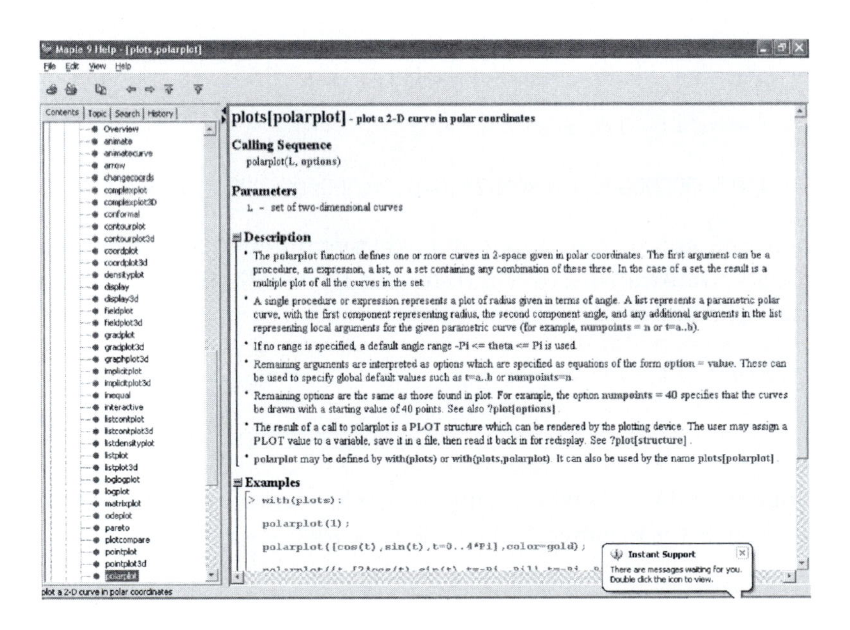

EXAMPLE 2.3.8 (The Unit Circle): The **unit circle** is the set of points (x, y) exactly 1 unit from the origin, $(0, 0)$, and, in rectangular coordinates, has equation $x^2 + y^2 = 1$. The unit circle is the classic example of a relation that is neither a function of x nor a function of y. The top half of the unit circle is given by $y = \sqrt{1 - x^2}$ and the bottom half is given by $y = -\sqrt{1 - x^2}$.

```
> plot([sqrt(1-x^2),-sqrt(1-x^2)],x=-1..1,
   view=[-3/2..3/2,-3/2..3/2],
> scaling=constrained,color=black);
```

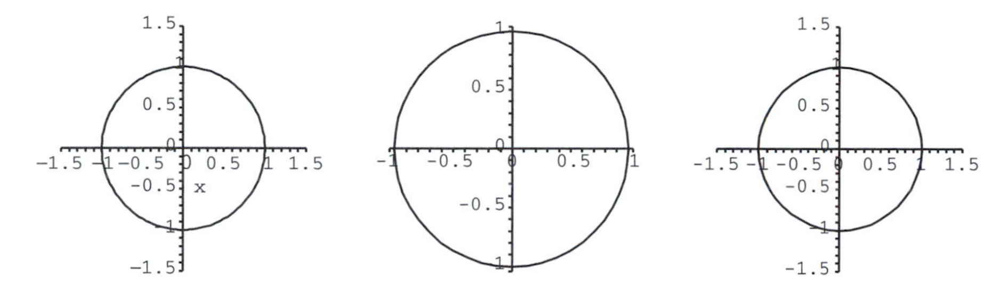

Figure 2-11 Three plots of the unit circle

Each point (x, y) on the unit circle is a function of the angle, t, that subtends the x-axis, which leads to a parametric representation of the unit circle, $\begin{cases} x = \cos t \\ y = \sin t \end{cases}$, $0 \le t \le 2\pi$, which we graph with `plot`.

```
> plot([cos(t),sin(t),t=0..2*Pi],color=black,
  scaling=constrained);
```

Using the change of variables $x = r\cos t$ and $y = r\sin t$ to convert from rectangular to polar coordinates, a polar equation for the unit circle is $r = 1$. We use `plot` together with the `coords=polar` to graph $r = 1$.

```
> plot(1,t=0..2*Pi,view=[-3/2..3/2,-3/2..3/2],
  scaling=constrained,color=black,
> coords=polar);
```

The three plots are shown side-by-side in Figure 2-11. Of course, they all look like unit circles.

EXAMPLE 2.3.9: Graph the parametric equations

$$\begin{cases} x = t + \sin 2t \\ y = t + \sin 3t \end{cases}, \quad -2\pi \le t \le 2\pi.$$

SOLUTION: After defining x and y, we use `plot` to graph the parametric equations in Figure 2-12.

```
> x:=t->t+sin(2*t):
> y:=t->t+sin(3*t):
```

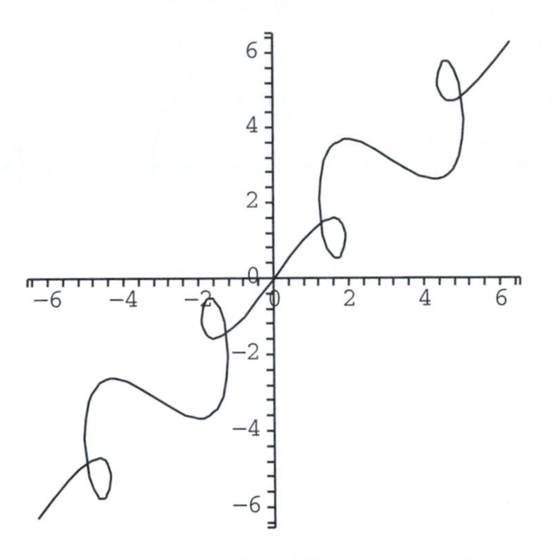

Figure 2-12 $(x(t), y(t)),\ -2\pi \le t \le 2\pi$

```
> plot([x(t),y(t),t=-2*Pi..2*Pi],color=black,
   scaling=constrained);
```

■

In the following example, the equations involve integrals.

Remark. Topics from calculus are discussed in Chapter 3. For now, we state that `int(f(x),x=a..b)` attempts to evaluate $\int_a^b f(x)\,dx$.

EXAMPLE 2.3.10 (Cornu Spiral): The **Cornu spiral** (or **clothoid**) (see [11] and [17]) has parametric equations

$$x = \int_0^t \sin\left(\frac{1}{2}u^2\right) du \qquad \text{and} \qquad y = \int_0^t \cos\left(\frac{1}{2}u^2\right) du.$$

Graph the Cornu spiral.

SOLUTION: We begin by defining x and y. Notice that Maple can evaluate these integrals, even though the results are in terms of the

`FresnelS` and `FresnelC` functions, which are defined in terms of integrals:

$$\texttt{FresnelS(t)} = \int_0^t \sin\left(\frac{\pi}{2}u^2\right)\,du \qquad \text{and}$$

$$\texttt{FresnelC(t)} = \int_0^t \cos\left(\frac{\pi}{2}u^2\right)\,du.$$

```
> x:=t->int(sin(u^2/2),u=0..t);
> x(t);
```

$$t \mapsto \sqrt{\pi}\mathit{FresnelS}\left(\frac{t}{\sqrt{\pi}}\right)$$

$$\sqrt{\pi}\mathit{FresnelS}\left(\frac{t}{\sqrt{\pi}}\right)$$

```
> x(1);
```

$$\mathit{FresnelS}\left(\frac{1}{\sqrt{\pi}}\right)\sqrt{\pi}$$

```
> y:=t->int(cos(u^2/2),u=0..t);
> y(t);
```

$$t \mapsto \sqrt{\pi}\mathit{FresnelC}\left(\frac{t}{\sqrt{\pi}}\right)$$

$$\sqrt{\pi}\mathit{FresnelC}\left(\frac{t}{\sqrt{\pi}}\right)$$

We use `plot` to graph the Cornu spiral in Figure 2-13. The option `scaling=constrained` instructs Maple to generate the plot to scale.

```
> plot([x(t),y(t),t=-10..10],color=black,
   scaling=constrained);
```

■

Observe that the graph of the polar equation $r = f(\theta)$, $\alpha \le \theta \le \beta$ is the same as the graph of the parametric equations

$$x = f(\theta)\cos\theta \qquad \text{and} \qquad y = f(\theta)\sin\theta, \qquad \alpha \le \theta \le \beta.$$

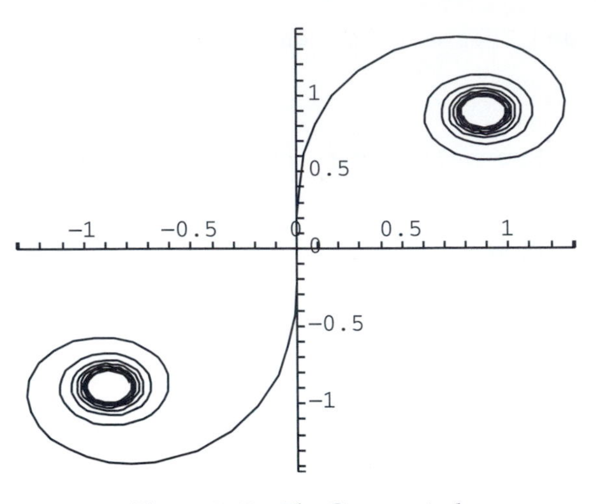

Figure 2-13 The Cornu spiral

EXAMPLE 2.3.11: Graph (a) $r = \sin(8\theta/7)$, $0 \le \theta \le 14\pi$; (b) $r = \theta\cos\theta$, $-19\pi/2 \le \theta \le 19\pi/2$; (c) ("The Butterfly") $r = e^{\cos\theta} - 2\cos 4\theta + \sin^5(\theta/12)$, $0 \le \theta \le 24\pi$; and (d) ("The Lituus") $r^2 = 1/\theta$, $0.1 \le \theta \le 10\pi$.

SOLUTION: For (a) and (b) we use `plot` together with the `coords=polar` option. First define r and then use `plot` to generate the graph of the polar curve.

```
> plot(sin(8*theta/7),theta=0..14*Pi,coords=polar,
   color=black,scaling=constrained);
```

```
> plot(theta*cos(theta),theta=-19*Pi/2..19*Pi/2,
   coords=polar,color=black,scaling=constrained);
```

You do not need to reload the `plots` package if you have already loaded it during your current Maple session.

For (c) and (d) we use `polarplot`. Using standard mathematical notation, we know that $\sin^5(\theta/12) = (\sin(\theta/12))^5$. However, when defining r with Maple, be sure you use the form `sin(θ/12)^5`, not `sin^5(θ/12)`, which Maple will not interpret in the way intended.

```
> with(plots):
> polarplot(exp(cos(theta))-2*cos(4*theta)
   +sin(theta/12)^5,theta=0..24*Pi,
> color=black,scaling=constrained);
```

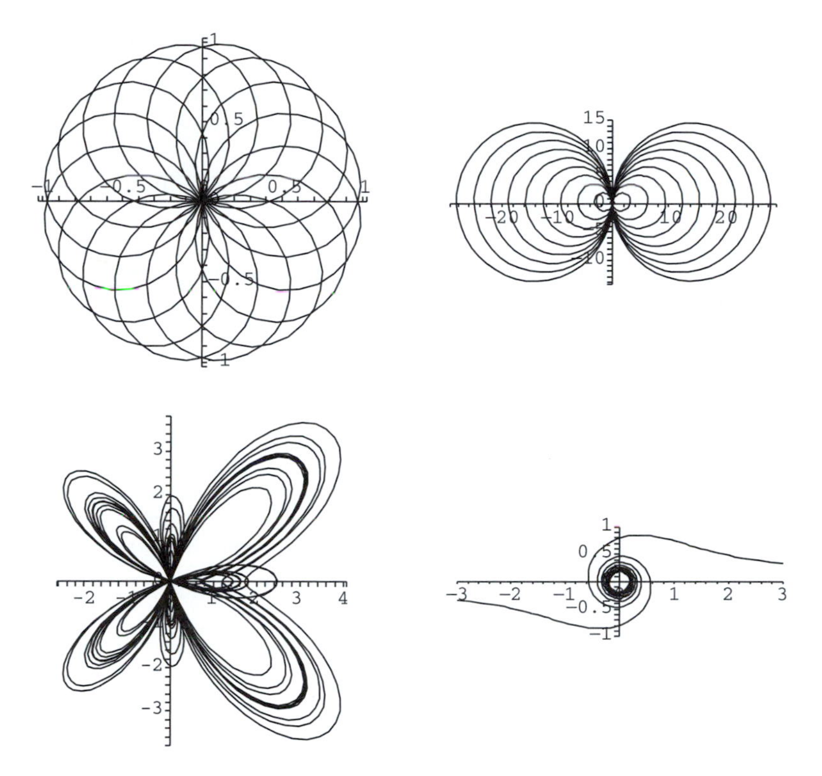

Figure 2-14 Graphs of four polar equations

For (d), we graph $r^2 = 1/\theta$ by graphing $r = 1/\sqrt{\theta}$ and $r = -1/\sqrt{\theta}$ together with `polarplot`.

```
> polarplot([sqrt(1/theta),-sqrt(1/theta)],
   theta=0..10*Pi,
> color=black,view=[-3..3,-1..1],scaling=constrained);
```

All four graphs are shown in Figure 2-14.
■

2.3.3 Three-Dimensional and Contour Plots; Graphing Equations

An elementary function of two variables, $z = f(x,y) =$ *expression in x and y*, is typically defined using the form

```
f:=(x,y)->expression in x and y
```

Once a function has been defined, a basic graph is generated with `plot3d`:

```
plot3d(f(x,y),x=a..b,y=c..d)
```

graphs $f(x,y)$ for $a \le x \le b$ and $c \le y \le d$.

For details regarding `plot3d` and its options enter `?plot3d` or `?options[plot3d]`, as we do here.

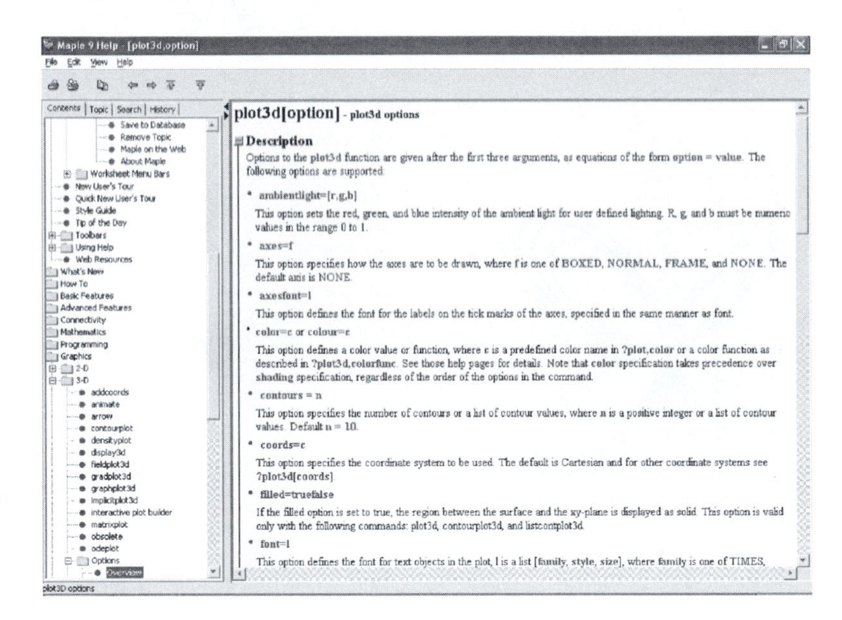

Graphs of several level curves of $z = f(x,y)$ are generated with

```
contourplot(f(x,y),x=a..b,y=c..d).
```

Note that `contourplot` is contained in the `plots` package so be sure to load the `plots` package first by entering `with(plots)` or enter the command in its long form, `plots[contourplot](...)`.

For details regarding `contourplot` and its options enter `?contourplot` or `?options[contourplot]`.

EXAMPLE 2.3.12: Let

$$f(x,y) = \frac{x^2 y}{x^4 + 4y^2}.$$

(a) Calculate $f(1,-1)$. (b) Graph $f(x,y)$ and several contour plots of $f(x,y)$ on a region containing $(0,0)$.

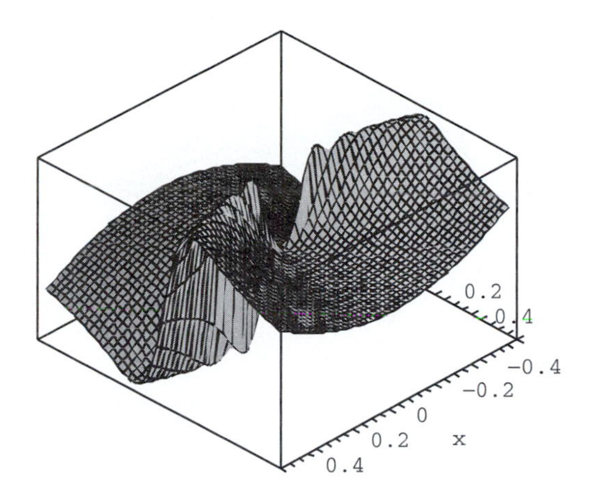

Figure 2-15 Three-dimensional plot of $f(x, y)$

SOLUTION: After defining $f(x, y)$, we evaluate $f(1, -1) = -1/5$.

```
> f:=(x,y)->x^2*y/(x^4+4*y^2);
```

$$(x, y) \mapsto \frac{x^2 y}{x^4 + 4y^2}$$

```
> f(1,-1);
```

$$-1/5$$

Next, we use `plot3d` to graph $f(x, y)$ for $-1/2 \le x \le 1/2$ and $-1/2 \le y \le 1/2$ in Figure 2-15. We illustrate the use of the `axes` and `grid` options.

```
> plot3d(f(x,y),x=-1/2..1/2,y=-1/2..1/2,axes=BOXED,
   grid=[50,50]);
```

Two contour plots are generated with `contourplot`. The second illustrates the use of the `grid`, `color`, and `scaling` options (Figure 2-16).

```
> with(plots):
> contourplot(f(x,y),x=-1/2..1/2,y=-1/2..1/2,
   grid=[50,50]);

> contourplot(f(x,y),x=-1/4..1/4,y=-1/4..1/4,
   grid=[60,60],
> color=BLACK,scaling=CONSTRAINED);
```

■

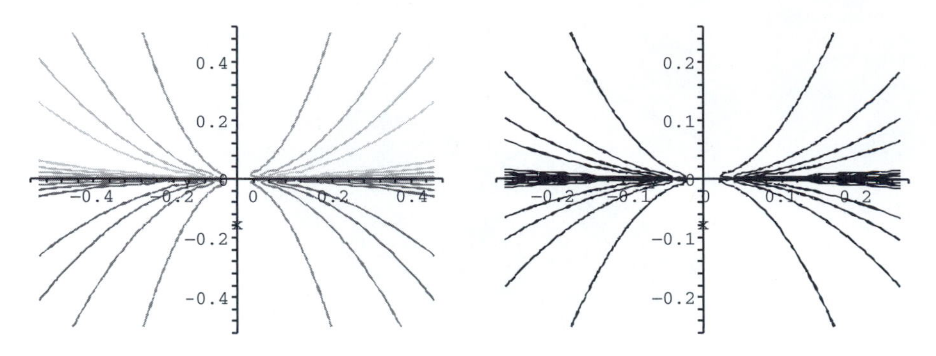

Figure 2-16 Two contour plots of $f(x, y)$

Various perspectives can be adjusted by clicking on the graphic and dragging the bounding box. Also, once you have selected the graphic, the **Plot** submenu becomes available from the Maple menu.

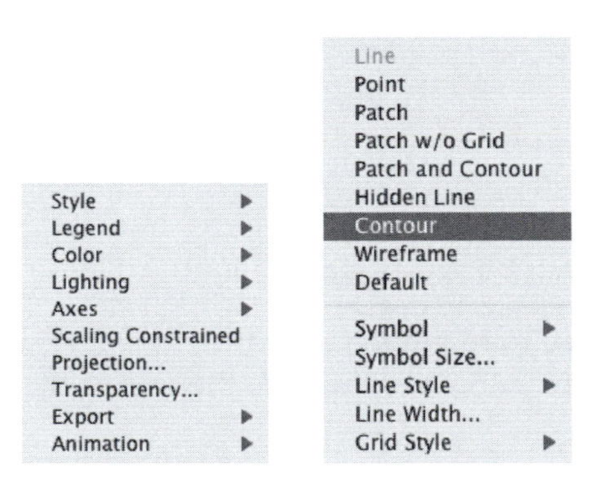

Figure 2-17 shows four different views of the graph of $g(x, y) = x \sin y + y \sin x$ for $0 \le x \le 5\pi$ and $0 \le y \le 5\pi$. In the first, we have slightly rotated the plot. In the second, we selected **Patch and Contour** from the **Style** submenu. In the third we selected **Wireframe** from the **Style** submenu and **Normal** from the **Axes** submenu. In the fourth, we selected **Hidden Line** from **Style** and **Graylevel** from **Color**. In subsequent examples, we will see that these options can be included in the `plot3d` command as well.

`contourplot` is especially useful when graphing equations. The graph of the equation $f(x, y) = C$, where C is a constant, is the same as the contour plot of $z = f(x, y)$ corresponding to C. That is, the graph of $f(x, y) = C$ is the same as the level curve of $z = f(x, y)$ corresponding to $z = C$.

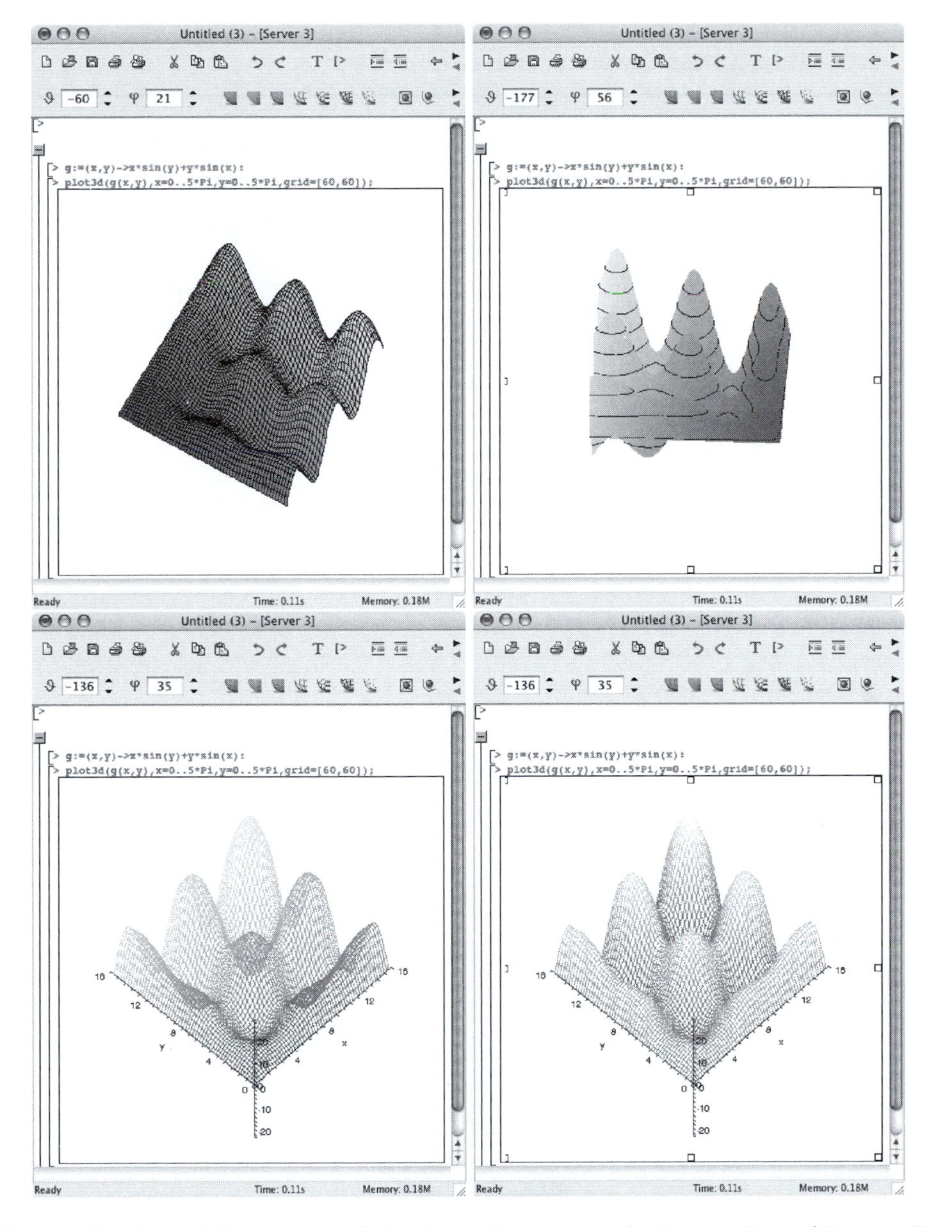

Figure 2-17 Four different plots of $g(x, y) = x \sin y + y \sin x$ for $0 \leq x \leq 5\pi$ and $0 \leq y \leq 5\pi$

EXAMPLE 2.3.13: Graph the unit circle, $x^2 + y^2 = 1$.

SOLUTION: We first graph $z = x^2 + y^2$ for $-4 \leq x \leq 4$ and $-4 \leq y \leq 4$ with `plot3d` in Figure 2-18.

```
> plot3d(x^2+y^2,x=-4..4,y=-4..4,axes=boxed);
```

The graph of $x^2 + y^2 = 1$ is the graph of $z = x^2 + y^2$ corresponding to $z = 1$. We use `contourplot` together with the `contours` option to graph this equation in Figure 2-19.

```
> with(plots):
> contourplot(x^2+y^2,x=-3/2..3/2,y=-3/2..3/2,
   contours=[1],color=black);
```

Multiple graphs can be generated as well. As an illustration, we graph $x^2 + y^2 = C$ for $C = 1, 4$, and 9 in Figure 2-20.

```
> contourplot(x^2+y^2,x=-4..4,y=-4..4,
> contours=[1,4,9],color=black,grid=[50,50]);
```

■

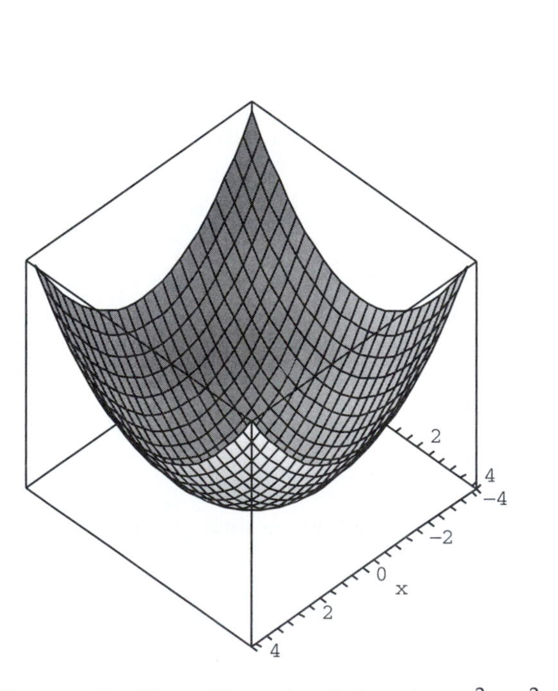

Figure 2-18 Three-dimensional plot of $z = x^2 + y^2$

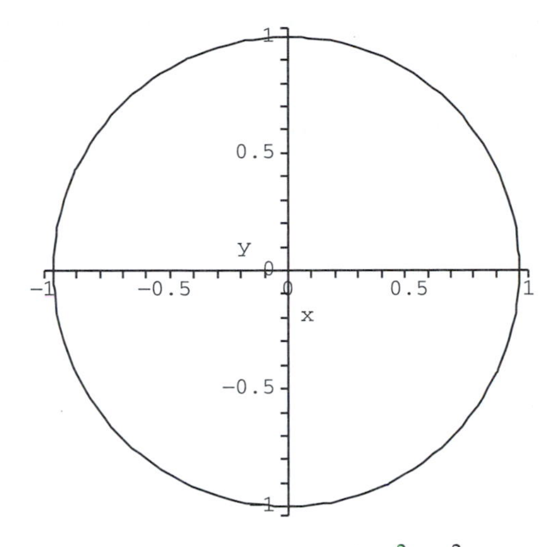

Figure 2-19 The unit circle, $x^2 + y^2 = 1$

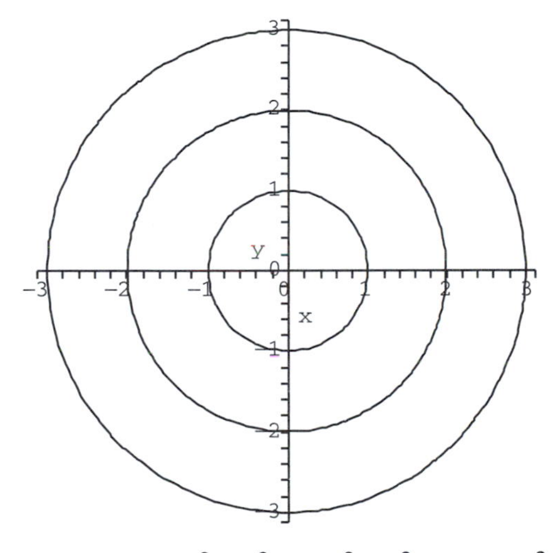

Figure 2-20 Graphs of $x^2 + y^2 = 1$, $x^2 + y^2 = 4$, and $x^2 + y^2 = 9$

As an alternative to using `contourplot` to graph equations, you can also use the `implicitplot` function, which is also contained in the `plots` package.

After loading the `plots` package by entering `with(plots)`, the command

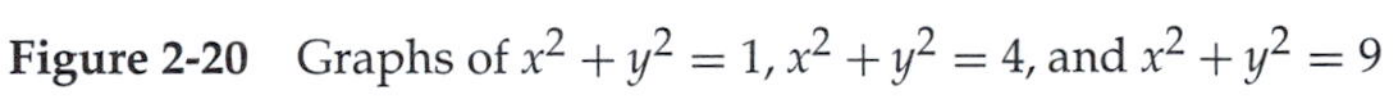

```
implicitplot(equation,x=a..b,y=c..d)
```

graphs the equation, `equation` for $a \leq x \leq b$ and $c \leq y \leq d$.

EXAMPLE 2.3.14: Graph the equation $y^2 - x^4 + 2x^6 - x^8 = 0$ for $-1.5 \leq x \leq 1.5$.

SOLUTION: After loading the `plots` package, we define `eq` to be the equation $y^2 - x^4 + 2x^6 - x^8 = 0$ and then use `implicitplot` to graph `eq` for $-1.5 \leq x \leq 1.5$ and $-1 \leq y \leq 1$ in Figure 2-21. We illustrate the use of the `grid` option (to increase the number of sample points) and the `color` option.

```
> with(plots):
> eq:=y^2-x^4+2*x^6-x^8=0:
> implicitplot(eq,x=-1.5..1.5,y=-1..1,
> grid=[90,90],color=black);
```

■

Equations can be plotted together, as with the `plot` command, with

```
implicitplot([eq1,eq2,...,eqn],x=a..b,y=c..d)
```

Any options included are passed to the plot of the respective equation.

EXAMPLE 2.3.15: Graph the equations $x^2 + y^2 = 1$ and $4x^2 - y^2 = 1$ for $-1.5 \leq x \leq 1.5$.

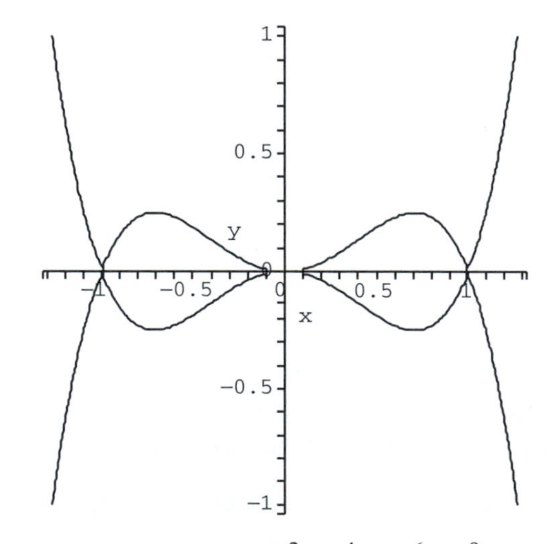

Figure 2-21 Plot of $y^2 - x^4 + 2x^6 - x^8 = 0$

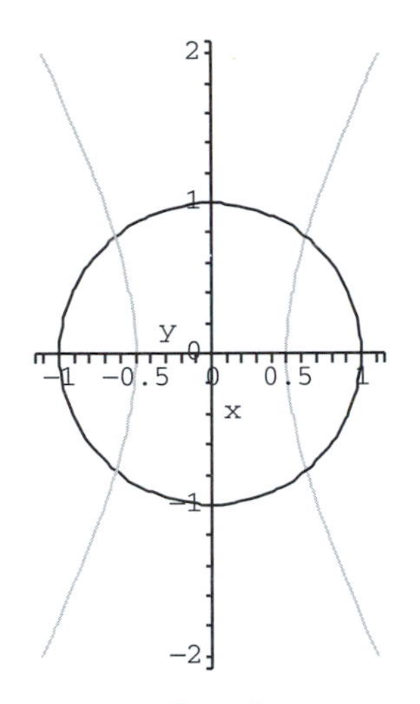

Figure 2-22 Plots of $x^2 + y^2 = 1$ and $4x^2 - y^2 = 1$

SOLUTION: We use `implicitplot` to graph the equations together on the same axes in Figure 2-22. The graph of $x^2 + y^2 = 1$ is the unit circle while the graph of $4x^2 - y^2 = 1$ is a hyperbola. With the included `color` option, the circle is in black and the hyperbola is in gray.

```
> with(plots):
> implicitplot([x^2+y^2=1,4*x^2-y^2=1],
   x=-1.5..1.5,y=-2..2,
> color=[black,gray], scaling=constrained);
```

■

Also see Example 2.3.19.

EXAMPLE 2.3.16 (Conic Sections): A **conic section** is a graph of the equation

$$Ax^2 + Bxy + Cy^2 + Dx + Ey + F = 0.$$

Except when the conic is degenerate, the conic $Ax^2 + Bxy + Cy^2 + Dx + Ey + F = 0$ is a (an)

1. **Ellipse** or **circle** if $B^2 - 4AC < 0$;
2. **Parabola** if $B^2 - 4AC = 0$; or
3. **Hyperbola** if $B^2 - 4AC > 0$.

Graph the conic section $ax^2 + bxy + cy^2 = 1$ for $-4 \le x \le 4$ and for various values of a, b, and c.

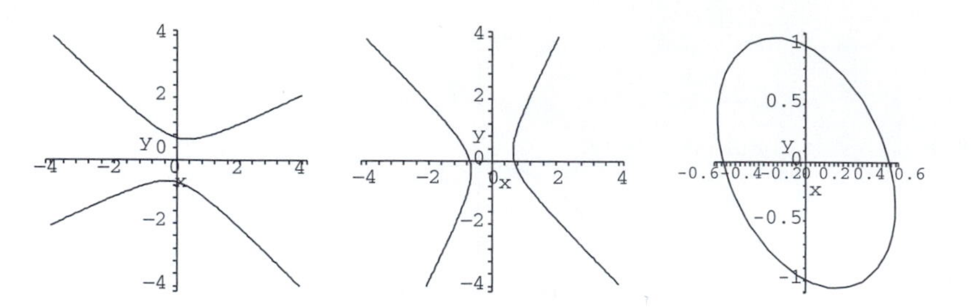

Figure 2-23 Plots of three conic sections

SOLUTION: We define the function `p`. Given `a`, `b`, and `c`, `p` plots the equation $ax^2 + bxy + cy^2 = 1$ for $-4 \le x \le 4$ and $-4 \le y \le 4$. The plot is displayed in black because we include the option `color=black` and is drawn to scale because we include the option `scaling=constrained`. We increase the number of points sampled by Maple with `grid=[60,60]`, which results in smoother plots.

```
> with(plots):
> p:=(a,b,c)->implicitplot(a*x^2+b*x*y+c*y^2=1,
   x=-4..4,y=-4..4,
> scaling=constrained,color=black,grid=[60,60]):
```

We then compute $p(-1, 1, 2)$, $p(2, -1, 1)$, and $p(2, 1, 1)$. The results are shown side-by-side in Figure 2-23.

```
> p(-1,1,2);
> p(2,1,-1);
> p(2,1,1);
```

■

2.3.4 Parametric Curves and Surfaces in Space

The command

```
spacecurve([x(t),y(t),z(t)],t=a..b)
```

generates the three-dimensional curve $\begin{cases} x = x(t) \\ y = y(t), a \le t \le b \\ z = z(t) \end{cases}$ and the command

```
plot3d([x(u,v),y(u,v),z(u,v)],u=a..b,v=c..d)
```

plots the surface $\begin{cases} x = x(u,v) \\ y = y(u,v), a \le u \le b, c \le v \le d. \\ z = z(u,v) \end{cases}$

As with the `implicitplot` and `contourplot` commands, `spacecurve` is contained in the `plots` package.

Thus,

```
> with(plots):
> x:=t->t*cos(2*t):
> y:=t->t*sin(2*t):
> z:=t->t/5:
> spacecurve([x(t),y(t),z(t)],t=0..8*Pi,
> numpoints=240,axes=NORMAL,color=black);
```

loads the `plots` package, defines $x(t) = t\cos 2t$, $y(t) = t\sin 2t$, and $z(t) = t/5$ and then graphs the parametric equations $\begin{cases} x = x(t) \\ y = y(t) \\ z = z(t) \end{cases}$ for $0 \le t \le 8\pi$. We have used the `numpoints` option to increase the number of sample points resulting in a smoother plot. The `axes=NORMAL` option instructs Maple to place axes on the plot, and the color of the graph is in black because of the option `color=black` (Figure 2-24).

Entering `?spacecurve` returns a description of the `spacecurve` command along with a list of options and their current settings.

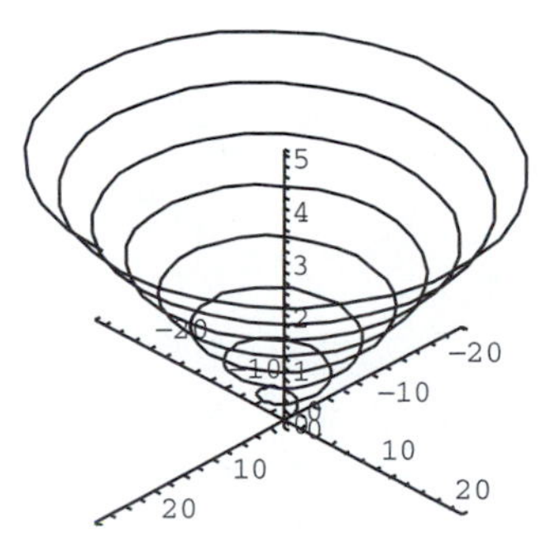

Figure 2-24 A curve in space

EXAMPLE 2.3.17 (Umbilic Torus NC): A parametrization of **Umbilic Torus NC** is given by $\mathbf{r}(s,t) = x(s,t)\mathbf{i} + y(s,t)\mathbf{j} + z(s,t)\mathbf{k}$, $-\pi \le s \le \pi$, $-\pi \le t \le \pi$, where

$$x = \left[7 + \cos\left(\frac{1}{3}s - 2t\right) + 2\cos\left(\frac{1}{3}s + t\right)\right]\sin s$$

$$y = \left[7 + \cos\left(\frac{1}{3}s - 2t\right) + 2\cos\left(\frac{1}{3}s + t\right)\right]\cos s$$

and

$$z = \sin\left(\frac{1}{3}s - 2t\right) + 2\sin\left(\frac{1}{3}s + t\right).$$

Graph the torus.

SOLUTION: We define x, y, and z.

```
> x:=(s,t)->(7+cos(1/3*s-2*t)+2*cos(1/3*s+t))*sin(s):
> y:=(s,t)->(7+cos(1/3*s-2*t)+2*cos(1/3*s+t))*cos(s):
> z:=(s,t)->sin(1/3*s-2*t)+2*sin(1/3*s+t):
```

The torus is then graphed with `plot3d` in Figure 2-25. We illustrate the use of the `grid`, `axes`, and `scaling` options.

```
> plot3d([x(s,t),y(s,t),z(s,t)],s=-Pi..Pi,t=-Pi..Pi,
> grid=[40,40],axes=boxed,scaling=constrained);
```

■

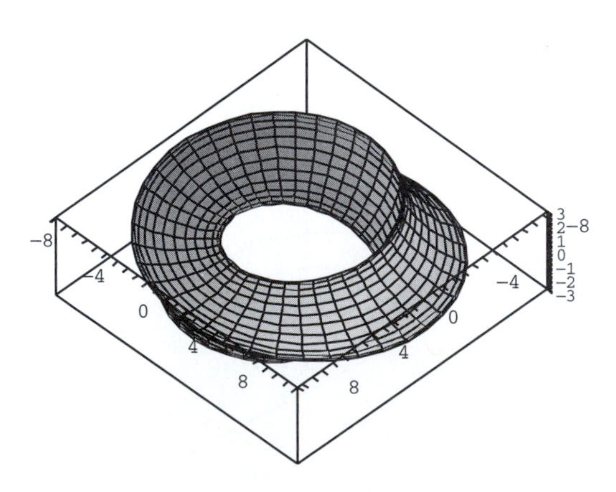

Figure 2-25 Umbilic torus

EXAMPLE 2.3.18 (Gray's Torus Example): A parametrization of an **elliptical torus** is given by

$$x = (a + b\cos v)\cos u, \quad y = (a + b\cos v)\sin u, \quad z = c\sin v.$$

This example is explored in detail in sections 8.2 and 11.4 of Gray's *Modern Differential Geometry of Curves and Surfaces* [11], an indispensable reference for those who use Maple's graphics extensively.

For positive integers p and q, the curve with parametrization

$$x = (a + b\cos qt)\cos pt, \quad y = (a + b\cos qt)\sin pt, \quad z = c\sin qt$$

winds around the elliptical torus and is called a **torus knot**.

Plot the torus if $a = 8$, $b = 3$, and $c = 5$ and then graph the torus knots for $p = 2$ and $q = 5$, $p = 1$ and $q = 10$, and $p = 2$ and $q = 3$.

SOLUTION: We begin by defining `torus` and `torusknot`. Given a, b, and c, `torus(a,b,c)` plots the torus. In the case of `torusknot`, we have used `proc` to define the "indexed function," `torusknot(a,b,c)(p,q)`.

```
> torus:=(a,b,c)->
> plot3d([(a+b*cos(u))*cos(v),(a+b*cos(u))*sin(v),
  c*sin(u)],
> u=0..2*Pi,v=0..2*Pi,
> grid=[60,60],scaling=constrained,axes=boxed):
> torusknot:=(a,b,c)->proc(p,q)
> spacecurve(
> [(a+b*cos(q*t))*cos(p*t),(a+b*cos(q*t))*sin(p*t),
> c*sin(q*t)],t=0..3*Pi,numpoints=300,color=black,
> scaling=constrained,axes=boxed)
> end proc:
```

Next, we use `torus` and `torusknot` to generate all four graphs

```
> torus(8,3,5);
> torusknot(8,3,5)(2,5);
> torusknot(8,3,5)(1,10);
> torusknot(8,3,5)(2,3);
```

and show the results in Figure 2-26.

■

EXAMPLE 2.3.19 (Quadric Surfaces): The **quadric surfaces** are the three-dimensional objects corresponding to the conic sections in two dimensions. A **quadric surface** is a graph of

Also see Example 2.3.16.

$$Ax^2 + By^2 + Cz^2 + Dxy + Exz + Fyz + Gx + Hy + Iz + J = 0,$$

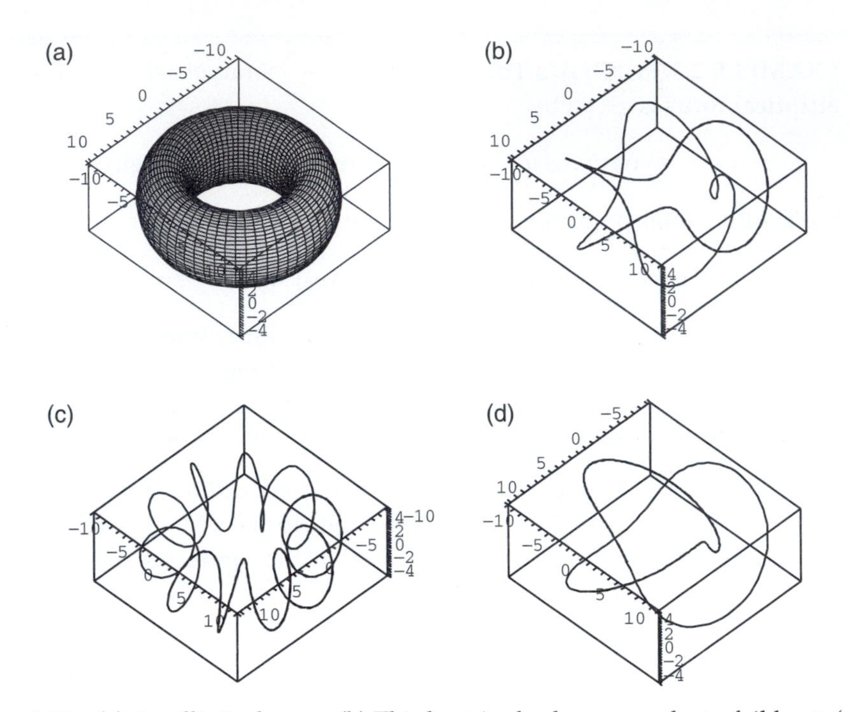

Figure 2-26 (a) An elliptical torus. (b) This knot is also known as the trefoil knot. (c) The curve generated by `torusknot(8,3,5)(1,10)` is not a knot. (d) The torus knot with $p = 2$ and $q = 3$

where $A, B, C, D, E, F, G, H, I,$ and J are constants.

The intersection of a plane and a quadric surface is a conic section.

Three of the basic quadric surfaces, in standard form, and a parametrization of the surface are listed in Table 2-1.

Graph the ellipsoid with equation $\frac{1}{16}x^2 + \frac{1}{4}y^2 + z^2 = 1$, the hyperboloid of one sheet with equation $\frac{1}{16}x^2 + \frac{1}{4}y^2 - z^2 = 1$, and the hyperboloid of two sheets with equation $\frac{1}{16}x^2 - \frac{1}{4}y^2 - z^2 = 1$.

SOLUTION: A parametrization of the ellipsoid with equation $\frac{1}{16}x^2 + \frac{1}{4}y^2 + z^2 = 1$ is given by

$$x = 4\cos t\cos r, \quad y = 2\cos t\sin r,$$

$$z = \sin t, \quad -\pi/2 \le t \le \pi/2, -\pi \le r \le \pi,$$

which is graphed with `plot3d` and shown in Figure 2-27(a).

```
> x:='x':y:='y':z:='z':
> x:=(t,r)->4*cos(t)*cos(r):
```

Table 2-1 Three of the basic quadric surfaces and their parametrizations

Name	Parametric Equations
Ellipsoid $\dfrac{x^2}{a^2}+\dfrac{y^2}{b^2}+\dfrac{z^2}{c^2}=1$	$\begin{cases} x=a\cos t\cos r \\ y=b\cos t\sin r, \quad -\pi/2\le t\le\pi/2, -\pi\le r\le\pi \\ z=c\sin t \end{cases}$
Hyperboloid of One Sheet $\dfrac{x^2}{a^2}+\dfrac{y^2}{b^2}-\dfrac{z^2}{c^2}=1$	$\begin{cases} x=a\sec t\cos r \\ y=b\sec t\sin r, \quad -\pi/2< t<\pi/2, -\pi\le r\le\pi \\ z=c\tan t \end{cases}$
Hyperboloid of Two Sheets $\dfrac{x^2}{a^2}-\dfrac{y^2}{b^2}-\dfrac{z^2}{c^2}=1$	$\begin{cases} x=a\sec t \\ y=b\tan t\cos r, -\pi/2< t<\pi/2 \text{ or } \pi/2<t<3\pi/2, -\pi\le r\le\pi \\ z=c\tan t\sin r \end{cases}$

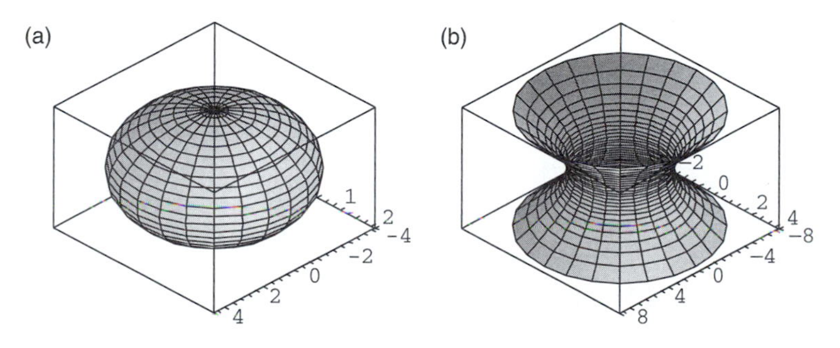

Figure 2-27 (a) Plot of $\frac{1}{16}x^2+\frac{1}{4}y^2+z^2=1$. (b) Plot of $\frac{1}{16}x^2+\frac{1}{4}y^2-z^2=1$

```
> y:=(t,r)->2*cos(t)*sin(r):
> z:=(t,r)->sin(t):
> plot3d([x(t,r),y(t,r),z(t,r)],
> t=-Pi/2..Pi/2,r=-Pi..Pi,axes=BOXED);
```

A parametrization of the hyperboloid of one sheet with equation $\frac{1}{16}x^2+\frac{1}{4}y^2-z^2=1$ is given by

$$x=4\sec t\cos r, \quad y=2\sec t\sin r,$$

$$z=\tan t, \quad -\pi/2<t<\pi/2, -\pi\le r\le\pi.$$

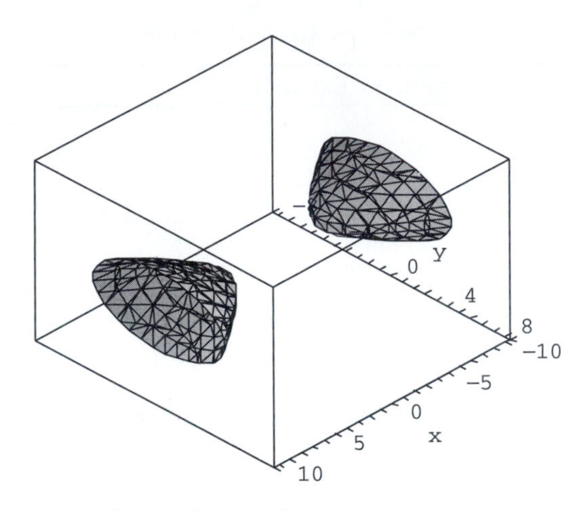

Figure 2-28 Plot of $\frac{1}{16}x^2 - \frac{1}{4}y^2 - z^2 = 1$ generated with `implicitplot3d`

Because $\sec t$ and $\tan t$ are undefined if $t = \pm\pi/2$, we use `plot3d` to graph these parametric equations on a subinterval of $[-\pi/2, \pi/2]$, $[-\pi/3, \pi/3]$ (Figure 2-27(b)).

```
> x:='x':y:='y':z:='z':
> x:=(t,r)->4*sec(t)*cos(r):
> y:=(t,r)->2*sec(t)*sin(r):
> z:=(t,r)->tan(t):
> plot3d([x(t,r),y(t,r),z(t,r)],
> t=-Pi/3..Pi/3,r=-Pi..Pi,axes=BOXED);
```

For (c), we take advantage of the `implicitplot3d` command, which is located in the `plots` package. After the `plots` package has been loaded by entering `with(plots)`, the command

```
implicitplot3d(f(x,y,z),x=a..b,y=c..d,z=u..v)
```

attempts to graph the level surface of $w = f(x, y, z)$ corresponding to $w = 0$.

After loading the `plots` package, we use `implicitplot3d` to graph the equation $\frac{1}{16}x^2 - \frac{1}{4}y^2 - z^2 - 1 = 0$ in Figure 2-28, illustrating the use of the `axes` and `grid` options.

```
> with(plots):
> implicitplot3d(x^2/16-y^2/4-z^2=1,
> x=-10..10,y=-8..8,z=-5..5,axes=BOXED,
   grid=[20,20,20]);
```

■

2.4 Solving Equations and Inequalities

2.4.1 Exact Solutions of Equations

Maple can find exact solutions to many equations and systems of equations, including exact solutions to polynomial equations of degree four or less. Equations in Maple are of the form

```
left-hand side=right-hand side
```

The equals sign '=' between the left-hand side and right-hand side specifies that the object is an equation. For example, to represent the equation $3x + 7 = 4$ in Maple, type `3*x+7=4`. The command `solve(lhs=rhs,x)` attempts to solve the equation `lhs` = `rhs` for x. If the only unknown in the equation `lhs` = `rhs` is x and Maple does not need to use inverse functions to solve for x, the command `solve(lhs=rhs)` solves the equation `lhs` = `rhs` for x. Hence, to solve the equation $3x + 7 = 4$, both the commands `solve(3*x+7=4)` and `solve(3*x+7=4,x)` return the same result.

EXAMPLE 2.4.1: Solve the equations $3x + 7 = 4$, $\dfrac{x^2 - 1}{x - 1} = 0$, and $x^3 + x^2 + x + 1 = 0$.

SOLUTION: In each case, we use `solve` to solve the indicated equation. Be sure to include the equals sign '=' between the left- and right-hand sides of each equation. Thus, the result of entering

Solving linear equations is discussed in more detail in Chapter 5.

```
> solve(3*x+7=4);
```

$$-1$$

means that the solution of $3x+7=4$ is $x=-1$ and the result of entering

```
> solve((x^2-1)/(x-1)=0,x);
```

$$-1$$

means that the solution of $\dfrac{x^2-1}{x-1}=0$ is $x=-1$. On the other hand, the equation $x^3+x^2+x+1=0$ has two imaginary roots. We see that entering

```
> solve(x^3+x^2+x+1=0);
```

$$-1,\ i,\ -i$$

yields all three solutions. Thus, the solutions of $x^3+x^2+x+1=0$ are $x=-1$ and $x=\pm i$. Remember that the Maple symbol `I` represents the complex number $i=\sqrt{-1}$. In general, Maple can find the exact solutions of any polynomial equation of degree four or less.

■

Lists and tables are discussed in more detail in Chapter 4.

Observe that the results of a `solve` command are a **list**.

Maple can also solve equations involving more than one variable for one variable in terms of other unknowns.

EXAMPLE 2.4.2: (a) Solve the equation $v=\pi r^2/h$ for h. (b) Solve the equation $a^2+b^2=c^2$ for a.

SOLUTION: These equations involve more than one unknown so we must specify the variable for which we are solving in the `solve` commands. Thus, entering

```
> solve(v=Pi*r^2/h,h);
```

$$\frac{\pi\, r^2}{v}$$

solves the equation $v = \pi r^2/h$ for h. (Be sure to include an asterisk, *, between π and r.) Similarly, entering

```
> solve(a^2+b^2=c^2,a);
```

$$\sqrt{-b^2+c^2},\ -\sqrt{-b^2+c^2}$$

solves the equation $a^2 + b^2 = c^2$ for a.

■

If Maple needs to use inverse functions to solve an equation, you must be sure to specify the variable(s) for which you want Maple to solve.

EXAMPLE 2.4.3: Find a solution of $\sin^2 x - 2\sin x - 3 = 0$.

SOLUTION: When the command `solve(sin(x)^2-2*sin(x)-3=0)` is entered, Maple solves the equation for x.

```
> solve(sin(x)^2-2*sin(x)-3=0);
```

$$\arcsin(3)\,,\ -1/2\,\pi$$

However, when we set `_EnvAllSolutions:=true`, Maple attempts to solve the equation for *all* values of x. In this case, the equation has infinitely many solutions of the form $x = \frac{1}{2}(4k-1)\pi, k = 0, \pm 1, \pm 2, \ldots$; $\sin x = 3$ has no solutions.

```
> _EnvAllSolutions:=true:
> solve(sin(x)^2-2*sin(x)-3=0);
```

$$\arcsin(3) - 2\arcsin(3)\,\_B1 + 2\pi\,\_Z1 + \pi\,\_B1,\ -1/2\,\pi + 2\pi\,\_Z2$$

■

The example indicates that it is especially important to be careful when dealing with equations involving trigonometric functions.

EXAMPLE 2.4.4: Let $f(\theta) = \sin 2\theta + 2\cos\theta$, $0 \le \theta \le 2\pi$. (a) Solve $f'(\theta) = 0$. (b) Graph $f(\theta)$ and $f'(\theta)$.

SOLUTION: After defining $f(\theta)$, we use `diff` to compute $f'(\theta)$ and then use `solve` to solve $f'(\theta) = 0$.

`diff(f(x),x)` computes $f'(x)$. Topics from calculus are discussed in more detail in Chapter 3.

```
> f:=theta->sin(2*theta)+2*cos(theta):
> df:=diff(f(theta),theta);
```

$$df := 2\cos(2\theta) - 2\sin(\theta)$$

```
> solve(df=0,theta);
```

$$-1/2\pi, 1/6\pi, 5/6\pi$$

Notice that $-\pi/2$ is not between 0 and 2π. Moreover, $\pi/6$ and $5\pi/6$ are *not* the only solutions of $f'(\theta) = 0$ between 0 and 2π. Proceeding by hand, we use the identity $\cos 2\theta = 1 - 2\sin^2\theta$ and factor:

$$2\cos 2\theta - 2\sin\theta = 0$$

$$1 - 2\sin^2\theta - \sin\theta = 0$$

$$2\sin^2\theta + \sin\theta - 1 = 0$$

$$(2\sin\theta - 1)(\sin\theta + 1) = 0$$

so $\sin\theta = 1/2$ or $\sin\theta = -1$. Because we are assuming that $0 \le \theta \le 2\pi$, we obtain the solutions $\theta = \pi/6, 5\pi/6$, or $3\pi/2$. We perform the same steps with Maple.

`subs(x=y,expression)` replaces all occurrences of x in *expression* by y.

```
> s1:=expand(df);
```

$$s1 := 4(\cos(\theta))^2 - 2 - 2\sin(\theta)$$

```
> s2:=subs(cos(theta)^2=1-sin(theta)^2,s1);
```

$$s2 := 2 - 4(\sin(\theta))^2 - 2\sin(\theta)$$

```
> factor(s2);
```

$$-2(\sin(\theta) + 1)(2\sin(\theta) - 1)$$

Finally, we graph $f(\theta)$ and $f'(\theta)$ with `plot` in Figure 2-29. Note that the plot is drawn to scale because we include the option `scaling=constrained`.

```
> plot([f(theta),df],theta=0..2*Pi,color=[black,gray],
> scaling=constrained);
```

■

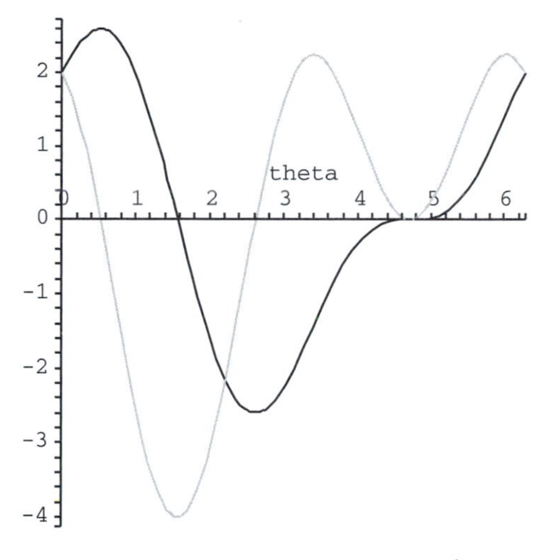

Figure 2-29 Graphs of $f(\theta)$ and $f'(\theta)$

We can also use `solve` to find the solutions, if any, of various types of systems of equations. Entering

```
solve([lhs=rhs1,lhs2=rhs2],{x,y})
```

solves a system of two equations for x and y while entering

```
solve([lhs=rhs1,lhs2=rhs2])
```

attempts to solve the system of equations for all unknowns. In general, `solve` can find the solutions to a system of linear equations. In fact, if the systems to be solved are inconsistent or dependent, Maple 's output indicates so.

Systems of linear equations are discussed in more detail in Chapter 5.

EXAMPLE 2.4.5: Solve each system: (a) $\begin{cases} 3x-y=4 \\ x+y=2 \end{cases}$; (b) $\begin{cases} 2x-3y+4z=2 \\ 3x-2y+z=0 \\ x+y-z=1 \end{cases}$;

(c) $\begin{cases} 2x-2y-2z=-2 \\ -x+y+3z=0 \\ -3x+3y-2z=1 \end{cases}$; and (d) $\begin{cases} -2x+2y-2z=-2 \\ 3x-2y+2z=2 \\ x+3y-3z=-3 \end{cases}$.

SOLUTION: In each case we use solve to solve the given system. For (a), the result of entering

```
> solve(3*x-y=4,x+y=2);
```

$$\{y = 1/2, x = 3/2\}$$

means that the solution of $\begin{cases} 3x - y = 4 \\ x + y = 2 \end{cases}$ is $(x, y) = (3/2, 1/2)$.

(b) We can verify that the results returned by Maple are correct. First, we name the system of equations sys and then use solve to solve the system of equations naming the result sols.

```
> sys:=2*x-3*y+4*z=2,3*x-2*y+z=0,x+y-z=1:
> sols:=solve(sys);
```

$$sols := \left\{x = \frac{7}{10}, y = 9/5, z = 3/2\right\}$$

We verify the result by substituting the values obtained with solve back into sys with subs.

```
> subs(sols,sys);
```

$$\{1 = 1, 0 = 0, 2 = 2\}$$

means that the solution of $\begin{cases} 2x - 3y + 4z = 2 \\ 3x - 2y + z = 0 \\ x + y - z = 1 \end{cases}$ is $(x, y, z) =$ $(7/10,\ 9/5,\ 3/2)$.

(c) When we use solve to solve this system, Maple returns nothing, which indicates that the system has no solution; the system is inconsistent.

```
> solve(2*x-2*y-2*z=-2,-x+y+3*z=0,
  -3*x+3*y-2*z=1,x,y,z);
```

(d) On the other hand, when we use solve to solve this system, Maple's result indicates that the system has infinitely many solutions. That is, all ordered triples of the form $\{(0, z-1, z)|z \text{ real}\}$ are solutions of the system.

```
> solve(-2*x+2*y-2*z=-2,3*x-2*y+2*z=2,
  x+3*y-3*z=-3);
```

$$\{x = 0, y = -1 + z, z = z\}$$

■

We can often use `solve` to find solutions of a nonlinear system of equations as well.

EXAMPLE 2.4.6: Solve the systems (a) $\begin{cases} 4x^2 + y^2 = 4 \\ x^2 + 4y^2 = 4 \end{cases}$ and (b) $\begin{cases} \dfrac{1}{a^2}x^2 + \dfrac{1}{b^2}y^2 = 1 \\ y = mx \end{cases}$ (a, b greater than zero) for x and y.

SOLUTION: (a) The graphs of the equations are both ellipses. We use `contourplot` to graph each equation, naming the results `cp1` and `cp2`, respectively, and then use `display` to display both graphs together in Figure 2-30. The solutions of the system correspond to the intersection points of the two graphs.

```
> with(plots):
> cp1:=contourplot(4*x^2+y^2,x=-4..4,y=-4..4,
   contours=[4],color=black):
> cp2:=contourplot(x^2+4*y^2,x=-4..4,y=-4..4,
   contours=[4],color=black):
> display(cp1,cp2);
```

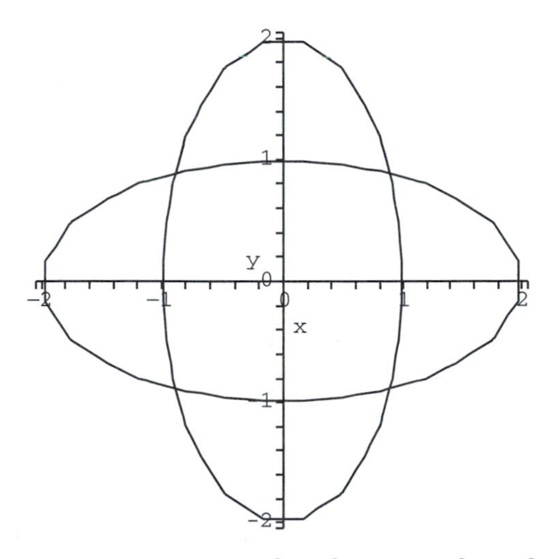

Figure 2-30 Graphs of $4x^2 + y^2 = 4$ and $x^2 + 4y^2 = 4$

Finally, we use `solve` to find the solutions of the system.

```
> sola:=solve(4*x^2+y^2=4,x^2+4*y^2=4);
```

$$sola := \left\{y = 2\,RootOf\left(5\_Z^2 - 1, label = \_L1\right),\right.$$
$$\left. x = 2\,RootOf\left(5\_Z^2 - 1, label = \_L1\right)\right\}$$

$$\left\{x = 2/5\sqrt{5}, y = 2/5\sqrt{5}\right\}, \left\{x = -2/5\sqrt{5}, y = -2/5\sqrt{5}\right\}$$

For (b), we also use `solve` to find the solutions of the system. However, because the unknowns in the equations are a, b, m, x, and y, we must specify that we want to solve for x and y in the `solve` command.

```
> solb:=solve(x^2/a^2+y^2/b^2=1,y=m*x,x,y);
```

$$solb := \{y = mRootOf((b^2 + m^2a^2)\_Z^2 - 1, label = \_L4)ba,$$
$$x = RootOf((b^2 + m^2a^2)\_Z^2 - 1, label = \_L4)ba\}$$

$$\left\{y = m\sqrt{(b^2 + m^2a^2)^{-1}}ba, x = \sqrt{(b^2 + m^2a^2)^{-1}}ba\right\},$$

$$\left\{y = -m\sqrt{(b^2 + m^2a^2)^{-1}}ba, x = -\sqrt{(b^2 + m^2a^2)^{-1}}ba\right\}$$

■

Although Maple can find the exact solution to every polynomial equation of degree four or less, exact solutions to some equations may not be meaningful. In those cases, Maple can provide approximations of the exact solutions using either the `evalf` command in conjunction with `solve` or `fsolve`.

EXAMPLE 2.4.7: Approximate the solutions to the equation $1 - x^2 = x^3$.

SOLUTION: This is a polynomial equation with degree less than five so `solve` will find the exact solutions of the equation. However, the solutions are quite complicated so we use `evalf` to obtain approximate solutions of it.

```
> solb:=solve(1-x^2=x^3,x);
```

$$solb := 1/6\sqrt[3]{100+12\sqrt{69}} + 2/3\frac{1}{\sqrt[3]{100+12\sqrt{69}}} - 1/3,$$

$$-1/12\sqrt[3]{100+12\sqrt{69}} - 1/3\frac{1}{\sqrt[3]{100+12\sqrt{69}}} - 1/3$$

$$+1/2i\sqrt{3}\left(1/6\sqrt[3]{100+12\sqrt{69}} - 2/3\frac{1}{\sqrt[3]{100+12\sqrt{69}}}\right),$$

$$-1/12\sqrt[3]{100+12\sqrt{69}} - 1/3\frac{1}{\sqrt[3]{100+12\sqrt{69}}} - 1/3$$

$$-1/2i\sqrt{3}\left(1/6\sqrt[3]{100+12\sqrt{69}} - 2/3\frac{1}{\sqrt[3]{100+12\sqrt{69}}}\right)$$

```
> evalf(solb);
```

$$0.7548776667, -0.8774388331 + 0.7448617670\,i,$$

$$-0.8774388331 - 0.7448617670\,i$$

■

To solve an identity, use `solve` together with `identity`.

EXAMPLE 2.4.8: Solve $(A+B)\cos x + (A-B+C)\sin x + (A+B-C+D+1)e^x + (A-B+C-D+2)x = 0$.

SOLUTION: In differential equations, we learn that if a linear combination of linearly independent functions is identically the zero function, the corresponding coefficients must be 0. Because the set $S = \{\cos x, \sin x, e^x, x\}$ is linearly independent, the coefficients must all be 0. After defining `eqn` and declaring it to be an identity in the variable x, we use `solve` to solve for A, B, C, and D. Note that we use lower-case letters to avoid any possible ambiguity with built-in Maple objects.

```
> eqn:=(a+b)*cos(x)+(a-b+c)*sin(x)+
> (a+b-c+d+1)*exp(x)+(a-b+c-d+2)*x=0;
```

$$eqn := (a+b)\cos(x) + (a-b+c)\sin(x)$$

$$+ (a+b-c+d+1)\,e^x + (a-b+c-d+2)\,x = 0$$

```
> sols:=solve(identity(eqn,x));
```

$$sols := \{d = 2, a = -3/2, c = 3, b = 3/2\}$$

We verify that these values result in an identity by substituting back into the equation with subs.

```
> subs(sols,eqn);
```

$$0 = 0$$

■

2.4.2 Solving Inequalities

You can also use solve to solve many inequalities. In Maple, the symbols <, <=, >, >=, and <>, represent "less than," "less than or equal to," "greater than," "greater than or equal to," and "not equal to," respectively.

EXAMPLE 2.4.9: Solve $x^3 - 2x^2 - x + 2 \geq 0$.

SOLUTION: We use solve to solve the inequality. We must be careful of our interpretation of the result. Looking back at the inequality, we see that the endpoints must be included. Thus, the solution is $-1 \leq x \leq 1$ and $x \geq 2$.

```
> solve(x^3-2*x^2-x+2>=0);
```

$$RealRange(-1, 1), RealRange(2, \infty)$$

We confirm the result by plotting $y = x^3 - 2x^2 - x + 2$ with plot in Figure 2-31.

```
> plot(x^3-2*x^2-x+2,x=-2..3);
```

■

EXAMPLE 2.4.10: Find the domain of $f(x) = \dfrac{\sqrt{-x^4 + 4x^3 + 4x^2 - 16x}}{x^2 - 2x - 3}$.

SOLUTION: The domain is the values of x for which $-x^4 + 4x^3 + 4x^2 - 16x \geq 0$ and $x^2 - 2x - 3 \neq 0$. We solve these two inequalities together with solve.

```
> solve(-x^4+4*x^3+4*x^2-16*x>=0,x^2-2*x-3<>0,x);
```

$$\{x < -1, -2 \leq x\}, \{-1 < x, x \leq 0\}, \{2 \leq x, x < 3\}, \{x \leq 4, 3 < x\}$$

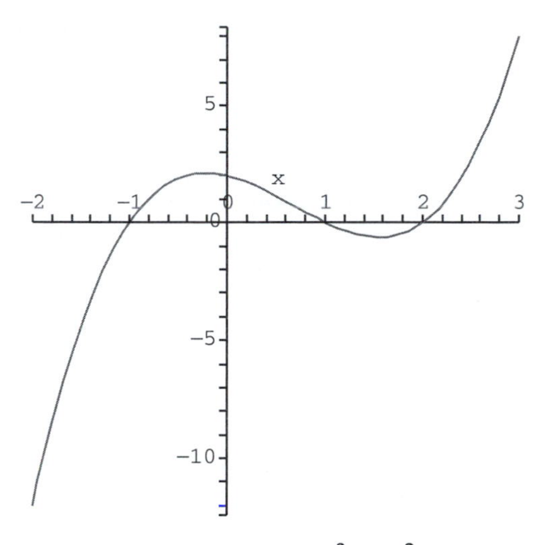

Figure 2-31 Plot of $y = x^3 - 2x^2 - x + 2$

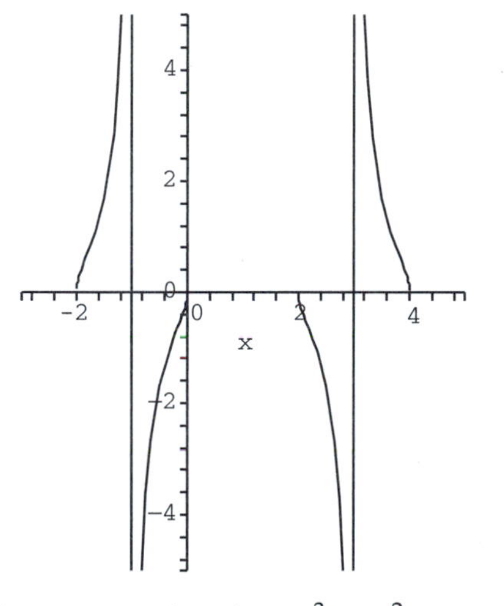

Figure 2-32 Plot of $y = x^3 - 2x^2 - x + 2$

Thus, the domain of $f(x)$ is

$$[-2, -1) \cup (-1, 0] \cup [2, 3) \cup (3, 4],$$

which we confirm by using `plot` to graph $f(x)$ in Figure 2-32.

```
> plot(sqrt(-x^4+4*x^3+4*x^2-16*x)/(x^2-2*x-3),x=-3..5,
   color=black,
> view=[-3..5,-5..5],scaling=constrained);
```

■

2.4.3 Approximate Solutions of Equations

When solving an equation is either impractical or impossible, Maple provides `fsolve` to approximate solutions of equations.

1. `fsolve(eqn,x)` attempts to find a solution of `eqn`.
2. `fsolve(eqn,x=a..b)` attempts to find a solution of `eqn` contained in the interval $[a, b]$.

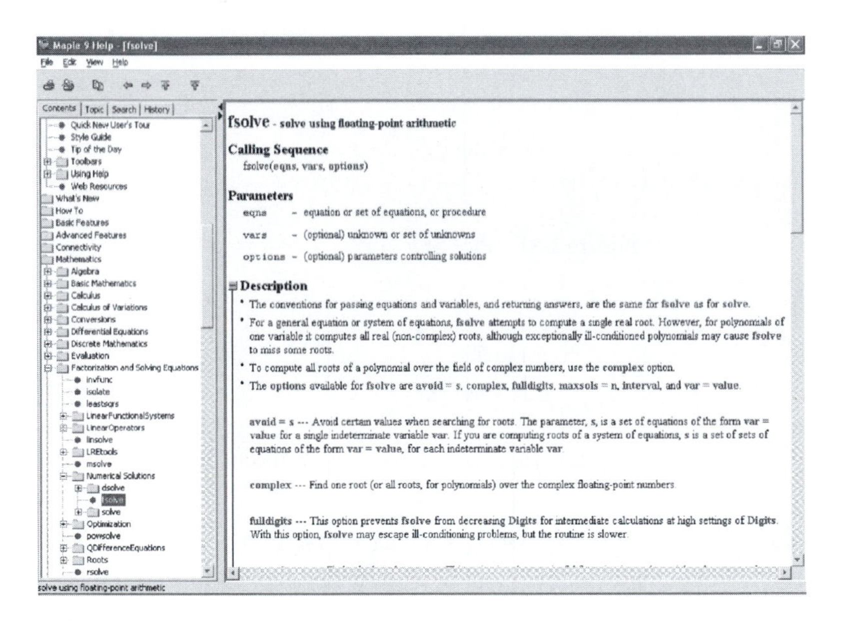

EXAMPLE 2.4.11: Approximate the solutions of $x^5 + x^4 - 4x^3 + 2x^2 - 3x - 7 = 0$.

SOLUTION: We use `fsolve` to approximate the solutions of the equation. Thus, entering

```
> fsolve(x^5+x^4-4*x^3+2*x^2-3*x-7=0);
```

$$-2.744632420, -0.8808584760, 1.796450526$$

approximates the real solutions of $x^5 + x^4 - 4x^3 + 2x^2 - 3x - 7 = 0$ while including the `complex` option in the `fsolve` command

```
> fsolve(x^5+x^4-4*x^3+2*x^2-3*x-7=0,x,complex);
```

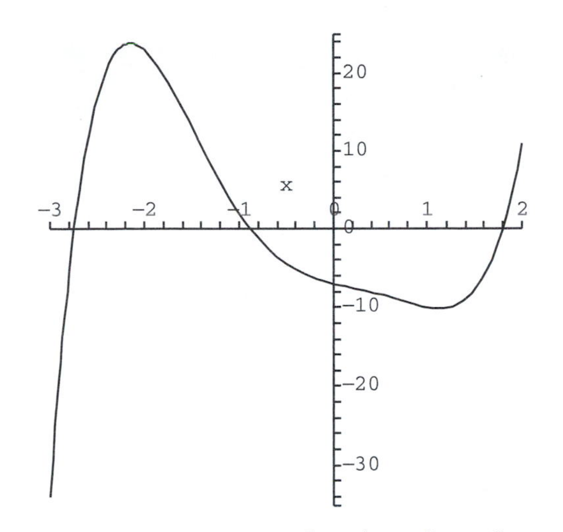

Figure 2-33 Graph of $f(x) = x^5 + x^4 - 4x^3 + 2x^2 - 3x - 7$

$$-2.744632420, -0.8808584760, 0.4145201849 - 1.199959840\,i,$$
$$0.4145201849 + 1.199959840\,i, 1.796450526$$

approximates all solutions.

`fsolve` may also be used to approximate each root of the equation if we supply an initial approximation of the solution that we wish to approximate. The real solutions of $x^5 + x^4 - 4x^3 + 2x^2 - 3x - 7 = 0$ correspond to the values of x where the graph of $f(x) = x^5 + x^4 - 4x^3 + 2x^2 - 3x - 7$ intersects the x-axis. We use `plot` to graph $f(x)$ in Figure 2-33.

```
> f:=x->x^5+x^4-4*x^3+2*x^2-3*x-7:
> plot(f(x),x=-3..2,color=black);
```

We see that the graph intersects the x-axis near $x \approx -2.5$, -1, and 1.5. We use these values as initial approximations of each solution. Thus, entering

```
> fsolve(f(x)=0,x=-1..-0.5);
```

$$-0.8808584760$$

approximates the solution near -1 and entering

```
> fsolve(f(x)=0,x=1.5..2);
```

$$1.796450526$$

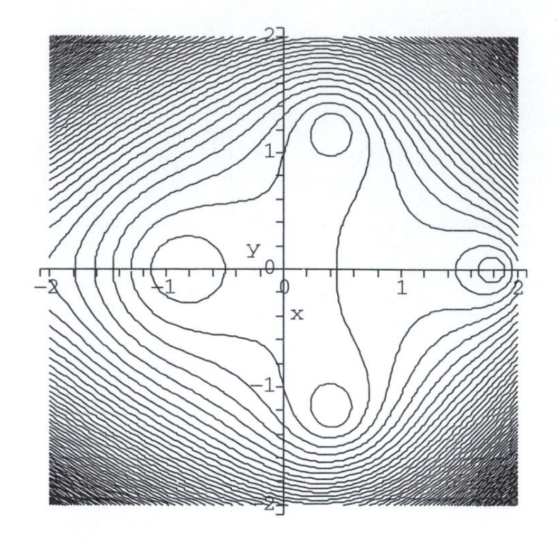

Figure 2-34 Level curves of $w = |f(x+iy)|$, $-2 \leq x \leq 2$, $-2 \leq y \leq 2$

approximates the solution near 1.5. Note that `fsolve` may be used to approximate complex solutions as well. To obtain initial guesses, observe that the solutions of $f(z) = 0$, $z = x + iy$, x, y real, are the level curves of $w = |f(z)|$ that are points. In Figure 2-34, we use `contourplot` to graph various level curves of $w = |f(x+iy)|$, $-2 \leq x \leq 2$, $-2 \leq y \leq 2$. In the plot, observe that the two complex solutions occur at $x \pm iy \approx 0.5 \pm 1.2i$.

```
> f:=z->z^5+z^4-4*z^3+2*z^2-3*z-7:
> with(plots):
> contourplot(abs(f(x+I*y)),x=-2..2,y=-2..2,
   contours=60,grid=[60,60],axes=normal,
> color=black);
```

Thus, entering

```
> fsolve(Re(f(x+I*y))=0,Im(f(x+I*y))=0,
   x=0..0.5,y=1.0..1.25);
```

$$\{y = 1.199959840, x = 0.4145201850\}$$

approximates the solution near $x+iy \approx 0.5+1.2i$. For polynomials with real coefficients, complex solutions occur in conjugate pairs so the other complex solution is approximately $0.41452 - 1.19996i$.

■

EXAMPLE 2.4.12: Find the first three non-negative solutions of $x = \tan x$.

SOLUTION: We attempt to solve $x = \tan x$ with `solve`.

```
> solve(x=tan(x),x);
```

$$RootOf\,(-\tan(\_Z) + \_Z)$$

We next graph $y = x$ and $y = \tan x$ together in Figure 2-35.

```
> plot([x,tan(x)],x=0..4*Pi,view=[0..4*Pi,-4*Pi..4*Pi],
> color=[black,gray]);
```

Remember that vertical lines are never the graphs of functions. In this case, they represent the vertical asymptotes at odd multiples of $\pi/2$.

In the graph, we see that $x = 0$ is a solution. This is confirmed with `fsolve`.

```
> fsolve(x=tan(x),x);
```

$$0.0$$

The second solution is near 4 while the third solution is near 7. Using `fsolve` together with these initial approximations locates the second two solutions.

```
> fsolve(x=tan(x),x=4..5);
```

$$4.493409458$$

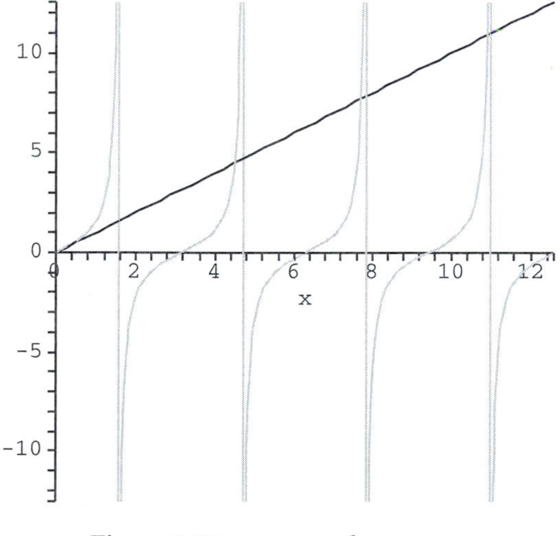

Figure 2-35 $y = x$ and $y = \tan x$

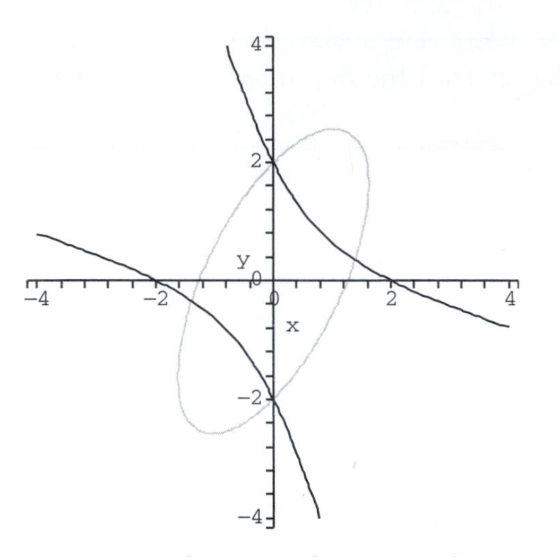

Figure 2-36 Graphs of $x^2 + 4xy + y^2 = 4$ and $5x^2 - 4xy + 2y^2 = 8$

```
> fsolve(x=tan(x),x=7..8);
```

$$7.725251837$$

■

`fsolve` can also be used to approximate solutions to systems of equations. When approximations of solutions of systems of equations are desired, use either `solve` and `evalf` together, when possible, or `fsolve`.

EXAMPLE 2.4.13: Approximate the solutions to the system of equations $\begin{cases} x^2 + 4xy + y^2 = 4 \\ 5x^2 - 4xy + 2y^2 = 8 \end{cases}$.

SOLUTION: We begin by using `contourplot` to graph each equation in Figure 2-36. From the resulting graph, we see that $x^2 + 4xy + y^2 = 4$ is a hyperbola, $5x^2 - 4xy + 2y^2 = 8$ is an ellipse, and there are four solutions to the system of equations.

```
> with(plots):
> cp1:=contourplot(x^2+4*x*y+y^2-4,x=-4..4,y=-4..4,
   contours=[0],
> grid=[60,60],color=black):
> cp2:=contourplot(5*x^2-4*x*y+2*y^2-8,
```

```
> x=-4..4,y=-4..4,contours=[0],
> grid=[60,60],color=gray):
> display(cp1,cp2);
```

From the graph we see that possible solutions are $(0, 2)$ and $(0, -2)$. In fact, substituting $x = 0$ and $y = -2$, and $x = 0$ and $y = 2$, into each equation verifies that these points are both exact solutions of the equation. The remaining two solutions are approximated with `fsolve`.

```
> fsolve(x^2+4*x*y+y^2=4,5*x^2-4*x*y+2*y^2=8,
  x=1..2,y=0..1);
```

$$\{x = 1.392621248, y = 0.3481553119\}$$

```
> fsolve(x^2+4*x*y+y^2=4,5*x^2-4*x*y+2*y^2=8,
  x=-1.5..-1,y=-1..0);
```

$$\{y = -0.3481553119, x = -1.392621248\}$$

■

Calculus 3

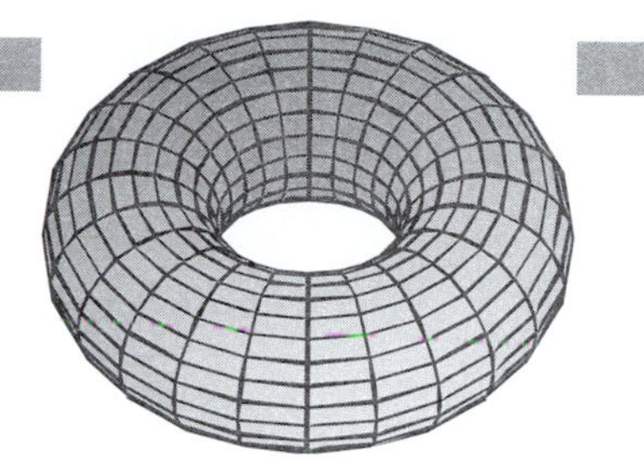

Chapter 3 introduces Maple's built-in calculus commands. The examples used to illustrate the various commands are similar to examples routinely done in first-year calculus courses.

3.1 Limits

Remember that pressing **Enter** or **Return** evaluates commands while pressing **Shift-Return** or **Shift-Enter** gives new lines so that you can continue typing Maple input.

One of the first topics discussed in calculus is that of limits. Maple can be used to investigate limits graphically and numerically. In addition, the Maple command

```
limit(f(x),x=a)
```

attempts to compute the limit of $y = f(x)$ as x approaches a, $\lim_{x\to a} f(x)$, where a can be a finite number, ∞ (`infinity`), or $-\infty$ (`-infinity`).

Remark. To define a function of a single variable, $f(x) = \textit{expression in } x$, enter `f:=x->expression in x`. To generate a basic plot of $y = f(x)$ for $a \le x \le b$, enter `plot(f(x),x=a..b)`.

3.1.1 Using Graphs and Tables to Predict Limits

EXAMPLE 3.1.1: Use a graph and table of values to investigate $\lim_{x\to 0} \dfrac{\sin 3x}{x}$.

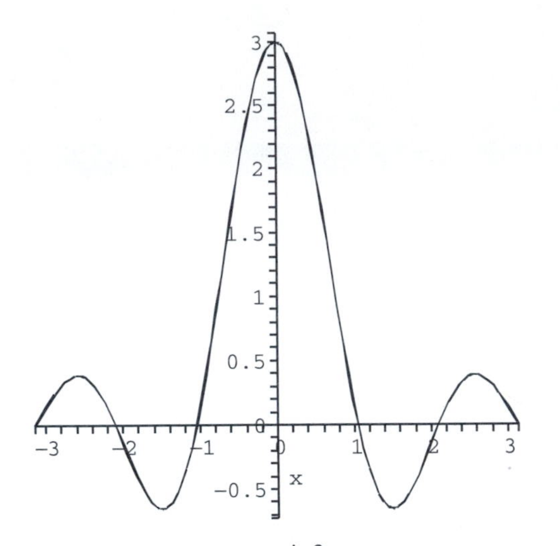

Figure 3-1 Graph of $f(x) = \frac{\sin 3x}{x}$ on the interval $[-\pi, \pi]$

`f:='f'`: clears all prior definitions of f, if any. Clearing function definitions before defining new ones helps eliminate any possible confusion and/or ambiguities.

SOLUTION: We clear all prior definitions of f, define $f(x) = \frac{\sin 3x}{x}$, and then graph $y = f(x)$ on the interval $[-\pi, \pi]$ with `plot`.

```
> f:='f':
> f:=x->sin(3*x)/x:
> plot(f(x),x=-Pi..Pi,color=black);
```

`rand()` returns a "random" 12-digit non-negative integer. Because we are generating "random" numbers, your results will differ from those obtained here.

From the graph shown in Figure 3-1, we might, correctly, conclude that $\lim_{x\to 0} \frac{\sin 3x}{x} = 3$. Further evidence that $\lim_{x\to 0} \frac{\sin 3x}{x} = 3$ can be obtained by computing the values of $f(x)$ for values of x "near" 0. In the following, we use `rand` to define `xvals` to be a table of six "random" real numbers. The first number in `xvals` is between -1 and 1, the second between $-1/10$ and $1/10$, and so on.

```
> xvals:=[seq((-1)^rand()*rand()*10.^(-12-n),n=0..5)];
```

$$xvals := [-0.3211106933, -0.04742561436, 0.007467538305,$$
$$-0.0007229741218, -0.00007455800374, mbox, -0.000003100754872]$$

`map(f,[x1,x2,...,xn])` returns the list $[f(x_1), f(x_2), \ldots, f(x_n)]$.

We then use `map` to compute the value of $f(x)$ for each x in `xvals`.

```
> fvals:=map(f,xvals);
```

$$fvals := [2.557056020, 2.989888890, 2.999749068,$$
$$2.999997647, 2.999999974, 3.0]$$

From these values, we might again correctly deduce that $\lim_{x\to 0} \frac{\sin 3x}{x} = 3$. Of course, these results do not prove that $\lim_{x\to 0} \frac{\sin 3x}{x} = 3$ but they are helpful in convincing us that $\lim_{x\to 0} \frac{\sin 3x}{x} = 3$.
■

3.1.2 Computing Limits

Some limits involving rational functions can be computed by factoring the numerator and denominator.

EXAMPLE 3.1.2: Compute $\lim_{x\to -9/2} \dfrac{2x^2 + 25x + 72}{72 - 47x - 14x^2}$.

SOLUTION: We define `frac1` to be the rational expression $\dfrac{2x^2 + 25x + 72}{72 - 47x - 14x^2}$. We then attempt to compute the value of `frac1` if $x = -9/2$ by using `eval` to evaluate `frac1` if $x = -9/2$ but see that it is undefined.

```
> frac1:=(2*x^2+25*x+72)/(72-47*x-14*x^2):
> eval(frac1,x=-9/2);

Error, numeric exception: division by zero
```

Factoring the numerator and denominator with `factor`, `numer`, and `denom`, we see that

$$\lim_{x\to -9/2} \frac{2x^2 + 25x + 72}{72 - 47x - 14x^2} = \lim_{x\to -9/2} \frac{(x+8)(2x+9)}{(8-7x)(2x+9)} = \lim_{x\to -9/2} \frac{x+8}{8-7x}.$$

The fraction $(x+8)/(8-7x)$ is named `frac2` and the limit is evaluated by computing the value of `frac2` if $x = -9/2$.

```
> factor(numer(frac1));
```

$$-(x+8)(2x+9)$$

```
> factor(denom(frac1));
```

$$(2x+9)(7x-8)$$

```
> frac2:=simplify(frac1);
```

$$frac2 := -\frac{x+8}{7x-8}$$

```
> eval(frac2,x=-9/2);
```

$$\frac{7}{79}$$

We conclude that

$$\lim_{x\to -9/2} \frac{2x^2+25x+72}{72-47x-14x^2} = \frac{7}{79}.$$

■

We can also use the `limit` command to evaluate frequently encountered limits:

```
limit(f(x),x=a)
```

attempts to compute $\lim_{x\to a} f(x)$. Thus, entering

```
> limit((2*x^2+25*x+72)/(72-47*x-14*x^2),x=-9/2);
```

$$\frac{7}{79}$$

computes $\lim_{x\to -9/2} \dfrac{2x^2+25x+72}{72-47x-14x^2} = \dfrac{7}{79}$.

EXAMPLE 3.1.3: Calculate each limit: (a) $\lim_{x\to -5/3} \dfrac{3x^2-7x-20}{21x^2+14x-35}$; (b) $\lim_{x\to 0} \dfrac{\sin x}{x}$; (c) $\lim_{x\to\infty} \left(1+\dfrac{1}{x}\right)^x$; (d) $\lim_{x\to 0} \dfrac{e^{3x}-1}{x}$; (e) $\lim_{x\to\infty} e^{-2x}\sqrt{x}$; and (f) $\lim_{x\to 1^+} \left(\dfrac{1}{\ln x} - \dfrac{1}{x-1}\right)$.

SOLUTION: In each case, we use `limit` to evaluate the indicated limit. Entering

```
> limit((3*x^2-7*x-20)/(21*x^2+14*x-35),x=-5/3);
```

$$\frac{17}{56}$$

computes

$$\lim_{x\to -5/3} \frac{3x^2-7x-20}{21x^2+14x-35} = \frac{17}{56};$$

and entering

```
> limit(sin(x)/x,x=0);
```

$$1$$

computes

$$\lim_{x \to 0} \frac{\sin x}{x} = 1.$$

Maple represents ∞ by `infinity`. Thus, entering

```
> limit((1+1/x)^x,x=infinity);
```

$$e^1$$

computes

$$\lim_{x \to \infty} \left(1 + \frac{1}{x}\right)^x = e.$$

Entering

```
> limit((exp(3*x)-1)/x,x=0);
```

$$3$$

computes

$$\lim_{x \to 0} \frac{e^{3x} - 1}{x} = 3.$$

Entering

```
> limit(exp(-2*x)*sqrt(x),x=infinity);
```

$$0$$

computes $\lim_{x \to \infty} e^{-2x}\sqrt{x} = 0$, and entering

```
> limit(1/ln(x)-1/(x-1),x=1);
```

$$1/2$$

computes

$$\lim_{x \to 1^+} \left(\frac{1}{\ln x} - \frac{1}{x - 1}\right) = \frac{1}{2}.$$

■

Because $\ln x$ is undefined for $x \leq 0$, a right-hand limit is mathematically necessary, even though Maple's `limit` function computes the limit correctly without the distinction.

We can often use the `limit` command to compute symbolic limits.

EXAMPLE 3.1.4: If $\$P$ is compounded n times per year at an annual interest rate of r, the value of the account, A, after t years is given by

$$A = P\left(1 + \frac{r}{n}\right)^{nt}.$$

The formula for continuously compounded interest is obtained by taking the limit of this expression as $t \to \infty$.

SOLUTION: The formula for continuously compounded interest, $A = Pe^{rt}$, is obtained using `limit`.

```
> limit(p*(1+r/n)^(n*t),n=infinity);
```

$$e^{rt}p$$

■

3.1.3 One-Sided Limits

In some cases, Maple can compute certain one-sided limits. The command

```
limit(f(x),x=a,left)
```

attempts to compute $\lim_{x\to a^-} f(x)$ while

```
limit(f(x),x=a,right)
```

attempts to compute $\lim_{x\to a^+} f(x)$.

EXAMPLE 3.1.5: Compute (a) $\lim_{x\to 0^+} |x|/x$; (b) $\lim_{x\to 0^-} |x|/x$; (c) $\lim_{x\to 0^+} 1/x$; (d) $\lim_{x\to 0^-} 1/x$; (e) $\lim_{x\to 0^+} e^{-1/x}$; and (f) $\lim_{x\to 0^-} e^{-1/x}$.

SOLUTION: Even though $\lim_{x\to 0} |x|/x$ does not exist, $\lim_{x\to 0^+} |x|/x = 1$ and $\lim_{x\to 0^-} |x|/x = -1$, as we see using `limit` together with the `left` and `right` options, respectively.

```
> limit(abs(x)/x,x=0);
```

$$undefined$$

```
> limit(abs(x)/x,x=0,right);
```

$$1$$

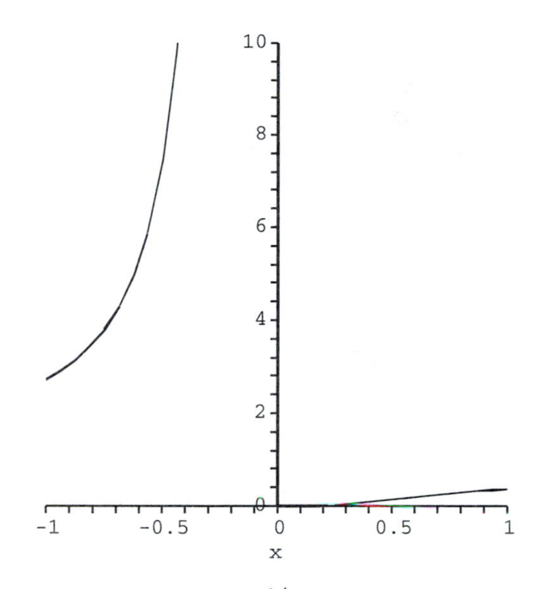

Figure 3-2 Graph of $y = e^{-1/x}$ on the interval $[-3/2, 3/2]$

```
> limit(abs(x)/x,x=0,left);
```

$$-1$$

The `right` and `left` options are used to calculate the correct values for (c) and (d), respectively. For (c), we have:

```
> limit(1/x,x=0);
> limit(1/x,x=0,right);
> limit(1/x,x=0,left);
```

undefined

$$\infty$$

$$-\infty$$

For (e) we see that $\lim_{x \to 0} e^{-1/x}$ does not exist (see Figure 3-2).

```
> limit(exp(-1/x),x=0);
```

undefined

Using `limit` together with the `left` and `right` options gives the correct left and right limits as well.

```
> limit(exp(-1/x),x=0,right);
```

$$0$$

```
> limit(exp(-1/x),x=0,left);
```

$$\infty$$

We confirm these results by graphing $y = e^{-1/x}$ with `plot` in Figure 3-2.

```
> plot(exp(-1/x),x=-3/2..3/2,view=[-1..1,0..10],
   color=black);
```

■

The `limit` command together with the `left` and `right` options is a "fragile" command and should be used with caution because its results are unpredictable, especially for the beginner. It is wise to check or confirm results using a different technique for nearly all problems faced by the beginner.

3.2 Differential Calculus

3.2.1 Definition of the Derivative

Definition 1 (The Derivative): The **derivative** of $y = f(x)$ is

$$y' = f'(x) = \frac{dy}{dx} = \lim_{h\to 0} \frac{f(x+h) - f(x)}{h}, \tag{3.1}$$

provided the limit exists.

The `limit` command can be used along with `simplify` to compute the derivative of a function using the definition of the derivative.

EXAMPLE 3.2.1: Use the definition of the derivative to compute the derivative of (a) $f(x) = x + 1/x$, (b) $g(x) = 1/\sqrt{x}$, and (c) $h(x) = \sin 2x$.

`Limit` is the *inert* form of the `limit` function: `Limit(f(x),x=a)` returns the symbols $\lim_{x\to a} f(x)$ while `limit(f(x),x=a)` attempts to calculate $\lim_{x\to a} f(x)$.

SOLUTION: For (a) and (b), we first define f and g, compute the difference quotient, $\left(f(x+h) - f(x)\right)/h$, simplify the difference quotient with `simplify`, and use `limit` to calculate the derivative.

```
> f:=x->x+1/x:
> s1:=(f(x+h)-f(x))/h;
```

$$s1 := \left(h + (x+h)^{-1} - x^{-1}\right)h^{-1}$$

```
> s2:=simplify(s1);
```

$$s2 := \frac{x^2 + xh - 1}{(x+h)\,x}$$

```
> Limit(s2,h=0)=limit(s2,h=0);
```

$$\lim_{h\to 0} \frac{x^2 + xh - 1}{(x+h)\,x} = \frac{x^2 - 1}{x^2}$$

```
> g:=x->1/sqrt(x):
> s1:=(g(x+h)-g(x))/h;
```

$$s1 := \left(\frac{1}{\sqrt{x+h}} - \frac{1}{\sqrt{x}}\right) h^{-1}$$

```
> s2:=simplify(s1);
```

$$s2 := -\frac{-\sqrt{x} + \sqrt{x+h}}{\sqrt{x+h}\sqrt{x}h}$$

```
> Limit(s2,h=0)=limit(s2,h=0);
```

$$\lim_{h\to 0} -\frac{-\sqrt{x} + \sqrt{x+h}}{\sqrt{x+h}\sqrt{x}h} = -1/2\,x^{-3/2}$$

For (c), we define h and then use `expand` to simplify the difference quotient. We use `limit` to compute the derivative. The result is written as a single trigonometric function using `combine` with the `trig` option and shows us that $\frac{d}{dx}(\sin 2x) = 2\cos 2x$.

```
> h:=x->sin(2*x):
> s1:=(h(x+h)-h(x))/h;
```

$$s1 := \frac{\sin(2\,x + 2\,h) - \sin(2\,x)}{h}$$

```
> s2:=expand(s1);
```

$$s2 := 4\,\frac{\sin(x)\cos(x)(\cos(h))^2}{h} - 4\,\frac{\sin(x)\cos(x)}{h} + 4\,\frac{(\cos(x))^2\sin(h)\cos(h)}{h} - 2\,\frac{\sin(h)\cos(h)}{h}$$

```
> s3:=limit(s2,h=0);
```

$$s3 := 4\,(\cos(x))^2 - 2$$

```
> s4:=combine(s3,trig);
```

$$s4 := 2\,\cos(2\,x)$$

■

If the derivative of $y = f(x)$ exists at $x = a$, a geometric interpretation of $f'(a)$ is that $f'(a)$ is the slope of the line tangent to the graph of $y = f(x)$ at the point $(a, f(a))$.

To motivate the definition of the derivative, many calculus texts choose a value of x, $x = a$, and then draw the graph of the secant line passing through the points $(a, f(a))$ and $(a + h, f(a + h))$ for "small" values of h to show that as h approaches 0, the secant line approaches the tangent line. An equation of the secant line passing through the points $(a, f(a))$ and $(a + h, f(a + h))$ is given by

$$y - f(a) = \frac{f(a+h) - f(a)}{(a+h) - a}(x - a) \quad \text{or} \quad y = \frac{f(a+h) - f(a)}{h}(x - a) + f(a).$$

EXAMPLE 3.2.2: If $f(x) = 9 - 4x^2$, graph $f(x)$ together with the secant line containing $(1, f(1))$ and $(1 + h, f(1 + h))$ for various values of h.

SOLUTION: We define $f(x) = 9 - 4x^2$ and $y(x, h)$ to be a function returning the line containing $(1, f(1))$ and $(1 + h, f(1 + h))$.

```
> with(plots):
> f:=x->9-4*x^2:
> y:=(x,h)->(f(1+h)-f(1))/h*(x-1)+f(1):
```

In the following, we use `animate` to show the graphs of $f(x)$ and $y(x, h)$ for $h = 1, 2, \ldots, 9$. The resulting animation can be played and controlled from the Maple menu (Figure 3-3).

```
> animate(f(x),y(x,1/h),x=-3..3,h=1..10,color=black,
> view=[-3..3,-10..10]);
```

If instead the command is entered as

```
> A:=animate(f(x),y(x,1/h),x=-3..3,h=1..10,
   color=black,
> view=[-3..3,-10..10]):
> display(A);
```

the result is displayed as a graphics array (Figure 3-4).

Greater control over the graphics is obtained by using `plot` directly as indicated with the following commands.

```
> hvals:=[seq((2/3)^i,i=-1..9)];
```

$$hvals := \left[3/2, 1, 2/3, 4/9, \frac{8}{27}, \frac{16}{81}, \frac{32}{243}, \frac{64}{729}, \frac{128}{2187}, \frac{256}{6561}, \frac{512}{19683}\right]$$

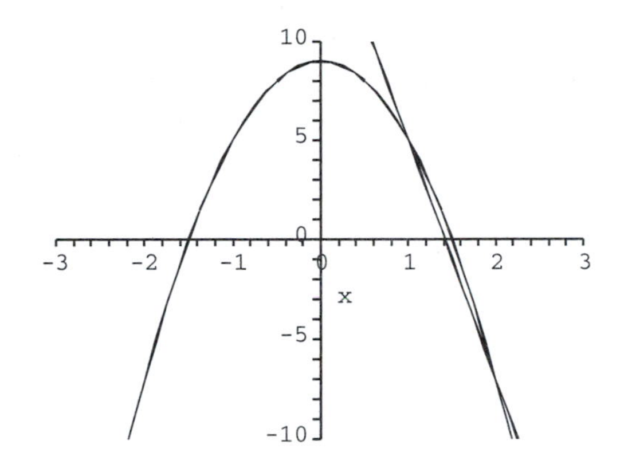

Figure 3-3 An animation

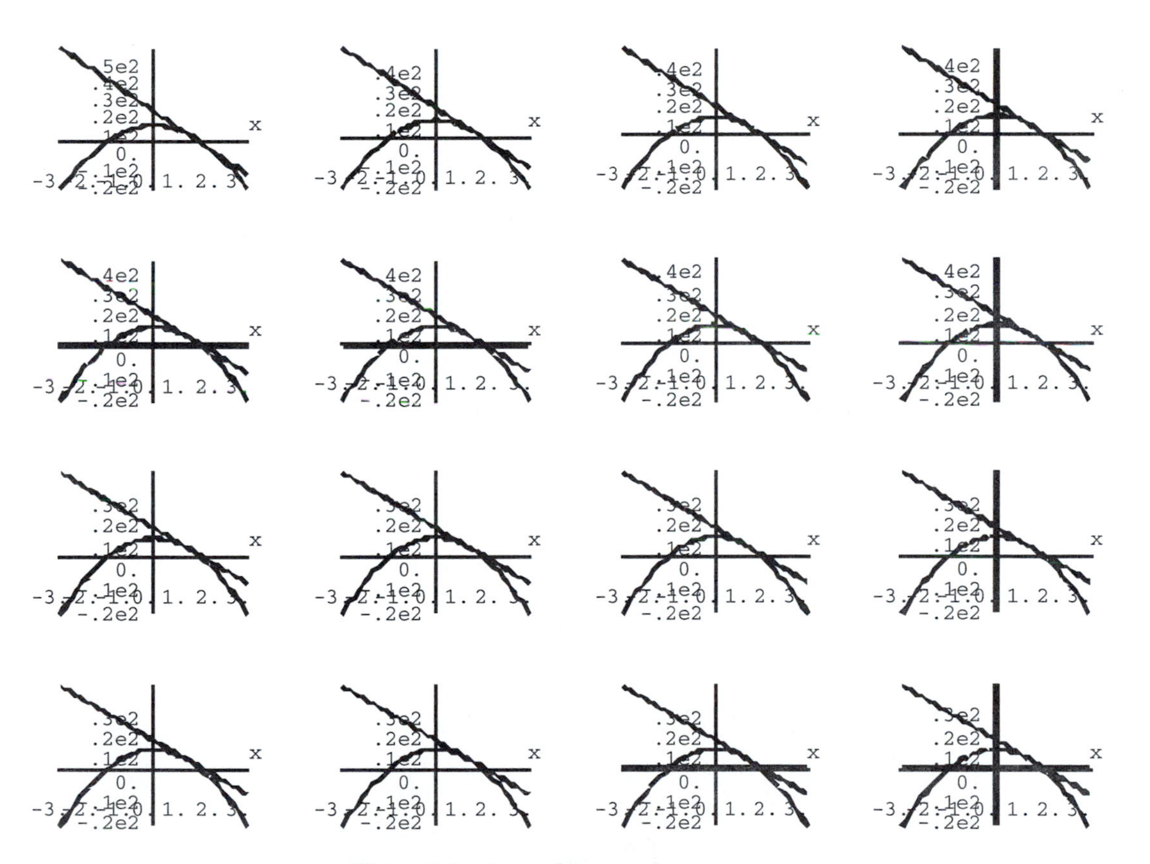

Figure 3-4 A graphics array

```
> toshow:=[seq(plot([f(x),y(x,h)],x=-3..3,
   color=[black,gray],
> view=[-3..3,-10..10]),h=hvals)]:
```

Entering

```
> display(toshow);
```

displays the result as a graphics array while

```
> display(toshow,insequence=true);
```

generates an animation.

■

3.2.2 Calculating Derivatives

The functions `D` and `diff` are used to differentiate functions. Assuming that $y = f(x)$ is differentiable,

1. `diff(f(x),x)` computes and returns $f'(x) = df/dx$,
2. `D(f)(x)` computes and returns $f'(x) = df/dx$,
3. `diff(f(x),x$n)` computes and returns $f^{(n)}(x) = d^n f/dx^n$, and
4. `(D@@n)(f)(x)` computes and returns $f^{(n)}(x) = d^n f/dx^n$.

Maple knows the numerous differentiation rules, including the product, quotient, and chain rules. Thus, entering

```
> f:='f':g:='g':
> diff(f(x)*g(x),x);
```

$$\left(\frac{d}{dx}f(x)\right)g(x) + f(x)\frac{d}{dx}g(x)$$

shows us that $\frac{d}{dx}(f(x)) \cdot g(x) = f'(x)g(x) + f(x)g'(x)$; entering

```
> (simplify@diff)(f(x)/g(x),x);
```

$$\frac{\left(\frac{d}{dx}f(x)\right)g(x) - f(x)\frac{d}{dx}g(x)}{(g(x))^2}$$

shows us that $\frac{d}{dx}(f(x)/g(x)) = (f'(x)g(x) - f(x)g'(x))/(g(x))^2$; and entering

```
> diff(f(g(x)),x);
```

$$\mathrm{D}(f)(g(x))\,\frac{d}{dx}g(x)$$

shows us that $\frac{d}{dx}(f(g(x))) = f'(g(x))\,g'(x)$.

EXAMPLE 3.2.3: Compute the first and second derivatives of (a) $y = x^4 + \frac{4}{3}x^3 - 3x^2$, (b) $f(x) = 4x^5 - \frac{5}{2}x^4 - 10x^3$, (c) $y = \sqrt{e^{2x} + e^{-2x}}$, and (d) $y = (1 + 1/x)^x$.

SOLUTION: For (a), we use `diff`.

```
> diff(x^4+4/3*x^3-3*x^2,x);
```

$$4x^3 + 4x^2 - 6x$$

```
> diff(x^4+4/3*x^3-3*x^2,x$2);
```

$$12x^2 + 8x - 6$$

For (b), we first define f and then use `D` to calculate $f'(x)$ and $f''(x)$.

```
> f:=x->4*x^5-5/2*x^4-10*x^3:
> factor(D(f)(x));
> factor((D@@2)(f)(x));
```

$$10x^2(x+1)(2x-3)$$

$$10x\left(8x^2 - 3x - 6\right)$$

For (c), we use `simplify` together with `diff` to calculate and simplify y' and y''.

```
> diff(sqrt(exp(2*x)+exp(-2*x)),x);
> (simplify@diff)(sqrt(exp(2*x)+exp(-2*x)),x$2);
```

$$1/2\,\frac{2e^{2x} - 2e^{-2x}}{\sqrt{e^{2x} + e^{-2x}}}$$

$$\frac{e^{4x} + 6 + e^{-4x}}{\left(e^{2x} + e^{-2x}\right)^{3/2}}$$

By hand, (d) would require logarithmic differentiation. The second derivative would be particularly difficult to compute by hand. Maple quickly computes and simplifies each derivative.

```
> simplify(diff((1+1/x)^x,x));
> simplify(diff((1+1/x)^x,x$2));
```

$$\left(\frac{x+1}{x}\right)^x \left(x \ln\left(\frac{x+1}{x}\right) + \ln\left(\frac{x+1}{x}\right) - 1\right)(x+1)^{-1}$$

$$\left(\frac{x+1}{x}\right)^x \left(x^3 \left(\ln\left(\frac{x+1}{x}\right)\right)^2 + 2x^2 \left(\ln\left(\frac{x+1}{x}\right)\right)^2 - 2x^2 \ln\left(\frac{x+1}{x}\right)\right.$$
$$\left. + x\left(\ln\left(\frac{x+1}{x}\right)\right)^2 - 2x\ln\left(\frac{x+1}{x}\right) + x - 1\right) x^{-1}(x+1)^{-2}$$

■

map and operations on lists are discussed in more detail in Chapter 4.

The command `map(f,list)` applies the function `f` to each element of the list `list`. Thus, if you are computing the derivatives of a large number of functions, you can use `map` together with `diff`.

EXAMPLE 3.2.4: Compute the first and second derivatives of $\sin x$, $\cos x$, $\tan x$, $\sin^{-1} x$, $\cos^{-1} x$, and $\tan^{-1} x$.

SOLUTION: Notice that lists are contained in brackets. Thus, entering

```
> map(diff,[sin(x),cos(x),tan(x),arcsin(x),arccos(x),
  arctan(x)],x);
```

$$\left[\cos(x), -\sin(x), 1+(\tan(x))^2, \frac{1}{\sqrt{1-x^2}}, -\frac{1}{\sqrt{1-x^2}}, \left(1+x^2\right)^{-1}\right]$$

and

```
> map(diff,[sin(x),cos(x),tan(x),arcsin(x),arccos(x),
  arctan(x)],x$2);
```

$$\left[-\sin(x), -\cos(x), 2\tan(x)\left(1+(\tan(x))^2\right), \frac{x}{\left(1-x^2\right)^{3/2}},\right.$$
$$\left. -\frac{x}{\left(1-x^2\right)^{3/2}}, -2\frac{x}{\left(1+x^2\right)^2}\right]$$

computes the first and second derivatives of the three trigonometric functions and their inverses.

■

3.2.3 Implicit Differentiation

If an equation contains two variables, x and y, implicit differentiation can be carried out by explicitly declaring y to be a function of x, $y = y(x)$, and using `diff` or by using the `implicitdiff` command.

EXAMPLE 3.2.5: Find $y' = dy/dx$ if (a) $\cos(e^{xy}) = x$ and (b) $\ln(x/y) + 5xy = 3y$.

SOLUTION: For (a) we illustrate the use of `diff`. Notice that we are careful to specifically indicate that $y = y(x)$. First we differentiate with respect to x

```
> s1:=diff(cos(exp(x*y(x)))=x,x);
```

$$s1 := -\sin\left(e^{xy(x)}\right)\left(y(x) + x\frac{d}{dx}y(x)\right)e^{xy(x)} = 1$$

and then we solve the resulting equation for $y' = dy/dx$ with

```
> s2:=solve(s1,diff(y(x),x));
```

$$s2 := -\frac{\sin\left(e^{xy(x)}\right)e^{xy(x)}y(x) + 1}{\sin\left(e^{xy(x)}\right)e^{xy(x)}x}$$

For (b), we use `implicitdiff`.

```
> implicitdiff(ln(x/y)+5*x*y=3*y,y,x);
```

$$-\frac{y(1 + 5xy)}{x(-1 + 5xy - 3y)}$$

shows us that if $\ln(x/y) + 5xy = 3y$,

$$y' = \frac{dy}{dx} = -\frac{(1 + 5xy)y}{(5xy - 3y - 1)x}$$

■

3.2.4 Tangent Lines

If $f'(a)$ exists, we interpret $f'(a)$ to be the slope of the line tangent to the graph of $y = f(x)$ at the point $(a, f(a))$. An equation of the tangent is given by

$$y - f(a) = f'(a)(x - a) \quad \text{or} \quad y = f'(a)(x - a) + f(a).$$

EXAMPLE 3.2.6: Find an equation of the line tangent to the graph of $f(x) = \sin x^{1/3} + \cos^{1/3} x$ at the point with x-coordinate $x = 5\pi/3$.

SOLUTION: We first define $f(x)$ and compute $f'(x)$.

```
> f:=x->sin(x^(1/3))+cos(x)^(1/3):
> D(f)(x);
```

$$1/3\,\frac{\cos\left(\sqrt[3]{x}\right)}{x^{2/3}} - 1/3\,\frac{\sin(x)}{(\cos(x))^{2/3}}$$

Then, the slope of the line tangent to the graph of $f(x)$ at the point with x-coordinate $x = 5\pi/3$ is

```
> D(f)(5*Pi/3);
> evalf(D(f)(5*Pi/3));
```

$$1/15\,\frac{\cos\left(1/3\,\sqrt[3]{5}3^{2/3}\sqrt[3]{\pi}\right)\sqrt[3]{5}3^{2/3}}{\pi^{2/3}} + 1/6\,2^{2/3}\sqrt{3}$$

$$0.4400126493$$

while the y-coordinate of the point is

```
> f(5*Pi/3);
> evalf(f(5*Pi/3));
```

$$\sin\left(1/3\,\sqrt[3]{5}3^{2/3}\sqrt[3]{\pi}\right) + 1/2\,2^{2/3}$$

$$1.780008715$$

Thus, an equation of the line tangent to the graph of $f(x)$ at the point with x-coordinate $x = 5\pi/3$ is

$$y - \left(\frac{1}{\sqrt[3]{2}} + \sin\sqrt[3]{5\pi/3}\right) = \left(\frac{\cos\sqrt[3]{5\pi/3}}{\sqrt[3]{3}\sqrt[3]{25\pi^2}} + \frac{1}{\sqrt[3]{2}\sqrt{3}}\right)\left(x - \frac{5\pi}{3}\right),$$

as shown in Figure 3-5. To generate the plot, notice that we redefine f using the `surd` function because computing $f(x)$ values requires taking odd roots of negative numbers.

surd(x,n) = $\begin{cases} x^{1/n}, & x \geq 0 \\ -(-x)^{1/n}, & x < 0 \end{cases}$

```
> fsurd:=x->sin(surd(x,3))+surd(cos(x),3):
> plot([fsurd(x),D(f)(5*Pi/3)*(x-5*Pi/3)+f(5*Pi/3)],
   x=0..4*Pi,
> color=[black,gray],scaling=constrained);
```

■

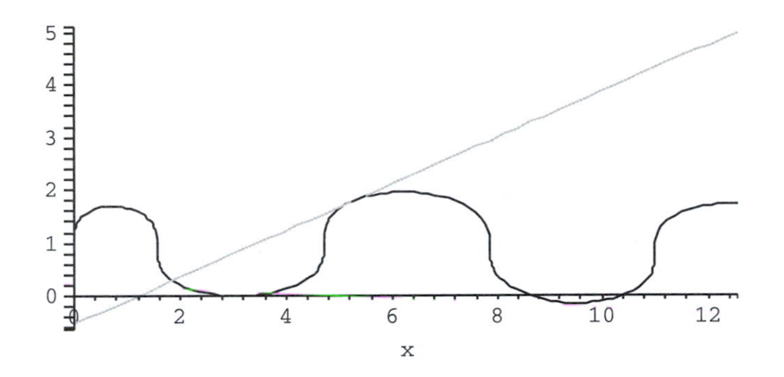

Figure 3-5 $f(x) = \sin x^{1/3} + \cos^{1/3} x$ together with its tangent at the point $(5\pi/3, f(5\pi/3))$

EXAMPLE 3.2.7: Find an equation of the line tangent to the graph of $f(x) = 9 - 4x^2$ at the point $(1, f(1))$.

SOLUTION: After defining f, we see that $f(1) = 5$ and $f'(1) = -8$

```
> f:=x->9-4*x^2:
> f(1);
> D(f)(1);
```

$$5$$

$$-8$$

so an equation of the line tangent to $y = f(x)$ at the point $(1,5)$ is $y - 5 = -8(x-1)$ or $y = -8x + 13$. We can visualize the tangent at $(1, f(1))$ with `showtangent`, which is contained in the `student` package, or `plot` (Figure 3-6).

```
> with(student):

> showtangent(f(x),x=1,color=[gray,black]);
> plot([f(x),D(f)(1)*(x-1)+f(1)],x=-3..3,
   color=[black,gray],
> view=[-3..3,-10..10]);
```

In addition, we can view a sequence of lines tangent to the graph of a function for a sequence of x values using `animate`. In the following, we use `animate` to generate graphs of $y = f(x)$ and $y = f'(a)(x-a) + f(a)$ for 50 equally spaced values of a between -3 and 3 (Figure 3-7).

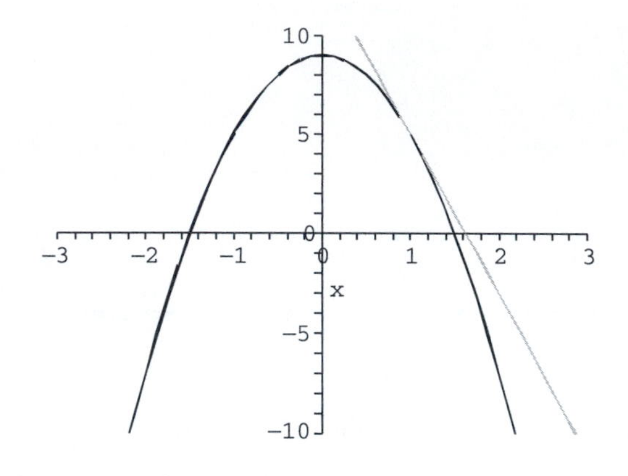

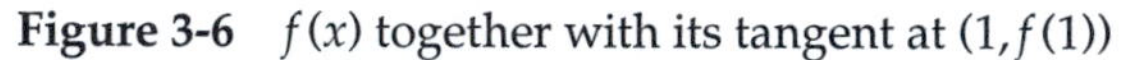
Figure 3-6 $f(x)$ together with its tangent at $(1, f(1))$

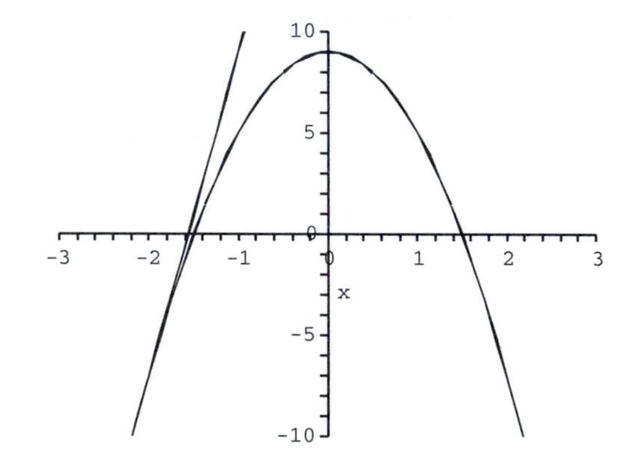

Figure 3-7 An animation

```
> animate(f(x),D(f)(a)*(x-a)+f(a),x=-3..3,a=-2..2,
   frames=50,
> color=black,view=[-3..3,-10..10]);
```

On the other hand,

```
> A:=animate(f(x),D(f)(a)*(x-a)+f(a),x=-3..3,a=-2..2,
   frames=9,
> color=black,view=[-3..3,-10..10]):
> display(A);
```

graphs $y = f(x)$ and $y = f'(a)(x - a) + f(a)$ for nine equally spaced values of a between -3 and 3 and displays the result as a graphics array (Figure 3-8).

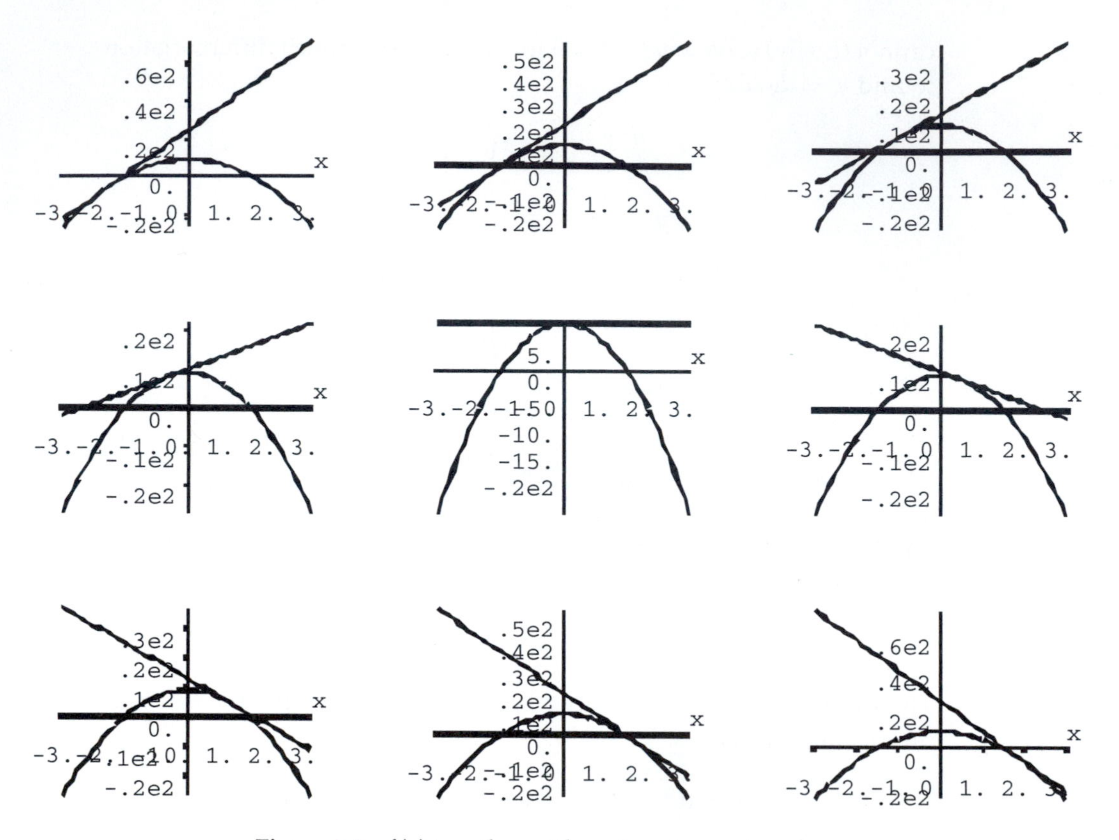

Figure 3-8 $f(x)$ together with various tangents

In the graphs, notice that where the tangent lines have positive slope ($f'(x) > 0$), $f(x)$ is increasing while where the tangent lines have negative slope ($f'(x) < 0$), $f(x)$ is decreasing.

■

Tangent Lines of Implicit Functions

EXAMPLE 3.2.8: Find equations of the tangent line and normal line to the graph of $x^2y - y^3 = 8$ at the point $(-3, 1)$. Find and simplify $y'' = d^2y/dx^2$.

SOLUTION: We will evaluate $y' = dy/dx$ if $x = -3$ and $y = 1$ to determine the slope of the tangent line at the point $(-3, 1)$. Note that we

By the product and chain rules, $\frac{d}{dx}(x^2y) = \frac{d}{dx}(x^2)y + x^2\frac{d}{dx}(y) = 2x \cdot y + x^2 \cdot \frac{dy}{dx} = 2xy + x^2y'$.

cannot (easily) solve $x^2y - y^3 = 8$ for y so we use implicit differentiation to find $y' = dy/dx$:

$$\frac{d}{dx}\left(x^2y - y^3\right) = \frac{d}{dx}(8)$$

$$2xy + x^2y - 3y^2y' = 0$$

$$y' = \frac{-2xy}{x^2 - 3y^2}.$$

```
> with(plots):
> eq:=x^2*y-y^3=8:
> s1:=implicitdiff(eq,y,x);
```

$$s1 := -2\frac{xy}{x^2 - 3y^2}$$

We then use `eval` to find that the slope of the tangent at $(-3, 1)$ is

```
> s2:=eval(s1,x=-3,y=1);
```

$$s2 := 1$$

The slope of the normal is $-1/1 = -1$. Equations of the tangent and normal are given by

$$y - 1 = 1(x + 3) \qquad \text{and} \qquad y - 1 = -1(x + 3),$$

respectively (Figure 3-9).

```
> cp1:=contourplot(x^2*y-y^3-8,x=-5..5,y=-5..5,
   grid=[50,50],
> color=black,contours=[0]):
```

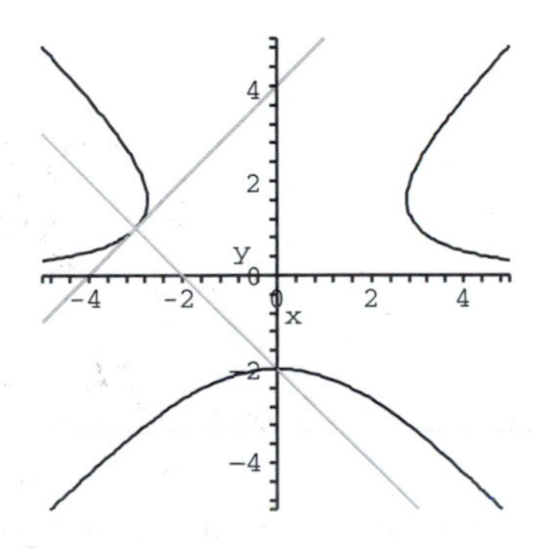

Figure 3-9 Graphs of $x^2y - y^3 = 8$ (in black) and the tangent and normal at $(-3, 1)$ (in gray)

```
> cp2:=plot([(x+3)+1,-(x+3)+1],x=-5..5,
   color=[gray,gray]):
> display(cp1,cp2,view=[-5..5,-5..5],
   scaling=constrained);
```

To find $y'' = d^2y/dx^2$, we also use `implicitdiff`. Both

```
> implicitdiff(eq,y,x,x);
```

$$6\,\frac{y\left(x^4+2\,x^2y^2-3\,y^4\right)}{x^6-9\,x^4y^2+27\,x^2y^4-27\,y^6}$$

and

```
> implicitdiff(eq,y,x$2);
```

$$6\,\frac{y\left(x^4+2\,x^2y^2-3\,y^4\right)}{x^6-9\,x^4y^2+27\,x^2y^4-27\,y^6}$$

find

$$y'' = \frac{d^2y}{dx^2} = \frac{6\left(x^2y-y^3\right)\left(x^2+3y^2\right)}{\left(x^2-3y^2\right)^3}.$$

Because $x^2y - y^3 = 8$, the second derivative is further simplified to

$$y'' = \frac{d^2y}{dx^2} = \frac{48\left(x^2+3y^2\right)}{\left(x^2-3y^2\right)^3}.$$

■

Parametric Equations and Polar Coordinates

For the parametric equations $\{x = f(t), y = g(t)\}$, $t \in I$,

$$y' = \frac{dy}{dx} = \frac{dy/dt}{dx/dt} = \frac{g'(t)}{f'(t)}$$

and

$$y'' = \frac{d^2y}{dx^2} = \frac{d}{dx}\frac{dy}{dx} = \frac{d/dt(dy/dx)}{dx/dt}.$$

If $\{x = f(t), y = g(t)\}$ has a tangent line at the point $(f(a), g(a))$, parametric equations of the tangent are given by

$$x = f(a) + tf'(a) \qquad \text{and} \qquad y = g(a) + tg'(a). \tag{3.2}$$

If $f'(a), g'(a) \neq 0$, we can eliminate the parameter from (3.2)

$$\frac{x - f(a)}{f'(a)} = \frac{y - g(a)}{g'(a)}$$

$$y - g(a) = \frac{g'(a)}{f'(a)}(x - f(a))$$

and obtain an equation of the tangent line in point-slope form.

```
> x:='x':y:='y':
> l:=solve(x(a)+t*D(x)(a)=X,t);
> r:=solve(y(a)+t*D(y)(a)=Y,t);
```

$$l := \frac{-x(a) + X}{D(x)(a)}$$

$$r := \frac{-y(a) + Y}{D(y)(a)}$$

EXAMPLE 3.2.9 (The Cycloid): The **cycloid** has parametric equations

$$x = t - \sin t \qquad \text{and} \qquad y = 1 - \cos t.$$

Graph the cycloid together with the line tangent to the graph of the cycloid at the point $(x(a), y(a))$ for various values of a between -2π and 4π.

SOLUTION: After defining x and y we use `diff` to compute dy/dt and dx/dt. We then compute $dy/dx = (dy/dt)/(dx/dt)$ and d^2y/dx^2.

```
> x:=t->t-sin(t):
> y:=t->1-cos(t):
> dx:=D(x)(t);
> dy:=D(y)(t);
> dydx:=dy/dx;
```

$$dx := 1 - \cos(t)$$

$$dy := \sin(t)$$

$$dydx := \frac{\sin(t)}{1 - \cos(t)}$$

```
> dypdt:=simplify(diff(dydx,t));
```

$$dypdt := (-1 + \cos(t))^{-1}$$

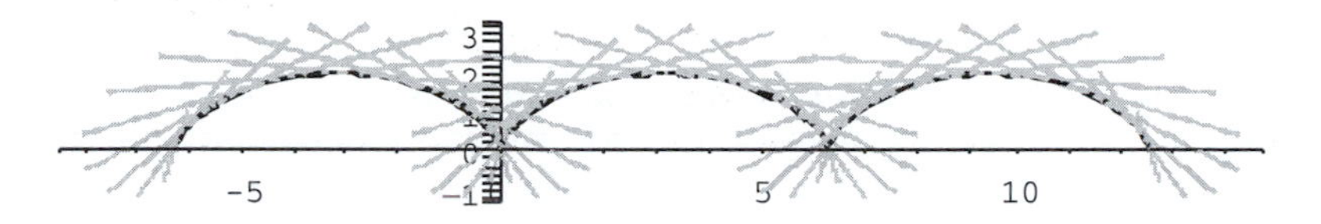

Figure 3-10 The cycloid with various tangents

```
> secondderiv:=simplify(dypdt/dx);
> factor(denom(secondderiv));
```

$$secondderiv := -\left(1 - 2\cos(t) + (\cos(t))^2\right)^{-1}$$

$$(-1 + \cos(t))^2$$

We then use `plot` to graph the cycloid for $-2\pi \le t \le 4\pi$, naming the resulting graph `p1`.

```
> with(plots):
> p1:=plot([x(t),y(t),t=-2*Pi..4*Pi],color=BLACK):
```

Next, we use `seq` and `plot` to graph 40 tangent lines, equation (3.2), for 40 equally spaced values of a between -2π and 4π and name the resulting graph `p2`. Finally, we show `p1` and `p2` together with the `display` function. The resulting plot is shown to scale because the lengths of the x- and y-axes are equal and we include the option `scaling=CONSTRAINED`. In the graphs, notice that on intervals for which dy/dx is defined, dy/dx is a decreasing function and, consequently, $d^2y/dx^2 < 0$ (Figure 3-10).

```
> avals:=[seq(-2*Pi+6*Pi*i/39,i=0..39)]:
> p2:=plot([seq([x(a)+t*D(x)(a),y(a)+t*D(y)(a),
   t=-2..2],a=avals)],
> color=gray):
> display(p1,p2,scaling=CONSTRAINED);
```

■

EXAMPLE 3.2.10 (Orthogonal Curves): Two lines L_1 and L_2 with slopes m_1 and m_2, respectively, are **orthogonal** if their slopes are negative reciprocals: $m_1 = -1/m_2$.

Extended to curves, we say that the curves C_1 and C_2 are **orthogonal** at a point of intersection if their respective tangent lines to the curves at that point are orthogonal.

Show that the family of curves with equation $x^2 + 2xy - y^2 = C$ is orthogonal to the family of curves with equation $y^2 + 2xy - x^2 = C$.

SOLUTION: We begin by defining `Eq1` and `Eq2` to be the equations $x^2 + 2xy - y^2 = C$ and $y^2 + 2xy - x^2 = C$, respectively.

```
> Eq1:=x^2+2*x*y-y^2=c:
> Eq2:=y^2+2*x*y-x^2=c:
```

We then use `implicitdiff` to find $y' = dy/dx$. Because the derivatives are negative reciprocals, we conclude that the curves are orthogonal. We confirm this graphically by graphing several members of each family with `contourplot` and showing the results together (Figure 3-11).

```
> dEq1:=implicitdiff(Eq1,y,x);
> dEq2:=implicitdiff(Eq2,y,x);
```

$$dEq1 := -\frac{x+y}{x-y}$$

$$dEq2 := \frac{x-y}{x+y}$$

```
> with(plots):
> cp1:=contourplot(x^2+2*x*y-y^2,x=-5..5,y=-5..5,
   color=black):
> cp2:=contourplot(y^2+2*x*y-x^2,x=-5..5,y=-5..5,
   color=gray):
> display(cp1,cp2,scaling=CONSTRAINED);
```

■

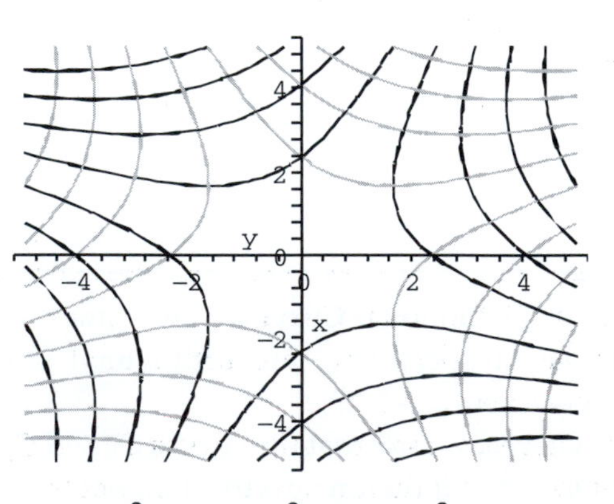

Figure 3-11 $x^2 + 2xy - y^2 = C$ and $y^2 + 2xy - x^2 = C$ for various values of C

EXAMPLE 3.2.11 (The Mean-Value Theorem for Derivatives):
Theorem 1 (The Mean-Value Theorem). *If $y = f(x)$ is continuous on $[a,b]$ and differentiable on (a,b) then there is at least one value of c between a and b for which*

$$f'(c) = \frac{f(b) - f(a)}{b - a} \quad \textit{or, equivalently,} \quad f(b) - f(a) = f'(c)(b - a). \tag{3.3}$$

Find all number(s) c that satisfy the conclusion of the Mean-Value theorem for $f(x) = x^2 - 3x$ on the interval $[0, 7/2]$.

SOLUTION: By the Power rule, $f'(x) = 2x - 3$. The slope of the secant containing $(0, f(0))$ and $(7/2, f(7/2))$ is

$$\frac{f(7/2) - f(0)}{7/2 - 0} = \frac{1}{2}.$$

Solving $2x - 3 = 1/2$ for x gives us $x = 7/4$.

```
> f:=x->x^2-3*x:
> solve(D(f)(x)=0);
> solve(D(f)(x)=(f(7/2)-f(0))/(7/2-0));
```

$$3/2$$

$$7/4$$

$x = 7/4$ satisfies the conclusion of the Mean-Value theorem for $f(x) = x^2 - 3x$ on the interval $[0, 7/2]$, as shown in Figure 3-12.

```
> with(plots):
> p1:=plot(f(x),x=-2..4,color=black):
> p2:=plot(f(x),x=0..7/2,color=black,thickness=5):
> p3:=plot([[0,f(0)],[7/4,f(7/4)]],style=point,
   color=black,
> symbol=circle,symbolsize=15):
> p4:=plot([D(f)(7/4)*(x-7/4)+f(7/4),
   (f(7/2)-f(0))/(7/2-0)*x],
> x=-2..4,color=black,style=point,symbol=point):
> display([p1,p2,p3,p4],view=[-2..4,-5..5],
   scaling=constrained);
```

■

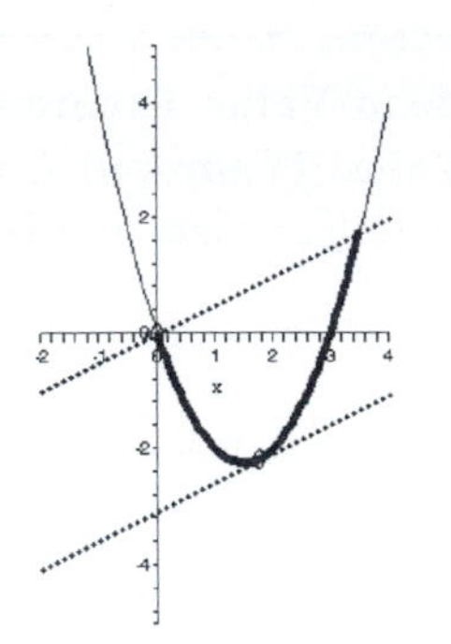

Figure 3-12 Graphs of $f(x) = x^2 - 3x$, the secant containing $(0, f(0))$ and $(7/2, f(7/2))$, and the tangent at $(7/4, f(7/4))$

3.2.5 The First Derivative Test and Second Derivative Test

Examples 3.2.12 and 3.2.13 illustrate the following properties of the first and second derivative.

Theorem 2. *Let $y = f(x)$ be continuous on $[a, b]$ and differentiable on (a, b).*

1. *If $f'(x) = 0$ for all x in (a, b), then $f(x)$ is constant on $[a, b]$.*
2. *If $f'(x) > 0$ for all x in (a, b), then $f(x)$ is increasing on $[a, b]$.*
3. *If $f'(x) < 0$ for all x in (a, b), then $f(x)$ is decreasing on $[a, b]$.*

For the second derivative, we have the following theorem.

Theorem 3. *Let $y = f(x)$ have a second derivative on (a, b).*

1. *If $f''(x) > 0$ for all x in (a, b), then the graph of $f(x)$ is concave up on (a, b).*
2. *If $f''(x) < 0$ for all x in (a, b), then the graph of $f(x)$ is concave down on (a, b).*

The **critical points** correspond to those points on the graph of $y = f(x)$ where the tangent line is horizontal or vertical; the number $x = a$ is a **critical number** if $f'(a) = 0$ or $f'(x)$ does not exist if $x = a$. The **inflection points** correspond to those points on the graph of $y = f(x)$ where the graph of $y = f(x)$ is neither concave up nor concave down. Theorems 2 and 3 help establish the First Derivative Test and Second Derivative Test.

Theorem 4 (First Derivative Test). *Let $x = a$ be a critical number of a function $y = f(x)$ continuous on an open interval I containing $x = a$. If $f(x)$*

is differentiable on I, except possibly at $x = a$, $f(a)$ can be classified as follows.

1. *If $f'(x)$ changes from positive to negative at $x = a$, then $f(a)$ is a* ***relative maximum****.*
2. *If $f'(x)$ changes from negative to positive at $x = a$, then $f(a)$ is a* ***relative minimum****.*

Theorem 5 (Second Derivative Test). *Let $x = a$ be a critical number of a function $y = f(x)$ and suppose that $f''(x)$ exists on an open interval containing $x = a$.*

1. *If $f''(a) < 0$, then $f(a)$ is a relative maximum.*
2. *If $f''(a) > 0$, then $f(a)$ is a relative minimum.*

EXAMPLE 3.2.12: Graph $f(x) = 3x^5 - 5x^3$.

SOLUTION: We begin by defining $f(x)$ and then computing and factoring $f'(x)$ and $f''(x)$.

```
> f:=x->3*x^5-5*x^3:
> d1:=factor(D(f)(x));
> d2:=factor((D@@2)(f)(x));
```

$$d1 := 15\,x^2\,(x-1)\,(x+1)$$

$$d2 := 30\,x\left(2\,x^2-1\right)$$

By inspection, we see that the critical numbers are $x = 0$, 1, and -1 while $f''(x) = 0$ if $x = 0$, $1/\sqrt{2}$, or $-1/\sqrt{2}$. Of course, these values can also be found with `solve` as done next in `cns` and `ins`, respectively.

```
> cns:=[solve(d1=0,x)];
> ins:=[solve(d2=0,x)];
```

$$cns := [-1, 1, 0, 0]$$

$$ins := [0, 1/2\sqrt{2}, -1/2\sqrt{2}]$$

We find the critical and inflection points by using `map` to compute $f(x)$ for each value of x in `cns` and `ins`, respectively. The result means that the critical points are $(0,0)$, $(1,-2)$, and $(-1,2)$; the inflection points are $(0,0)$, $(1/\sqrt{2}, -7\sqrt{2}/8)$, and $(-1/\sqrt{2}, 7\sqrt{2}/8)$. We also see that $f''(0) = 0$ so Theorem 5 cannot be used to classify $f(0)$.

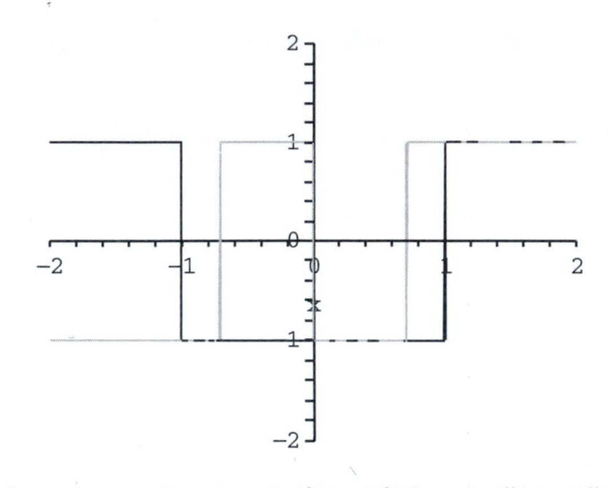

Figure 3-13 Graphs of $|f'(x)|/f'(x)$ and $|f''(x)|/f''(x)$

On the other hand, $f''(1) = 30 > 0$ and $f''(-1) = -30 < 0$ so by Theorem 5, $f(1) = -2$ is a relative minimum and $f(-1) = 2$ is a relative maximum.

```
> cps:=map(f,cns);
> ips:=map(f,ins);
> map((D@@2)(f),cns);
```

$$cps := [2, -2, 0, 0]$$

$$ips := [0, -\frac{7}{8}\sqrt{2}, \frac{7}{8}\sqrt{2}]$$

$$[-30, 30, 0, 0]$$

We can graphically determine the intervals of increase and decrease by noting that if $f'(x) > 0$ ($f'(x) < 0$), $|f'(x)|/f'(x) = 1$ ($|f'(x)|/f'(x) = -1$). Similarly, the intervals for which the graph is concave up and concave down can be determined by noting that if $f''(x) > 0$ ($f''(x) < 0$), $|f''(x)|/f''(x) = 1$ ($|f''(x)|/f''(x) = -1$). We use `plot` to graph $|f'(x)|/f'(x)$ and $|f''(x)|/f''(x)$ in Figure 3-13.

```
> plot([abs(d1)/d1,abs(d2)/d2],x=-2..2,
   color=[black,gray],view=[-2..2,-2..2]);
```

From the graph, we see that $f'(x) > 0$ for x in $(-\infty, -1) \cup (1, \infty)$, $f'(x) < 0$ for x in $(-1, 1)$, $f''(x) > 0$ for x in $(-1/\sqrt{2}, 0) \cup (1/\sqrt{2}, \infty)$, and $f''(x) < 0$ for x in $(-\infty, -1/\sqrt{2}) \cup (0, 1/\sqrt{2})$. Thus, the graph of $f(x)$ is

- increasing and concave down for x in $(-\infty, -1)$,
- decreasing and concave down for x in $(-1, -1/\sqrt{2})$,
- decreasing and concave up for x in $(-1/\sqrt{2}, 0)$,

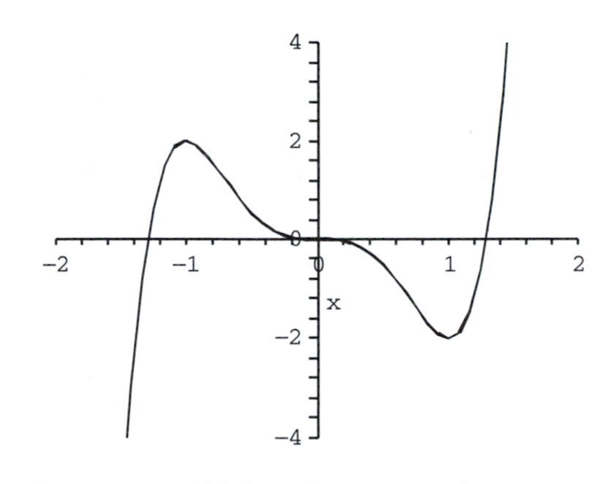

Figure 3-14 $f(x)$ for $-2 \le x \le 2$ and $-4 \le y \le 4$

- decreasing and concave down for x in $(0, 1\sqrt{2})$,
- decreasing and concave up for x in $(1/\sqrt{2}, 1)$, and
- increasing and concave up for x in $(1, \infty)$.

We also see that $f(0) = 0$ is neither a relative minimum nor maximum. To see all points of interest, our domain must contain -1 and 1 while our range must contain -2 and 2. We choose to graph $f(x)$ for $-2 \le x \le 2$; we choose the range displayed to be $-4 \le y \le 4$ (Figure 3-14).

```
> plot(f(x),x=-2..2,view=[-2..2,-4..4],color=black);
```

■

Remember to be especially careful when working with functions that involve odd roots.

EXAMPLE 3.2.13: Graph $f(x) = (x-2)^{2/3}(x+1)^{1/3}$.

SOLUTION: We begin by defining $f(x)$ and then computing and simplifying $f'(x)$ and $f''(x)$ with `D` and `simplify`.

```
> f:=x->(x-2)^(2/3)*(x+1)^(1/3):
> d1:=simplify(D(f)(x));
> d2:=simplify((D@@2)(f)(x));
```

$$d1 := \frac{x}{\sqrt[3]{x-2}\,(x+1)^{2/3}}$$

$$d2 := -2\,\frac{1}{(x-2)^{4/3}\,(x+1)^{5/3}}$$

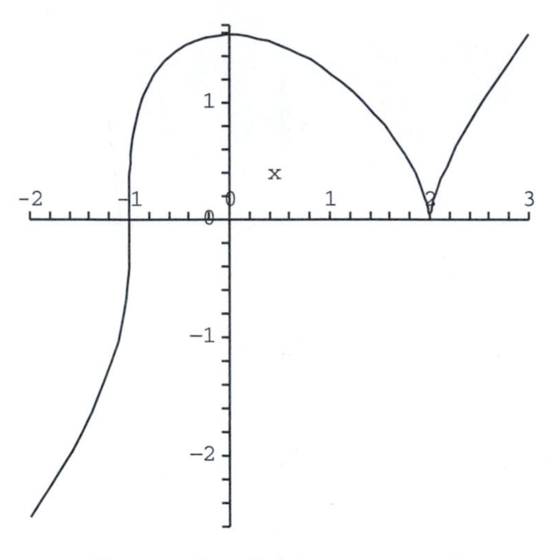

Figure 3-15 $f(x)$ for $-2 \leq x \leq 3$

By inspection, we see that the critical numbers are $x = 0$, 2, and -1. We cannot use Theorem 5 to classify $f(2)$ and $f(-1)$ because $f''(x)$ is undefined if $x = 2$ or -1. On the other hand, $f''(0) < 0$ so $f(0) = 2^{2/3}$ is a relative maximum. By hand, we make a sign chart to see that the graph of $f(x)$ is

- increasing and concave up on $(-\infty, -1)$,
- increasing and concave down on $(-1, 0)$,
- decreasing and concave down on $(0, 2)$, and
- increasing and concave down on $(2, \infty)$.

Hence, $f(-1) = 0$ is neither a relative minimum nor maximum while $f(2) = 0$ is a relative minimum by Theorem 4. To graph $f(x)$, redefine $f(x)$ using `surd` and then use `plot` to graph $f(x)$ for $-2 \leq x \leq 3$ in Figure 3-15.

```
> f:=x->surd((x-2)^2,3)*surd(x+1,3):
plot(f(x),x=-2..3,color=black);
```

■

The previous examples illustrate that if $x = a$ is a critical number of $f(x)$ and $f'(x)$ makes a *simple change in sign* from positive to negative at $x = a$, then $(a, f(a))$ is a relative maximum. If $f'(x)$ makes a *simple change in sign* from negative to positive at $x = a$, then $(a, f(a))$ is a relative minimum. Maple is especially useful in investigating interesting functions for which this may not be the case.

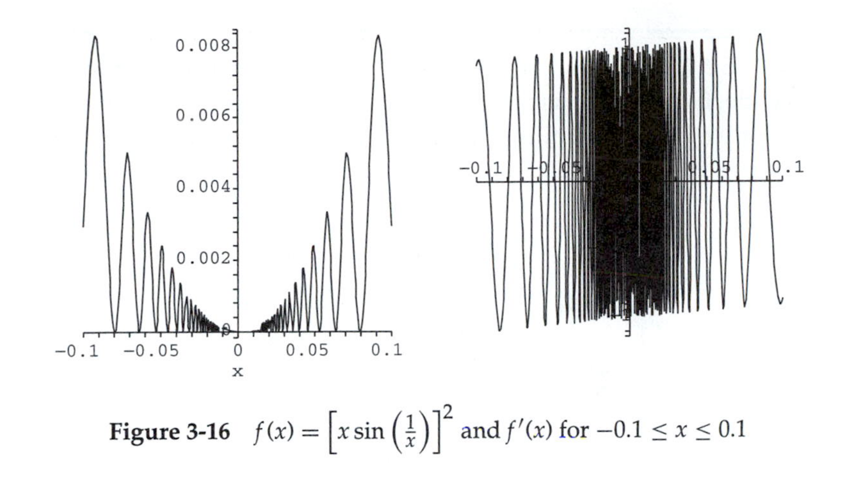

Figure 3-16 $f(x) = \left[x \sin\left(\frac{1}{x}\right)\right]^2$ and $f'(x)$ for $-0.1 \le x \le 0.1$

EXAMPLE 3.2.14: Consider

$$f(x) = \begin{cases} x^2 \sin^2\left(\dfrac{1}{x}\right), & x \neq 0 \\ 0, & x = 0 \end{cases}.$$

$x = 0$ is a critical number because $f'(x)$ does not exist if $x = 0$. $(0, 0)$ is both a relative and absolute minimum but $f'(x)$ does not make a simple change in sign at $x = 0$, as illustrated in Figure 3-16.

```
> f:=x->x^2*sin(1/x)^2:
> factor(D(f)(x));
```

$$2 \sin\left(x^{-1}\right)\left(x \sin\left(x^{-1}\right) - \cos\left(x^{-1}\right)\right)$$

```
> plot(f(x),x=-0.1..0.1,color=black);
> plot(D(f)(x),x=-0.1..0.1,color=black);
```

In the figure, notice that the derivative "oscillates" infinitely many times near $x = 0$, so the first derivative test cannot be used to classify $(0, 0)$.

3.2.6 Applied Max/Min Problems

Maple can be used to assist in solving maximization/minimization problems encountered in a differential calculus course.

EXAMPLE 3.2.15: A woman is located on one side of a body of water 4 miles wide. Her position is directly across from a point on the other side of the body of water 16 miles from her house, as shown in the figure.

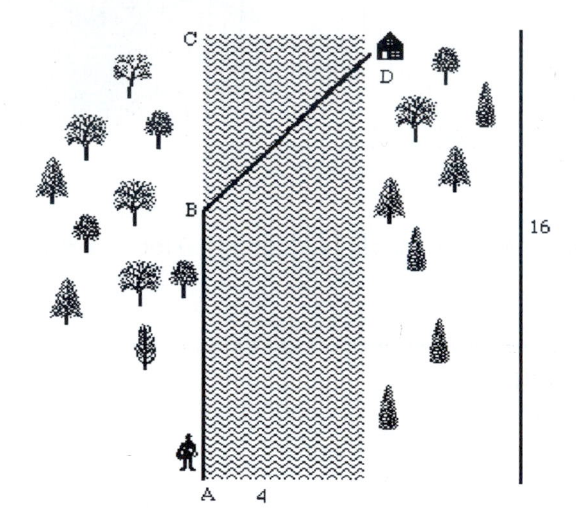

If she can move across land at a rate of 10 miles per hour and move over water at a rate of 6 miles per hour, find the least amount of time for her to reach her house.

SOLUTION: From the figure, we see that the woman will travel from A to B by land and then from B to D by water. We wish to find the least time for her to complete the trip.

Let x denote the distance BC, where $0 \le x \le 16$. Then, the distance AB is given by $16 - x$ and, by the Pythagorean theorem, the distance BD is given by $\sqrt{x^2 + 4^2}$. Because rate $\times$ time $=$ distance, time $=$ distance/rate. Thus, the time to travel from A to B is $\frac{1}{10}(16 - x)$, the time to travel from B to D is $\frac{1}{6}\sqrt{x^2 + 16}$, and the total time to complete the trip, as a function of x, is

$$\mathit{time}(x) = \frac{1}{10}(16 - x) + \frac{1}{6}\sqrt{x^2 + 16}, \quad 0 \le x \le 16.$$

We must minimize the function *time*. First, we define $y = t(x)$ (to avoid conflict with the built-in function `time`) and then verify that

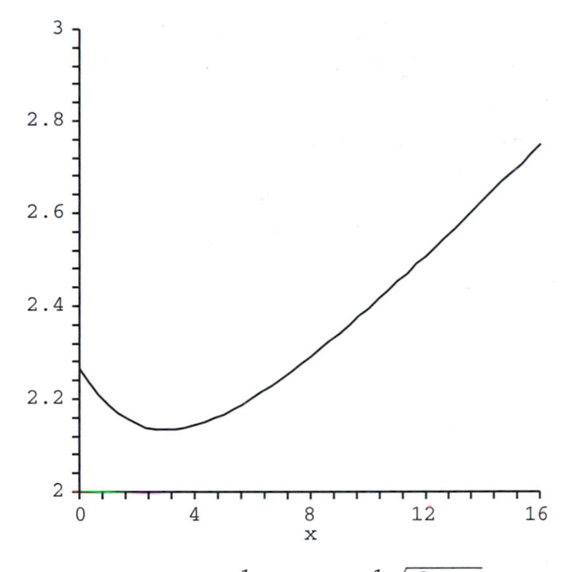

Figure 3-17 Plot of $t(x) = \frac{1}{10}(16 - x) + \frac{1}{6}\sqrt{x^2 + 16}$, $0 \le x \le 16$

$y = t(x)$ has a minimum by graphing $y = t(x)$ on the interval $[0, 16]$ in Figure 3-17.

```
> t:=x->(16-x)/10+1/6*sqrt(x^2+16):
> plot(t(x),x=0..16,view=[0..16,2..3],color=black);
```

Next, we compute the derivative of `t` and find the values of x for which the derivative is 0 with `solve`.

```
> simplify(D(t)(x));
```

$$1/30\,\frac{-3\sqrt{x^2+16}+5\,x}{\sqrt{x^2+16}}$$

```
> solve(D(t)(x)=0);
```

$$3$$

At this point, we can calculate the minimum time by calculating `t(3)`.

```
> t(3);
```

$$\frac{32}{15}$$

Alternatively, we demonstrate how to find the value of `t(x)` with `subs` and `simplify`.

```
> s1:=subs(x=3,t(x));
```

$$s1 := \frac{13}{10} + 1/6\sqrt{25}$$

```
> simplify(s1);
```

$$\frac{32}{15}$$

Regardless of our evaluation method, we see that the minimum time to complete the trip is 32/15 hours.

■

One of the more interesting applied max/min problems is the *beam problem*. We present two solutions.

EXAMPLE 3.2.16 (The Beam Problem): Find the exact length of the longest beam that can be carried around a corner from a hallway 2 feet wide to a hallway that is 3 feet wide (Figure 3-18).

SOLUTION: We assume that the beam has negligible thickness. Our first approach is algebraic. Using Figure 3-18 and the Pythagorean

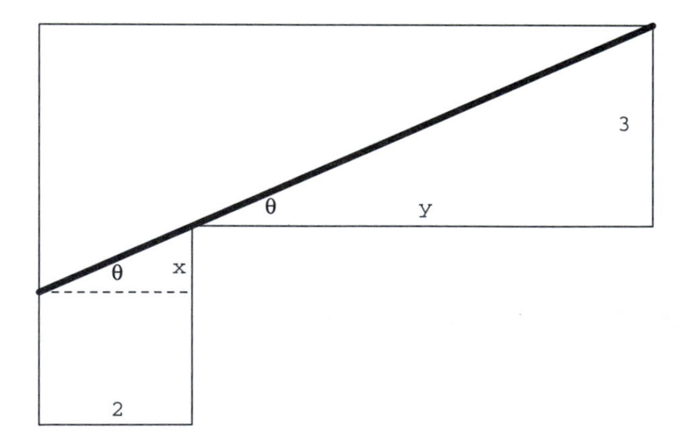

Figure 3-18 The length of the beam is found using similar triangles

theorem, the total length of the beam is

$$L = \sqrt{2^2 + x^2} + \sqrt{y^2 + 3^2}.$$

By similar triangles,

$$\frac{y}{3} = \frac{2}{x} \qquad \text{so} \qquad y = \frac{6}{x}$$

and the length of the beam, L, becomes

$$L(x) = \sqrt{4 + x^2} + \sqrt{9 + \frac{36}{x^2}}, \quad 0 < x < \infty.$$

```
> l:=x->sqrt(2^2+x^2)+sqrt((6/x)^2+3^2);
```

$$l := x \mapsto \sqrt{4 + x^2} + 3\sqrt{4x^{-2} + 1}$$

Observe that the length of the longest beam is obtained by *minimizing* L. (Why?)

We ignore negative values because length must be non-negative.

Differentiating gives us

```
> D(l)(x);
```

$$\frac{x}{\sqrt{4 + x^2}} - 12x^{-3}\frac{1}{\sqrt{4x^{-2} + 1}}$$

and solving $L'(x) = 0$ for x results in

```
> cns:=solve(D(l)(x)=0);
```

$$cns := -\sqrt{2}\sqrt[6]{18}, \sqrt{2}\sqrt[6]{18}$$

so $x = 2^{2/3}3^{1/3} \approx 2.29$.

```
> evalf(cns);
```

$$-2.289428485, 2.289428485$$

```
> l(cns[2]);
```

$$\sqrt{4 + 2\sqrt[3]{18}} + \sqrt{18^{2/3} + 9}$$

```
> evalf(l(cns[2]));
```

$$7.023482380$$

It follows that the length of the beam is

$$\begin{aligned} L(2^{2/3}3^{1/3}) &= \sqrt{9 + 3 \cdot 2^{2/3} \cdot 3^{1/3}} + \sqrt{4 + 2 \cdot 2^{1/3} \cdot 3^{2/3}} \\ &= \sqrt{13 + 9 \cdot 2^{2/3} \cdot 3^{1/3} + 6 \cdot 2^{1/3} \cdot 3^{2/3}} \approx 7.02. \end{aligned}$$

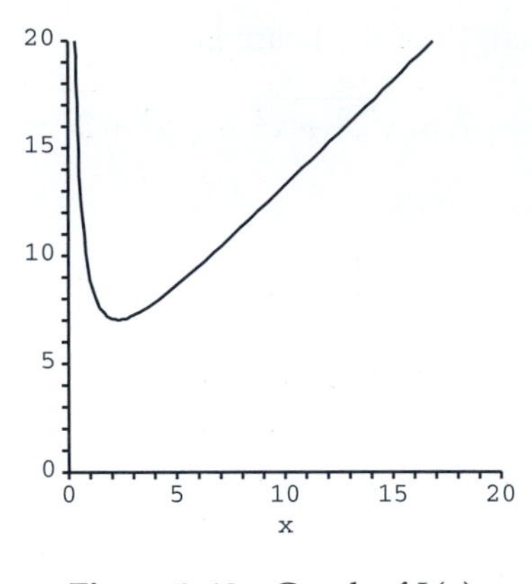

Figure 3-19 Graph of $L(x)$

See Figure 3-19.

```
> plot(l(x),x=0..20,color=black,view=[0..20,0..20],
   scaling=constrained);
```

Our second approach uses right triangle trigonometry. In terms of θ, the length of the beam is given by

$$L(\theta) = 3\csc\theta + 2\sec\theta, \quad 0 < \theta < \pi/2.$$

Differentiating gives us

$$L'(\theta) = -3\csc\theta\cot\theta + 2\sec\theta\tan\theta.$$

To avoid typing the θ symbol, we define L as a function of t.

```
> l:=t->2*sec(t)+3*csc(t):
> D(l)(t);
```

$$2\sec(t)\tan(t) - 3\csc(t)\cot(t)$$

We now solve $L'(\theta) = 0$. First multiply through by $\sin\theta$ and then by $\tan\theta$.

$$2\sec\theta\tan\theta = 3\csc\theta\cot\theta$$

$$\tan^2\theta = \frac{3}{2}\cot\theta$$

$$\tan^3\theta = \frac{3}{2}$$

$$\tan\theta = \sqrt[3]{\frac{3}{2}}.$$

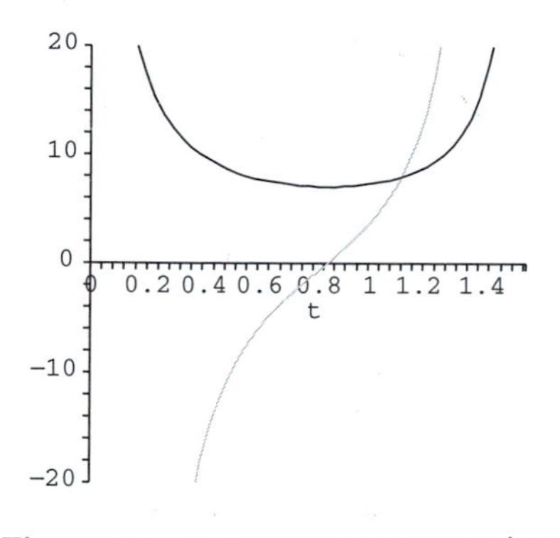

Figure 3-20 Graph of $L(\theta)$ and $L'(\theta)$

In this case, observe that we cannot compute θ exactly. However, we do not need to do so. Let $0 < \theta < \pi/2$ be the unique solution of $\tan\theta = \sqrt[3]{3/2}$ (Figure 3-20). Using the identity $\tan^2\theta + 1 = \sec^2\theta$, we find that $\csc\theta = \sqrt{1 + \sqrt[3]{4/9}}$. Similarly, because $\tan\theta = \sqrt[3]{3/2}$ and $\cot^2\theta + 1 = \csc^2\theta$, $\sec\theta = \sqrt[3]{3/2}\sqrt{1 + \sqrt[3]{4/9}}$. Hence, the length of the beam is

$$L(\theta) = 2\sqrt[3]{\frac{3}{2}}\sqrt{1 + \sqrt[3]{\frac{4}{9}}} + 3\sqrt{1 + \sqrt[3]{\frac{4}{9}}} \approx 7.02.$$

```
> plot([l(t),D(l)(t)],t=0..Pi/2,color=[black,gray],
> view=[0..Pi/2,-20..20]);
```

■

In the next two examples, the constants do not have specific numerical values.

EXAMPLE 3.2.17: Find the volume of the right circular cone of maximum volume that can be inscribed in a sphere of radius R.

SOLUTION: Try to avoid three-dimensional figures unless they are absolutely necessary. For this problem, a cross-section of the situation is sufficient (Figure 3-21).

The volume, V, of a right circular cone with radius r and height h is $V = \frac{1}{3}\pi r^2 h$. Using the notation in Figure 3-21, the volume is given by

$$V = \frac{1}{3}\pi x^2 (R + y). \tag{3.4}$$

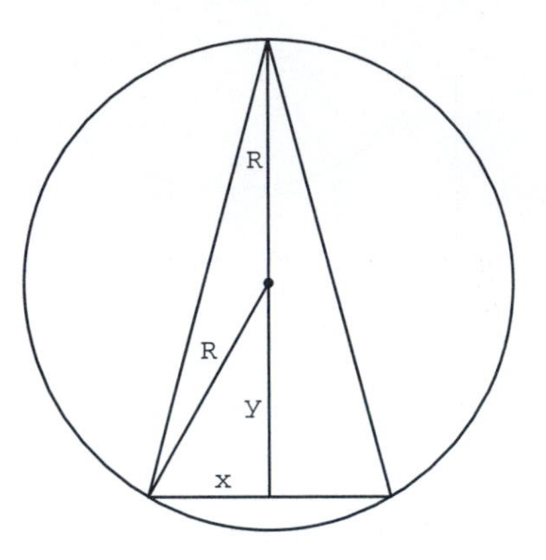

Figure 3-21 Cross-section of a right circular cone inscribed in a sphere

However, by the Pythagorean theorem, $x^2 + y^2 = R^2$ so $x^2 = R^2 - y^2$ and (3.4) becomes

$$V = \frac{1}{3}\pi\left(R^2 - y^2\right)(R + y) = \frac{1}{3}\pi\left(R^3 + R^2 y - Ry^2 - y^3\right), \tag{3.5}$$

```
> y:='y':
> s1:=expand((r^2-y^2)*(r+y));
```

$$s1 := r^3 + r^2 y - y^2 r - y^3$$

where $0 \le y \le R$. $V(y)$ is continuous on $[0, R]$ so it will have a minimum and maximum value on this interval. Moreover, the minimum and maximum values occur either at the endpoints of the interval or at the critical numbers on the interior of the interval. Differentiating (3.5) with respect to y gives us

Remember that R is a constant.

$$\frac{dV}{dy} = \frac{1}{3}\pi\left(R^2 - 2Ry - 3y^2\right) = \frac{1}{3}\pi(R - 3y)(R + y)$$

```
> s2:=diff(s1,y);
```

$$s2 := r^2 - 2\,yr - 3\,y^2$$

and we see that $dV/dy = 0$ if $y = \frac{1}{3}R$ or $y = -R$.

```
> factor(s2);
> solve(s2=0,y);
```

$$(r + y)\,(r - 3\,y)$$

$$-r,\ 1/3\,r$$

We ignore $y = -R$ because $-R$ is not in the interval $[0, R]$. Note that $V(0) = V(R) = 0$. The maximum volume of the cone is

$$V\left(\frac{1}{3}R\right) = \frac{1}{3}\pi \cdot \frac{32}{27}R^3 = \frac{32}{81}\pi R^2 \approx 1.24R^3.$$

```
> s3:=subs(y=r/3,s1);
```

$$s3 := \frac{32}{27}r^3$$

```
> s3*Pi/3;
```

$$\frac{32}{81}r^3\pi$$

```
> evalf(s3*Pi/3);
```

$$1.241123024\,r^3$$

■

EXAMPLE 3.2.18 (The Stayed-Wire Problem): Two poles D feet apart with heights L_1 feet and L_2 feet are to be stayed by a wire as shown in Figure 3-22. Find the minimum amount of wire required to stay the poles, as illustrated in Figure 3-22.

SOLUTION: Using the notation in Figure 3-22, the length of the wire, L, is

$$L(x) = \sqrt{L_1{}^2 + x^2} + \sqrt{L_2{}^2 + (D-x)^2}, \qquad 0 \le x \le D. \qquad (3.6)$$

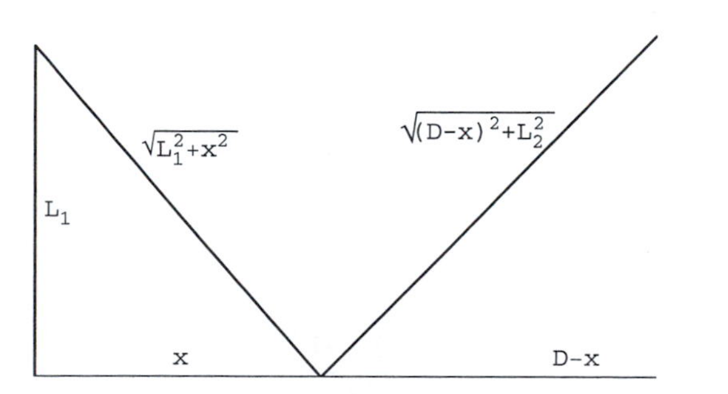

Figure 3-22 When the wire is stayed to minimize the length, the result is two similar triangles

In the special case that $L_1 = L_2$, the length of the wire to stay the poles is minimized when the wire is placed halfway between the two poles, at a distance $D/2$ from each pole. Thus, we assume that the lengths of the poles are different; we assume that $L_1 < L_2$, as illustrated in Figure 3-22. We compute $L'(x)$ and then solve $L'(x) = 0$.

```
> l:='l':
> l:=x->sqrt(x^2+l1^2)+sqrt((d-x)^2+l2^2);
```

$$l := x \mapsto \sqrt{x^2 + l1^2} + \sqrt{d^2 - 2\,dx + x^2 + l2^2}$$

```
> simplify(D(l)(x));
```

$$\frac{x\sqrt{d^2 - 2\,dx + x^2 + l2^2} - \sqrt{x^2 + l1^2}d + \sqrt{x^2 + l1^2}x}{\sqrt{x^2 + l1^2}\sqrt{d^2 - 2\,dx + x^2 + l2^2}}$$

```
> l(0);
```

$$\sqrt{l1^2} + \sqrt{d^2 + l2^2}$$

```
> l(d);
```

$$\sqrt{d^2 + l1^2} + \sqrt{l2^2}$$

```
> solve(D(l)(x)=0,x);
```

$$\frac{dl1}{l2 + l1}, \frac{dl1}{-l2 + l1}$$

The result indicates that $x = L_1D/\,(L_1 + L_2)$ minimizes $L(x)$. Moreover, the triangles formed by minimizing L are similar triangles.

```
> l1/(d*l1/(l1+l2));
```

$$\frac{l2 + l1}{d}$$

```
> simplify(l2/(d-d*l1/(l1+l2)));
```

$$\frac{l2 + l1}{d}$$

■

3.2.7 Antidifferentiation

Antiderivatives

$F(x)$ is an **antiderivative** of $f(x)$ if $F'(x) = f(x)$. The symbol

$$\int f(x)\,dx$$

means "find all antiderivatives of $f(x)$." Because all antiderivatives of a given function differ by a constant, we usually find an antiderivative, $F(x)$, of $f(x)$ and then write

$$\int f(x)\,dx = F(x) + C,$$

where C represents an arbitrary constant. The commands

```
int(f(x),x)
```

and

```
integrate(f(x),x)
```

attempt to find an antiderivative, $F(x)$, of $f(x)$. Maple does not include the "$+C$" that we include when writing $\int f(x)\,dx = F(x) + C$. In the same way as `diff` can differentiate many functions, `int` (or `integrate`) can antidifferentiate many functions. However, antidifferentiation is a fundamentally difficult procedure so it is not difficult to find functions $f(x)$ for which the command `int(f(x),x)` returns unevaluated.

EXAMPLE 3.2.19: Evaluate each of the following antiderivatives: (a) $\int \frac{1}{x^2}e^{1/x}dx$, (b) $\int x^2 \cos x\,dx$, (c) $\int x^2\sqrt{1+x^2}\,dx$, (d) $\int \frac{x^2-x+2}{x^3-x^2+x-1}dx$, and (e) $\int \frac{\sin x}{x}dx$.

SOLUTION: Entering

```
> int(1/x^2*exp(1/x),x);
```

$$-e^{x^{-1}}$$

shows us that $\int \frac{1}{x^2}e^{1/x}dx = -e^{1/x} + C$. Notice that Maple does not automatically include the arbitrary constant, C. When computing several antiderivatives, you can use map to apply int to a list of antiderivatives, which we illustrate to compute (b), (c), and (d).

```
> map(int,[x^2*cos(x),x^2*sqrt(1+x^2),
  (x^2-x+2)/(x^3-x^2+x-1)],x);
```

$$\left[x^2 \sin(x) - 2\sin(x) + 2x\cos(x),\right.$$
$$1/4x\left(1+x^2\right)^{3/2} - 1/8x\sqrt{1+x^2} - 1/8\,arcsinh(x),$$
$$\left.\ln(x-1) - \arctan(x)\right]$$

For (e), we see that there is not a "closed form" antiderivative of $\int \frac{\sin x}{x}dx$ and the result is given in terms of a definite integral, the **sine integral function**:

$$Si(x) = \int_0^x \frac{\sin t}{t}dt.$$

```
> int(sin(x)/x,x);
```

$$Si(x)$$

■

u-Substitutions

Usually, the first antidifferentiation technique discussed is the method of u **substitution**. Suppose that $F(x)$ is an antiderivative of $f(x)$. Given

$$\int f(g(x))\, g'(x)\, dx,$$

we let $u = g(x)$ so that $du = g'(x)\, dx$. Then,

$$\int f(g(x))\, g'(x)\, dx = \int f(u)\, du = F(u) + C = F(g(x)) + C,$$

where $F(x)$ is an antiderivative of $f(x)$. After mastering u-substitutions, the **integration by parts formula**,

$$\int u\, dv = uv - \int v\, du, \tag{3.7}$$

is introduced.

EXAMPLE 3.2.20: Evaluate $\int 2^x\sqrt{4^x - 1}\, dx$.

SOLUTION: We use `int` to evaluate the antiderivative. Notice that the result is *very* complicated.

```
> int(2^x*sqrt(4^x-1),x);
```

$$1/2\,\frac{e^{x\ln(2)}\sqrt{\left(e^{x\ln(2)}\right)^2-1}}{\ln(2)}-1/2\,\frac{\ln\left(e^{x\ln(2)}+\sqrt{\left(e^{x\ln(2)}\right)^2-1}\right)}{\ln(2)}$$

Proceeding by hand, we let $u = 2^x$. Then, $du = 2^x \ln 2\, dx$ or, equivalently, $\frac{1}{\ln 2}du = 2^x\, dx$

```
> diff(2^x,x);
```

$$2^x \ln(2)$$

so $\int 2^x\sqrt{4^x-1}\,dx = \frac{1}{\ln 2}\int\sqrt{u^2-1}\,du$. We now use `int` to evaluate $\int\sqrt{u^2-1}\,du$

```
> s1:=int(sqrt(u^2-1),u);
```

$$s1 := 1/2\,u\sqrt{u^2-1}-1/2\,\ln\left(u+\sqrt{u^2-1}\right)$$

and then `subs` to replace u with 2^x.

```
> s2:=subs(u=2^x,s1);
```

$$s2 := 1/2\,2^x\sqrt{(2^x)^2-1}-1/2\,\ln\left(2^x+\sqrt{(2^x)^2-1}\right)$$

Clearly, proceeding by hand results in a significantly simpler antiderivative than using `int` directly.

You can also use the `changevar` command contained in the `student` package to perform the change of variables as illustrated with the following commands.

```
> with(student):
> s1:=changevar(2^x=u,Int(2^x*sqrt(4^x-1),x));
```

$$s1 := \int\sqrt{4^{\frac{\ln(u)}{\ln(2)}}-1}\,(\ln(2))^{-1}\,du$$

```
> s2:=value(s1);
```

$$s2 := 1/2\,\frac{u\sqrt{u^2-1}}{\ln(2)}-1/2\,\frac{\ln\left(u+\sqrt{u^2-1}\right)}{\ln(2)}$$

```
> subs(u=2^x,s2);
```

$$1/2\frac{2^x\sqrt{(2^x)^2-1}}{\ln(2)}-1/2\frac{\ln\left(2^x+\sqrt{(2^x)^2-1}\right)}{\ln(2)}$$

■

3.3 Integral Calculus

3.3.1 Area

In integral calculus courses, the definite integral is frequently motivated by investigating the area under the graph of a positive continuous function on a closed interval. Let $y=f(x)$ be a non-negative continuous function on an interval $[a,b]$ and let n be a positive integer. If we divide $[a,b]$ into n subintervals of equal length and let $[x_{k-1},x_k]$ denote the kth subinterval, the length of each subinterval is $(b-a)/n$ and $x_k=a+k\frac{b-a}{n}$. The area bounded by the graphs of $y=f(x)$, $x=a$, $x=b$, and the y-axis can be approximated with the sum

$$\sum_{k=1}^{n} f\left(x_k{}^*\right)\frac{b-a}{n}, \tag{3.8}$$

where $x_k{}^* \in [x_{k-1},x_k]$. Typically, we take $x_k{}^*=x_{k-1}=a+(k-1)\frac{b-a}{n}$ (the left endpoint of the kth subinterval), $x_k{}^*=x_k=a+k\frac{b-a}{n}$ (the right endpoint of the kth subinterval), or $x_k{}^*=\frac{1}{2}\left(x_{k-1}+x_k\right)=a+\frac{1}{2}(2k-1)\frac{b-a}{n}$ (the midpoint of the kth subinterval). For these choices of $x_k{}^*$, (3.8) becomes

$$\frac{b-a}{n}\sum_{k=1}^{n} f\left(a+(k-1)\frac{b-a}{n}\right) \tag{3.9}$$

$$\frac{b-a}{n}\sum_{k=1}^{n} f\left(a+k\frac{b-a}{n}\right), \quad \text{and} \tag{3.10}$$

$$\frac{b-a}{n}\sum_{k=1}^{n} f\left(a+\frac{1}{2}(2k-1)\frac{b-a}{n}\right), \tag{3.11}$$

respectively. If $y=f(x)$ is increasing on $[a,b]$, (3.9) is an under-approximation and (3.10) is an upper approximation: (3.9) corresponds to an approximation of the area using n inscribed rectangles; (3.10) corresponds to an approximation of the area using n circumscribed rectangles. If $y = f(x)$ is decreasing on $[a, b]$, (3.10) is an under-approximation and (3.9) is an upper approximation: (3.10) corresponds to an approximation of the area using n inscribed rectangles; (3.9) corresponds to an approximation of the area using n circumscribed rectangles.

The functions `leftsum(f(x),x=a..b,n)`, `middlesum(f(x),x=a..b,n)`, and `rightsum(f(x),x=a..b,n)`, which are contained in the `student` package, compute (3.9), (3.11), and (3.10), respectively, and `leftbox(f(x),x=a..b,n)`, `middlebox(f(x),x=a..b,n)`, and `rightbox(f(x),x=a..b,n)`, which are also contained in the `student` package, generate the corresponding graphs.

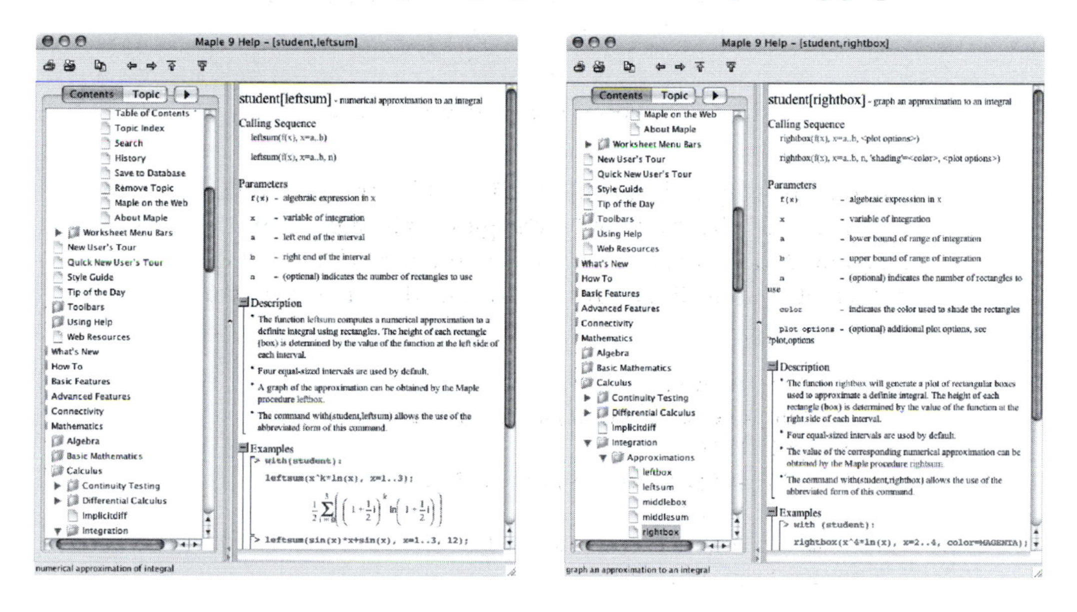

EXAMPLE 3.3.1: Let $f(x) = 9 - 4x^2$. Approximate the area bounded by the graph of $y = f(x)$, $x = 0$, $x = 3/2$, and the y-axis using (a) 100 inscribed and (b) 100 circumscribed rectangles. (c) What is the exact value of the area?

SOLUTION: We begin by defining and graphing $y = f(x)$ in Figure 3-23.

```
> f:=x->9-4*x^2:
> plot(f(x),x=0..3/2,color=black);
```

The first derivative, $f'(x) = -8x$, is negative on the interval so $f(x)$ is decreasing on $[0, 3/2]$. Thus, an approximation of the area using 100

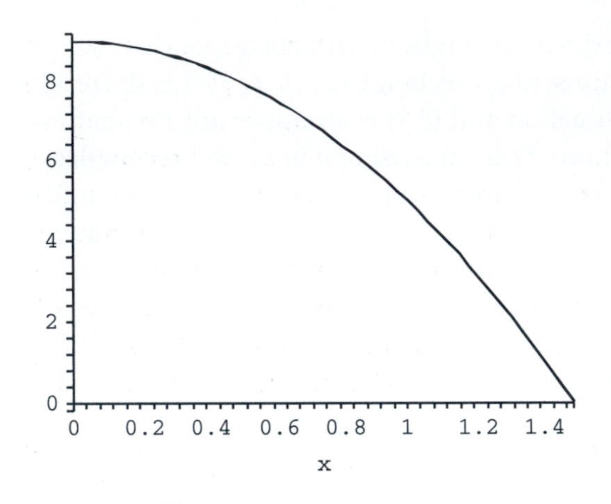

Figure 3-23 $f(x)$ for $0 \le x \le 3/2$

inscribed rectangles is given by (3.10) while an approximation of the area using 100 circumscribed rectangles is given by (3.9). These values are computed using `leftsum` and `rightsum`. The use of `middlesum` is illustrated as well. Approximations of the sums are obtained with `evalf`.

`evalf(number)` returns a numerical approximation of `number`.

```
> with(student):
> l100:=leftsum(f(x),x=0..3/2,100);
> evalf(l100);
> r100:=rightsum(f(x),x=0..3/2,100);
> evalf(r100);
> m100:=middlesum(f(x),x=0..3/2,100);
> evalf(m100);
```

$$l100 := \frac{3}{200} \sum_{i=0}^{99} 9 - \frac{9}{10000} i^2$$

9.067275000

$$r100 := \frac{3}{200} \sum_{i=1}^{100} 9 - \frac{9}{10000} i^2$$

8.932275000

$$m100 := \frac{3}{200} \sum_{i=0}^{99} 9 - 4\left(\frac{3}{200} i + \frac{3}{400}\right)^2$$

9.000112500

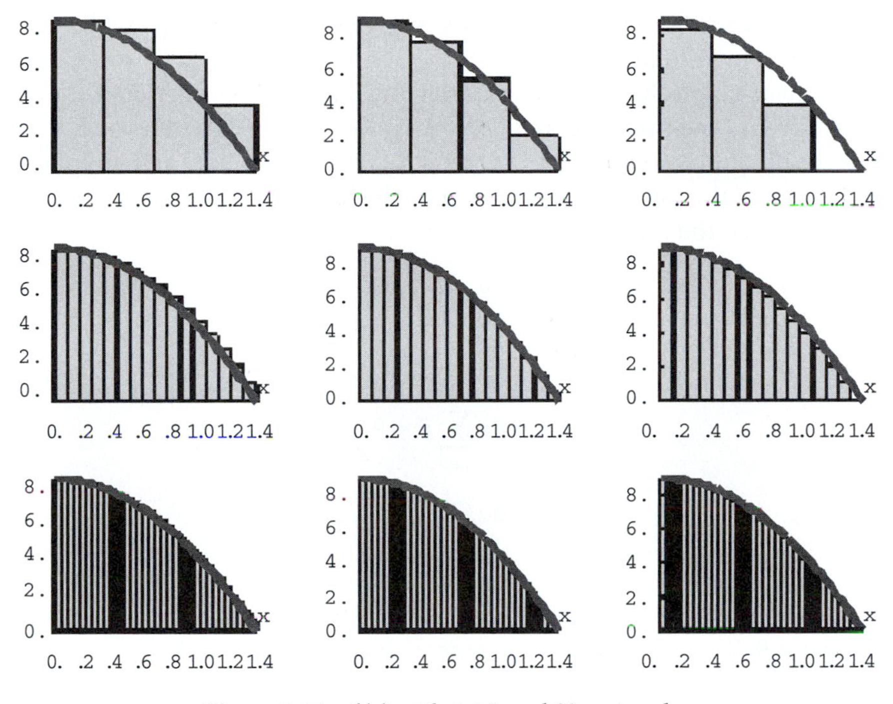

Figure 3-24 $f(x)$ with 4, 16, and 32 rectangles

Observe that these three values appear to be close to 9. In fact, 9 is the exact value of the area of the region bounded by $y = f(x)$, $x = 0$, $x = 3/2$, and the y-axis. To help us see why this is true, we use `leftbox`, `middlebox`, and `rightbox` to visualize the situation using $n = 4$, 16, and 32 rectangles in Figure 3-24.

```
> with(plots):
> A:=array(1..3,1..3):
> A[1,1]:=leftbox(f(x),x=0..3/2,4):
> A[1,2]:=middlebox(f(x),x=0..3/2,4):
> A[1,3]:=rightbox(f(x),x=0..3/2,4):
> A[2,1]:=leftbox(f(x),x=0..3/2,16):
> A[2,2]:=middlebox(f(x),x=0..3/2,16):
> A[2,3]:=rightbox(f(x),x=0..3/2,16):
> A[3,1]:=leftbox(f(x),x=0..3/2,32):
> A[3,2]:=middlebox(f(x),x=0..3/2,32):
> A[3,3]:=rightbox(f(x),x=0..3/2,32):
> display(A);
```

Notice that as n increases, the under-approximations increase while the upper approximations decrease.

These graphs help convince us that the limit of the sum as $n \to \infty$ of the areas of the inscribed and circumscribed rectangles is the same. We compute the exact value of (3.9) with `leftsum`, evaluate and simplify the sum with `simplify`, and compute the limit as $n \to \infty$ with `limit`. We see that the limit is 9.

```
> ls:=leftsum(f(x),x=0..3/2,n);
> ls2:=(simplify@value)(ls);
> limit(ls2,n=infinity);
```

$$ls := 3/2 \sum_{i=0}^{n-1} 9 - 9 \frac{i^2}{n^2} n^{-1}$$

$$ls2 := 9/4 \frac{4\,n^2 + 3\,n - 1}{n^2}$$

$$9$$

Similar calculations are carried out for (3.10) and again we see that the limit is 9. We conclude that the exact value of the area is 9.

```
> rs:=rightsum(f(x),x=0..3/2,n);
> rs2:=(simplify@value)(rs);
> limit(rs2,n=infinity);
```

$$rs := 3/2 \sum_{i=1}^{n} 9 - 9 \frac{i^2}{n^2} n^{-1}$$

$$rs2 := 9/4 \frac{4\,n^2 - 1 - 3\,n}{n^2}$$

$$9$$

For illustrative purposes, we confirm this result with `middlesum`.

```
> ms:=middlesum(f(x),x=0..3/2,n);
> ms2:=(simplify@value)(ms);
> limit(ms2,n=infinity);
```

$$ms := 3/2 \sum_{i=0}^{n-1} 9 - 9 \frac{(i + 1/2)^2}{n^2} n^{-1}$$

$$ms2 := \frac{9}{8} \frac{8\,n^2 + 1}{n^2}$$

$$9$$

■

3.3.2 The Definite Integral

In integral calculus courses, we formally learn that the **definite integral** of the function $y = f(x)$ from $x = a$ to $x = b$ is

$$\int_a^b f(x)\,dx = \lim_{|P|\to 0} \sum_{k=1}^{n} f\left(x_k{}^*\right)\Delta x_k, \tag{3.12}$$

provided that the limit exists. In (3.12), $P = \{a = x_0 < x_1 < x_2 < \ldots < x_n = b\}$ is a partition of $[a, b]$, $|P|$ is the **norm** of P,

$$|P| = \max\{x_k - x_{k-1} | k = 1, 2, \ldots, n\},$$

$\Delta x_k = x_k - x_{k-1}$, and $x_k{}^* \in [x_{k-1}, x_k]$.

The Fundamental Theorem of Calculus provides the fundamental relationship between differentiation and integration.

Theorem 6 (The Fundamental Theorem of Calculus). *Suppose that $y = f(x)$ is continuous on $[a, b]$.*

1. *If $F(x) = \int_a^x f(t)\,dt$, then F is an antiderivative of f: $F'(x) = f(x)$.*
2. *If G is any antiderivative of f, then $\int_a^b f(x)\,dx = G(b) - G(a)$.*

Maple's `int` and `integrate` commands can compute many definite integrals. The commands

```
int(f(x),x=a..b)
```

and

```
integrate(f(x),x=a..b)
```

attempt to compute $\int_a^b f(x)\,dx$. Because integration is a fundamentally difficult procedure, it is easy to create integrals for which the exact value cannot be found explicitly. In those cases, use `evalf` to obtain an approximation of the integral's value.

EXAMPLE 3.3.2: Evaluate (a) $\int_1^4 \left(x^2 + 1\right) / \sqrt{x}\,dx$; (b) $\int_0^{\sqrt{\pi/2}} x \cos x^2\,dx$; (c) $\int_0^{\pi} e^{2x} \sin^2 2x\,dx$; (d) $\int_0^1 \frac{2}{\sqrt{\pi}} e^{-x^2}\,dx$; and (e) $\int_{-1}^{0} \sqrt[3]{u}\,du$.

SOLUTION: We evaluate (a)–(c) directly with `int`.

```
> int((x^2+1)/sqrt(x),x=1..4);
```

$$\frac{72}{5}$$

```
> int(x*cos(x^2),x=0..sqrt(Pi/2));
```

$$1/2$$

```
> integrate(exp(2*x)*sin(2*x)^2,x=0..Pi);
```

$$1/5\,e^{2\pi} - 1/5$$

For (d), the result returned is in terms of the **error function**, `erf(x)`, which is defined by the integral

$$\texttt{erf}(x) = \frac{2}{\sqrt{\pi}} \int_0^x e^{-t^2}\, dt.$$

```
> int(2/sqrt(Pi)*exp(-x^2),x=0..1);
```

$$erf\,(1)$$

We use `evalf` to obtain an approximation of the value of the definite integral.

```
> evalf(int(2/sqrt(Pi)*exp(-x^2),x=0..1));
```

$$0.8427007929$$

(e) Recall that Maple does not return a real number when we compute odd roots of negative numbers so the following result would be surprising to many students in an introductory calculus course because it is complex.

See Chapter 2, Example 2.1.3.

```
> int(u^(1/3),u=-1..0);
```

$$3/4\,\sqrt[3]{-1}$$

Therefore, we use `surd` when typing the integrand so that Maple returns the real-valued third root of u.

```
> int(surd(u,3),u=-1..0);
```

$$-3/4$$

■

Improper integrals are computed using `int` in the same way as other definite integrals.

EXAMPLE 3.3.3: Evaluate (a) $\int_0^1 \frac{\ln x}{\sqrt{x}}\, dx$; (b) $\int_{-\infty}^{\infty} \frac{2}{\sqrt{\pi}} e^{-x^2}\, dx$; (c) $\int_1^{\infty} \frac{1}{x\sqrt{x^2-1}} dx$; (d) $\int_0^{\infty} \frac{1}{x^2+x^4}\, dx$; (e) $\int_2^4 \frac{1}{\sqrt[3]{(x-3)^2}}\, dx$; and (f) $\int_{-\infty}^{\infty} \frac{1}{x^2+x-6}\, dx$.

SOLUTION: (a) This is an improper integral because the integrand is discontinuous on the interval $[0, 1]$, but we see that the improper integral converges to -4.

```
> int(ln(x)/sqrt(x),x=0..1);
```

$$-4$$

(b) This is an improper integral because the interval of integration is infinite, but we see that the improper integral converges to 2.

```
> int(2/sqrt(Pi)*exp(-x^2),x=-infinity..infinity);
```

$$2$$

(c) This is an improper integral because the integrand is discontinuous on the interval of integration and because the interval of integration is infinite, but we see that the improper integral converges to $\pi/2$.

```
> int(1/(x*sqrt(x^2-1)),x=1..infinity);
```

$$1/2\,\pi$$

(d) As with (c), this is an improper integral because the integrand is discontinuous on the interval of integration and because the interval of integration is infinite, but we see that the improper integral diverges to ∞.

```
> int(1/(x^2+x^4),x=0..infinity);
```

$$\infty$$

(e) Recall that Maple does not return a real number when we compute odd roots of negative numbers so the following result would be surprising to many students in an introductory calculus course because it contains imaginary numbers. We use `surd` to carefully define the integrand so that the returned result is the expected one.

```
> int(1/(x-3)^(2/3),x=2..4);
```

$$-3/2\,i\sqrt{3}+3/2$$

```
> int((1/(x-3)^2)^(1/3),x=2..4);
```

$$\int_2^4 \sqrt[3]{(x-3)^{-2}}\,dx$$

```
> int(surd(1/(x-3),3)^2,x=2..4);
```

$$6$$

(f) In this case, Maple warns us that the improper integral diverges.

```
> factor(x^2+x-6);
```

$$(x+3)\,(x-2)$$

```
> int(1/(x^2+x-6),x=-infinity..-4);
```

$$1/5\ln(2)+1/5\ln(3)$$

```
> int(1/(x^2+x-6),x=-4..-3);
```

$$\infty$$

```
> int(1/(x^2+x-6),x=-3..0);
```

$$-\infty$$

```
> int(1/(x^2+x-6),x=0..2);
```

$$-\infty$$

```
> int(1/(x^2+x-6),x=2..3);
```

$$\infty$$

```
> int(1/(x^2+x-6),x=3..infinity);
```

$$1/5\ln(2)+1/5\ln(3)$$

```
> s1:=int(1/(x^2+x-6),x=-infinity..infinity);
```

$$s1 := \mathit{undefined}$$

To help us understand why the improper integral diverges, we note that

$$\frac{1}{x^2+x-6}=\frac{1}{5}\left(\frac{1}{x-2}-\frac{1}{x+3}\right)$$

and

$$\int \frac{1}{x^2+x-6}\,dx=\int \frac{1}{5}\left(\frac{1}{x-2}-\frac{1}{x+3}\right)dx=\frac{1}{5}\ln\left(\frac{x-2}{x+3}\right)+C.$$

```
> convert(1/(x^2+x-6),parfrac,x);
```

$$-1/5\,(x+3)^{-1}+1/5\,(x-2)^{-1}$$

```
> int(1/(x^2+x-6),x);
```

$$1/5\ln(x-2)-1/5\ln(x+3)$$

```
> s1:=int(1/(x^2+x-6),x);
```

$$s1 := 1/5\ln(x-2)-1/5\ln(x+3)$$

Hence the integral is improper because the interval of integration is infinite and because the integrand is discontinuous on the interval of integration so

$$\begin{aligned}\int_{-\infty}^{\infty}\frac{1}{x^2+x-6}\,dx &= \int_{-\infty}^{-4}\frac{1}{x^2+x-6}\,dx+\int_{-4}^{-3}\frac{1}{x^2+x-6}\,dx\\ &+\int_{-3}^{0}\frac{1}{x^2+x-6}\,dx+\int_{0}^{2}\frac{1}{x^2+x-6}\,dx\\ &+\int_{2}^{3}\frac{1}{x^2+x-6}\,dx+\int_{3}^{\infty}\frac{1}{x^2+x-6}\,dx.\end{aligned}\tag{3.13}$$

We conclude that the improper integral diverges because at least one of the improper integrals in (3.13) diverges.

■

In some cases, Maple can help illustrate the steps carried out when computing integrals using standard methods of integration like u-substitutions and integration by parts.

EXAMPLE 3.3.4: Evaluate (a) $\int_e^{e^3}\frac{1}{x\sqrt{\ln x}}\,dx$ and (b) $\int_0^{\pi/4}x\sin 2x\,dx$.

SOLUTION: (a) We let $u = \ln x$. Then, $du = \frac{1}{x}dx$ so $\int_e^{e^3} \frac{1}{x\sqrt{\ln x}}\,dx = \int_1^3 \frac{1}{\sqrt{u}}du = \int_1^3 u^{-1/2}du$. We use changevar, which is contained in the student package, to perform this change of variables.

The new lower limit of integration is 1 because if $x = e$, $u = \ln e = 1$. The new upper limit of integration is 3 because if $x = e^3$, $u = \ln e^3 = 3$. Int represents the inert form of the int command. That is, Int(f(x),x=a..b) returns the symbols $\int_a^b f(x)\,dx$ while int(f(x),x=a..b) attempts to evaluate $\int_a^b f(x)\,dx$.

```
> with(student):
> s1:=changevar(ln(x)=u,Int(1/(x*sqrt(ln(x))),
   x=exp(1)..exp(3)));
```

$$s1 := \int_1^3 \frac{1}{\sqrt{u}}\,du$$

The value of the definite integral is obtained with value.

```
> value(s1);
```

$$2\,\frac{-\sqrt{\pi}+\sqrt{\pi}\sqrt{3}}{\sqrt{\pi}}$$

To evaluate (b), we use integration by parts and let $u = x \Rightarrow du = dx$ and $dv = \sin 2x\,dx \Rightarrow v = -\frac{1}{2}\cos 2x$. We carry out the calculation using intparts, which is also contained in the student package.

```
> s1:=intparts(Int(x*sin(2*x),x=0..Pi/4),x);
```

$$s1 := -\int_0^{1/4\,\pi} -1/2\,\cos(2\,x)\,dx$$

The results mean that

$$\int_0^{\pi/4} x\sin 2x\,dx = -\frac{1}{2}x\cos 2x\bigg]_0^{\pi/4} + \frac{1}{2}\int_0^{\pi/4}\cos 2x\,dx$$

$$= 0 + \frac{1}{2}\int_0^{\pi/4}\cos 2x\,dx.$$

The result is evaluated with value.

```
> value(s1);
```

$$1/4$$

■

3.3.3 Approximating Definite Integrals

Because integration is a fundamentally difficult procedure, Maple is unable to compute a "closed form" of the value of many definite integrals. In these cases, numerical integration can be used to obtain an approximation of the definite integral using `evalf` together with `int` or `Int`.

EXAMPLE 3.3.5: Evaluate $\int_0^{\sqrt[3]{\pi}} e^{-x^2} \cos x^3 \, dx$.

SOLUTION: In this case, Maple is unable to evaluate the integral with `int`.

```
> i1:=int(exp(-x^2)*cos(x^3),x=0..Pi^(1/3));
```

$$i1 := \int_0^{\sqrt[3]{\pi}} e^{-x^2} \cos\left(x^3\right) dx$$

An approximation is obtained with `evalf`.

```
> evalf(i1);
```

$$0.7015656956$$

■

In some cases, you may wish to investigate particular numerical methods that can be used to approximate integrals. If needed you can use the functions `leftsum`, `middlesum`, and `rightsum` that are contained in the `student` package and were discussed previously. In addition the `student` package contains the functions `simpson`, which implements Simpson's rule, and `trapezoid`, which implements the trapezoidal rule.,

EXAMPLE 3.3.6: Let $f(x) = e^{-(x-3)^2 \cos(4(x-3))}$. (a) Graph $y = f(x)$ on the interval $[1,5]$. Use (b) Simpson's rule with $n = 4$, (c) the trapezoidal rule with $n = 4$, and (d) the midpoint rule with $n = 4$ to approximate $\int_1^5 f(x)\, dx$.

SOLUTION: We define f, and then graph $y = f(x)$ on the interval $[1,5]$ with `plot` in Figure 3-25.

```
> with(student):
> f:=x->exp(-(x-3)^2*cos(4*(x-3))):
> plot(f(x),x=1..5,color=black);
```

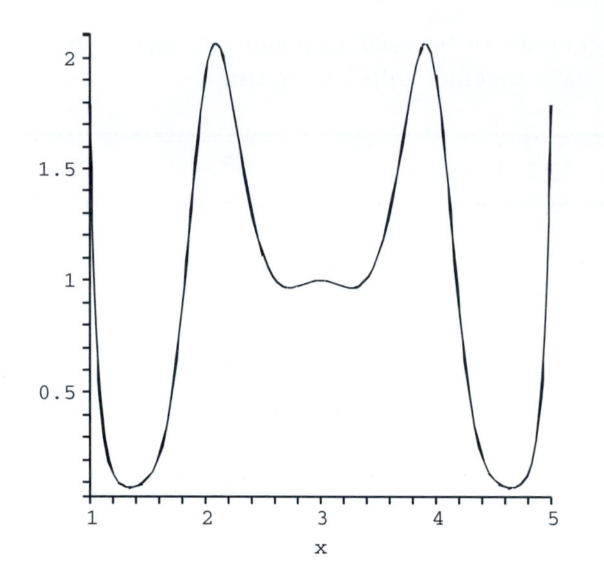

Figure 3-25 $f(x)$ for $1 \le x \le 5$

We use the functions `simpson`, `trapezoid`, and `middlesum` to approximate $\int_1^5 f(x)\,dx$ using $n = 4$ rectangles. In each case, `evalf` is used to evaluate the sum.

```
> s1:=simpson(f(x),x=1..5,4);
> evalf(s1);
> t1:=trapezoid(f(x),x=1..5,4);
> evalf(t1);
> m1:=middlesum(f(x),x=1..5,4);
> evalf(m1);
```

$$s1 := 2/3\,e^{-4\cos(8)} + 4/3\sum_{i=1}^{2} e^{-(2i-3)^2\cos(8i-12)} + 2/3\sum_{i=1}^{1} e^{-(-2+2i)^2\cos(-8+8i)}$$

$$6.986497720$$

$$t1 := e^{-4\cos(8)} + \sum_{i=1}^{3} e^{-(-2+i)^2\cos(-8+4i)}$$

$$6.634680453$$

$$m1 := \sum_{i=0}^{3} e^{-(-3/2+i)^2} \cos(-6+4\,i)$$

2.449844263

We obtain an accurate approximation of the value of the integral using `evalf` together with `int`.

```
> evalf(int(f(x),x=1..5));
```

3.761001249

Notice that with $n = 4$ rectangles, the midpoint rule gives the best approximation. However, as n increases, Simpson's rule gives a better approximation, as we see using $n = 50$ rectangles.

```
> s1:=simpson(f(x),x=1..5,50);
> evalf(s1);
> t1:=trapezoid(f(x),x=1..5,50);
> evalf(t1);
> m1:=middlesum(f(x),x=1..5,50);
> evalf(m1);
```

$$s1 := \frac{4}{75} e^{-4} \cos(8) + \frac{8}{75} \sum_{i=1}^{25} e^{-\left(-\frac{52}{25}+\frac{4}{25}\,i\right)^2} \cos\left(-\frac{208}{25}+\frac{16}{25}\,i\right) mbox$$
$$+ \frac{4}{75} \sum_{i=1}^{24} e^{-\left(-2+\frac{4}{25}\,i\right)^2} \cos\left(-8+\frac{16}{25}\,i\right)$$

3.764454020

$$t1 := \frac{2}{25} e^{-4} \cos(8) + \frac{2}{25} \sum_{i=1}^{49} e^{-\left(-2+\frac{2}{25}\,i\right)^2} \cos\left(-8+\frac{8}{25}\,i\right)$$

3.791301168

$$m1 := \frac{2}{25} \sum_{i=0}^{49} e^{-\left(-\frac{49}{25}+\frac{2}{25}\,i\right)^2} \cos\left(-\frac{196}{25}+\frac{8}{25}\,i\right)$$

3.746232810

■

3.3.4 Area

Suppose that $y = f(x)$ and $y = g(x)$ are continuous on $[a, b]$ and that $f(x) \geq g(x)$ for $a \leq x \leq b$. The **area** of the region bounded by the graphs of $y = f(x)$, $y = g(x)$, $x = a$, and $x = b$ is

$$A = \int_a^b \left[f(x) - g(x)\right] dx. \tag{3.14}$$

EXAMPLE 3.3.7: Find the area between the graphs of $y = \sin x$ and $y = \cos x$ on the interval $[0, 2\pi]$.

SOLUTION: We graph $y = \sin x$ and $y = \cos x$ on the interval $[0, 2\pi]$ in Figure 3-26. The graph of $y = \cos x$ is gray.

```
> plot([sin(x),cos(x)],x=0..2*Pi,color=[black,gray],
> scaling=CONSTRAINED);
```

To find the upper and lower limits of integration, we must solve the equation $\sin x = \cos x$ for x. In this case, we set `_EnvAllSolutions := true` to force Maple to try to find all solutions to the equation.

```
> _EnvAllSolutions := true:
> s1:=solve(sin(x)=cos(x),x);
```

$$s1 := 1/4\pi + \pi\_Z1$$

Thus, for $0 \leq x \leq 2\pi$, $\sin x = \cos x$ if $x = \pi/4$ or $x = 5\pi/4$. Hence, the area of the region between the graphs is given by

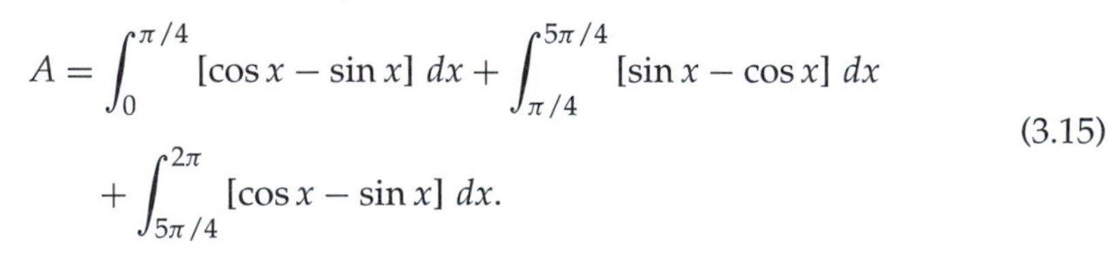

$$A = \int_0^{\pi/4} [\cos x - \sin x]\, dx + \int_{\pi/4}^{5\pi/4} [\sin x - \cos x]\, dx + \int_{5\pi/4}^{2\pi} [\cos x - \sin x]\, dx. \tag{3.15}$$

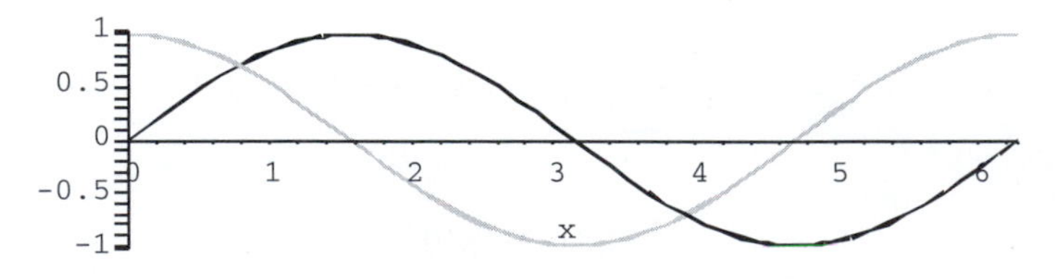

Figure 3-26 $y = \sin x$ and $y = \cos x$ on the interval $[0, 2\pi]$

Notice that if we take advantage of symmetry we can simplify (3.15) to

$$A = 2\int_{\pi/4}^{5\pi/4} [\sin x - \cos x]\, dx. \tag{3.16}$$

We evaluate (3.16) with `int` to see that the area is $4\sqrt{2}$.

```
> 2*int(sin(x)-cos(x),x=Pi/4..5*Pi/4);
```

$$4\sqrt{2}$$

■

In cases when we cannot calculate the points of intersection of two graphs exactly, we can frequently use `fsolve` to approximate the points of intersection.

EXAMPLE 3.3.8: Let

$$p(x) = \frac{3}{10}x^5 - 3x^4 + 11x^3 - 18x^2 + 12x + 1$$

and

$$q(x) = -4x^3 + 28x^2 - 56x + 32.$$

Approximate the area of the region bounded by the graphs of $y = p(x)$ and $y = q(x)$.

SOLUTION: After defining p and q, we graph them on the interval $[-1, 5]$ in Figure 3-27.

```
> p:='p':q:='q':
> p:=3*x^5/10-3*x^4+11*x^3-18*x^2+12*x+1:
> q:=-4*x^3+28*x^2-56*x+32:
> plot([p,q],x=-1..5,-15..20,color=[black,gray]);
```

The x-coordinates of the three intersection points are the solutions of the equation $p(x) = q(x)$. Although Maple can solve this equation exactly, approximate solutions are more useful for the problem and are obtained with `fsolve`.

```
> intpts:=fsolve(p=q,x);
```

$$intpts := 0.7720583045,\ 2.291819211,\ 3.865127100$$

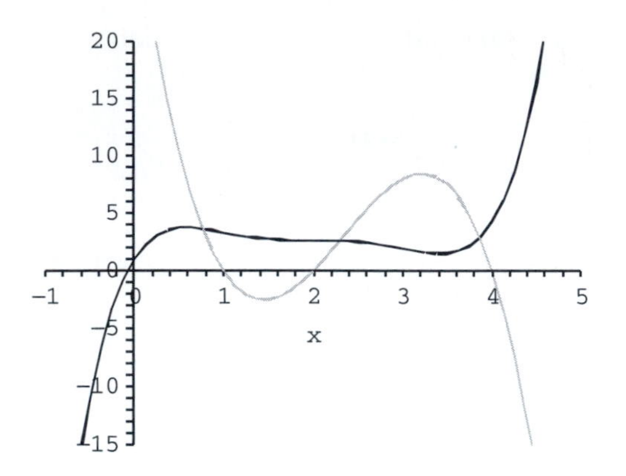

Figure 3-27 p and q on the interval $[-1, 5]$

Using the roots to the equation $p(x) = q(x)$ and the graph we see that $p(x) \geq q(x)$ for $0.772 \leq x \leq 2.292$ and $q(x) \geq p(x)$ for $2.292 \leq x \leq 3.865$. Hence, an approximation of the area bounded by p and q is given by the sum

$$\int_{0.772}^{2.292} \left[p(x) - q(x)\right] dx + \int_{2.292}^{3.865} \left[q(x) - p(x)\right] dx.$$

These two integrals are computed with `evalf` and `Int`.

```
> intone:=evalf(Int(p-q,x=intpts[1]..intpts[2]));
> inttwo:=evalf(Int(q-p,x=intpts[2]..intpts[3]));
```

$$intone := 5.269124281$$

$$inttwo := 6.925994162$$

and added to see that the area is approximately 12.195.

```
> intone+inttwo;
```

$$12.19511844$$

■

Parametric Equations

If the curve, C, defined parametrically by $x = x(t)$, $y = y(t)$, $a \leq t \leq b$ is a non-negative continuous function of x and $x(a) < x(b)$ the area under the graph of C and above the x-axis is

Graphically, y is a function of x, $y = y(x)$ if the graph of $y = y(x)$ passes the vertical line test.

$$\int_{x(a)}^{x(b)} y\, dx = \int_a^b y(t)x'(t)\, dt.$$

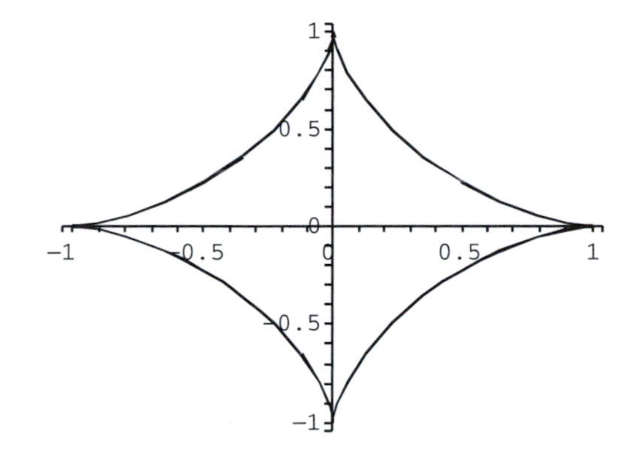

Figure 3-28 The astroid $x = \sin^3 t, y = \cos^3 t, 0 \le t \le 2\pi$

EXAMPLE 3.3.9 (The Astroid): Find the area enclosed by the **astroid** $x = \sin^3 t, y = \cos^3 t, 0 \le t \le 2\pi$.

SOLUTION: We begin by defining x and y and then graphing the asteroid with `plot` in Figure 3-28.

```
> x:=t->sin(t)^3:
> y:=t->cos(t)^3:
```

```
> plot([sin(t)^3,cos(t)^3,t=0..2*Pi],
   color=black,scaling=CONSTRAINED);
```

Observe that $x(0) = 0$ and $x(\pi/2) = 1$ and the graph of the astroid in the first quadrant is given by $x = \sin^3 t, y = \cos^3 t, 0 \le t \le \pi/2$. Hence, the area of the astroid in the first quadrant is given by

$$\int_0^{\pi/2} y(t)x'(t)\, dt = 3 \int_0^{\pi/2} \sin^2 t \cos^4 t\, dt$$

and the total area is given by

$$A = 4 \int_0^{\pi/2} y(t)x'(t)\, dt = 3 \int_0^{\pi/2} \sin^2 t \cos^4 t\, dt = \frac{3}{8}\pi \approx 1.178,$$

which is computed with `int` and then approximated with `evalf`.

```
> area:=4*int(y(t)*D(x)(t),t=0..Pi/2);
```

$$area := 3/8\,\pi$$

```
> evalf(area);
```

$$1.178097245$$

■

Polar Coordinates

For problems involving "circular symmetry" it is often easier to work in polar coordinates. The relationship between (x, y) in rectangular coordinates and (r, θ) in polar coordinates is given by

$$x = r\cos\theta \qquad y = r\sin\theta$$

and

$$r^2 = x^2 + y^2 \qquad \tan\theta = \frac{y}{x}.$$

If $r = f(\theta)$ is continuous and non-negative for $\alpha \le \theta \le \beta$, then the **area** A of the region enclosed by the graphs of $r = f(\theta)$, $\theta = \alpha$, and $\theta = \beta$ is

$$A = \frac{1}{2}\int_{\alpha}^{\beta} [f(\theta)]^2 \, d\theta = \frac{1}{2}\int_{\alpha}^{\beta} r^2 \, d\theta.$$

EXAMPLE 3.3.10 (Lemniscate of Bernoulli): The **Lemniscate of Bernoulli** is given by

$$\left(x^2 + y^2\right)^2 = a^2\left(x^2 - y^2\right),$$

where a is a constant. (a) Graph the Lemniscate of Bernoulli if $a = 2$. (b) Find the area of the region bounded by the Lemniscate of Bernoulli.

SOLUTION: This problem is much easier solved in polar coordinates so we first convert the equation from rectangular to polar coordinates with subs and then solve for r with solve.

```
> lofb:=(x^2+y^2)^2=a^2*(x^2-y^2):
> topolar:=subs(x=r*cos(theta),y=r*sin(theta),lofb);
```

$$topolar := \left(r^2(\cos(\theta))^2 + r^2(\sin(\theta))^2\right)^2 = a^2\left(r^2(\cos(\theta))^2 - r^2(\sin(\theta))^2\right)$$

```
> solve(topolar,r);
```

$$0, 0, \sqrt{1-2(\sin(\theta))^2}a, -\sqrt{1-2(\sin(\theta))^2}a$$

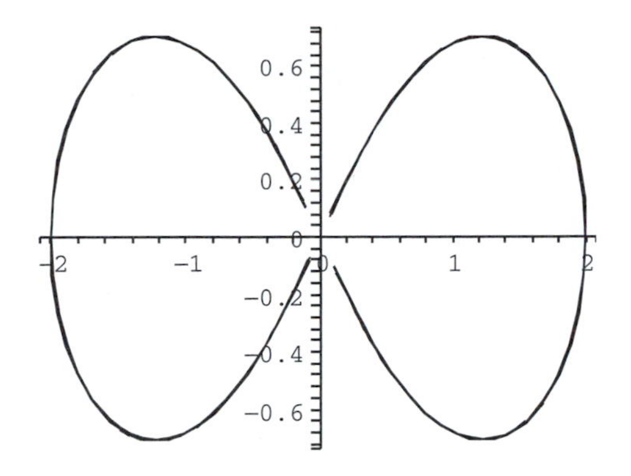

Figure 3-29 The Lemniscate

```
> s1:=isolate(topolar,r^2);
```

$$s1 := r^2 = \frac{a^2\left((\cos(\theta))^2 - (\sin(\theta))^2\right)}{2(\cos(\theta))^2(\sin(\theta))^2 + (\sin(\theta))^4 + (\cos(\theta))^4}$$

```
> s2:=simplify(s1,trig);
```

$$s2 := r^2 = \left(2(\cos(\theta))^2 - 1\right)a^2$$

```
> combine(s2,trig);
```

$$r^2 = \cos(2\theta)a^2$$

These results indicate that an equation of the Lemniscate in polar coordinates is $r^2 = a^2\cos 2\theta$. The graph of the Lemniscate is then generated in Figure 3-29 using `plot` together with the `coords=polar` option.

```
> plot(2*sqrt(cos(2*theta)),-2*sqrt(cos(2*theta)),
> theta=0..2*Pi,coords=polar,color=[black,black]);
```

The portion of the Lemniscate in quadrant 1 is obtained by graphing $r = 2\cos 2\theta, 0 \leq \theta \leq \pi/4$ (Figure 3-30).

```
> plot(2*sqrt(cos(2*theta)),theta=0..Pi/4,
> theta=coords=polar,color=black);
```

Then, taking advantage of symmetry, the area of the Lemniscate is given by

$$A = 2\cdot\frac{1}{2}\int_{-\pi/4}^{\pi/4} r^2\,d\theta = 2\int_0^{\pi/4} r^2\,d\theta = 2\int_0^{\pi/4} a^2\cos 2\theta\,d\theta = a^2,$$

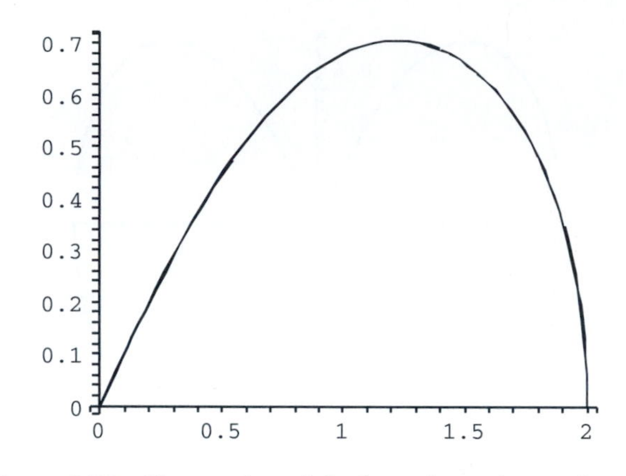

Figure 3-30 The portion of the Lemniscate in quadrant 1

which we calculate with `int`.

```
> int(2*a^2*cos(2*theta),theta=0..Pi/4);
```

$$a^2$$

■

3.3.5 Arc Length

Let $y = f(x)$ be a function for which $f'(x)$ is continuous on an interval $[a, b]$. Then the **arc length** of the graph of $y = f(x)$ from $x = a$ to $x = b$ is given by

$$L = \int_a^b \sqrt{\left(\frac{dy}{dx}\right)^2 + 1}\, dx. \tag{3.17}$$

The resulting definite integrals used for determining arc length are usually difficult to compute because they involve a radical. In these situations, Maple is helpful with approximating solutions to these types of problems.

EXAMPLE 3.3.11: Find the length of the graph of

$$y = \frac{x^4}{8} + \frac{1}{4x^2}$$

from (a) $x = 1$ to $x = 2$ and (b) $x = -2$ to $x = -1$.

SOLUTION: With no restrictions on the value of x, $\sqrt{x^2} = |x|$. Notice that Maple does not automatically algebraically simplify $\sqrt{\left(\frac{dy}{dx}\right)^2 + 1}$ because Maple does not know if x is positive or negative unless we use assume to instruct Maple to make the desired choice.

```
> y:=x->x^4/8+1/(4*x^2):
> simplify(sqrt(D(y)(x)^2+1));
```

$$1/2\sqrt{\frac{(x^6+1)^2}{x^6}}$$

In fact, for (b), x is negative so

$$\frac{1}{2}\sqrt{\frac{(x^6+1)^2}{x^6}} = -\frac{1}{2}\frac{x^6+1}{x^3}.$$

Maple simplifies

$$\frac{1}{2}\sqrt{\frac{(x^6+1)^2}{x^6}} = \frac{1}{2}\frac{x^6+1}{x^3}$$

and correctly evaluates the arc length integral (3.17) for (a).

```
> assume(x>0):
> simplify(sqrt(D(y)(x)^2+1));
```

$$1/2\sqrt{\frac{(x^6+1)^2}{x^6}}$$

```
> int(sqrt(D(y)(x)^2+1),x=1..2);
```

$$\frac{33}{16}$$

For (b), we compute the arc length integral (3.17).

```
> assume(x<0):
> int(sqrt(D(y)(x)^2+1),x=-2..-1);
```

$$\frac{33}{16}$$

As we expect due to symmetry, both values are the same.

■

Parametric Equations

C is **smooth** if both $x'(t)$ and $y'(t)$ are continuous on (a, b) and not simultaneously zero for $t \in (a, b)$.

If the smooth curve, C, defined parametrically by $x = x(t)$, $y = y(t)$, $t \in [a, b]$ is traversed exactly once as t increases from $t = a$ to $t = b$, the arc length of C is given by

$$L = \int_a^b \sqrt{\left(\frac{dx}{dt}\right)^2 + \left(\frac{dy}{dt}\right)^2}\, dt. \tag{3.18}$$

EXAMPLE 3.3.12: Find the length of the graph of $x = \sqrt{2}t^2$, $y = 2t - \frac{1}{2}t^3$, $-2 \le t \le 2$.

SOLUTION: For illustrative purposes, we graph $x = \sqrt{2}t^2$, $y = 2t - \frac{1}{2}t^3$ for $-3 \le t \le 3$ (in black) and $-2 \le t \le 2$ (in thick black) in Figure 3-31.

```
> x:=t->t^2*sqrt(2):
> y:=t->2*t-1/2*t^3:
> with(plots):
> p1:=plot([x(t),y(t),t=-3..3],color=black):
> p2:=plot([x(t),y(t),t=-2..2],color=black,
   thickness=4):
> display(p1,p2);
```

Maple is able to compute the exact value of the arc length (3.18) although the result is quite complicated and not displayed here for length considerations.

```
> factor(D(x)(t)^2+D(y)(t)^2);
```

$$1/4\left(3t^2 + 4t + 4\right)\left(3t^2 - 4t + 4\right)$$

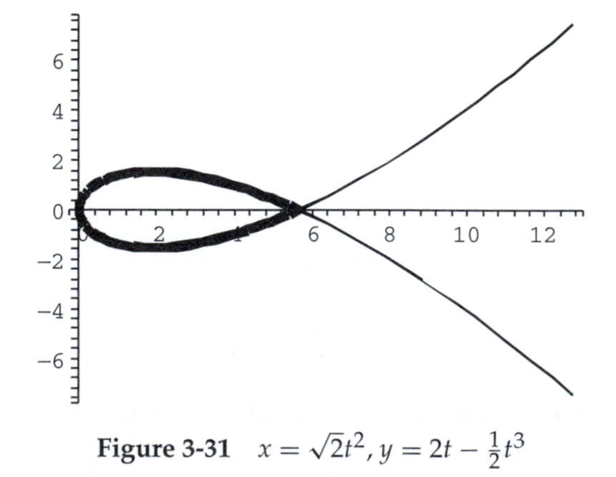

Figure 3-31 $x = \sqrt{2}t^2, y = 2t - \frac{1}{2}t^3$

```
> i1:=int(2*sqrt(D(x)(t)^2+D(y)(t)^2),t=0..2):
```

A more meaningful approximation is obtained with `evalf` or using `Int` together with `evalf`.

```
> evalf(i1);
```

$$13.70985196 - 0.000000002720174721\, i$$

```
> evalf(Int(2*sqrt(D(x)(t)^2+D(y)(t)^2),t=0..2));
```

$$13.70985196$$

We conclude that the arc length is approximately 13.71.

■

Polar Coordinates

If the smooth polar curve C given by $r = f(\theta)$, $\alpha \leq \theta \leq \beta$ is traversed exactly once as θ increases from α to β, the arc length of C is given by

$$L = \int_{\alpha}^{\beta} \sqrt{\left(\frac{dr}{d\theta}\right)^2 + r^2}\, d\theta. \tag{3.19}$$

EXAMPLE 3.3.13: Find the length of the graph of $r = \theta, 0 \leq \theta \leq 10\pi$.

SOLUTION: We begin by defining r and then graphing r with `plot` using the `coords=polar` option in Figure 3-32.

```
> r:=theta->theta:
> plot(r(theta),theta=0..10*Pi,coords=polar,
   color=black,
> scaling=CONSTRAINED);
```

Using (3.19), the length of the graph of r is given by $\int_0^{10\pi} \sqrt{1+\theta^2}\, d\theta$. The exact value is computed with `int`

```
> ev:=int(sqrt(D(r)(theta)^2+r(theta)^2),
   theta=0..10*Pi):
```

and then approximated with `evalf`.

```
> evalf(ev);
```

$$495.8005145$$

We conclude that the length of the graph is approximately 495.8.

■

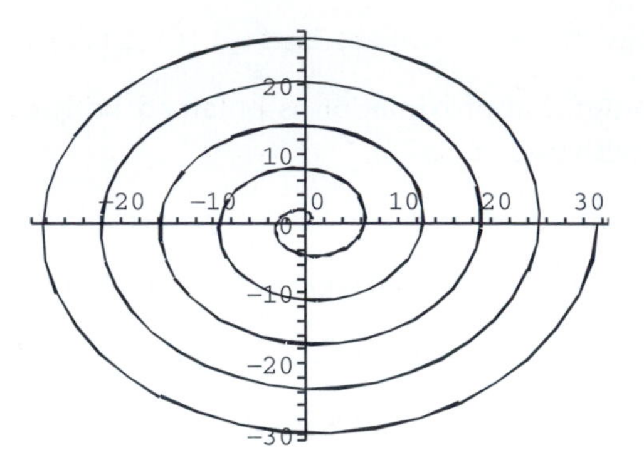

Figure 3-32 $r = \theta$ for $0 \le \theta \le 10\pi$

3.3.6 Solids of Revolution

Volume

Let $y = f(x)$ be a non-negative continuous function on $[a, b]$. The **volume** of the solid of revolution obtained by revolving the region bounded by the graphs of $y = f(x)$, $x = a$, $x = b$, and the x-axis about the x-axis is given by

$$V = \pi \int_a^b [f(x)]^2 \, dx. \tag{3.20}$$

If $0 \le a < b$, the **volume** of the solid of revolution obtained by revolving the region bounded by the graphs of $y = f(x)$, $x = a$, $x = b$, and the x-axis about the y-axis is given by

$$V = 2\pi \int_a^b x f(x) \, dx. \tag{3.21}$$

EXAMPLE 3.3.14: Let $g(x) = x \sin^2 x$. Find the volume of the solid obtained by revolving the region bounded by the graphs of $y = g(x)$, $x = 0$, $x = \pi$, and the x-axis about (a) the x-axis and (b) the y-axis.

SOLUTION: After defining g, we graph g on the interval $[0, \pi]$ in Figure 3-33.

```
> g:='g':
> g:=x->x*sin(x)^2:
> plot(g(x),x=0..Pi,color=black,scaling=CONSTRAINED);
```

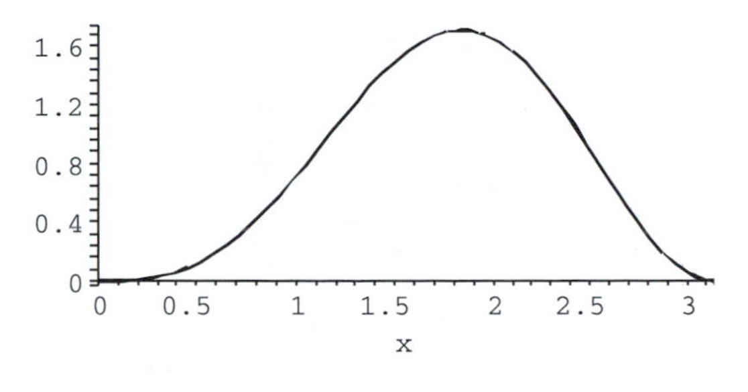

Figure 3-33 $g(x)$ for $0 \le x \le \pi$

The volume of the solid obtained by revolving the region about the x-axis is given by (3.20), while the volume of the solid obtained by revolving the region about the y-axis is given by (3.21). These integrals are computed with `int` and named `xvol` and `yvol`, respectively. `evalf` is used to approximate each volume.

```
> xvol:=int(Pi*g(x)^2,x=0..Pi);
> evalf(xvol);
```

$$xvol := 1/8\,\pi^4 - \frac{15}{64}\,\pi^2$$

$$9.862947848$$

```
> yvol:=int(2*Pi*x*g(x),x=0..Pi);
> evalf(yvol);
```

$$yvol := -1/2\,\pi^2 + 1/3\,\pi^4$$

$$27.53489482$$

We can use `plot3d` to visualize the resulting solids by parametrically graphing the equations given by

$$\begin{cases} x = r\cos t \\ y = r\sin t \\ z = g(r) \end{cases}$$

for r between 0 and π and t between $-\pi$ and π to visualize the graph of the solid obtained by revolving the region about the y-axis, and by parametrically graphing the equations given by

$$\begin{cases} x = r \\ y = g(r)\cos t \\ z = g(r)\sin t \end{cases}$$

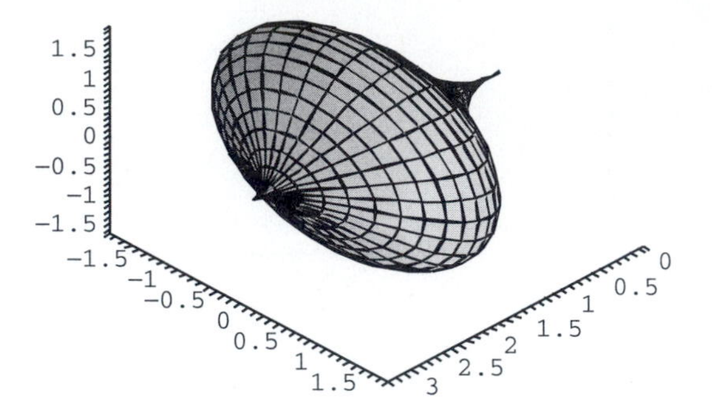

Figure 3-34 $g(x)$ revolved about the x-axis

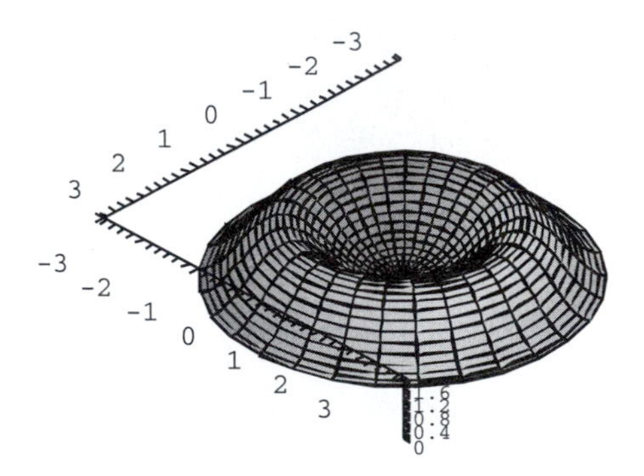

Figure 3-35 $g(x)$ revolved about the y-axis

for r between 0 and π and t between $-\pi$ and π to visualize the graph of the solid obtained by revolving the region about the x-axis (Figures 3-34 and 3-35). In this case, we identify the z-axis as the y-axis. Notice that we are simply using polar coordinates for the x- and y-coordinates, and the height above the x,y-plane is given by $z = g(r)$ because r is replacing x in the new coordinate system.

```
> plot3d([r,g(r)*cos(t),g(r)*sin(t)],
> r=0..Pi,t=0..2*Pi,grid=[30,30],axes=FRAME,
   scaling=CONSTRAINED);
```

```
> plot3d([r*cos(t),r*sin(t),g(r)],
> r=0..Pi,t=0..2*Pi,grid=[30,30],axes=FRAME,
   scaling=CONSTRAINED);
```

■

We now demonstrate a volume problem that requires the method of disks.

EXAMPLE 3.3.15: Let $f(x) = e^{-(x-3)\cos[4(x-3)]}$. Approximate the volume of the solid obtained by revolving the region bounded by the graphs of $y = f(x)$, $x = 1$, $x = 5$, and the x-axis about the x-axis.

SOLUTION: Proceeding as in the previous example, we first define and graph f on the interval $[1, 5]$ in Figure 3-36.

```
> f:='f':
> f:=x->exp(-(x-3)^2*cos(4*(x-3))):
> plot(f(x),x=1..5,color=black,scaling=CONSTRAINED);
```

In this case, an approximation is desired so we use `evalf` together with `Int` to approximate the integral $V = \int_1^5 \pi \left[f(x)\right]^2 \, dx$.

```
> evalf(Int(Pi*f(x)^2,x=1..5));
```

$$16.07615213$$

In the same manner as in the previous example, `plot3d` can be used to visualize the resulting solid by graphing the set of equations given parametrically by

$$\begin{cases} x = r \\ y = f(r)\cos t \\ z = f(r)\sin t \end{cases}$$

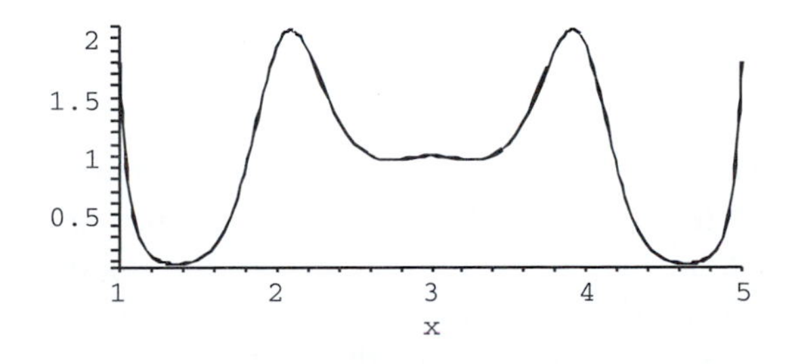

Figure 3-36 $f(x)$ for $1 \le x \le 5$

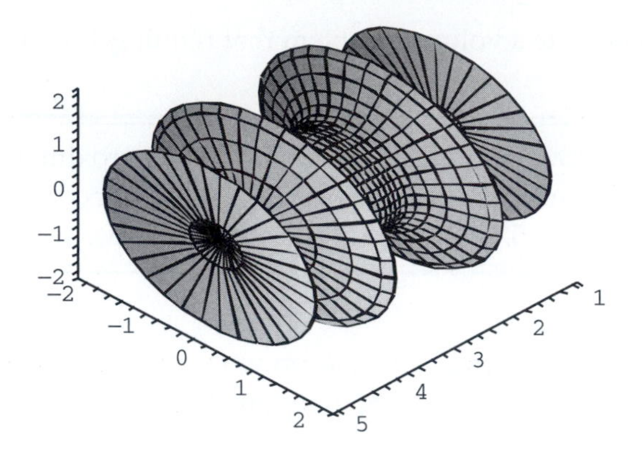

Figure 3-37 $f(x)$ revolved about the x-axis

for r between 1 and 5 and t between 0 and 2π. In this case, polar coordinates are used in the y,z-plane with the distance from the x-axis given by $f(x)$. Because r replaces x in the new coordinate system, $f(x)$ becomes $f(r)$ in these equations (Figure 3-37).

```
> plot3d([r,f(r)*cos(t),f(r)*sin(t)],
> r=1..5,t=0..2*Pi,grid=[45,35],axes=FRAME,
  scaling=CONSTRAINED);
```

■

Surface Area

Let $y = f(x)$ be a non-negative function for which $f'(x)$ is continuous on an interval $[a, b]$. Then the **surface area** of the solid of revolution obtained by revolving the region bounded by the graphs of $y = f(x)$, $x = a$, $x = b$, and the x-axis about the x-axis is given by

$$SA = 2\pi \int_a^b f(x)\sqrt{1 + [f'(x)]^2}\, dx. \tag{3.22}$$

EXAMPLE 3.3.16 (Gabriel's Horn): Gabriel's Horn is the solid of revolution obtained by revolving the area of the region bounded by $y = 1/x$ and the x-axis for $x \geq 1$ about the x-axis. Show that the surface area of Gabriel's Horn is infinite but that its volume is finite.

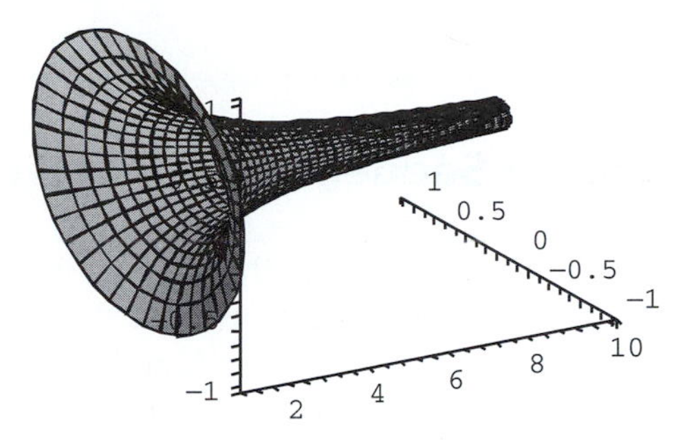

Figure 3-38 A portion of Gabriel's Horn

SOLUTION: After defining $f(x) = 1/x$, we use `plot3d` to visualize a portion of Gabriel's Horn in Figure 3-38.

```
> f:='f':
> f:=x->1/x:
> plot3d([r,f(r)*cos(t),f(r)*sin(t)],
> r=1..10,t=0..2*Pi,grid=[40,40],axes=FRAME,
  orientation=[-120,64]);
```

Using (3.22), the surface area of Gabriel's Horn is given by the improper integral

$$SA = 2\pi \int_1^\infty \frac{1}{x}\sqrt{1+\frac{1}{x^4}}\,dx = 2\pi \lim_{L\to\infty} \int_1^L \frac{1}{x}\sqrt{1+\frac{1}{x^4}}\,dx.$$

```
> int(2*Pi*f(x)*sqrt(1+D(f)(x)^2),x=1..infinity);
```

$$\infty$$

On the other hand, using (3.20) the volume of Gabriel's Horn is given by the improper integral

$$SA = 2\pi \int_1^\infty \frac{1}{x^2}\,dx = \pi \lim_{L\to\infty} \int_1^L \frac{1}{x^2}\,dx,$$

which converges to π.

```
> int(Pi*f(x)^2,x=1..infinity);
```

$$\pi$$

■

3.4 Series

3.4.1 Introduction to Sequences and Series

Sequences and series are usually discussed in the third quarter or second semester of introductory calculus courses. Most students find that it is one of the most difficult topics covered in calculus. A **sequence** is a function with domain consisting of the positive integers. The **terms** of the sequence $\{a_n\}$ are $a_1, a_2, a_3, \ldots$. The nth term is a_n; the $(n+1)$th term is a_{n+1}. If $\lim_{n\to\infty} a_n = L$, we say that $\{a_n\}$ **converges** to L. If $\{a_n\}$ does not converge, $\{a_n\}$ **diverges**. We can sometimes prove that a sequence converges by applying the following theorem.

A sequence $\{a_n\}$ is monotonic if $\{a_n\}$ is increasing ($a_{n+1} \geq a_n$ for all n) or decreasing ($a_{n+1} \leq a_n$ for all n).

Theorem 7. *Every bounded monotonic sequence converges.*

In particular, Theorem 7 gives us the following special cases.

1. If $\{a_n\}$ has positive terms and is eventually decreasing, $\{a_n\}$ converges.
2. If $\{a_n\}$ has negative terms and is eventually increasing, $\{a_n\}$ converges.

After you have defined a sequence, use `seq` to compute the first few terms of the sequence.

1. `seq(a(n),n=1..m)` returns the list $[a_1, a_2, a_3, \ldots, a_m]$.
2. `seq(a(n),n=k..m)` returns $[a_k, a_{k+1}, a_{k+2}, \ldots, a_m\}$.

EXAMPLE 3.4.1: If $a_n = \dfrac{50^n}{n!}$, show that $\lim_{n\to\infty} a_n = 0$.

SOLUTION: We remark that the symbol $n!$ in the denominator of a_n represents the **factorial sequence**:

$$n! = n \cdot (n-1) \cdot (n-2) \cdot \cdots \cdot 2 \cdot 1.$$

We begin by defining a_n and then computing the first few terms of the sequence with `seq`.

```
> a:=n->50^n/n!:
> afewterms:=[seq(a(n),n=1..10)];
> evalf(afewterms);
```

$$[50.0, 1250.0, 20833.33333, 260416.6667, 2604166.667, 21701388.89,$$
$$155009920.6, 968812004.0, 5382288911.0, 26911444550.0]$$

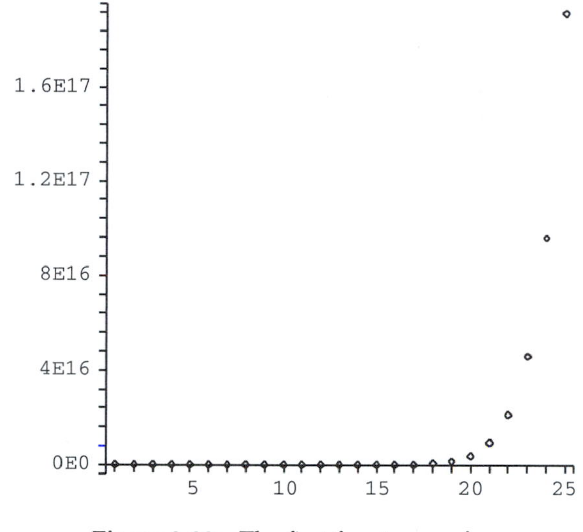

Figure 3-39 The first few terms of a_n

The first few terms increase in magnitude. In fact, this is further confirmed by graphing the first few terms of the sequence using `plot` together with the `style=point` option in Figure 3-39. Based on the graph and the values of the first few terms we might incorrectly conclude that the sequence diverges.

```
> toplot1:=[seq([k,a(k)],k=1..25)]:
> plot(toplot1,style=point,color=black);
```

However, notice that

$$a_{n+1} = \frac{50}{n+1} a_n \Rightarrow \frac{a_{n+1}}{a_n} = \frac{50}{n+1}.$$

Because $50/(n+1) < 1$ for $n > 49$, we conclude that the sequence is decreasing for $n > 49$. Because it has positive terms, it is bounded below by 0 so the sequence converges by Theorem 7. Let $L = \lim_{n\to\infty} a_n$. Then,

$$\begin{aligned} \lim_{n\to\infty} a_{n+1} &= \lim_{n\to\infty} \frac{50}{n+1} a_n \\ L &= \lim_{n\to\infty} \frac{50}{n+1} \cdot L \\ L &= 0. \end{aligned}$$

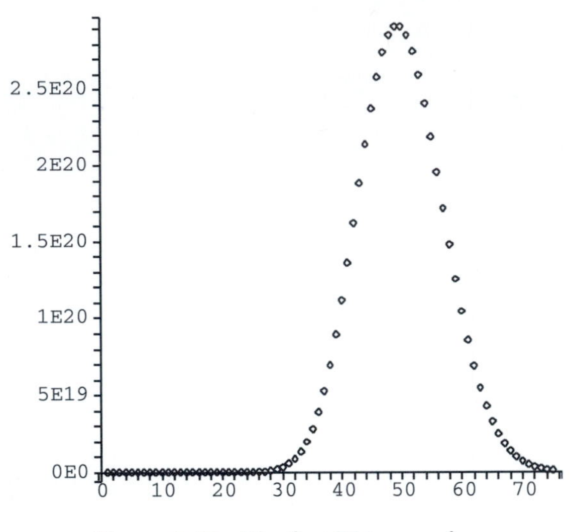

Figure 3-40 The first 75 terms of a_n

When we graph a larger number of terms, it is clear that the limit is 0 (Figure 3-40). It is a good exercise to show that for any real value of x, $\lim_{n\to\infty} \frac{x^n}{n!} = 0$.

```
> toplot2:=[seq([k,a(k)],k=1..75)]:
> plot(toplot2,style=point,color=black);
```

■

An **infinite series** is a series of the form

$$\sum_{k=1}^{\infty} a_k \tag{3.23}$$

where $\{a_n\}$ is a sequence. The nth **partial sum** of (3.23) is

$$s_n = \sum_{k=1}^{n} a_k = a_1 + a_2 + \cdots + a_n. \tag{3.24}$$

Notice that the partial sums of the series (3.23) form a sequence $\{s_n\}$. Hence, we say that the infinite series (3.23) **converges** to L if the sequence of partial sums $\{s_n\}$ converges to L, and write

$$\sum_{k=1}^{\infty} a_k = L.$$

The infinite series (3.23) **diverges** if the sequence of partial sums diverges. Given the infinite series (3.23),

```
sum(a(k),k=1..n)
```

calculates the nth partial sum (3.24). In *some* cases, if the infinite series (3.23) converges,

```
sum(a(k),k=1..infinity)
```

can compute the value of the infinite sum. You should think of the `sum` function as a "fragile" command and be certain to examine its results carefully.

EXAMPLE 3.4.2: Determine whether each of the following series converges or diverges. If the series converges, find its sum. (a) $\sum_{k=1}^{\infty}(-1)^{k+1}$; (b) $\sum_{k=2}^{\infty}\frac{2}{k^2-1}$; (c) $\sum_{k=0}^{\infty} ar^k$.

SOLUTION: For (a), we compute the nth partial sum (3.24) in `sn` with `sum`.

```
> sn:=sum((-1)^(k+1),k=1..n);
```

$$sn := 1/2\,(-1)^{n+1} + 1/2$$

Notice that the odd partial sums are 1:

$$s_{2n+1} = \frac{1}{2}\left((-1)^{2n+1+1}+1\right) = \frac{1}{2}(1+1) = 1$$

while the even partial sums are 0:

$$s_{2n} = \frac{1}{2}\left((-1)^{2n+1}+1\right) = \frac{1}{2}(-1+1) = 0.$$

We confirm that the limit of the partial sums does not exist with `limit`. Maple's result indicates that it cannot determine the limit uniquely.

```
> limit(sn,n=infinity);
```

$$0 \ldots 1$$

However, when we attempt to compute the infinite sum with `sum`, Maple is able to determine that the sum diverges.

```
> sum((-1)^(k+1),k=1..infinity);
```

undefined

Thus, the series diverges.

For (b), we have a **telescoping series**. Using partial fractions,

$$\begin{aligned}\sum_{k=2}^{\infty}\frac{2}{k^2-1} &= \sum_{k=2}^{\infty}\left(\frac{1}{k-1}-\frac{1}{k+1}\right)\\ &= \left(1-\frac{1}{3}\right)+\left(\frac{1}{2}-\frac{1}{4}\right)+\left(\frac{1}{3}-\frac{1}{5}\right)+\cdots+\left(\frac{1}{n-2}-\frac{1}{n}\right)\\ &\quad+\left(\frac{1}{n-1}-\frac{1}{n+1}\right)+\cdots\end{aligned}$$

we see that the nth partial sum is given by

$$s_n = \frac{3}{2}-\frac{1}{n}-\frac{1}{n+1}$$

and $s_n \to 3/2$ as $n \to \infty$ so the series converges to $3/2$:

$$\sum_{k=2}^{\infty}\frac{2}{k^2-1} = \frac{3}{2}.$$

We perform the same steps with Maple using `sum`, `convert` with the `parfrac` option, and `limit`.

`convert` with the `parfrac` option computes the partial fraction decomposition of a rational expression.

```
> sn:=sum(1/(k-1)-1/(k+1),k=2..n);
```

$$sn := -\frac{1+2n}{(n+1)\,n}+3/2$$

```
> convert(sn,parfrac,n);
```

$$3/2-(n+1)^{-1}-n^{-1}$$

```
> limit(sn,n=infinity);
```

$$3/2$$

In this case, you can use `sum` to find the sum of the infinite series.

```
> sum(1/(k-1)-1/(k+1),k=2..infinity);
```

$$3/2$$

(c) A series of the form $\sum_{k=0}^{\infty} ar^k$ is called a **geometric series**. We compute the nth partial sum of the geometric series with sum.

```
> sn:=sum(a*r^k,k=0..n);
```

$$sn := \frac{ar^{n+1}}{r-1} - \frac{a}{r-1}$$

When using limit to determine the limit of s_n as $n \to \infty$, we see that Maple returns the limit unevaluated because it does not know the value of r.

```
> limit(sn,n=infinity);
```

$$\lim_{n\to\infty} \frac{ar^{n+1}}{r-1} - \frac{a}{r-1}$$

In fact, the geometric series diverges if $|r| \geq 1$ and converges if $|r| < 1$.

```
> assume(r<1,r>-1);

> limit(sn,n=infinity);
```

$$\lim_{n\to\infty} \frac{ar^{n+1}}{r-1} - \frac{a}{r-1}$$

Observe that if we simply compute the sum with sum, Maple returns $a/(1-r)$, which is correct if $|r| < 1$ but incorrect if $|r| \geq 1$.

```
> r:='r':
> sum(a*r^k,k=0..infinity);
```

$$-\frac{a}{r-1}$$

However, the result of entering

```
> sum((-5/3)^k,k=0..infinity);
```

$$\textit{undefined}$$

is correct because the series $\sum_{k=0}^{\infty} \left(-\frac{5}{3}\right)^k$ is geometric with $|r| = 5/3 \geq 1$ and, consequently, diverges. Similarly,

```
> sum(9*(1/10)^k,k=1..infinity);
```

$$1$$

is correct because $\sum_{k=1}^{\infty} 9\left(\frac{1}{10}\right)^k$ is geometric with $a = 9/10$ and $r = 1/10$ so the series converges to

$$\frac{a}{1-r} = \frac{9/10}{1-1/10} = 1.$$

■

3.4.2 Convergence Tests

Frequently used convergence tests are stated in the following theorems.

Theorem 8 (The Divergence Test). *Let $\sum_{k=1}^{\infty} a_k$ be an infinite series. If $\lim_{k\to\infty} a_k \neq 0$, then $\sum_{k=1}^{\infty} a_k$ diverges.*

Theorem 9 (The Integral Test). *Let $\sum_{k=1}^{\infty} a_k$ be an infinite series with positive terms. If $f(x)$ is a decreasing continuous function for which $f(k) = a_k$ for all k, then $\sum_{k=1}^{\infty} a_k$ and $\int_1^{\infty} f(x)\,dx$ either both converge or both diverge.*

Theorem 10 (The Ratio Test). *Let $\sum_{k=1}^{\infty} a_k$ be an infinite series with positive terms and let $\rho = \lim_{k\to\infty} \frac{a_{k+1}}{a_k}$.*

1. *If $\rho < 1$, $\sum_{k=1}^{\infty} a_k$ converges.*
2. *If $\rho > 1$, $\sum_{k=1}^{\infty} a_k$ diverges.*
3. *If $\rho = 1$, the Ratio Test is inconclusive.*

Theorem 11 (The Root Test). *Let $\sum_{k=1}^{\infty} a_k$ be an infinite series with positive terms and let $\rho = \lim_{k\to\infty} \sqrt[k]{a_k}$.*

1. *If $\rho < 1$, $\sum_{k=1}^{\infty} a_k$ converges.*
2. *If $\rho > 1$, $\sum_{k=1}^{\infty} a_k$ diverges.*
3. *If $\rho = 1$, the Root Test is inconclusive.*

Theorem 12 (The Limit Comparison Test). *Let $\sum_{k=1}^{\infty} a_k$ and $\sum_{k=1}^{\infty} b_k$ be infinite series with positive terms and let $L = \lim_{k\to\infty} \frac{a_k}{b_k}$. If $0 < L < \infty$, then either both series converge or both series diverge.*

EXAMPLE 3.4.3: Determine whether each of the following series converges or diverges: (a) $\sum_{k=1}^{\infty}\left(1+\frac{1}{k}\right)^k$; (b) $\sum_{k=1}^{\infty}\frac{1}{k^p}$; (c) $\sum_{k=1}^{\infty}\frac{k}{3^k}$; (d) $\sum_{k=1}^{\infty}\frac{(k!)^2}{(2k)!}$; (e) $\sum_{k=1}^{\infty}\left(\frac{k}{4k+1}\right)^k$; and (f) $\sum_{k=1}^{\infty}\frac{2\sqrt{k}+1}{(\sqrt{k}+1)(2k+1)}$.

SOLUTION: (a) Using `limit`, we see that the limit of the terms is $e \neq 0$ so the series diverges by the Divergence Test, Theorem 8.

```
> limit((1+1/k)^k,k=infinity);
```

$$e^1$$

It is a very good exercise to show that the limit of the terms of the series is e by hand. Let $L = \lim_{k\to\infty}\left(1+\frac{1}{k}\right)^k$. Take the logarithm of each side of this equation and apply L'Hôpital's rule:

$$\begin{aligned}
\ln L &= \lim_{k\to\infty} \ln\left(1+\frac{1}{k}\right)^k \\
\ln L &= \lim_{k\to\infty} k \ln\left(1+\frac{1}{k}\right) \\
\ln L &= \lim_{k\to\infty} \frac{\ln\left(1+\frac{1}{k}\right)}{\frac{1}{k}} \\
\ln L &= \lim_{k\to\infty} \frac{\frac{1}{1+\frac{1}{k}}\cdot -\frac{1}{k^2}}{-\frac{1}{k^2}} \\
\ln L &= 1.
\end{aligned}$$

Exponentiating yields $L = e^{\ln L} = e^1 = e$.
(b) A series of the form $\sum_{k=1}^{\infty}\frac{1}{k^p}$ is called a ***p*-series**. Let $f(x) = x^{-p}$. Then, $f(x)$ is continuous and decreasing for $x \geq 1$, $f(k) = k^{-p}$ and

$$\int_1^{\infty} x^{-p}dx = \begin{cases} \infty, & \text{if } p \leq 1 \\ 1/(p-1), & \text{if } p > 1 \end{cases}$$

so the p-series converges if $p > 1$ and diverges if $p \leq 1$. If $p = 1$, the series $\sum_{k=1}^{\infty} \frac{1}{k}$ is called the **harmonic series.**

```
> p:='p':
> s1:=int(x^(-p),x=1..infinity);
```

$$s1 := \lim_{x \to \infty} -\frac{x^{-p+1} - 1}{p - 1}$$

```
> assume(p>1):
> value(s1);
```

$$(p - 1)^{-1}$$

(c) Let $f(x) = x \cdot 3^{-x}$. Then, $f(k) = k \cdot 3^{-k}$ and $f(x)$ is decreasing for $x > 1/\ln 3$.

```
> f:=x->x*3^(-x):
> factor(D(f)(x));
```

$$-3^{-x}\left(-1 + x\ln(3)\right)$$

```
> solve(-1+x*ln(3)=0);
```

$$(\ln(3))^{-1}$$

int and integrate can be used interchangeably.

Using integrate, we see that the improper integral $\int_1^{\infty} f(x)\,dx$ converges.

```
> ival:=integrate(f(x),x=1..infinity);
> evalf(ival);
```

$$ival := 1/3\,\frac{\ln(3) + 1}{(\ln(3))^2}$$

$$0.5795915583$$

Thus, by the Integral Test, Theorem 9, we conclude that the series converges. Note that when applying the Integral Test, if the improper integral converges its value is *not* the value of the sum of the series. In this case, we see that Maple is able to evaluate the sum with sum and the series converges to 3/4.

```
> sum(k*3^(-k),k=1..infinity);
```

$$3/4$$

(d) If a_k contains factorials, the Ratio Test, Theorem 10, is a good first test to try. After defining a_k we compute

$$\lim_{k\to\infty}\frac{a_{k+1}}{a_k}=\lim_{k\to\infty}\frac{\dfrac{[(k+1)!]^2}{[2(k+1)]!}}{\dfrac{(k!)^2}{(2k)!}}$$

$$=\lim_{k\to\infty}\frac{(k+1)!\cdot(k+1)!}{k!\cdot k!}\frac{(2k)!}{(2k+2)!}$$

$$=\lim_{k\to\infty}\frac{(k+1)^2}{(2k+2)(2k+1)}=\lim_{k\to\infty}\frac{(k+1)}{2(2k+1)}=\frac{1}{4}.$$

Because $1/4<1$, the series converges by the Ratio Test. We confirm these results with Maple.

Remark. Use `simplify` to simplify expressions involving factorials.

```
> a:=k->(k!)^2/(2*k)!:
> s1:=simplify(a(k+1)/a(k));
```

$$s1:=1/2\,\frac{k+1}{2k+1}$$

```
> limit(s1,k=infinity);
```

$$1/4$$

We illustrate that we can approximate the sum using `evalf` and `sum` as follows.

```
> evalf(sum(a(k),k=1..infinity));
```

$$0.7363998585$$

(e) Because

$$\lim_{k\to\infty}\sqrt[k]{\left(\frac{k}{4k+1}\right)^k}=\lim_{k\to\infty}\frac{k}{4k+1}=\frac{1}{4}<1,$$

the series converges by the Root Test, Theorem 11.

```
> a:=k->(k/(4*k+1))^k:
> limit(a(k)^(1/k),k=infinity);
```

$$1/4$$

As with (d), we can approximate the sum with `evalf` and `sum`.

```
> evalf(sum(a(k),k=1..infinity));
```

$$0.2657572097$$

(f) We use the Limit Comparison Test, Theorem 12, and compare the series to $\sum_{k=1}^{\infty} \frac{\sqrt{k}}{k\sqrt{k}} = \sum_{k=1}^{\infty} \frac{1}{k}$, which diverges because it is a p-series with $p = 1$. Because

$$0 < \lim_{k\to\infty} \frac{\dfrac{2\sqrt{k}+1}{(\sqrt{k}+1)(2k+1)}}{\dfrac{1}{k}} = 1 < \infty$$

and the harmonic series diverges, the series diverges by the Limit Comparison Test.

```
> a:=k->(2*sqrt(k)+1)/((sqrt(k)+1)*(2*k+1)):
> b:=k->1/k:
> limit(a(k)/b(k),k=infinity);
```

$$1$$

■

3.4.3 Alternating Series

An **alternating series** is a series of the form

$$\sum_{k=1}^{\infty}(-1)^k a_k \quad \text{or} \quad \sum_{k=1}^{\infty}(-1)^{k+1} a_k \tag{3.25}$$

where $\{a_k\}$ is a sequence with positive terms.

Theorem 13 (Alternating Series Test). *If* $\{a_k\}$ *is decreasing and* $\lim_{k\to\infty} a_k = 0$, *the alternating series* (3.25) *converges.*

The alternating series (3.25) **converges absolutely** if $\sum_{k=1}^{\infty} a_k$ converges.

Theorem 14. *If the alternating series* (3.25) *converges absolutely, it converges.*

If the alternating series (3.25) converges but does not converge absolutely, we say that it **conditionally converges**.

EXAMPLE 3.4.4: Determine whether each of the following series converges or diverges. If the series converges, determine whether the convergence is conditional or absolute. (a) $\sum_{k=1}^{\infty} \frac{(-1)^{k+1}}{k}$; (b) $\sum_{k=1}^{\infty} (-1)^{k+1} \frac{(k+1)!}{4^k (k!)^2}$; (c) $\sum_{k=1}^{\infty} (-1)^{k+1} \left(1 + \frac{1}{k}\right)^k$.

SOLUTION: (a) Because $\{1/k\}$ is decreasing and $1/k \to 0$ as $k \to \infty$, the series converges. The series does not converge absolutely because the harmonic series diverges. Hence, $\sum_{k=1}^{\infty} \frac{(-1)^{k+1}}{k}$, which is called the **alternating harmonic series**, converges conditionally. We see that this series converges to ln 2 with `sum`.

```
> a:=k->(-1)^(k+1)/k:
> sum(a(k),k=1..infinity);
```

$$\ln(2)$$

(b) We test for absolute convergence first using the Ratio Test. Because

$$\lim_{k\to\infty} \frac{\dfrac{((k+1)+1)!}{4^{k+1}[(k+1)!]^2}}{\dfrac{(k+2)!}{4^k (k!)^2}} = \lim_{k\to\infty} \frac{k+1}{4(k+1)^2} = 0 < 1,$$

```
> a:=k->(k+1)!/(4^k*(k!)^2):
> simplify(a(k+1)/a(k));
> limit(a(k+1)/a(k),k=infinity);
```

$$1/4\,\frac{k+2}{(k+1)^2}$$

$$0$$

the series converges absolutely by the Ratio Test. Absolute convergence implies convergence so the series converges.

(c) Because $\lim_{k\to\infty} \left(1 + \frac{1}{k}\right)^k = e$, $\lim_{k\to\infty} (-1)^{k+1} \left(1 + \frac{1}{k}\right)^k$ does not exist, so the series diverges by the Divergence Test. We confirm that the limit of the terms is not zero with `limit`.

```
> a:=k->(-1)^(k+1)*(1+1/k)^k:
> sum(a(k),k=1..infinity);
```

$$\sum_{k=1}^{\infty}(-1)^{k+1}\left(k^{-1}+1\right)^{k}$$

```
> limit(a(k),k=infinity);
```

$$-e^{1}\ldots e^{1}$$

■

3.4.4 Power Series

Let x_0 be a number. A **power series** in $x - x_0$ is a series of the form

$$\sum_{k=0}^{\infty} a_k (x - x_0)^k . \tag{3.26}$$

A fundamental problem is determining the values of x, if any, for which the power series converges.

Theorem 15. *For the power series* (3.26), *exactly one of the following is true.*

1. *The power series converges absolutely for all values of* x. *The interval of convergence is* $(-\infty, \infty)$.
2. *There is a positive number* r *so that the series converges absolutely if* $x_0 - r < x < x_0 + r$. *The series may or may not converge at* $x = x_0 - r$ *and* $x = x_0 + r$. *The interval of convergence will be one of* $(x_0 - r, x_0 + r)$, $[x_0 - r, x_0 + r)$, $(x_0 - r, x_0 + r]$, *or* $[x_0 - r, x_0 + r]$.
3. *The series converges only if* $x = x_0$. *The interval of convergence is* $\{x_0\}$.

EXAMPLE 3.4.5: Determine the interval of convergence for each of the following power series: (a) $\sum_{k=0}^{\infty} \frac{(-1)^k}{(2k+1)!} x^{2k+1}$; (b) $\sum_{k=0}^{\infty} \frac{k!}{1000^k}(x-1)^k$; (c) $\sum_{k=1}^{\infty} \frac{2^k}{\sqrt{k}}(x-4)^k$.

SOLUTION: (a) We test for absolute convergence first using the Ratio Test. Because

$$\lim_{k\to\infty}\left|\frac{\dfrac{(-1)^{k+1}}{(2(k+1)+1)!}x^{2(k+1)+1}}{\dfrac{(-1)^k}{(2k+1)!}x^{2k+1}}\right| = \lim_{k\to\infty}\frac{1}{2(k+1)(2k+3)}x^2 = 0 < 1$$

```
> a:=k->(-1)^k/(2*k+1)!*x^(2*k+1):
> s1:=simplify(a(k+1)/a(k));
> limit(s1,k=infinity);
```

$$s1 := -1/2\,\frac{x^2}{(k+1)\,(2\,k+3)}$$

$$0$$

for all values of x, we conclude that the series converges absolutely for all values of x; the interval of convergence is $(-\infty,\infty)$. In fact, we will see later that this series converges to $\sin x$:

$$\sin x = \sum_{k=0}^{\infty}\frac{(-1)^{k+1}}{(2k+1)!}x^{2k+1} = x - \frac{1}{3!}x^3 + \frac{1}{5!}x^5 - \frac{1}{7!}x^7 + \cdots,$$

which means that the partial sums of the series converge to $\sin x$. Graphically, we can visualize this by graphing partial sums of the series together with the graph of $y = \sin x$. Note that the partial sums of a series are a recursively defined function: $s_n = s_{n-1} + a_n$, $s_0 = a_0$. We use this observation to define p to be the nth partial sum of the series.

```
> p:=(x,n)->sum(a(k),k=0..n);
```

$$p := (x,n) \mapsto \sum_{k=0}^{n} a\,(k)$$

```
> p(x,1);
```

$$x - 1/6\,x^3$$

In Figure 3-41 we graph $p_n(x) = \sum_{k=0}^{n}\frac{(-1)^k}{(2k+1)!}x^{2k+1}$ together with $y = \sin x$ for $n = 1$, 5, and 10. In the graphs, notice that as n increases, the graphs of $p_n(x)$ more closely resemble the graph of $y = \sin x$.

```
> plot([sin(x),p(x,1),p(x,5),p(x,10)],x=-2*Pi..2*Pi,
> view=[-2*Pi..2*Pi,-Pi..Pi],scaling=CONSTRAINED,
> color=[black,COLOR(RGB,.4,.4,.4),
   COLOR(RGB,.6,.6,.6)]);
```

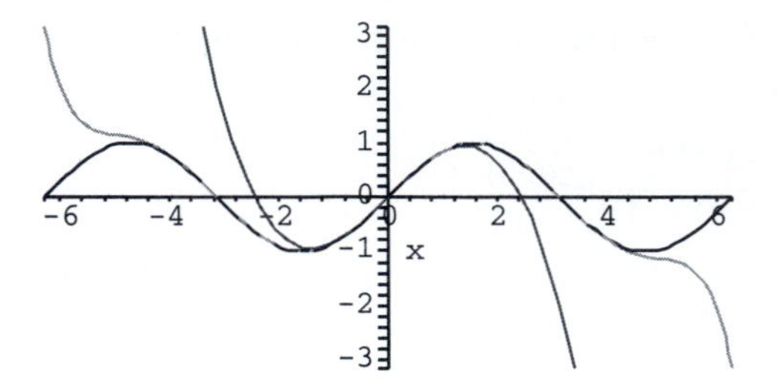

Figure 3-41 $y = \sin x$ together with the graphs of $p_1(x)$, $p_5(x)$, and $p_{10}(x)$

(b) As in (a), we test for absolute convergence first using the Ratio Test:

$$\lim_{k\to\infty}\left|\frac{\frac{(k+1)k!}{1000^{k+1}}(x-1)^{k+1}}{\frac{k!}{1000^k}(x-1)^k}\right| = \frac{1}{1000}(k+1)|x-1| = \begin{cases}0, & \text{if } x = 1\\ \infty, & \text{if } x \neq 1\end{cases}.$$

```
> a:=k->k!/1000^k*(x-1)^k:
> s1:=simplify(abs(a(k+1)/a(k)));
> limit(s1,k=infinity);
```

$$s1 := \frac{1}{1000}\,|(k+1)\,(x-1)|$$

$$\infty$$

Be careful of your interpretation of the result of the `limit` command because Maple does not consider the case $x = 1$ separately: if $x = 1$ the limit is 0. Because $0 < 1$ the series converges by the Ratio Test if $x = 1$.

The series converges only if $x = 1$; the interval of convergence is $\{1\}$. You should observe that if you graph several partial sums for "small" values of n, you might incorrectly conclude that the series converges.
(c) Use the Ratio Test to check absolute convergence first:

$$\lim_{k\to\infty}\left|\frac{\frac{2^{k+1}}{\sqrt{k+1}}(x-4)^{k+1}}{\frac{2^k}{\sqrt{k}}(x-4)^k}\right| = \lim_{k\to\infty} 2\sqrt{\frac{k}{k+1}}|x-4| = 2|x-4|.$$

By the Ratio Test, the series converges absolutely if $2|x-4| < 1$. We solve this inequality for x with `solve` to see that $2|x-4| < 1$ if $7/2 < x < 9/2$.

```
> a:=k->2^k/sqrt(k)*(x-4)^k:
> s1:=simplify(abs(a(k+1)/a(k)));
> s2:=limit(s1,k=infinity);
```

$$s1 := 2 \left| \frac{(x-4)\sqrt{k}}{\sqrt{k+1}} \right|$$

$$s2 := 2\,|x-4|$$

```
> solve(s2<1,x);
```

$$RealRange\,(Open\,(7/2)\,, Open\,(9/2))$$

We check $x = 7/2$ and $x = 9/2$ separately. If $x = 7/2$, the series becomes $\sum_{k=1}^{\infty}(-1)^k \frac{1}{\sqrt{k}}$, which converges conditionally.

```
> simplify(subs(x=7/2,a(k)));
```

$$\frac{(-1)^k}{\sqrt{k}}$$

On the other hand, if $x = 9/2$, the series is $\sum_{k=1}^{\infty} \frac{1}{\sqrt{k}}$, which diverges. We conclude that the interval of convergence is $[7/2, 9/2)$.

```
> simplify(subs(x=9/2,a(k)));
```

$$\frac{1}{\sqrt{k}}$$

■

3.4.5 Taylor and Maclaurin Series

Let $y = f(x)$ be a function with derivatives of all orders at $x = x_0$. The **Taylor series** for $f(x)$ about $x = x_0$ is

$$\sum_{k=0}^{\infty} \frac{f^{(k)}(x_0)}{k!}(x - x_0)^k. \tag{3.27}$$

The **Maclaurin series** for $f(x)$ is the Taylor series for $f(x)$ about $x = 0$. If $y = f(x)$ has derivatives up to at least order n at $x = x_0$, the nth degree **Taylor polynomial** for $f(x)$ about $x = x_0$ is

$$p_n(x) = \sum_{k=0}^{n} \frac{f^{(k)}(x_0)}{k!}(x - x_0)^k. \tag{3.28}$$

The nth degree **Maclaurin polynomial** for $f(x)$ is the nth degree Taylor polynomial for $f(x)$ about $x = 0$. Generally, finding Taylor and Maclaurin series using the definition is a tedious task at best.

EXAMPLE 3.4.6: Find the first few terms of (a) the Maclaurin series and (b) the Taylor series about $x = \pi/4$ for $f(x) = \tan x$.

SOLUTION: (a) After defining $f(x) = \tan x$, we use `seq` together with `D` and `simplify` to compute $f^{(k)}(0)/k!$ for $k = 0, 1, \ldots, 8$.

```
> f:=x->tan(x):
> [seq([k,simplify((D@@k)(f)(x)),(D@@k)(f)(0)/k!],
   k=0..8)];
```

$$\left[[0, \tan(x), 0], \right.$$
$$\left[1, 1 + (\tan(x))^2, 1\right],$$
$$\left[2, 2\tan(x)\left(1 + (\tan(x))^2\right), 0\right],$$
$$\left[3, 2\left(1 + (\tan(x))^2\right)\left(1 + 3(\tan(x))^2\right), 1/3\right],$$
$$\left[4, 8\tan(x)\left(1 + (\tan(x))^2\right)\left(2 + 3(\tan(x))^2\right), 0\right],$$
$$\left[5, 8\left(1 + (\tan(x))^2\right)\left(15(\tan(x))^2 + 15(\tan(x))^4 + 2\right), 2/15\right],$$
$$\left[6, 16\tan(x)\left(1 + (\tan(x))^2\right)\left(60(\tan(x))^2 + 45(\tan(x))^4 + 17\right), 0\right],$$
$$\left[7, 16\left(1 + (\tan(x))^2\right)\left(525(\tan(x))^4 + 315(\tan(x))^6 \right.\right.$$
$$\left.\left. + 231(\tan(x))^2 + 17\right), \frac{17}{315}\right],$$
$$\left[8, 128\left(1 + (\tan(x))^2\right)\tan(x)\left(630(\tan(x))^4 + 315(\tan(x))^6 \right.\right.$$
$$\left.\left.\left. + 378(\tan(x))^2 + 62\right), 0\right]\right]$$

Using the values in the table, we apply the definition to see that the Maclaurin series is

$$\sum_{k=0}^{\infty} \frac{f^{(k)}(0)}{k!} x^k = x + \frac{1}{3}x^3 + \frac{2}{15}x^5 + \frac{17}{315}x^7 + \cdots$$

For (b), we repeat (a) using $x = \pi/4$ instead of $x = 0$

```
> [seq([k,simplify((D@@k)(f)(x)),(D@@k)(f)(Pi/4)/k!],
   k=0..8)];
```

$$\Big[[0, \tan(x), 1],$$
$$\Big[1, 1+(\tan(x))^2, 2\Big],$$
$$\Big[2, 2\tan(x)\left(1+(\tan(x))^2\right), 2\Big],$$
$$\Big[3, 2\left(1+(\tan(x))^2\right)\left(1+3(\tan(x))^2\right), 8/3\Big],$$
$$\Big[4, 8\tan(x)\left(1+(\tan(x))^2\right)\left(2+3(\tan(x))^2\right), 10/3\Big],$$
$$\left[5, 8\left(1+(\tan(x))^2\right)\left(15(\tan(x))^2+15(\tan(x))^4+2\right), \frac{64}{15}\right],$$
$$\left[6, 16\tan(x)\left(1+(\tan(x))^2\right)\left(60(\tan(x))^2+45(\tan(x))^4+17\right), \frac{244}{45}\right],$$
$$\Big[7, 16\left(1+(\tan(x))^2\right)\Big(525(\tan(x))^4+315(\tan(x))^6$$
$$+231(\tan(x))^2+17\Big), \frac{2176}{315}\Big],$$
$$\Big[8, 128\left(1+(\tan(x))^2\right)\tan(x)\Big(630(\tan(x))^4+315(\tan(x))^6$$
$$+378(\tan(x))^2+62\Big), \frac{554}{63}\Big]\Big]$$

and then apply the definition to see that the Taylor series about $x=\pi/4$ is

$$\sum_{k=0}^{\infty}\frac{f^{(k)}(x_0)}{k!}(x-x_0)^k = 1+2\left(x-\frac{\pi}{4}\right)+2\left(x-\frac{\pi}{4}\right)^2+\frac{8}{3}\left(x-\frac{\pi}{4}\right)^3$$
$$+\frac{10}{3}\left(x-\frac{\pi}{4}\right)^4+\frac{64}{15}\left(x-\frac{\pi}{4}\right)^5+\frac{244}{45}\left(x-\frac{\pi}{4}\right)^6+\cdots$$

From the series, we can see various Taylor and Maclaurin polynomials. For example, the third Maclaurin polynomial is

$$p_3(x) = x+\frac{1}{3}x^3$$

and the fourth-degree Taylor polynomial about $x=\pi/4$ is

$$p_4(x) = 1+2\left(x-\frac{\pi}{4}\right)+2\left(x-\frac{\pi}{4}\right)^2+\frac{8}{3}\left(x-\frac{\pi}{4}\right)^3+\frac{10}{3}\left(x-\frac{\pi}{4}\right)^4.$$

■

The command

```
series(f(x),x=x0,n)
```

computes (3.27) to (at least) order $n-1$. Because of the O-term in the result that represents the terms omitted from the power series for $f(x)$ expanded about the point $x = x_0$, the result of entering a `series` command is not a function that can be evaluated if x is a particular number. We remove the remainder (O-) term of the power series `series(f(x),x=x0,n)` using the command `convert` with the `polynom` option. The resulting polynomial can then be evaluated for particular values of x.

EXAMPLE 3.4.7: Find the first few terms of the Taylor series for $f(x)$ about $x = x_0$: (a) $f(x) = \cos x$, $x = 0$; (b) $f(x) = 1/x^2$, $x = 1$.

SOLUTION: Entering

```
> series(cos(x),x=0);
```

$$series\left(1 - 1/2\,x^2 + 1/24\,x^4 + O\left(x^6\right), x, 6\right)$$

computes the Maclaurin series to order 6. Entering

```
> series(cos(x),x=0,14);
```

$$series\left(1 - 1/2\,x^2 + 1/24\,x^4 - \frac{1}{720}x^6 + \frac{1}{40320}x^8 - \frac{1}{3628800}x^{10} + \frac{1}{479001600}x^{12} + O\left(x^{14}\right), x, 14\right)$$

computes the Maclaurin series to order 14. In this case, the Maclaurin series for $\cos x$ converges to $\cos x$ for all real x. To see this graphically, we define the function `p`. Given n, `p(n)` returns the Maclaurin polynomial of degree n for $\cos x$.

```
> p:=proc(n)
> convert(series(cos(x),x=0,n+1),polynom)
> end:
```

For example, $p_8(x)$ is given by

```
> p(8);
```

$$1 - 1/2\,x^2 + 1/24\,x^4 - \frac{1}{720}x^6 + \frac{1}{40320}x^8$$

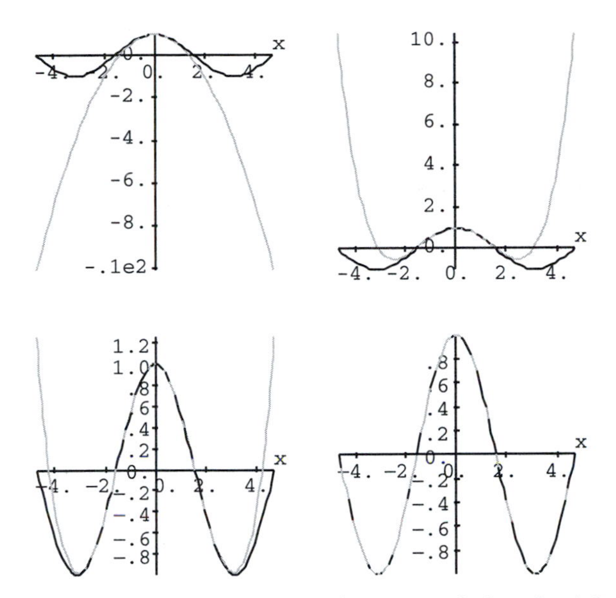

Figure 3-42 Graphs of $y = \cos x$ together with its second, fourth, eighth, and sixteenth Maclaurin polynomials

We then graph $\cos x$ together with the Maclaurin polynomial of degree $n = 2, 4, 8$, and 16 on the interval $[-3\pi/2, 3\pi/2]$ in Figure 3-42. Notice that as n increases, the graph of the Maclaurin polynomial more closely resembles the graph of $\cos x$. We would see the same pattern if we increased the length of the interval and the value of n.

```
> with(plots):
> A:=array(1..2,1..2):
> A[1,1]:=plot([cos(x),p(2)],x=-3*Pi/2..3*Pi/2,
   color=[black,gray]):
> A[1,2]:=plot([cos(x),p(4)],x=-3*Pi/2..3*Pi/2,
   color=[black,gray]):
> A[2,1]:=plot([cos(x),p(8)],x=-3*Pi/2..3*Pi/2,
   color=[black,gray]):
> A[2,2]:=plot([cos(x),p(16)],x=-3*Pi/2..3*Pi/2,
   color=[black,gray]):
> display(A);
```

(b) After defining $f(x) = 1/x^2$, we compute the first 10 terms of the Taylor series for $f(x)$ about $x = 1$ with `series`.

```
> f:=x->1/x^2:
> p10:=series(f(x),x=1,10);
```

$$p10 := series\big(3 - 2\,x + 3\,(x-1)^2 - 4\,(x-1)^3 + 5\,(x-1)^4 - 6\,(x-1)^5 + 7\,(x-1)^6 - 8\,(x-1)^7 + 9\,(x-1)^8 - 10\,(x-1)^9 + O(x-1)\,, x-1, 10\big)$$

In this case, the pattern for the series is relatively easy to see: the Taylor series for $f(x)$ about $x = 1$ is

$$\sum_{k=0}^{\infty}(-1)^k(k+1)(x-1)^k.$$

This series converges absolutely if

$$\lim_{k\to\infty}\left|\frac{(-1)^{k+1}(k+2)(x-1)^{k+1}}{(-1)^k(k+1)(x-1)^k}\right| = |x-1| < 1$$

or $0 < x < 2$. The series diverges if $x = 0$ and $x = 2$. In this case, the series converges to $f(x)$ on the interval $(0, 2)$.

```
> a:=k->(-1)^k*(k+1)*(x-1)^k:
> s1:=simplify(abs(a(k+1)/a(k)));
```

$$s1 := \left|\frac{(k+2)\,(x-1)}{k+1}\right|$$

```
> limit(s1,k=infinity);
```

$$|x-1|$$

```
> solve(abs(x-1)<1,x);
```

$$RealRange\,(Open(0), Open(2))$$

To see this, we graph $f(x)$ together with the Taylor polynomial for $f(x)$ about $x = 1$ of degree n for large n. Regardless of the size of n, the graphs of $f(x)$ and the Taylor polynomial closely resemble each other on the interval $(0, 2)$ – but not at the endpoints or outside the interval (Figure 3-43).

```
> f:=x->1/x^2:
> p:=proc(n)
> convert(series(f(x),x=1,n+1),polynom)
> end:
```

```
> plot([f(x),p(16)],x=0..2,color=[black,gray],
   view=[0..2,-5..45]);
```

■

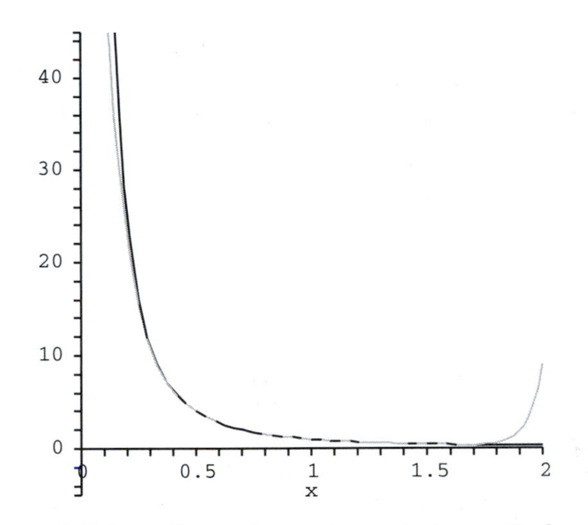

Figure 3-43 Graph of $f(x)$ together with the sixteenth-degree Taylor polynomial about $x = 1$

3.4.6 Taylor's Theorem

Taylor's theorem states the relationship between $f(x)$ and the Taylor series for $f(x)$ about $x = x_0$.

Theorem 16 (Taylor's Theorem). *Let $y = f(x)$ have (at least) $n+1$ derivatives on an interval I containing $x = x_0$. Then, for every number $x \in I$, there is a number z between x and x_0 so that*

$$f(x) = p_n(x) + R_n(x),$$

where $p_n(x)$ is given by (3.28) *and*

$$R_n(x) = \frac{f^{(n+1)}(z)}{(n+1)!}(x - x_0)^{n+1}. \tag{3.29}$$

EXAMPLE 3.4.8: Use Taylor's theorem to show that

$$\sin x = \sum_{k=0}^{\infty} \frac{(-1)^k}{(2k+1)!} x^{2k+1}.$$

SOLUTION: Let $f(x) = \sin x$. Then, for each value of x, there is a number z between 0 and x so that $\sin x = p_n(x) + R_n(x)$ where $p_n(x) = \sum_{k=0}^{n} \frac{f^{(k)}(0)}{k!} x^k$ and $R_n(x) = \frac{f^{(n+1)}(z)}{(n+1)!} x^{n+1}$. Regardless of the value of n, $f^{(n+1)}(z)$ is one of $\sin z$, $-\sin z$, $\cos z$, or $-\cos z$, which are all bounded by 1. Then,

$$\left|\sin x - p_n(x)\right| = \left|\frac{f^{(n+1)}(z)}{(n+1)!} x^{n+1}\right|$$

$$\left|\sin x - p_n(x)\right| \le \frac{1}{(n+1)!} |x|^{n+1}$$

and $\frac{x^n}{n!} \to 0$ as $n \to \infty$ for all real values of x.

You should remember that the number z in $R_n(x)$ is guaranteed to exist by Taylor's theorem. However, from a practical point of view, you would rarely (if ever) need to compute the z value for a particular x value.

For illustrative purposes, we show the difficulties. Suppose we wish to approximate $\sin(\pi/180)$ using the Maclaurin polynomial of degree 4, $p_4(x) = x - \frac{1}{6}x^3$, for $\sin x$. The fourth remainder is

The Maclaurin polynomial of degree 4 for $\sin x$ is $\sum_{k=0}^{4} \frac{f^{(k)}(0)}{k!} x^4 = 0 + x + 0 \cdot x^2 + \frac{-1}{3!} x^3 + 0 \cdot x^4$.

$$R_4(x) = \frac{1}{120} \cos z\, x^5.$$

```
> f:=x->sin(x):
> r5:=(D@@5)(f)(z)/5!*x^5;
```

$$r5 := \frac{1}{120} \cos(z)\, x^5$$

If $x = \pi/180$ there is a number z between 0 and $\pi/180$ so that

$$\left|R_4\left(\frac{\pi}{180}\right)\right| = \frac{1}{120} \cos z \left(\frac{\pi}{180}\right)^5$$
$$\le \frac{1}{120}\left(\frac{\pi}{180}\right)^5 \approx 0.135 \times 10^{-10},$$

which shows us that the maximum the error can be is $\frac{1}{120}\left(\frac{\pi}{180}\right)^5 \approx 0.135 \times 10^{-10}$.

```
> maxerror:=evalf(1/120*(Pi/180)^5);
```

$$maxerror := 1.349601624 \times 10^{-11}$$

Abstractly, the exact error can be computed. By Taylor's theorem, z satisfies

$$f\left(\frac{\pi}{180}\right) = p_4\left(\frac{\pi}{180}\right) + R_4\left(\frac{\pi}{180}\right)$$

$$\sin\frac{\pi}{180} = \frac{1}{180}\pi - \frac{1}{34992000}\pi^3 + \frac{1}{22674816000000}\pi^5\cos z$$

$$0 = \frac{1}{180}\pi - \frac{1}{34992000}\pi^3 + \frac{1}{22674816000000}\pi^5\cos z - \sin\frac{\pi}{180}.$$

We graph the right-hand side of this equation with `plot` in Figure 3-44. The exact value of z is the z-coordinate of the point where the graph intersects the z-axis.

```
> p4:=convert(series(f(x),x=0,5),polynom);
```

$$p4 := x - 1/6x^3$$

```
> exval:=sin(Pi/180);
> p4b:=subs(x=Pi/180,p4);
> r5b:=subs(x=Pi/180,r5);
```

$$exval := \sin\left(\frac{1}{180}\pi\right)$$

$$p4b := \frac{1}{180}\pi - \frac{1}{34992000}\pi^3$$

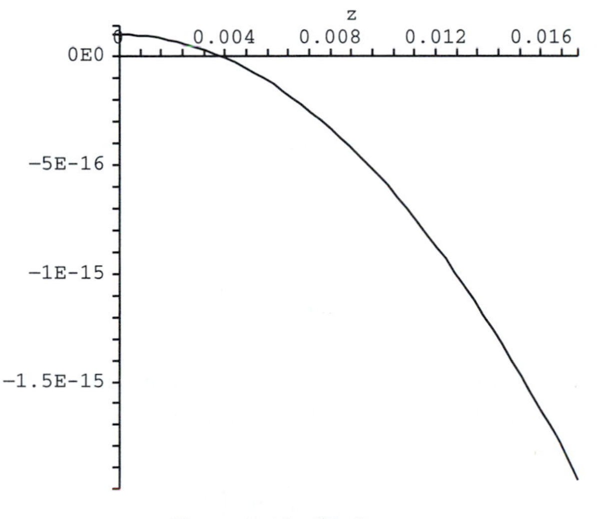

Figure 3-44 Finding z

$$r5b := \frac{1}{22674816000000}\cos(z)\,\pi^5$$

```
> toplot:=r5b+p4b-exval:
> plot(toplot,z=0..Pi/180,color=black);
```

We can use `fsolve` to approximate z, if we increase the number of digits carried in floating point calculations with `Digits`.

```
> Digits:=20:
> exz:=fsolve(toplot=0,z,0..0.01,fulldigits);
```

$$exz := 0.0038086149165541606417$$

Alternatively, we can compute the exact value of z with `solve`

```
> cz:=solve(toplot,z);
```

$$cz := \arccos\left(648000\,\frac{-194400\,\pi + \pi^3 + 34992000\sin\left(\frac{1}{180}\pi\right)}{\pi^5}\right)$$

and then approximate the result with `evalf`.

```
> czn:=evalf(cz);
```

$$czn := 0.0038086261175999712083$$

■

3.4.7 Other Series

In calculus, we learn that the power series $f(x) = \sum_{k=0}^{\infty} a_k\,(x - x_0)^k$ is differentiable and integrable on its interval of convergence. However, for series that are not power series this result is not generally true. For example, in more advanced courses, we learn that the function

$$f(x) = \sum_{k=0}^{\infty} \frac{1}{2^k}\sin\left(3^k x\right)$$

is continuous for all values of x but nowhere differentiable. We can use Maple to help us see why this function is not differentiable. Let

$$f_n(x) = \sum_{k=0}^{n} \frac{1}{2^k}\sin\left(3^k x\right).$$

Notice that $f_n(x)$ is defined recursively by $f_0(x) = \sin x$ and $f_n(x) = f_{n-1}(x) + \frac{1}{2^n}\sin(3^n x)$. We use Maple to recursively define $f_n(x)$.

```
> f:='f':
> f:=proc(n) option remember;
> f(n-1)+sin(3^n*x)/2^n end:
> f(0):=sin(x):
```

We define $f_n(x)$ using the form with `proc` using the `remember` option so that Maple "remembers" the values it computes. Thus, to compute `f(5)`, Maple uses the previously computed values, namely `f(4)`.
Next, we use `seq` to generate $f_3(x)$, $f_6(x)$, $f_9(x)$, and $f_{12}(x)$.

```
> ints:=seq(3*i,i=1..4);
> seq(f(n),n=ints);
```

$$\sin(x) + 1/2\sin(3x) + 1/4\sin(9x) + 1/8\sin(27x),$$

$$\sin(x) + 1/2\sin(3x) + 1/4\sin(9x) + 1/8\sin(27x) + 1/16\sin(81x) + 1/32\sin(243x) + \frac{1}{64}\sin(729x),$$

$$\sin(x) + 1/2\sin(3x) + 1/4\sin(9x) + 1/8\sin(27x) + 1/16\sin(81x) + 1/32\sin(243x) + \frac{1}{64}\sin(729x) + \frac{1}{128}\sin(2187x) + \frac{1}{256}\sin(6561x) + \frac{1}{512}\sin(19683x),$$

$$\sin(x) + 1/2\sin(3x) + 1/4\sin(9x) + 1/8\sin(27x) + 1/16\sin(81x) + 1/32\sin(243x) + \frac{1}{64}\sin(729x) + \frac{1}{128}\sin(2187x) + \frac{1}{256}\sin(6561x) + \frac{1}{512}\sin(19683x) + \frac{1}{1024}\sin(59049x) + \frac{1}{2048}\sin(177147x) + \frac{1}{4096}\sin(531441x)$$

We now graph each of these functions and show the results as a graphics array with `display` in Figure 3-45.

```
> with(plots):
> A:=array(1..2,1..2):
> A[1,1]:=plot(f(3),x=0..3*Pi,color=black):
> A[1,2]:=plot(f(6),x=0..3*Pi,color=black):
> A[2,1]:=plot(f(9),x=0..3*Pi,color=black):
```

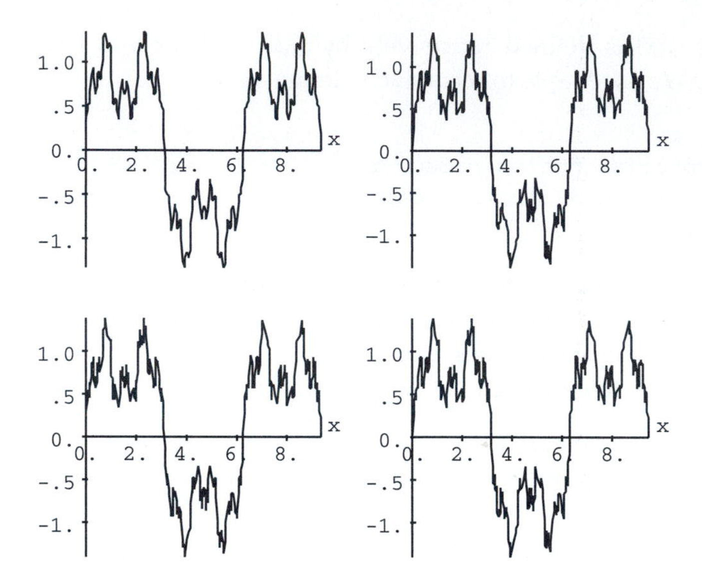

Figure 3-45 Approximating a function that is continuous everywhere but nowhere differentiable

```
> A[2,2]:=plot(f(12),x=0..3*Pi,color=black):
> display(A);
```

From these graphs, we see that for large values of n, the graph of $f_n(x)$, although actually smooth, appears "jagged" and thus we might suspect that $f(x) = \lim_{n\to\infty} f_n(x) = \sum_{k=0}^{\infty} \frac{1}{2^k} \sin\left(3^k x\right)$ is indeed continuous everywhere but nowhere differentiable.

3.5 Multi-Variable Calculus

Maple is useful in investigating functions involving more than one variable. In particular, the graphical analysis of functions that depend on two (or more) variables is enhanced with the help of Maple's graphics capabilities.

3.5.1 Limits of Functions of Two Variables

Maple's graphics and numerical capabilities are helpful in investigating limits of functions of two variables.

EXAMPLE 3.5.1: Show that the limit $\lim_{(x,y)\to(0,0)} \dfrac{x^2 - y^2}{x^2 + y^2}$ does not exist.

SOLUTION: We begin by defining $f(x,y) = \dfrac{x^2 - y^2}{x^2 + y^2}$. Next, we use `plot3d` to graph $z = f(x,y)$ for $-1/2 \le x \le 1/2$ and $-1/2 \le y \le 1/2$. `contourplot` is used to graph several level curves on the same rectangle (Figure 3-46). (To define a function of two variables, $f(x,y) =$ *expression in x and y*, enter `f:=(x,y)->expression in x and y`. `plot3d(f(x,y),x=a..b,y=c..d)` generates a basic graph of $z=f(x,y)$ for $a \le x \le b$ and $c \le y \le d$. `contourplot(f(x,y), x=a..b,y=c..d)` generates a basic plot of the level curves for $z = f(x,y)$.)

```
> with(plots):
> f:=(x,y)->(x^2-y^2)/(x^2+y^2):
> plot3d(f(x,y),x=-.5..0.5,y=-.5..0.5,axes=BOXED,
   color=gray,
> grid=[40,40],orientation=[-45,30]);
> contourplot(f(x,y),x=-.5..0.5,y=-.5..0.5,
   axes=FRAME,color=black,
> grid=[40,40],scaling=CONSTRAINED);
```

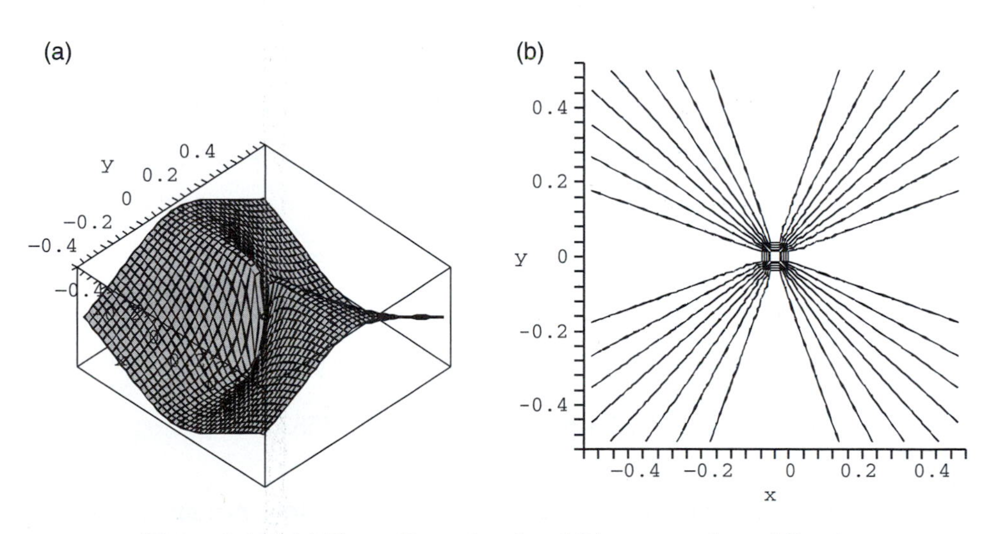

Figure 3-46 (a) Three-dimensional and (b) contour plots of $f(x,y)$

From the graph of the level curves, we suspect that the limit does not exist because we see that near $(0,0)$, $z = f(x,y)$ attains many different values. We obtain further evidence that the limit does not exist by computing the value of $z = f(x,y)$ for various points chosen randomly near $(0,0)$. We use `seq` and `rand` to generate 13 ordered triples $(x,y,f(x,y))$ for x and y "close to" 0. Because `rand` is included in the calculation, your results will almost certainly be different from those here. The first column corresponds to the x-coordinate, the second column to the y-coordinate, and the third column to the value of $z = f(x,y)$.

```
> r:='r':
> r:=proc(n)
> local a,b;
> a:=evalf(rand()*12^(-12-n));
> b:=evalf(rand()*12^(-12-n));
> [a,b,f(a,b)]
> end:
> seq(r(n),n=0..12);
```

[0.04793796027, 0.03601470117, 0.2784310478],

[0.003211728002, 0.004432581881, −0.3114705452],

[0.0004349643177, 0.0005816209138, −0.2826454490],

[0.000002081013642, 0.00004692497627, −0.9960742878],

[0.000003268561734, 0.000004032685324, −0.2070454032],

[0.0000001171056756, 0.0000001397610813, −0.1750364920],

[0.00000002994288814, 0.000000001471250566, 0.9951830886],

[0.0000000002767953294, 0.000000003006444334, −0.9831897010],

[0.0000000002120426056, 0.0000000001183556144, 0.5249101775],

[$1.399912437 \times 10^{-11}$, $2.001135850 \times 10^{-11}$, −0.3428382246],

[$1.722731318 \times 10^{-12}$, $2.653442115 \times 10^{-13}$, 0.9536519637],

[$2.348632968 \times 10^{-14}$, $6.481655897 \times 10^{-14}$, −0.7678806689],

[$6.609425775 \times 10^{-15}$, $3.429074467 \times 10^{-15}$, 0.5758332410]

From the third column, we see that $z = f(x,y)$ does not appear to approach any particular value for points chosen randomly near $(0,0)$.

In fact, along the line $y = mx$ we see that

$$f(x,y) = f(x,mx) = \frac{1-m^2}{1+m^2}.$$

We choose lines of the form $y = mx$ because near $(0,0)$ the level curves of $z = f(x,y)$ look like lines of the form $y = mx$.

Hence as $(x,y) \to (0,0)$ along $y = mx$, $f(x,y) = f(x,mx) \to \frac{1-m^2}{1+m^2}$. Thus, $f(x,y)$ does not have a limit as $(x,y) \to (0,0)$ because the value depends on the choice of m.

```
> v1:=simplify(f(x,m*x));
> subs(m=0,v1);
> subs(m=1,v1);
> subs(m=1/2,v1);
```

$$v1 := -\frac{-1+m^2}{1+m^2}$$

$$1$$

$$0$$

$$3/5$$

■

In some cases, you can establish that a limit does not exist by converting to polar coordinates. For example, in polar coordinates, $f(x,y) = \frac{x^2-y^2}{x^2+y^2}$ becomes $f(r\cos\theta, r\sin\theta) = 2\cos^2\theta - 1$

```
> simplify(f(r*cos(theta),r*sin(theta)));
```

$$2\,(\cos(\theta))^2 - 1$$

and

$$\lim_{(x,y)\to(0,0)} f(x,y) = \lim_{r\to 0} f(r\cos\theta, r\sin\theta) = \lim_{r\to 0} 2\cos^2\theta - 1 = 2\cos^2\theta - 1 = \cos 2\theta$$

depends on θ.

3.5.2 Partial and Directional Derivatives

Partial derivatives of functions of two or more variables are computed with Maple using `diff` or `D`. For $z = f(x,y)$,

1. `diff(f(x,y),x)` computes $\frac{\partial f}{\partial x} = f_x(x,y)$,
2. `diff(f(x,y),y)` computes $\frac{\partial f}{\partial y} = f_y(x,y)$,

3. `diff(f(x,y),x$n)` computes $\frac{\partial^n f}{\partial x^n}$,
4. `diff(f(x,y),x,y)` computes $\frac{\partial^2 f}{\partial x \partial y} = f_{yx}(x,y)$, and
5. `diff(f(x,y),x$n,y$m)` computes $\frac{\partial^{n+m} f}{\partial^n x \partial^m y}$.

You can also use `D` to compute partial derivatives. For example, `D[1](f)(x,y)` computes $\frac{\partial f}{\partial x} = f_x(x,y)$ and `D[2](f)(x,y)` computes $\frac{\partial f}{\partial y} = f_y(x,y)$. The calculations are carried out similarly for functions of more than two variables.

EXAMPLE 3.5.2: Calculate $f_x(x,y)$, $f_y(x,y)$, $f_{xy}(x,y)$, $f_{yx}(x,y)$, $f_{xx}(x,y)$, and $f_{yy}(x,y)$ if $f(x,y) = \sin\sqrt{x^2+y^2+1}$.

SOLUTION: After defining $f(x,y) = \sin\sqrt{x^2+y^2+1}$,

```
> f:=(x,y)->sin(sqrt(x^2+y^2+1)):
```

we illustrate the use of `D` and `diff` to compute the partial derivatives. Entering

```
> diff(f(x,y),x);
> D[1](f)(x,y);
```

$$\frac{\cos\left(\sqrt{x^2+y^2+1}\right)x}{\sqrt{x^2+y^2+1}}$$

$$\frac{\cos\left(\sqrt{x^2+y^2+1}\right)x}{\sqrt{x^2+y^2+1}}$$

computes $f_x(x,y)$. Entering

```
> diff(f(x,y),y);
> D[2](f)(x,y);
```

$$\frac{\cos\left(\sqrt{x^2+y^2+1}\right)y}{\sqrt{x^2+y^2+1}}$$

$$\frac{\cos\left(\sqrt{x^2+y^2+1}\right)y}{\sqrt{x^2+y^2+1}}$$

computes $f_y(x,y)$. Entering

```
> simplify(diff(f(x,y),x,y));
> simplify(D[1,2](f)(x,y));
```

$$-\frac{\left(\sin\left(\sqrt{x^2+y^2+1}\right)\sqrt{x^2+y^2+1}+\cos\left(\sqrt{x^2+y^2+1}\right)\right)yx}{\left(x^2+y^2+1\right)^{3/2}}$$

$$-\frac{\left(\sin\left(\sqrt{x^2+y^2+1}\right)\sqrt{x^2+y^2+1}+\cos\left(\sqrt{x^2+y^2+1}\right)\right)yx}{\left(x^2+y^2+1\right)^{3/2}}$$

computes $f_{yx}(x,y)$. Entering

```
> simplify(diff(f(x,y),y,x));
> simplify(D[2,1](f)(x,y));
```

$$-\frac{\left(\sin\left(\sqrt{x^2+y^2+1}\right)\sqrt{x^2+y^2+1}+\cos\left(\sqrt{x^2+y^2+1}\right)\right)yx}{\left(x^2+y^2+1\right)^{3/2}}$$

$$-\frac{\left(\sin\left(\sqrt{x^2+y^2+1}\right)\sqrt{x^2+y^2+1}+\cos\left(\sqrt{x^2+y^2+1}\right)\right)yx}{\left(x^2+y^2+1\right)^{3/2}}$$

computes $f_{xy}(x,y)$. Remember that under appropriate assumptions, $f_{xy}(x,y) = f_{yx}(x,y)$. Entering

```
> simplify(diff(f(x,y),x$2));
> simplify(D[1$2](f)(x,y));
```

$$-\frac{\sin\left(\sqrt{x^2+y^2+1}\right)x^2\sqrt{x^2+y^2+1}-\cos\left(\sqrt{x^2+y^2+1}\right)-\cos\left(\sqrt{x^2+y^2+1}\right)y^2}{\left(x^2+y^2+1\right)^{3/2}}$$

$$-\frac{\sin\left(\sqrt{x^2+y^2+1}\right)x^2\sqrt{x^2+y^2+1}-\cos\left(\sqrt{x^2+y^2+1}\right)-\cos\left(\sqrt{x^2+y^2+1}\right)y^2}{\left(x^2+y^2+1\right)^{3/2}}$$

computes $f_{xx}(x,y)$. Entering

```
> simplify(diff(f(x,y),y$2));
> simplify(D[2$2](f)(x,y));
```

$$\frac{\cos\left(\sqrt{x^2+y^2+1}\right)x^2-\sin\left(\sqrt{x^2+y^2+1}\right)y^2\sqrt{x^2+y^2+1}+\cos\left(\sqrt{x^2+y^2+1}\right)}{\left(x^2+y^2+1\right)^{3/2}}$$

$$\frac{\cos\left(\sqrt{x^2+y^2+1}\right)x^2-\sin\left(\sqrt{x^2+y^2+1}\right)y^2\sqrt{x^2+y^2+1}+\cos\left(\sqrt{x^2+y^2+1}\right)}{\left(x^2+y^2+1\right)^{3/2}}$$

computes $f_{yy}(x,y)$.

■

The vectors **i** and **j** are defined by $\mathbf{i} = \langle 1, 0\rangle$ and $\mathbf{j} = \langle 0, 1\rangle$.

The **directional derivative** of $z = f(x,y)$ in the direction of the unit vector $\mathbf{u} = \cos\theta\,\mathbf{i} + \sin\theta\,\mathbf{j}$ is

$$D_{\mathbf{u}}f(x,y) = f_x(x,y)\cos\theta + f_y(x,y)\sin\theta,$$

provided that $f_x(x,y)$ and $f_y(x,y)$ both exist.

Calculus of vector-valued functions is discussed in more detail in Chapter 5.

If $f_x(x,y)$ and $f_y(x,y)$ both exist, the **gradient** of $f(x,y)$ is the vector-valued function

$$\nabla f(x,y) = f_x(x,y)\mathbf{i} + f_y(x,y)\mathbf{j} = \left\langle f_x(x,y), f_y(x,y)\right\rangle.$$

Notice that if $\mathbf{u} = \langle\cos\theta, \sin\theta\rangle$,

$$D_{\mathbf{u}}f(x,y) = \nabla f(x,y)\cdot\langle\cos\theta, \sin\theta\rangle.$$

Use the `grad` command, which is contained in the `linalg` package,

```
grad(f(x,y),[x,y])
```

to compute $\nabla f(x,y)$.

EXAMPLE 3.5.3: Let $f(x,y) = 6x^2y - 3x^4 - 2y^3$. (a) Find $D_{\mathbf{u}}f(x,y)$ in the direction of $\mathbf{v} = \langle 3, 4\rangle$. (b) Compute

$$D_{\langle 3/5,4/5\rangle}f\left(\frac{1}{3}\sqrt{9+3\sqrt{3}}, 1\right).$$

(c) Find an equation of the line tangent to the graph of $6x^2y-3x^4-2y^3 = 0$ at the point $\left(\frac{1}{3}\sqrt{9+3\sqrt{3}}, 1\right)$.

SOLUTION: After defining $f(x,y) = 6x^2y - 3x^4 - 2y^3$, we graph $z = f(x,y)$ with `plot3d` in Figure 3-47, illustrating the `grid`, `gridstyle`, and `axes` options.

```
> f:=(x,y)->6*x^2*y-3*x^4-2*y^3:
> plot3d(f(x,y),x=-2..2,y=-2..3,
   view=[-2..2,-2..3,-2..2],grid=[35,35],
> gridstyle=triangular,axes=BOXED);
```

(a) A unit vector, **u**, in the same direction as **v** is

$$\mathbf{v} = \left\langle \frac{3}{\sqrt{3^2+4^2}}, \frac{4}{\sqrt{3^2+4^2}} \right\rangle = \left\langle \frac{3}{5}, \frac{4}{5} \right\rangle.$$

`norm(v,frobenius)` computes the Euclidean norm of the vector **v**. `norm` is contained in the `linalg` package.

```
> with(linalg):
> v:=[3,4]:
> u:=v/norm(v,frobenius);
```

$$u := [3/5, 4/5]$$

Then, $D_{\mathbf{u}}f(x,y) = \langle f_x(x,y), f_y(x,y) \rangle \cdot \mathbf{u}$, calculated in `du`.

```
> du:=dotprod([D[1](f)(x,y),D[2](f)(x,y)],u);
```

$$du := \frac{36}{5}yx - \frac{36}{5}x^3 + \frac{24}{5}x^2 - \frac{24}{5}y^2$$

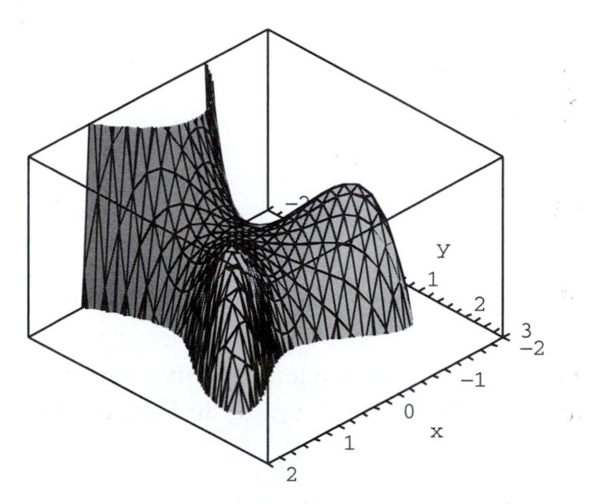

Figure 3-47 $f(x,y) = 6x^2y - 3x^4 - 2y^3$ for $-2 \le x \le 2$ and $-2 \le y \le 3$

(b) $D_{\langle 3/4,4/5\rangle}f\left(\frac{1}{3}\sqrt{9+3\sqrt{3}},1\right)$ is calculated by evaluating du if $x = \frac{1}{3}\sqrt{9+3\sqrt{3}}$ and $y = 1$.

```
> du1:=simplify(subs(x=1/3*sqrt(9+3*sqrt(3)),y=1,du));
> evalf(du1);
```

$$du1 := -4/5\sqrt{9+3\sqrt{3}}\sqrt{3} + 8/5\sqrt{3}$$

$$-2.449505301$$

(c) The gradient is evaluated if $x = \frac{1}{3}\sqrt{9+3\sqrt{3}}$ and $y = 1$.

```
> gradf:=grad(f(x,y),[x,y]);
> nvec:=simplify(subs(x=1/3*sqrt(9+3*sqrt(3)),y=1,
   eval(gradf)));
```

$$gradf := vector\left([12\,yx - 12\,x^3, 6\,x^2 - 6\,y^2]\right)$$

$$\left[-4/3\sqrt{9+3\sqrt{3}}\sqrt{3}, 2\sqrt{3}\right] := vector\left(\left[-4/3\sqrt{9+3\sqrt{3}}\sqrt{3}, 2\sqrt{3}\right]\right)$$

Generally, $\nabla f(x,y)$ is perpendicular to the level curves of $z = f(x,y)$, so

$$\text{nvec} = \nabla f\left(\frac{1}{3}\sqrt{9+3\sqrt{3}},1\right) = \left\langle f_x\left(\frac{1}{3}\sqrt{9+3\sqrt{3}},1\right), f_y\left(\frac{1}{3}\sqrt{9+3\sqrt{3}},1\right)\right\rangle$$

An equation of the line L containing (x_0, y_0) and perpendicular to $\mathbf{n} = \langle a, b\rangle$ is $a(x - x_0) + b(y - y_0) = 0$.

is perpendicular to $f(x,y) = 0$ at the point $\left(\frac{1}{3}\sqrt{9+3\sqrt{3}},1\right)$. Thus, an equation of the line tangent to the graph of $f(x,y) = 0$ at the point $\left(\frac{1}{3}\sqrt{9+3\sqrt{3}},1\right)$ is

$$f_x\left(\frac{1}{3}\sqrt{9+3\sqrt{3}},1\right)\left(x - \frac{1}{3}\sqrt{9+3\sqrt{3}}\right) + f_y\left(\frac{1}{3}\sqrt{9+3\sqrt{3}},1\right)(y-1) = 0,$$

which we solve for y with solve. We confirm this result by graphing $f(x,y) = 0$ using contourplot with the contours=[0] option in conf and then graphing the tangent line in tanplot. tanplot and conf are shown together with display in Figure 3-48.

```
> with(plots):
> conf:=contourplot(f(x,y),x=-2..2,y=-2..3,
   contours=[0],
```

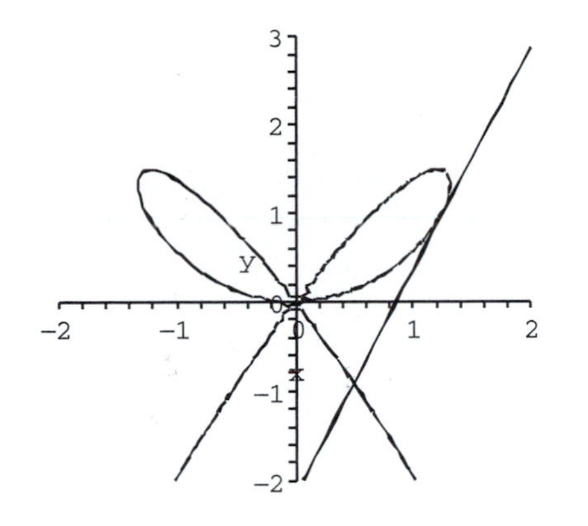

Figure 3-48 Level curves of $f(x, y)$

```
> color=black,grid=[40,40],scaling=CONSTRAINED)
> tanline:=solve(nvec[1]*(x-1/3*sqrt(9+3*sqrt(3)))
   +nvec[2]*(y-1)=0,y);
```

$$tanline := 1/9\left(2\sqrt{9+3\sqrt{3}}\sqrt{3}x - 3\sqrt{3} - 6\right)\sqrt{3}$$

```
> tanplot:=plot(tanline,x=-2..2,color=black,
   view=[-2..2,-2..3]):
> display([conf,tanplot],scaling=CONSTRAINED);
```

■

EXAMPLE 3.5.4: Let

$$f(x,y) = (y-1)^2 e^{-(x+1)^2-y^2} - \frac{10}{3}\left(-x^5 + \frac{1}{5}y - y^3\right)e^{-x^2-y^2} - \frac{1}{9}e^{-x^2-(y+1)^2}.$$

Calculate $\nabla f(x, y)$ and then graph $\nabla f(x, y)$ together with several level curves of $f(x, y)$.

SOLUTION: We begin by defining and graphing $z = f(x, y)$ with `plot3d` in Figure 3-49.

```
> f:=(x,y)->(y-1)^2*exp(-(x+1)^2-y^2)
   -10/3*(-x^5+1/5*y-y^3)*
> exp(-x^2-y^2)-1/9*exp(-x^2-(y+1)^2):
```

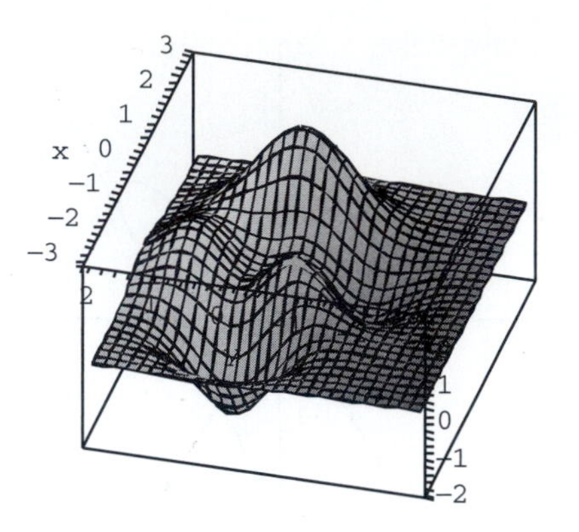

Figure 3-49 $f(x,y)$ for $-3 \le x \le 3$ and $-3 \le y \le 2$

Remember that with most operating systems, **Enter** and **Return** evaluate a Maple command; **shift-Enter** and **shift-Return** give a new line.

```
> with(plots):
> plot3d(f(x,y),x=-3..3,y=-3..2,orientation=[-165,45],
   axes=boxed,
> scaling=CONSTRAINED,grid=[30,30]);
> conf:=contourplot(f(x,y),x=-3..3,y=-3..3,
   grid=[50,50],color=black,
> scaling=CONSTRAINED,contours=25):
```

In the three-dimensional plot, notice that z appears to have six relative extrema: three relative maxima and three relative minima. We also graph several level curves of $f(x,y)$ with `contourplot` and name the resulting graphic `conf`. The graphic is not displayed because we include a colon at the end of the `contourplot` command.

Next we calculate $f_x(x,y)$ and $f_y(x,y)$ using `simplify` and `diff`. The gradient is the vector-valued function $\langle f_x(x,y), f_y(x,y)\rangle$, computed with the `grad` function that is contained in the `linalg` package.

```
> fx:=simplify(diff(f(x,y),x));
> fy:=simplify(diff(f(x,y),y));
```

$$fx := -2\,e^{-x^2-2x-1-y^2}y^2x - 2\,e^{-x^2-2x-1-y^2}y^2 + 4\,e^{-x^2-2x-1-y^2}yx + 4\,e^{-x^2-2x-1-y^2}y - 2\,e^{-x^2-2x-1-y^2}x - 2\,e^{-x^2-2x-1-y^2} + \frac{50}{3}x^4e^{-x^2-y^2} - \frac{20}{3}x^6e^{-x^2-y^2} + 4/3\,xe^{-x^2-y^2}y - \frac{20}{3}xe^{-x^2-y^2}y^3 + 2/9\,xe^{-x^2-y^2-2y-1}$$

$$fy := -2e^{-x^2-2x-1-y^2} - 2e^{-x^2-2x-1-y^2}y^3 + 4e^{-x^2-2x-1-y^2}y^2$$
$$-2/3e^{-x^2-y^2} + \frac{34}{3}e^{-x^2-y^2}y^2 - \frac{20}{3}ye^{-x^2-y^2}x^5 - \frac{20}{3}y^4e^{-x^2-y^2}$$
$$+2/9e^{-x^2-y^2-2y-1}y + 2/9e^{-x^2-y^2-2y-1}$$

```
> with(linalg):
> grad(f(x,y),[x,y]);
```

$$vector\left(\left[(y-1)^2(-2x-2)e^{-(x+1)^2-y^2} + \frac{50}{3}x^4e^{-x^2-y^2}\right.\right.$$
$$+\frac{20}{3}\left(-x^5+1/5y-y^3\right)xe^{-x^2-y^2} + 2/9xe^{-x^2-(y+1)^2},$$
$$2(y-1)e^{-(x+1)^2-y^2} - 2(y-1)^2ye^{-(x+1)^2-y^2}$$
$$-10/3\left(1/5-3y^2\right)e^{-x^2-y^2} + \frac{20}{3}\left(-x^5+1/5y-y^3\right)ye^{-x^2-y^2}$$
$$\left.\left.-1/9(-2y-2)e^{-x^2-(y+1)^2}\right]\right)$$

To graph the gradient, we use `gradplot`, which is contained in the `plots` package. We use `gradplot` to graph the gradient, naming the resulting graphic `gradf`. `gradf` and `conf` are displayed together using `display`.

Use `gradplot` in the same way that you use `contourplot`.

```
> gradf:=gradplot(f(x,y),x=-3..3,y=-3..3,
   grid=[40,40],arrows=THICK,
> color=black,scaling=CONSTRAINED):
> display(conf,gradf);
```

In the result (Figure 3-50), notice that the gradient is perpendicular to the level curves; the gradient is pointing in the direction of maximal increase of $z = f(x, y)$.

■

Classifying Critical Points

Let $z = f(x, y)$ be a real-valued function of two variables with continuous second-order partial derivatives. A **critical point** of $z = f(x, y)$ is a point (x_0, y_0) in the interior of the domain of $z = f(x, y)$ for which

$$f_x(x_0, y_0) = 0 \quad \text{and} \quad f_y(x_0, y_0) = 0.$$

Critical points are classified by the *Second Derivatives* (or *Partials*) *Test*.

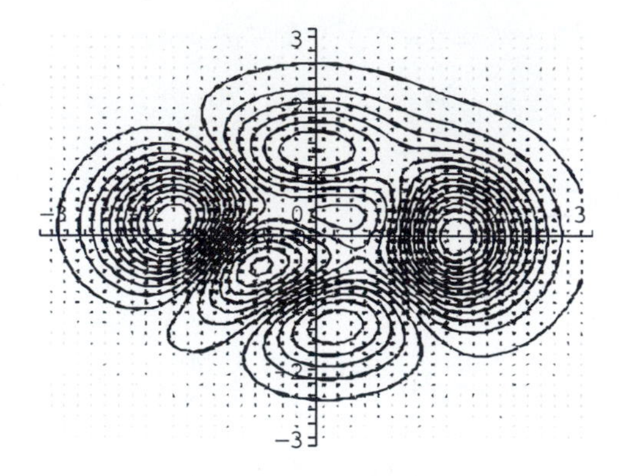

Figure 3-50 Contour plot of $f(x,y)$ along with several gradient vectors

Theorem 17 (Second Derivatives Test). *Let (x_0, y_0) be a critical point of a function $z = f(x,y)$ of two variables and let*

$$d = f_{xx}(x_0, y_0)\, f_{yy}(x_0, y_0) - \left[f_{xy}(x_0, y_0)\right]^2. \tag{3.30}$$

1. *If $d > 0$ and $f_{xx}(x_0, y_0) > 0$, then $z = f(x,y)$ has a **relative** (or **local**) **minimum** at (x_0, y_0).*
2. *If $d > 0$ and $f_{xx}(x_0, y_0) < 0$, then $z = f(x,y)$ has a **relative** (or **local**) **maximum** at (x_0, y_0).*
3. *If $d < 0$, then $z = f(x,y)$ has a **saddle point** at (x_0, y_0).*
4. *If $d = 0$, no conclusion can be drawn and (x_0, y_0) is called a **degenerate critical point**.*

EXAMPLE 3.5.5: Find the relative maximum, relative minimum, and saddle points of $f(x,y) = -2x^2 + x^4 + 3y - y^3$.

SOLUTION: After defining $f(x,y)$, the critical points are found with `solve` and named `critpts`.

```
> f:=(x,y)->-2*x^2+x^4+3*y-y^3:
> critpts:=[solve(D[1](f)(x,y)=0,
   D[2](f)(x,y)=0,x,y)];
```

$$critpts := \left[\{x = 0, y = 1\}, \{y = -1, x = 0\}, \{x = 1, y = 1\}, \{x = 1, y = -1\}, \{x = -1, y = 1\}, \{x = -1, y = -1\}\right]$$

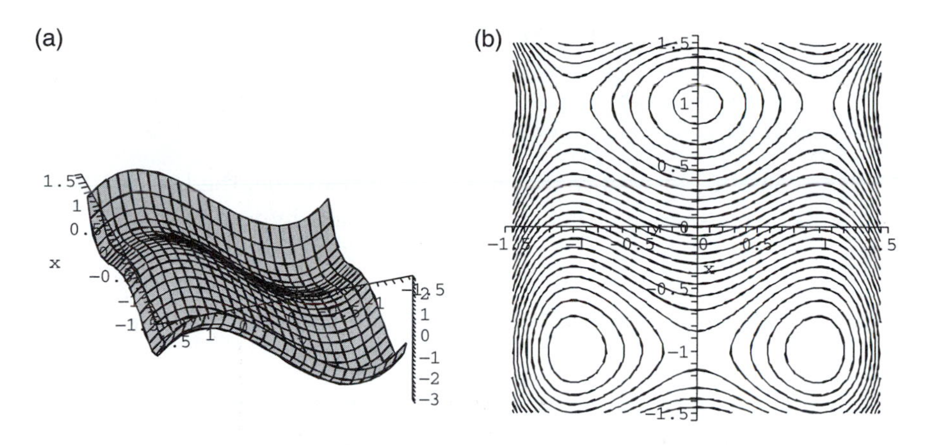

Figure 3-51 (a) Three-dimensional and (b) contour plots of $f(x, y)$

We then define dfxx. Given (x_0, y_0), dfxx (x_0, y_0) returns the ordered quadruple x_0, y_0, (3.30) evaluated at (x_0, y_0), and $f_{xx}(x_0, y_0)$.

```
> dfxx:=(x,y)->[x,y,D[1$2](f)(x,y)*D[2$2](f)(x,y)-
> D[1,2](f)(x,y)^2,D[1$2](f)(x,y)]:
```

For example,

```
> dfxx(0,1);
```

$$[0, 1, 24, -4]$$

shows us that a relative maximum occurs at $(0, 1)$. We then use subs to substitute the values in each element of critpts into dfxx.

```
> map(subs,critpts,dfxx(x,y));
```

$$[[0, 1, 24, -4], [0, -1, -24, -4], [1, 1, -48, 8],$$
$$[1, -1, 48, 8], [-1, 1, -48, 8], [-1, -1, 48, 8]]$$

From the result, we see that $(0, 1)$ results in a relative maximum, $(0, -1)$ results in a saddle, $(1, 1)$ results in a saddle, $(1, -1)$ results in a relative minimum, $(-1, 1)$ results in a saddle, and $(-1, -1)$ results in a relative minimum. We confirm these results graphically with a three-dimensional plot generated with plot3d and a contour plot generated with contourplot in Figure 3-51.

```
> with(plots):
> plot3d(f(x,y),x=-3/2..3/2,y=-3/2..3/2,axes=FRAMED,
```

```
> orientation=[162,38],color=gray);
> contourplot(f(x,y),x=-3/2..3/2,y=-3/2..3/2,
  contours=25,
> grid=[45,45],color=black,scaling=constrained);
```

In the contour plot, notice that near relative extrema, the level curves look like circles while near saddles they look like hyperbolas.
■

If the Second Derivatives Test fails, graphical analysis is especially useful.

EXAMPLE 3.5.6: Find the relative maximum, relative minimum, and saddle points of $f(x,y) = x^2 + x^2y^2 + y^4$.

SOLUTION: Initially we proceed in the same manner as in the previous example: we define $f(x,y)$ and compute the critical points. Several complex solutions are returned, which we ignore.

```
> f:=(x,y)->x^2+x^2*y^2+y^4:
> critpts:=[solve(D[1](f)(x,y)=0,
  D[2](f)(x,y)=0,x,y)];
```

$$critpts := \left[\{x=0, y=0\}, \left\{y = RootOf\left(\_Z^2+1, label=\_L3\right),\right.\right.$$
$$\left.\left. x = RootOf\left(\_Z^2-2, label=\_L4\right)\right\}\right]$$

```
> critpts:=evalf(map(allvalues,critpts));
```

$$critpts := \left[\{x=0.0, y=0.0\}, \{x=1.414213562, y=1.0i\},\right.$$
$$\{x=-1.414213562, y=1.0i\}, \{x=1.414213562, y=-1.0i\},$$
$$\left.\{x=-1.414213562, y=-1.0i\}\right]$$

We then compute the value of (3.30) at the real critical point, and the value of $f_{xx}(x,y)$ at this critical point.

```
> dfxx:=(x,y)->[x,y,D[1$2](f)(x,y)*D[2$2](f)(x,y)
  -D[1,2](f)(x,y)^2,D[1$2](f)(x,y)]:
> map(subs,critpts,dfxx(x,y));
```

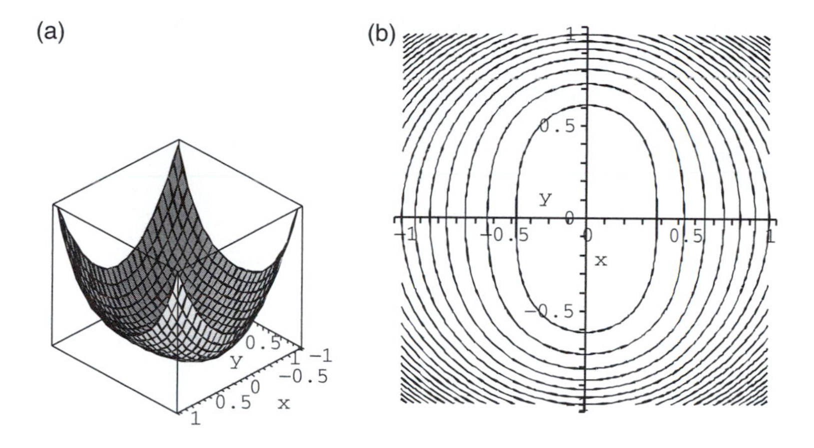

Figure 3-52 (a) Three-dimensional and (b) contour plots of $f(x,y)$

$$\left[[0.0,0.0,0.0,2.0],[1.414213562,1.0\,i,31.99999998,0],\right.$$
$$[-1.414213562,1.0\,i,31.99999998,0],$$
$$[1.414213562,-1.0\,i,31.99999998,0],$$
$$\left.[-1.414213562,-1.0\,i,31.99999998,0]\right]$$

The result shows us that the Second Derivatives Test fails at $(0,0)$.

```
> ?allvalues
> with(plots):
> plot3d(f(x,y),x=-1..1,y=-1..1,axes=BOXED,
   scaling=CONSTRAINED);
> contourplot(f(x,y),x=-1..1,y=-1..1,contours=20,
   color=black,grid=[50,50],scaling=CONSTRAINED);
```

However, the contour plot of $f(x,y)$ near $(0,0)$ indicates that an extreme value occurs at $(0,0)$. The three-dimensional plot shows that $(0,0)$ is a relative minimum (Figure 3-52).

■

Tangent Planes

Let $z = f(x,y)$ be a real-valued function of two variables. If both $f_x(x_0,y_0)$ and $f_y(x_0,y_0)$ exist, then an equation of the tangent plane to the graph of $z = f(x,y)$ at the point $(x_0,y_0,f(x_0,y_0))$ is given by

$$f_x(x_0,y_0)(x-x_0)+f_y(x_0,y_0)(y-y_0)-(z-z_0)=0, \tag{3.31}$$

where $z_0 = f(x_0, y_0)$. Solving for z yields the function (of two variables)

$$z = f_x(x_0, y_0)(x - x_0) + f_y(x_0, y_0)(y - y_0) + z_0. \tag{3.32}$$

Symmetric equations of the line perpendicular to the surface $z = f(x, y)$ at the point (x_0, y_0, z_0) are given by

$$\frac{x - x_0}{f_x(x_0, y_0)} = \frac{y - y_0}{f_y(x_0, y_0)} = \frac{z - z_0}{-1} \tag{3.33}$$

and parametric equations are

$$\begin{cases} x = x_0 + f_x(x_0, y_0)\,t \\ y = y_0 + f_y(x_0, y_0)\,t \\ z = z_0 - t. \end{cases} \tag{3.34}$$

The plane tangent to the graph of $z = f(x, y)$ at the point $(x_0, y_0, f(x_0, y_0))$ is the "best" linear approximation of $z = f(x, y)$ near $(x, y) = (x_0, y_0)$ in the same way as the line tangent to the graph of $y = f(x)$ at the point $(x_0, f(x_0))$ is the "best" linear approximation of $y = f(x)$ near $x = x_0$.

EXAMPLE 3.5.7: Find an equation of the plane tangent and normal line to the graph of $f(x, y) = 4 - \frac{1}{4}(2x^2 + y^2)$ at the point $(1, 2, 5/2)$.

SOLUTION: We define $f(x, y)$ and compute $f_x(1, 2)$ and $f_y(1, 2)$.

```
> f:=(x,y)->4-1/4*(2*x^2+y^2):
> f(1,2);
> dx:=D[1](f)(1,2);
> dy:=D[2](f)(1,2);
```

$$5/2$$

$$dx := -1$$

$$dy := -1$$

Using (3.32), an equation of the tangent plane is $z = -1(x - 1) - 1(y - 2) + f(1, 2)$. Using (3.34), parametric equations of the normal line are $x = 1 - t, y = 2 - t, z = f(1, 2) - t$. We confirm the result graphically by graphing $f(x, y)$ together with the tangent plane in `p1` using `plot3d`. We use `spacecurve`, which is contained in the `plots` package, to

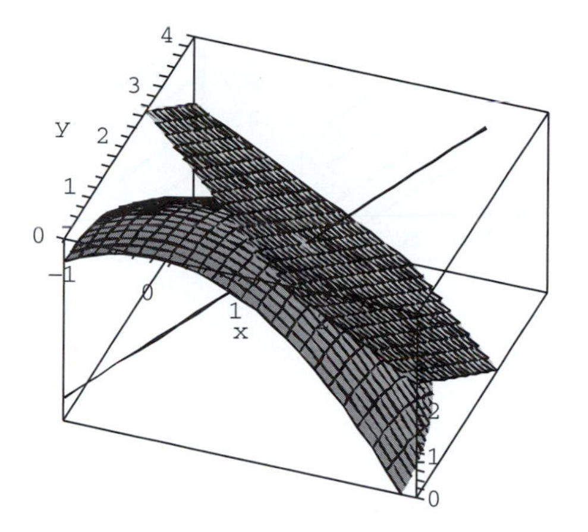

Figure 3-53 Graph of $f(x, y)$ with a tangent plane and normal line

graph the normal line in p2 and then display p1 and p2 together with display in Figure 3-53.

```
> with(plots):
> p1:=plot3d(f(x,y),dx*(x-1)+dy*(y-2)+f(1,2),
   x=-1..3,y=0..4,
> axes=BOXED,view=[-1..3,0..4,0..4],
   orientation=[-70,40]):
> p2:=spacecurve([1+dx*t,2+dy*t,f(1,2)-t],t=-4..4,
   color=black):
> display([p1,p2],scaling=CONSTRAINED);
```

Because $z = -1(x-1)-1(y-2)+f(1,2)$ is the "best" linear approximation of $f(x, y)$ near $(1, 2)$, the graphs are very similar near $(1, 2)$ as shown in the three-dimensional plot. We also expect the level curves of each near $(1, 2)$ to be similar, which is confirmed with contourplot in Figure 3-54.

```
> A:=array(1..2):
> A[1]:=contourplot(f(x,y),x=0.75..1.15,y=1.75..2.25,
   color=black,
> scaling=CONSTRAINED):
> A[2]:=contourplot(dx*(x-1)+dy*(y-2)+f(1,2),
   x=0.75..1.25,y=1.75..2.25,
> color=black,scaling=CONSTRAINED):
> display(A);
```

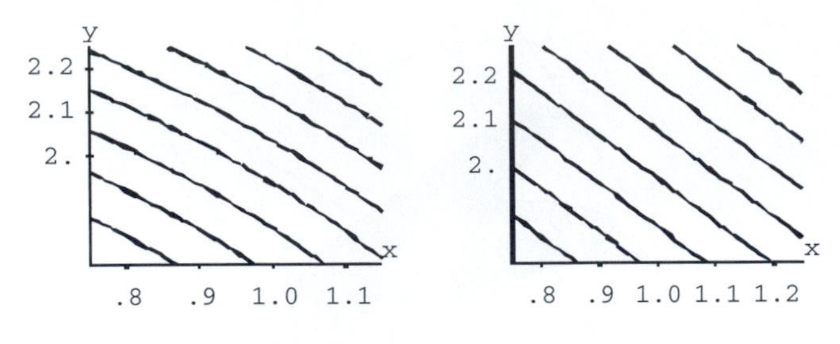

Figure 3-54 Zooming in near $(1,2)$

Lagrange Multipliers

Certain types of optimization problems can be solved using the method of *Lagrange multipliers* that is based on the following theorem.

Theorem 18 (Lagrange's Theorem). *Let $z = f(x,y)$ and $z = g(x,y)$ be real-valued functions with continuous partial derivatives and let $z = f(x,y)$ have an extreme value at a point (x_0, y_0) on the smooth constraint curve $g(x,y) = 0$. If $\nabla g\,(x_0, y_0) \neq \mathbf{0}$, then there is a real number λ satisfying*

$$\nabla f\,(x_0, y_0) = \lambda \,\nabla\, g\,(x_0, y_0)\,. \tag{3.35}$$

Graphically, the points (x_0, y_0) at which the extreme values occur correspond to the points where the level curves of $z = f(x,y)$ are tangent to the graph of $g(x,y) = 0$.

EXAMPLE 3.5.8: Find the maximum and minimum values of $f(x,y) = xy$ subject to the constraint $\frac{1}{4}x^2 + \frac{1}{9}y^2 = 1$.

SOLUTION: For this problem, $f(x,y) = xy$ and $g(x,y) = \frac{1}{4}x^2 + \frac{1}{9}y^2 - 1$. Observe that parametric equations for $\frac{1}{4}x^2 + \frac{1}{9}y^2 = 1$ are $x = 2\cos t$, $y = 3\sin t$, $0 \leq t \leq 2\pi$. In Figure 3-55, we use `spacecurve` to parametrically graph $g(x,y) = 0$ and $f(x,y)$ for x- and y-values on the curve $g(x,y) = 0$ by graphing

$$\begin{cases} x = 2\cos t \\ y = 3\sin t \\ z = 0 \end{cases} \quad \text{and} \quad \begin{cases} x = 2\cos t \\ y = 3\sin t \\ z = x \cdot y = 6\cos t \sin t \end{cases}$$

for $0 \leq t \leq 2\pi$. Our goal is to find the minimum and maximum values in Figure 3-55 and the points at which they occur.

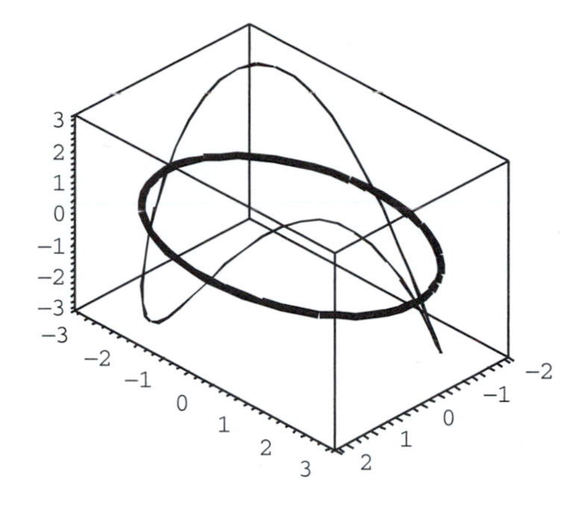

Figure 3-55 $f(x,y)$ on $g(x,y)=0$

```
> with(student):
> with(linalg):
> f:=(x,y)->x*y:
> g:=(x,y)->x^2/4+y^2/9-1:
> with(plots):
> s1:=spacecurve([2*cos(t),3*sin(t),0],t=0..2*Pi,
   color=black,thickness=3):
> s2:=spacecurve([2*cos(t),3*sin(t),6*cos(t)*sin(t)],
   t=0..2*Pi,color=black):
> display(s1,s2,scaling=CONSTRAINED,axes=BOXED);
```

To implement the method of Lagrange multipliers, we compute $f_x(x,y)$, $f_y(x,y)$, $g_x(x,y)$, and $g_y(x,y)$ with `diff`.

```
> fx:=diff(f(x,y),x);
> fy:=diff(f(x,y),y);
> gx:=diff(g(x,y),x);
> gy:=diff(g(x,y),y);
```

$$fx := y$$

$$fy := x$$

$$gx := 1/2\,x$$

$$gy := 2/9\,y$$

`solve` is used to solve the system of equations (3.35):

$$f_x(x,y) = \lambda g_x(x,y)$$

$$f_y(x,y) = \lambda g_y(x,y)$$
$$g(x,y) = 0$$

for x, y, and λ.

```
> vals:=solve(fx=lambda*gx,fy=lambda*gy,g(x,y)=0,
  x,y,lambda);
```

$$vals := \left\{y=3/2\,RootOf\left(\_Z^2-2\right),\lambda=3,x=RootOf\left(\_Z^2-2\right)\right\},$$
$$\left\{y=-3/2\,RootOf\left(\_Z^2-2\right),\lambda=-3,x=RootOf\left(\_Z^2-2\right)\right\}$$

The corresponding values of $f(x,y)$ are found using `subs` and `seq`.

```
> n1:=seq(subs(vals[i],[x,y,f(x,y)]),i=1..2);
```

$$n1 := \left[RootOf\left(\_Z^2-2\right),3/2\,RootOf\left(\_Z^2-2\right),\right.$$
$$\left.3/2\left(RootOf\left(\_Z^2-2\right)\right)^2\right],\left[RootOf\left(\_Z^2-2\right),\right.$$
$$\left.-3/2\,RootOf\left(\_Z^2-2\right),-3/2\left(RootOf\left(\_Z^2-2\right)\right)^2\right]$$

```
> allvalues(n1[1]);
> evalf(%);
> allvalues(n1[2]);
> evalf(%);
```

$$\left[\sqrt{2},3/2\sqrt{2},3\right],\left[-\sqrt{2},-3/2\sqrt{2},3\right]$$

$$[1.414213562, 2.121320343, 3.0], [-1.414213562, -2.121320343, 3.0]$$

$$\left[\sqrt{2},-3/2\sqrt{2},-3\right],\left[-\sqrt{2},3/2\sqrt{2},-3\right]$$

$$[1.414213562, -2.121320343, -3.0], [-1.414213562, 2.121320343, -3.0]$$

We conclude that the maximum value $f(x,y)$ subject to the constraint $g(x,y) = 0$ is 3 and occurs at $\left(\sqrt{2},\frac{3}{2}\sqrt{2}\right)$ and $\left(-\sqrt{2},-\frac{3}{2}\sqrt{2}\right)$. The minimum value is -3 and occurs at $\left(-\sqrt{2},\frac{3}{2}\sqrt{2}\right)$ and $\left(\sqrt{2},-\frac{3}{2}\sqrt{2}\right)$. We graph several level curves of $f(x,y)$ and the graph of $g(x,y) = 0$ with `contourplot` and show the graphs together with `display`. The minimum and maximum values of $f(x,y)$ subject to the constraint $g(x,y) = 0$

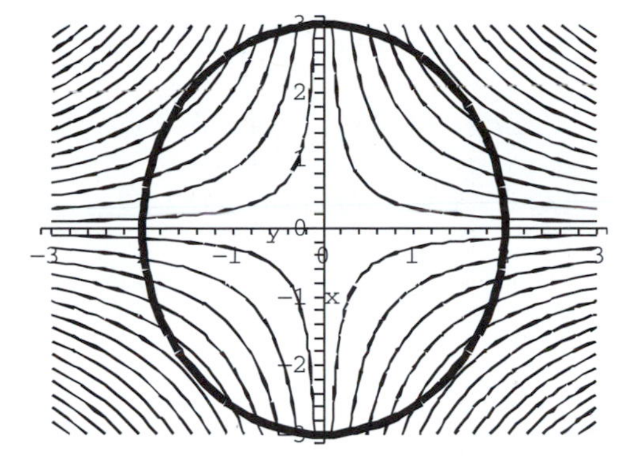

Figure 3-56 Level curves of $f(x, y)$ together with $g(x, y) = 0$

occur at the points where the level curves of $f(x, y)$ are tangent to the graph of $g(x, y) = 0$ as illustrated in Figure 3-56.

```
> with(plots):
> cp1:=contourplot(f(x,y),x=-3..3,y=-3..3,
   contours=30,
> color=BLACK):
> cp2:=contourplot(g(x,y),x=-3..3,y=-3..3,
   contours=[0],
> color=BLACK,thickness=3):
> display(cp1,cp2,scaling=CONSTRAINED);
```

You can also use the `extrema` function to find the maximum and minimum values of a function with appropriate constraints as illustrated in the following commands.

```
> extrema(x*y,x^2/4+y^2/9=1,x,y,'s');
```

$$\{-3, 3\}$$

```
> s;
```

$$\left\{\left\{y = 3/2\,RootOf\left(\_Z^2 - 2\right), x = RootOf\left(\_Z^2 - 2\right)\right\},\right.$$
$$\left.\left\{y = -3/2\,RootOf\left(\_Z^2 - 2\right), x = RootOf\left(\_Z^2 - 2\right)\right\}\right\}$$

```
> allvalues(s);
```

$$\left\{\left\{x=\sqrt{2}, y=-3/2\sqrt{2}\right\},\left\{y=3/2\sqrt{2}, x=\sqrt{2}\right\}\right\},$$
$$\left\{\left\{y=-3/2\sqrt{2}, x=-\sqrt{2}\right\},\left\{y=3/2\sqrt{2}, x=-\sqrt{2}\right\}\right\}$$

■

3.5.3 Iterated Integrals

The `int` and `integrate` commands, used to compute single integrals, are used to compute iterated integrals. The command

```
int(int(f(x,y),x=a..b),y=c..d)
```

attempts to compute the iterated integral

$$\int_c^d \int_a^b f(x,y)\,dx\,dy. \tag{3.36}$$

The `integrate` command works in the exact same way as the `int` command. If Maple cannot compute the exact value of the integral, it is returned unevaluated, in which case numerical results may be more useful. The iterated integral (3.36) is numerically evaluated with the command `evalf`. The `student` package contains the commands `Doubleint` and `Tripleint` that can be used to form double and triple iterated integrals, respectively. For example, after the `student` package has been loaded

```
Doubleint(f(x,y),x,y)
```

returns the unevaluated indefinite integral $\int\int f(x,y)\,dx\,dy$; the command

```
Doubleint(f(x,y),x=a..b,y=c..d)
```

returns the unevaluated definite integral $\int_c^d \int_a^b f(x,y)\,dx\,dy$. If the integral can be evaluated exactly, the exact value is obtained with `value`; numerical evaluation is obtained with `evalf`.

EXAMPLE 3.5.9: Evaluate each integral: (a) $\int_2^4 \int_1^2 \left(2xy^2+3x^2y\right)\,dx\,dy$; (b) $\int_0^2 \int_{y^2}^{2y} \left(3x^2+y^3\right)\,dx\,dy$; (c) $\int_0^\infty \int_0^\infty xye^{-x^2-y^2}\,dy\,dx$; (d) $\int_0^\pi \int_0^\pi e^{\sin xy}\,dx\,dy$.

SOLUTION: (a) First we compute $\iint (2xy^2 + 3x^2y)\,dx\,dy$ with `int`. Second, we compute $\int_2^4 \int_1^2 (2xy^2 + 3x^2y)\,dx\,dy$ with `int` and `Doubleint`.

```
> int(int(2*x*y^2+3*x^2*y,x),y);
> int(int(2*x*y^2+3*x^2*y,x=1..2),y=2..4);
> a1:=Doubleint(2*x*y^2+3*x^2*y,x=1..2,y=2..4);
> value(a1);
```

$$1/3\,x^2y^3 + 1/2\,x^3y^2$$

$$98$$

$$a1 := \int_2^4 \int_1^2 2\,xy^2 + 3\,x^2y dx\,dy$$

$$98$$

(b) We illustrate the same commands as in (a), except we are integrating over a nonrectangular region.

```
> int(int(3*x^2+y^3,x),y);
> int(int(3*x^2+y^3,x=y^2..2*y),y=0..2);
> a1:=Doubleint(3*x^2+y^3,x=y^2..2*y,y=0..2);
> value(a1);
```

$$x^3y + 1/4\,y^4x$$

$$\frac{1664}{105}$$

$$a1 := \int_0^2 \int_{y^2}^{2\,y} 3\,x^2 + y^3dx\,dy$$

$$\frac{1664}{105}$$

(c) Improper integrals can be handled in the same way as proper integrals.

```
> int(int(x*y*exp(-x^2-y^2),y),x);
> int(int(x*y*exp(-x^2-y^2),y=0..infinity),
   x=0..infinity);
> a1:=Doubleint(x*y*exp(-x^2-y^2),y=0..infinity,
   x=0..infinity);
> value(a1);
```

$$1/4\,e^{-x^2-y^2}$$

$$1/4$$

$$a1 := \int_0^{\infty}\int_0^{\infty} xye^{-x^2-y^2}\,dy\,dx$$

$$1/4$$

(d) In this case, Maple cannot evaluate the integral exactly so we use

```
> a1:=int(int(exp(sin(x*y)),x=0..Pi),y=0..Pi);
```

$$a1 := \int_0^{\pi}\int_0^{\pi} e^{\sin(xy)}\,dx\,dy$$

```
> evalf(a1);
```

$$15.50915577$$

```
> evalf(Int(Int(exp(sin(x*y)),x=0..Pi),y=0..Pi));
```

$$15.50915577$$

■

Area, Volume, and Surface Area

Typical applications of iterated integrals include determining the area of a planar region, the volume of a region in three-dimensional space, or the surface area of a region in three-dimensional space. The area of the planar region R is given by

$$A = \iint_R dA. \tag{3.37}$$

If $z = f(x,y)$ has continuous partial derivatives on a closed region R, then the surface area of the portion of the surface that projects onto R is given by

$$SA = \iint_R \sqrt{\left(\frac{\partial f}{\partial x}\right)^2 + \left(\frac{\partial f}{\partial y}\right)^2 + 1}\,dA. \tag{3.38}$$

If $f(x,y) \geq g(x,y)$ on R, the volume of the region between the graphs of $f(x,y)$ and $g(x,y)$ is

$$V = \iint_R \left(f(x,y) - g(x,y)\right)\,dA. \tag{3.39}$$

EXAMPLE 3.5.10: Find the area of the region R bounded by the graphs of $y = 2x^2$ and $y = 1 + x^2$.

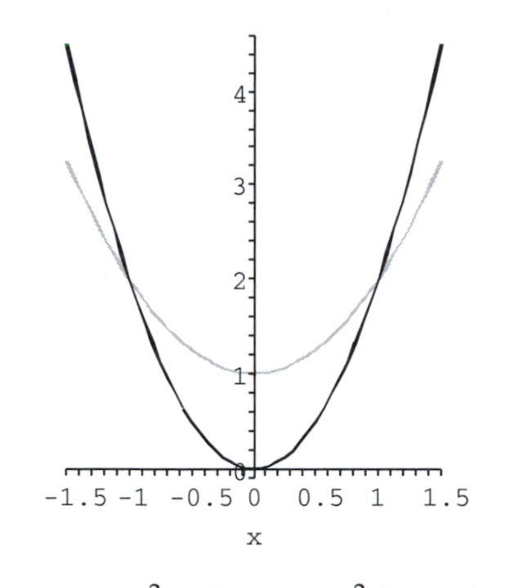

Figure 3-57 $y = 2x^2$ and $y = 1 + x^2$ for $-3/2 \le x \le 3/2$

SOLUTION: We begin by graphing $y = 2x^2$ and $y = 1 + x^2$ with `plot` in Figure 3-57. The x-coordinates of the intersection points are found with `solve`.

```
> plot([2*x^2,1+x^2],x=-3/2..3/2,color=[black,gray],
> scaling=CONSTRAINED);
```

```
> solve(2*x^2=1+x^2);
```

$$1, -1$$

Using (3.37) and taking advantage of symmetry, the area of R is given by

$$A = \iint_R dA = 2\int_0^1 \int_{2x^2}^{1+x^2} dy\, dx,$$

which we compute with `int`.

```
> 2*int(int(1,y=2*x^2..1+x^2),x=0..1);
```

$$4/3$$

We conclude that the area of R is $4/3$.

■

If the problem exhibits "circular symmetry," changing to polar coordinates is often useful. If $R = \{(r,\theta)\,|a \le r \le b, \alpha \le \theta \le \beta\}$, then

$$\iint_R f(x,y)\, dA = \int_\alpha^\beta \int_a^b f(r\cos\theta, r\sin\theta)\, r\, dr\, d\theta.$$

EXAMPLE 3.5.11: Find the surface area of the portion of $f(x,y) = \sqrt{4-x^2-y^2}$ that lies above the region $R = \{(x,y)\,|x^2+y^2 \le 1\}$.

SOLUTION: First, observe that the domain of $f(x,y)$ is

$$\left\{(x,y)\left|-\sqrt{4-y^2} \le x \le \sqrt{4-y^2}, -2 \le y \le 2\right.\right\} = \{(r,\theta)|0 \le r \le 2, 0 \le \theta \le 2\pi\}.$$

Similarly,

$$R=\left\{(x,y)\left|-\sqrt{1-y^2} \le x \le \sqrt{1-y^2}, -1 \le y \le 1\right.\right\} = \{(r,\theta)|0 \le r \le 1, 0 \le \theta \le 2\pi\}.$$

With this observation, we use `plot3d` to graph $f(x,y)$ in `p1` and the portion of the graph of $f(x,y)$ above R in `p2` and show the two graphs together with `display`. We wish to find the area of the black region in Figure 3-58.

```
> with(plots):
> f:=(x,y)->sqrt(4-x^2-y^2):
> p1:=plot3d(f(x,y),x=-sqrt(4-y^2)..sqrt(4-y^2),
   y=-2..2,
> style=wireframe,color=gray,gridstyle=triangular):
> p2:=plot3d(f(x,y),x=-sqrt(1-y^2)..sqrt(1-y^2),
   y=-1..1,
> color=gray,gridstyle=triangular):
```

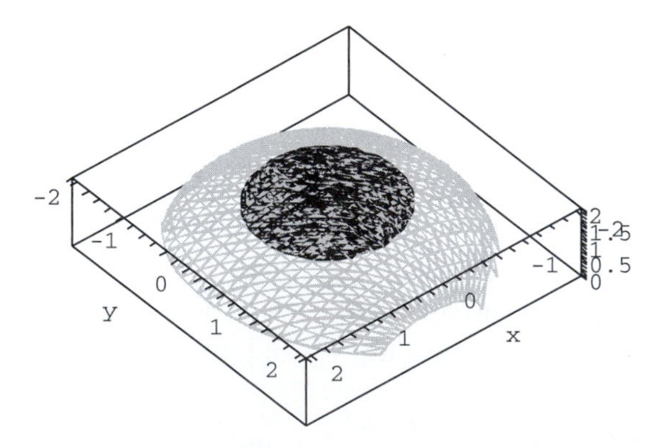

Figure 3-58 The portion of the graph of $f(x,y)$ above R

```
> display(p1,p2,axes=boxed,scaling=CONSTRAINED,
> orientation=[50,30]);
```

We compute $f_x(x,y)$, $f_y(x,y)$ and $\sqrt{[f_x(x,y)]^2+[f_y(x,y)]^2+1}$ with `diff` and `simplify`.

```
> fx:=diff(f(x,y),x);
> fy:=diff(f(x,y),y);
```

$$fx := -\frac{x}{\sqrt{4-x^2-y^2}}$$

$$fy := -\frac{y}{\sqrt{4-x^2-y^2}}$$

Then, using (3.38), the surface area is given by

$$\begin{aligned} SA &= \iint_R \sqrt{\left(\frac{\partial f}{\partial x}\right)^2+\left(\frac{\partial f}{\partial y}\right)^2+1}\,dA \\ &= \iint_R \frac{2}{\sqrt{4-x^2-y^2}}\,dA \\ &= \int_{-1}^{1}\int_{-\sqrt{1-y^2}}^{\sqrt{1-y^2}} \frac{2}{\sqrt{4-x^2-y^2}}\,dx\,dy. \end{aligned} \tag{3.40}$$

However, notice that in polar coordinates,

$$R=\left\{(r,\theta)\,|0\le r\le 1, 0\le\theta\le 2\pi\right\}$$

so in polar coordinates the surface area is given by

$$SA=\int_0^{2\pi}\int_0^1 \frac{2}{\sqrt{4-r^2}}\,r\,dr\,d\theta,$$

```
> s1:=simplify(sqrt(1+fx^2+fy^2));
```

$$s1 := 2\sqrt{-\left(-4+x^2+y^2\right)^{-1}}$$

```
> s2:=simplify(subs(x=r*cos(theta),
  y=r*sin(theta),s1));
```

$$s2 := 2\sqrt{-\left(-4+r^2\right)^{-1}}$$

which is much easier to evaluate than (3.40). We evaluate the iterated integral with `int`

```
> s3:=int(int(s2*r,r=0..1),theta=0..2*Pi);
```

$$s3 := -4\pi\sqrt{3} + 8\pi$$

```
> evalf(s3);
```

$$3.36714885$$

and conclude that the surface area is $\left(8 - 4\sqrt{3}\right)\pi \approx 3.367$.

■

EXAMPLE 3.5.12: Find the volume of the region between the graphs of $z = 4 - x^2 - y^2$ and $z = 2 - x$.

SOLUTION: We begin by graphing $z = 4 - x^2 - y^2$ and $z = 2 - x$ together with `plot3d` in Figure 3-59.

```
> with(plots):
> p1:=plot3d(4-x^2-y^2,x=-2..2,y=-2..2,
   style=WIREFRAME,
> gridstyle=triangular,color=black):
```

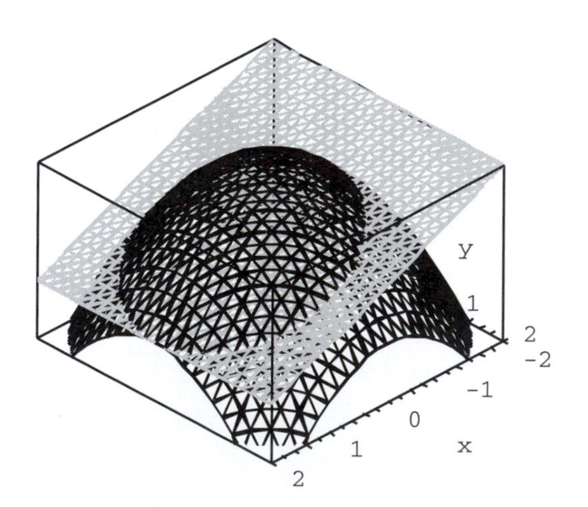

Figure 3-59 $z = 4 - x^2 - y^2$ and $z = 2 - x$ for $-2 \leq x \leq 2$ and $-2 \leq y \leq 2$

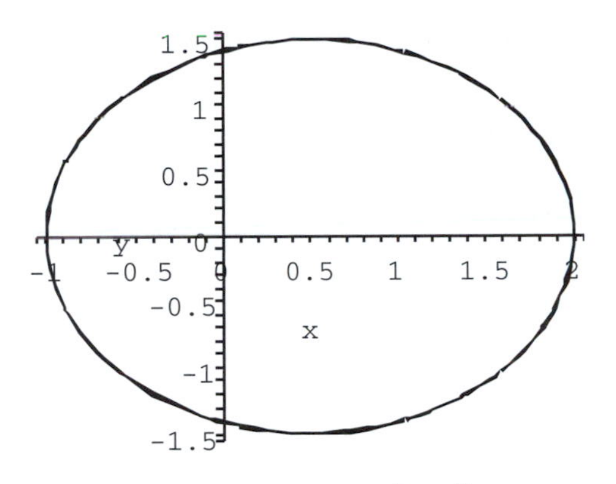

Figure 3-60 Graph of $4 - x^2 - y^2 = 2 - x$

```
> p2:=plot3d(2-x,x=-2..2,y=-2..2,style=WIREFRAME,
   color=gray):
> display(p1,p2,axes=BOXED,view=[-2..2,-2..2,-2..4]);
```

The region of integration, R, is determined by graphing $4-x^2-y^2 = 2-x$ with `implicitplot` in Figure 3-60.

```
> implicitplot(4-x^2-y^2=2-x,x=-2..2,y=-2..2,
> scaling=CONSTRAINED,color=black);
```

Completing the square with `completesquare` shows us that

$$R = \left\{ (x,y) \left| \left(x - \frac{1}{2}\right)^2 + y^2 \le \frac{9}{4} \right. \right\}$$
$$= \left\{ (x,y) \left| \frac{1}{2} - \frac{1}{2}\sqrt{9-4y^2} \le x \le \frac{1}{2} + \frac{1}{2}\sqrt{9-4y^2}, -\frac{3}{2} \le y \le \frac{3}{2} \right. \right\}.$$

```
> with(student):
> c1:=completesquare(4-x^2-y^2-(2-x),[x,y]);
```

$$c1 := -(x-1/2)^2 + 9/4 - y^2$$

```
> solve(c1=0,x);
```

$$1/2 + 1/2\sqrt{9-4y^2},\ 1/2 - 1/2\sqrt{9-4y^2}$$

Thus, using (3.39), the volume of the solid is given by

$$V = \iint_R \left[\left(4 - x^2 - y^2\right) - (2 - x)\right] dA$$
$$= \int_{-\frac{3}{2}}^{\frac{3}{2}} \int_{\frac{1}{2}-\frac{1}{2}\sqrt{9-4y^2}}^{\frac{1}{2}+\frac{1}{2}\sqrt{9-4y^2}} \left[\left(4 - x^2 - y^2\right) - (2 - x)\right] dx\, dy,$$

which we evaluate with `int`.

```
> i1:=int(int((4-x^2-y^2)-(2-x),
> x=1/2-1/2*sqrt(9-4*y^2)..1/2+1/2*sqrt(9-4*y^2)),
  y=-3/2..3/2);
```

$$i1 := \frac{81}{32}\pi$$

```
> evalf(i1);
```

$$7.952156405$$

We conclude that the volume is $\frac{81}{32}\pi \approx 7.952$.

■

Triple Iterated Integrals

Triple iterated integrals are calculated in the same manner as double iterated integrals.

EXAMPLE 3.5.13: Evaluate

$$\int_0^{\pi/4} \int_0^y \int_0^{y+z} (x + 2z) \sin y \, dx\, dz\, dy.$$

SOLUTION: Entering

```
> i1:=int(int(int(sin(y)*(x+2*z),x=0..y+z),
  z=0..y),y=0..Pi/4);
```

$$i1 := \frac{17}{8}\sqrt{2}\pi - 17/2\sqrt{2} - \frac{17}{768}\sqrt{2}\pi^3 + \frac{17}{64}\sqrt{2}\pi^2$$

calculates the triple integral exactly with `int`.

An approximation of the exact value is found with `evalf`.

```
> evalf(i1);
```

$$0.157205682$$

■

We illustrate how triple integrals can be used to find the volume of a solid when using spherical coordinates.

EXAMPLE 3.5.14: Find the volume of the torus with equation in spherical coordinates $\rho = \sin\phi$.

SOLUTION: We proceed by graphing the torus using `plot3d` with the `coords=spherical` option in Figure 3-61.

```
> plot3d(sin(phi),theta=0..2*Pi,phi=0..2*Pi,
> coords=spherical,axes=BOXED,grid=[30,30],
   gridstyle=triangular,
> scaling=CONSTRAINED);
```

In general, the volume of the solid region D is given by

$$V = \iiint_D dV.$$

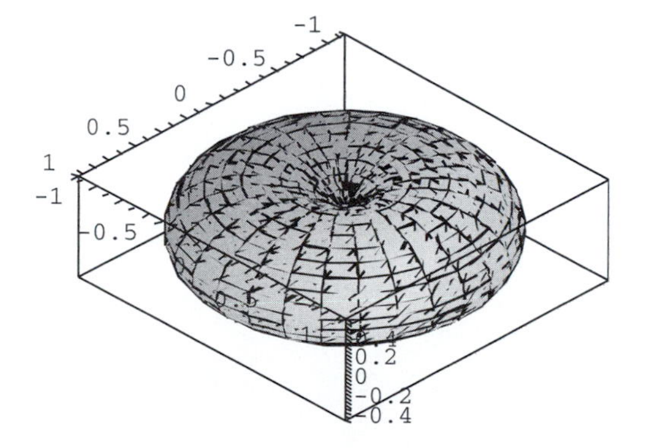

Figure 3-61 A graph of the torus

Thus, the volume of the torus is given by the triple iterated integral

$$V = \int_0^{2\pi} \int_0^{\pi} \int_0^{\sin\phi} \rho^2 \sin\phi \, d\rho \, d\phi \, d\theta,$$

```
> i1:=int(int(int(rho^2*sin(phi),rho=0..sin(phi)),
> phi=0..Pi),theta=0..2*Pi);
```

$$i1 := 1/4\,\pi^2$$

```
> evalf(i1);
```

$$2.467401101$$

which we evaluate with `int`. We conclude that the volume of the torus is $\frac{1}{4}\pi^2 \approx 2.467$.
■

Introduction to Lists and Tables

4

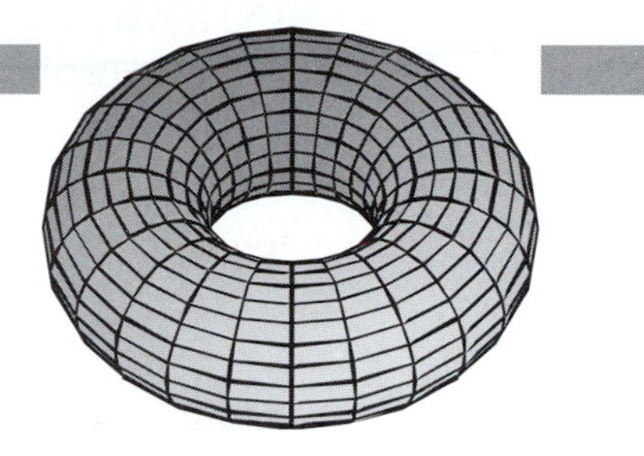

Chapter 4 introduces operations on lists and tables. The examples used to illustrate the various commands in this chapter are taken from calculus, business, dynamical systems, and engineering applications.

4.1 Lists and List Operations

4.1.1 Defining Lists

A **list** of n elements is a Maple object of the form

`list:=[a1,a2,a3,...,an].`

The ith element of the list is extracted from `list` with `list[i]`.

Elements of a list are separated by commas. Lists are always enclosed in brackets `[...]` and each element of a list may be (almost any) Maple object – even other lists. Because lists are Maple objects, they can be named. For easy reference, we will usually name lists.

Lists can be defined in a variety of ways: they may be completely typed in, imported from other programs and text files, or they may be created with either the `seq` or `array` commands. Given a function $f(x)$ and a number n, the command

Maple distinguishes between lists (order matters) and sets (order does not matter). Lists are contained in brackets; sets are contained in braces.

1. `[seq(f(i),i=1..n)]` creates the list `[f(1),...,f(n)]`;
2. `[seq(f(i),i=0..n)]` creates the list `[f(0),...,f(n)]`; and

3. `[seq(f(i),i=n..m)]` creates the list

```
[f(n),f(n+1),...,f(m)].
```

The `array` command will be discussed in Chapter 5.

In particular,

```
avals:=seq(a+(b-a)/(n-1),i=0..n-1)
```

returns a sequence of n equally spaced numbers between a and b so

```
seq(f(x),x=avals)
```

returns a sequence of $f(x)$ values for n equally spaced values of x between a and b and

```
seq([x,f(x)],x=avals)
```

returns a sequence of points $(x, f(x))$ for n equally spaced values of x between a and b.

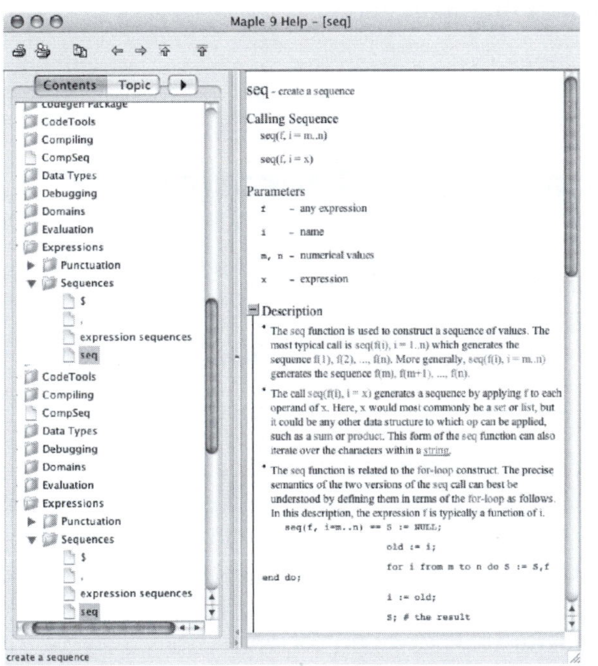

Once you have defined a list, elements are extracted with brackets (`[...]`) or `op`.

1. `list[i]` and `op(i,list)` return the ith element of `list`.
2. `list[i..j]` and `op(i..j,list)` return the ith through jth elements of `list`.

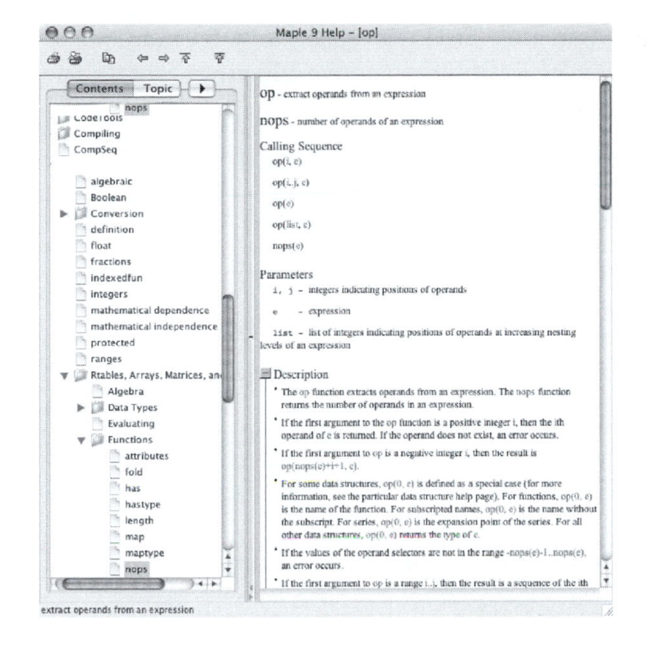

EXAMPLE 4.1.1: Use Maple to generate the list `[1,2,3,4,5,6,7,8,9,10]`.

SOLUTION: Generally, a given list can be constructed in several ways. Each of the following commands generates the list `[1,2,3,4,5,6,7,8,9,10]`.

```
> [1,2,3,4,5,6,7,8,9,10];
```

$$[1,2,3,4,5,6,7,8,9,10]$$

```
> [seq(i,i=1..10)];
```

$$[1, 2, 3, 4, 5, 6, 7, 8, 9, 10]$$

■

EXAMPLE 4.1.2: Use Maple to define `listone` to be the list of numbers [1,3/2,2,5/2,3,7/2,4].

SOLUTION: In this case, we generate a list and name the result `listone`. As in Example 4.1.1, we illustrate that `listone` can be created in several ways.

```
> listone:=[1,3/2,2,5/2,3,7/2,4];
```

$$listone := [1, 3/2, 2, 5/2, 3, 7/2, 4]$$

```
> listone:=seq(1+i/2,i=0..6);
```

$$listone := 1,\ 3/2,\ 2,\ 5/2,\ 3,\ 7/2,\ 4$$

Once you have defined a list, elements are extracted with `op` or `[...]`. Thus,

```
> listone[4];
```

$$5/2$$

returns the fourth element of `listone` while

```
> listone[4..6];
```

$$5/2,\ 3,\ 7/2$$

returns the fourth through sixth elements of `listone`.
■

EXAMPLE 4.1.3: Create a list of the first 25 prime numbers. What is the fifteenth prime number?

SOLUTION: The command `ithprime(n)` yields the nth prime number. We use `seq` to generate a list of the ordered pairs `[n,ithprime(n)]` for $n = 1, 2, 3, \ldots, 25$.

```
> prime_list:=[seq([i,ithprime(i)],i=1..25)];
```

$$prime\_list := [[1,2],[2,3],[3,5],[4,7],[5,11],[6,13],[7,17],[8,19], [9,23],[10,29],[11,31],[12,37],[13,41],[14,43],[15,47], [16,53],[17,59],[18,61],[19,67],[20,71],[21,73],[22,79], [23,83],[24,89],[25,97]]$$

The ith element of a list `list` is extracted from `list` with `list[i]`. From the resulting output, we see that the fifteenth prime number is 47.

```
> prime_list[15];
```

$$[15, 47]$$

■

In addition, we can use seq to generate lists consisting of the same or similar objects.

EXAMPLE 4.1.4: (a) Generate a list consisting of five copies of the letter a. (b) Generate a list consisting of 10 random integers between -10 and 10.

SOLUTION: Entering

```
> seq(a,i=1..5);
```

$$a, a, a, a, a$$

generates a list consisting of five copies of the letter a. For (b), we use the command rand to generate the desired list. Because we are using rand, your results will certainly differ from those obtained here.

```
> g:=rand(-10..10):
> seq(g(),i=1..10);
```

$$5, -10, 1, 1, -3, -10, 5, -4, -8, -1$$

■

4.1.2 Plotting Lists of Points

Lists are plotted using plot together with the style=point option. If you do not include the style=point option, successive points are connected with line segments.

1. plot([[x1,y1],[x2,y2],...,[xn,yn]],style=point) plots the list of points

$$[(x_1, y_1), (x_2, y_2), \ldots, (x_n, y_n)].$$

   The point symbol in the resulting plot is controlled with the option symbol=w, where w is one of BOX, CROSS, CIRCLE, POINT, and DIAMOND.

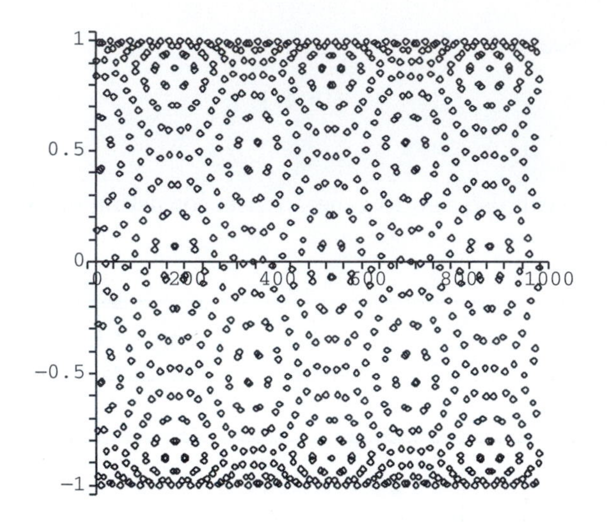

Figure 4-1 Plot of $(n, \sin n)$ for $n = 1, 2, \ldots, 1000$

2. `plot([y1,y2,..,yn],style=point)` plots the list of points

$$[(1, y_1), (2, y_2), \ldots, (n, y_n)].$$

You can also use the `listplot` command, which is contained in the `plots` package, in the exact same way to plot lists.

When a colon is included at the end of a command, the resulting output is suppressed.

EXAMPLE 4.1.5: Entering

```
> t1:=[seq([n,sin(n)],n=1..1000)]:
> plot(t1,style=point,color=black);
```

creates a list consisting of $(n, \sin n)$ for $n = 1, 2, \ldots, 1000$ and then graphs the list of points $(n, \sin n)$ for $n = 1, 2, \ldots, 1000$ (Figure 4-1).

EXAMPLE 4.1.6 (The Prime Difference Function and the Prime Number Theorem): In `t1`, we use `ithprime` and `seq` to compute a list of the first 3000 prime numbers.

```
> t1:=[seq(ithprime(n),n=1..3000)]:
```

`nops(list)` returns the number of elements of `list`.

We use `nops` to verify that `t1` has 3000 elements and `[..]` to see an abbreviated portion of `t1`.

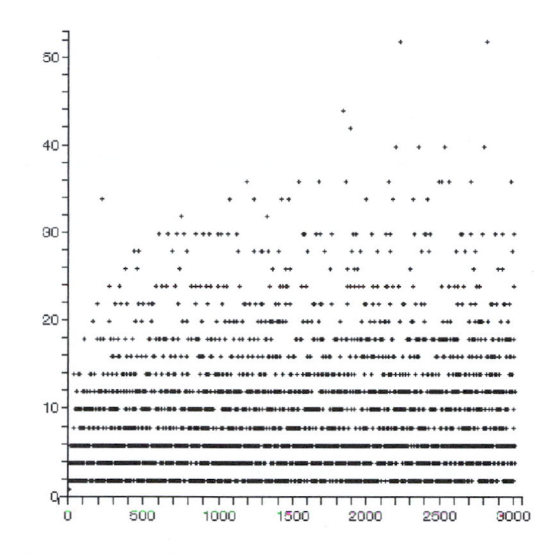

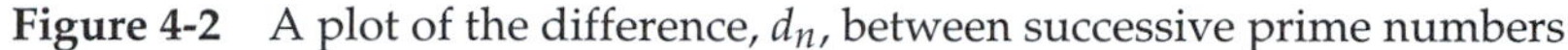

Figure 4-2 A plot of the difference, d_n, between successive prime numbers

```
> nops(t1);
```

$$3000$$

```
> t1[1..5];
> t1[2996..3000];
```

$$[2,3,5,7,11]$$

$$[27409,27427,27431,27437,27449]$$

In `t2`, we compute the difference, d_n, between the successive prime numbers in `t1`. The result is plotted with `plot` in Figure 4-2.

`list[i]` returns the ith element of `list` so `list[i+1]-list[i]` computes the difference between the $(i+1)$th and ith elements of `list`.

```
> t2:=[seq([i,t1[i+1]-t1[i]],i=1..2999)]:
> t2[1..5];
> t2[2995..2999];
> plot(t2,style=point,symbol=point,color=black);
```

$$[[1,1],[2,2],[3,2],[4,4],[5,2]]$$

$$[[2995,2],[2996,18],[2997,4],[2998,6],[2999,12]]$$

Let $\pi(n)$ denote the number of primes less than n and $Li(x)$ denote the **logarithmic integral**:

$$Li(x) = \int_0^x \frac{1}{\ln t}\,dt - \ln 2.$$

After defining $Li(x)$,

```
> li:=x->int(1/ln(t),t=2..x)-ln(2);
```

$$li := x \mapsto \int_2^x (\ln(t))^{-1}\,dt - \ln(2)$$

we compute $Li(3)$ and then approximate the result with `evalf`.

```
> li(3);
> evalf(li(3));
```

$$Ei(1, -\ln(2)) - Ei(1, -\ln(3)) - \ln(2)$$

$$0.425277635 + 0.0\,i$$

The **Prime Number theorem** states that

$$\pi(n) \approx Li(n).$$

(See [17].) In the following, we use `select` and `nops` to define $\pi(n)$. `select(list,criteria)` returns the elements of `list` for which `criteria` is true. Thus, given n, `select(isless(n),t1)` returns a list of the elements of `t1` less than n; `nops(select(isless(n),t1))` returns the number of elements in the list.

```
> isless:=x->proc(y) evalb(y<x) end proc:
```

For example,

```
> smallpi:=n->nops(select(isless(n),t1)):
> smallpi(100);
```

$$25$$

shows us that $\pi(100) = 25$. Note that because `t1` contains the first 3000 primes, `smallpi(n)` is valid for $1 \leq n \leq N$ where $\pi(N) = 3000$. In `t3`, we compute $\pi(n)$ for $n = 1, 2, \ldots, 5000$

```
> t3:=[seq([n,smallpi(n)],n=1..5000)]:
> p1:=plot(t3,style=point,symbol=point,color=gray):
```

and plot the resulting list with `plot`. In Figure 4-3, we display the plots of $\pi(n)$ and $Li(x)$ together.

```
> with(plots):
> p2:=plot(eval(li(x)),x=2..5000,color=black):
> display(p1,p2);
```

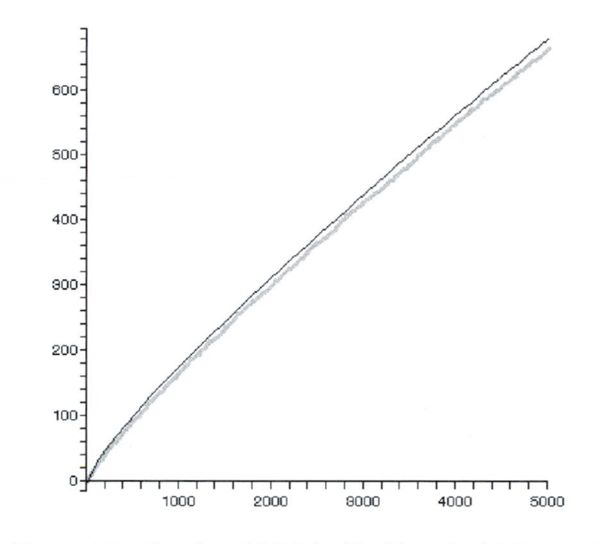

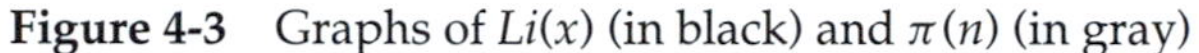
Figure 4-3 Graphs of $Li(x)$ (in black) and $\pi(n)$ (in gray)

You can iterate recursively with `seq`. Thus,

```
> seq(seq(a[i,2*j],j=1..5),i=1..5);
```

$$a_{1,2}, a_{1,4}, a_{1,6}, a_{1,8}, a_{1,10},$$

$$a_{2,2}, a_{2,4}, a_{2,6}, a_{2,8}, a_{2,10},$$

$$a_{3,2}, a_{3,4}, a_{3,6}, a_{3,8}, a_{3,10},$$

$$a_{4,2}, a_{4,4}, a_{4,6}, a_{4,8}, a_{4,10},$$

$$a_{5,2}, a_{5,4}, a_{5,6}, a_{5,8}, a_{5,10}$$

computes a list of a_{ij} values. For example,

```
> t1:=[seq(seq([sin(x+y),cos(x-y)],x=1..5),y=1..5)];
```

$$t1 := [[\sin(2), 1], [\sin(3), \cos(1)], [\sin(4), \cos(2)], [\sin(5), \cos(3)], [\sin(6), \cos(4)],$$
$$[\sin(3), \cos(1)], [\sin(4), 1], [\sin(5), \cos(1)], [\sin(6), \cos(2)], [\sin(7), \cos(3)],$$
$$[\sin(4), \cos(2)], [\sin(5), \cos(1)], [\sin(6), 1], [\sin(7), \cos(1)], [\sin(8), \cos(2)],$$
$$[\sin(5), \cos(3)], [\sin(6), \cos(2)], [\sin(7), \cos(1)], [\sin(8), 1], [\sin(9), \cos(1)],$$
$$[\sin(6), \cos(4)], [\sin(7), \cos(3)], [\sin(8), \cos(2)], [\sin(9), \cos(1)], [\sin(10), 1]]$$

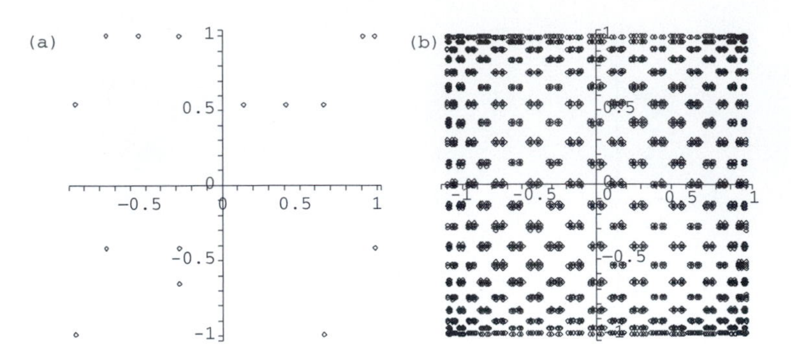

Figure 4-4 Plotting lists of ordered pairs (a) and (b)

returns a list of 25 points. The third point is extracted with `[...]`.

```
> t1[3];
```

$$[\sin(4), \cos(2)]$$

The points are plotted in Figure 4-4(a). In `t2`, we increase the number of points.

```
> t2:=[seq(seq([sin(x+y),cos(x-y)],x=1..75),y=1..75)]:
```

These are plotted with `plot` in Figure 4-4(b). We also illustrate the use of the `style`, `symbol`, and `color` options in the `plot` command.

```
> plot(t1,style=point,color=black);
> plot(t2,style=point,color=black,symbol=diamond);
```

Remark. Maple is very flexible and most calculations can be carried out in more than one way. Depending on how you think, some sequences of calculations may make more sense to you than others, even if they are less efficient than the most efficient way to perform the desired calculations. Often, the difference in time required for Maple to perform equivalent – but different – calculations is quite small. For the beginner, we think it is wisest to work with familiar calculations first and then efficiency.

EXAMPLE 4.1.7 (Dynamical Systems): A sequence of the form $x_{n+1} = f(x_n)$ is called a **dynamical system**. Sometimes, unusual behavior can be observed when working with dynamical systems. For example, consider the dynamical system with $f(x) = x + 2.5x(1 - x)$ and $x_0 = 1.2$. Note that we define x_n using `proc` with the `remember` option so that Maple remembers the functional values it computes and

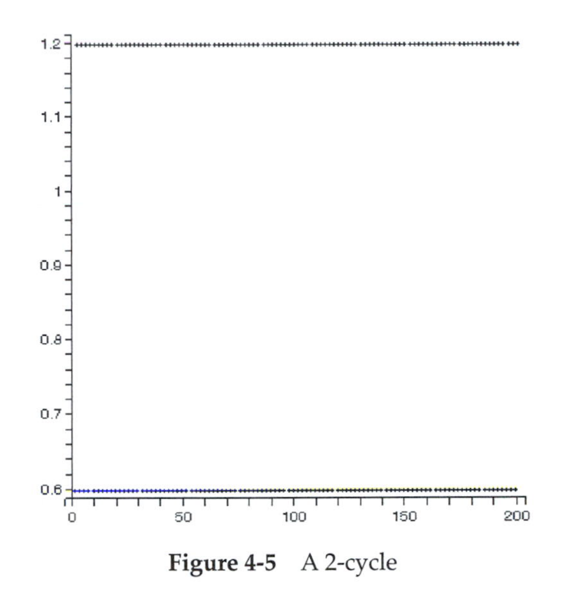

Figure 4-5 A 2-cycle

thus avoids recomputing functional values previously computed. This is particularly advantageous when we compute the value of x_n for large values of n.

Observe that $x_{n+1} = f(x_n)$ can also be computed with $x_{n+1} = f^n(x_0)$.

```
> f:=x->x+2.5*x*(1-x):
> x:=proc(n) option remember;
> f(x(n-1))
> end proc:
> x(0):=1.2:
```

In Figure 4-5, we see that the sequence oscillates between 0.6 and 1.2. We say that the dynamical system has a **2-cycle** because the values of the sequence oscillate between two numbers.

```
> tb:=[seq([n,x(n)],n=1..200)]:
> tb[1..5];
```

$$[[1, 0.600], [2, 1.2000000], [3, 0.6000000000],$$
$$[4, 1.200000000], [5, 0.6000000000]]$$

```
> plot(tb,style=point,symbol=point,color=black);
```

In Figure 4-6, we see that changing x_0 from 1.2 to 1.201 results in a 4-cycle.

```
> f:='f':
> x:='x':
```

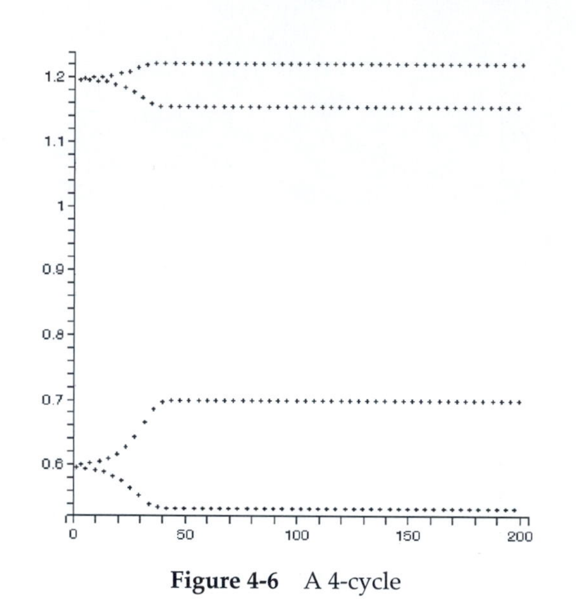

Figure 4-6 A 4-cycle

```
> f:=x->x+2.5*x*(1-x):
> x:=proc(n) option remember;
> f(x(n-1))
> end proc:
> x(0):=1.201:
> tb:=[seq([n,x(n)],n=1..200)]:
> plot(tb,style=point,symbol=point,color=black);
```

The calculations indicate that the behavior of the system can change considerably for small changes in x_0. With the following, we adjust the definition of x so that x depends on $x_0 = c$: $x(c, 0) = c$.

```
> f:='f':x:='x':
> f:=x->x+2.5*x*(1-x):
> x:=proc(c,n) option remember;
> if n=0 then c else f(x(c,n-1)) end if
> end proc:
```

In tb, we create a list of ordered pairs of the form $\{(c, x(c, n))|n = 100, \ldots, 150\}$ for 100 equally spaced values of c between 0 and 1.5, which are then graphed with plot in Figure 4-7(a).

```
> cvals:=seq(1.5/99*i,i=0..99):
> tb:=[seq(seq([c,x(c,n)],c=cvals),n=100..150)]:
> plot(tb,style=point,color=black,symbol=point);
```

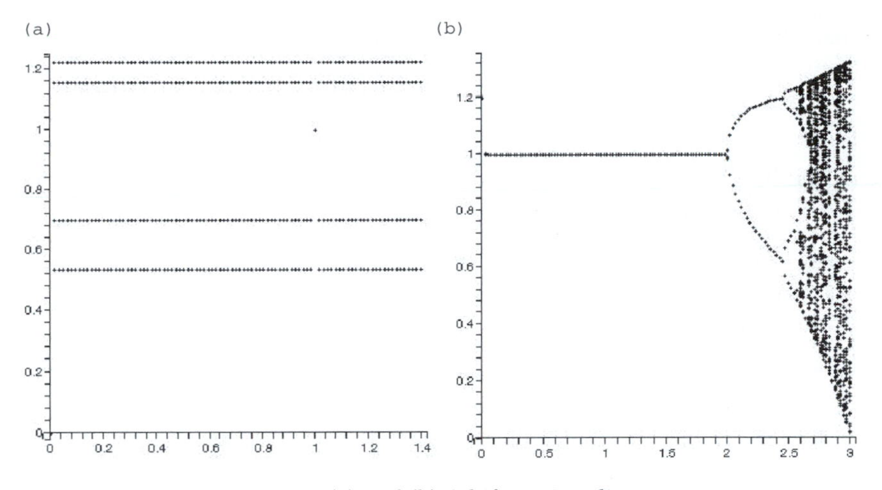

Figure 4-7 (a) and (b) A bifurcation diagram

Another interesting situation occurs if we fix x_0 and let c vary in $f(x) = x + cx(1 - x)$.

With the following we set $x_0 = 1.2$ and adjust the definition of f so that f depends on c: $f(x) = x + cx(1 - x)$.

```
> f:='f':x:='x':
> f:=(c,x)->x+c*x*(1-x):
> x:=proc(c,n) option remember;
> if n=0 then 1.2 else f(c,x(c,n-1)) end if
> end proc:
```

In tb, we create a list of ordered pairs of the form $\{(c, x(c,n))|n = 200, \ldots, 300\}$ for 150 equally spaced values of c between 0 and 3.5, which is then graphed with plot in Figure 4-7(b).

```
> x(2,3);
```

$$0.84644352$$

```
> cvals:=seq(3.5/149*i,i=0..149):
> tb:=[seq(seq([c,x(c,n)],c=cvals),n=200..300)]:

> plot(tb,style=point,color=black,symbol=point);
```

This plot is called a **bifurcation diagram**.

As indicated earlier, elements of lists can be numbers, ordered pairs, functions, and even other lists. You can also use Maple to manipulate lists in numerous ways. Most importantly, the map function is used to apply a function to a list:

```
map(f,[x1,x2,...,xn])
```

returns the list $[f(x_1), f(x_2), \ldots, f(x_n)]$. We will discuss other operations that can be performed on lists in the following sections.

EXAMPLE 4.1.8 (Hermite Polynomials): The **Hermite polynomials**, $H_n(x)$, satisfy the differential equation $y'' - 2xy' + 2ny = 0$ and the orthogonality relation $\int_{-\infty}^{\infty} H_n(x)H_m(x)e^{-x^2}\,dx = \delta_{mn}2^n n!\sqrt{\pi}$. The Maple command `H(n,x)`, which is contained in the `orthopoly` package, yields the Hermite polynomial $H_n(x)$. (a) Create a table of the first five Hermite polynomials. (b) Evaluate each Hermite polynomial if $x = 1$. (c) Compute the derivative of each Hermite polynomial in the table. (d) Compute an antiderivative of each Hermite polynomial in the table. (e) Graph the five Hermite polynomials on the interval $[-1, 1]$. (f) Verify that $H_n(x)$ satisfies $y'' - 2xy' + 2ny = 0$ for $n = 1, 2, \ldots, 5$.

SOLUTION: (a) After loading the `orthopoly` package, we proceed by using `H` together with `seq` to define `hermitetable` to be the list consisting of the first five Hermite polynomials.

```
> with(orthopoly);
```

$$[G, H, L, P, T, U]$$

```
> hermitetable:=[seq(H(n,x),n=1..5)];
```

$$hermitetable := \left[2\,x, -2+4\,x^2, 8\,x^3-12\,x, 12+16\,x^4-48\,x^2, 32\,x^5-160\,x^3+120\,x\right]$$

(b) We then use `subs` to evaluate each member of `hermitetable` if x is replaced by 1.

```
> subs(x=1,hermitetable);
```

$$[2, 2, -4, -20, -8]$$

(c) Both `diff(list,x)` and `map(diff,list,x)` differentiate each element of `list` with respect to x.

```
> diff(hermitetable,x);
```

$$\left[2, 8\,x, 24\,x^2-12, 64\,x^3-96\,x, 160\,x^4-480\,x^2+120\right]$$

```
> map(diff,hermitetable,x);
```

$$[2, 8x, 24x^2 - 12, 64x^3 - 96x, 160x^4 - 480x^2 + 120]$$

(d) `int` does not work in the same way as `diff`: we use `map(int, hermitetable,x)` to antidifferentiate each member of `hermitetable` with respect to x. Remember that Maple does not automatically include the "+C" that we include when we antidifferentiate.

```
> int(hermitetable,x);
```

Error, (in int) wrong number (or type) of arguments

```
> map(int,hermitetable,x);
```

$$[x^2, -2x+4/3x^3, 2x^4-6x^2, 12x+\frac{16}{5}x^5-16x^3, 16/3x^6-40x^4+60x^2]$$

(e) To graph the list `hermitetable`, we use `plot` to plot each function in the set `hermitetable` on the interval $[-2, 2]$ in Figure 4-8. In this case, we specify that the displayed y-values correspond to the interval $[-20,20]$. The plots of the Hermite polynomials are then shaded according to `grays`. The graph of $H_1(x)$ is in black and successive plots are lighter, with the graph of $H_5(x)$ being the lightest gray.

```
> plot(hermitetable,x=-1..1,view=[-1..1,-20..20]);
```

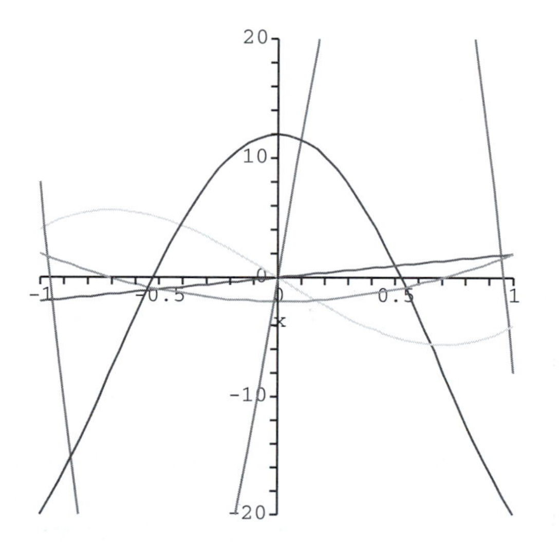

Figure 4-8 Graphs of $H_1(x)$ (in black), $H_2(x)$, $H_3(x)$, $H_4(x)$, and $H_5(x)$ (in light gray)

`hermitetable[n]` returns the nth element of hermitetable, which corresponds to $H_n(x)$. Thus,

```
> verifyde:=[seq(simplify(diff(hermitetable[n],x$2)-
> 2*x*diff(hermitetable[n],x)+2*n*hermitetable[n]),
  n=1..5)];
```

$$verifyde := [0, 0, 0, 0, 0]$$

computes and simplifies $H_n'' - 2xH_n' + 2nH_n$ for $n = 1, 2, \ldots, 5$. We use `seq` and `int` to compute $\int_{-\infty}^{\infty} H_n(x)H_m(x)e^{-x^2}\,dx$ for $n = 1, 2, \ldots, 5$ and $m = 1, 2, \ldots, 5$.

```
> verifyortho:=[seq([seq(int(hermitetable[n]
  *hermitetable[m]*exp(-x^2),
> x=-infinity..infinity),n=1..5)],m=1..5)];
```

$$verifyortho := \left[[2\sqrt{\pi}, 0, 0, 0, 0], [0, 8\sqrt{\pi}, 0, 0, 0], [0, 0, 48\sqrt{\pi}, 0, 0], [0, 0, 0, 384\sqrt{\pi}, 0], [0, 0, 0, 0, 3840\sqrt{\pi}]\right]$$

To view a table in traditional row-and-column form use `array`.

```
> array(verifyortho);
```

$$\begin{bmatrix} 2\sqrt{\pi} & 0 & 0 & 0 & 0 \\ 0 & 8\sqrt{\pi} & 0 & 0 & 0 \\ 0 & 0 & 48\sqrt{\pi} & 0 & 0 \\ 0 & 0 & 0 & 384\sqrt{\pi} & 0 \\ 0 & 0 & 0 & 0 & 3840\sqrt{\pi} \end{bmatrix}$$

Be careful when using `array`: `array(table)` is no longer a list and cannot be manipulated like a list.

■

4.2 Manipulating Lists: More on op and map

Often, Maple's output is given to us as a list that we need to use in subsequent calculations. Elements of a list are extracted with `op([...])`. `list[i]` returns the ith element of `list`; `list[i,j]` (or `list[i][j]`) returns the jth element of the ith element of `list`, and so on.

EXAMPLE 4.2.1: Let $f(x) = 3x^4 - 8x^3 - 30x^2 + 72x$. Locate and classify the critical points of $y = f(x)$.

SOLUTION: We begin by clearing all prior definitions of f and then defining f. The critical numbers are found by solving the equation $f'(x) = 0$. The resulting list is named `critnums`.

```
> f:='f':
> f:=x->3*x^4-8*x^3-30*x^2+72*x:
> critnums:=solve(D(f)(x)=0);
```

$$critnums := 1, -2, 3$$

`critnums` is a list.

```
> critnums[1];
```

$$1$$

We locate and classify the points by evaluating $f(x)$ and $f''(x)$ for each of the numbers in `critnums`. `seq(g(x),x=avals)` computes $g(x)$ for each value of x in `avals`

```
> seq([x,f(x),D(f)(x),(D@@2)(f)(x)],x=critnums);
```

$$[1, 37, 0, -72], [-2, -152, 0, 180], [3, -27, 0, 120]$$

replaces each x in the list $\{x, f(x), f''(x)\}$ by each of the x-values in `critnums`.

By the Second Derivative Test, we conclude that $y = f(x)$ has relative minima at the points $(-2, -152)$ and $(3, -27)$ while $f(x)$ has a relative maximum at $(1, 37)$. In fact, because $\lim_{x\to\pm\infty} = \infty$, -152 is the absolute minimum value of $f(x)$. These results are confirmed by the graph of $y = f(x)$ in Figure 4-9.

```
> plot(f(x),x=-4..4);
```

■

`map` is a very powerful and useful function: `map(f,list)` creates a list consisting of elements obtained by evaluating `f` for each element of `list`, provided that each member of `list` is an element of the domain of `f`.

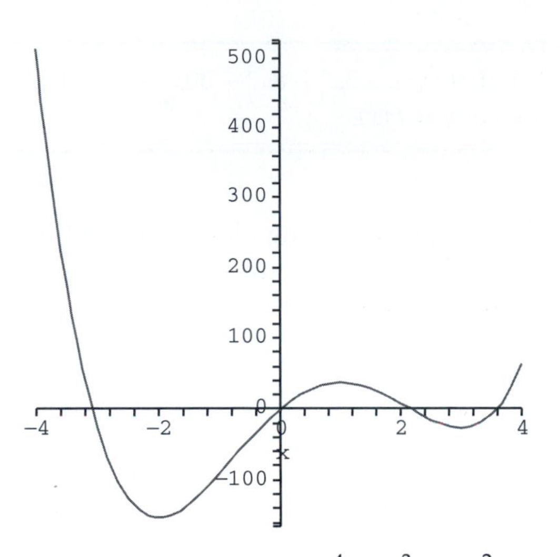

Figure 4-9 Graph of $f(x) = 3x^4 - 8x^3 - 30x^2 + 72x$

EXAMPLE 4.2.2: Entering

```
> t1:=[seq(n,n=1..100)];
```

$t1 := [1, 2, 3, 4, 5, 6, 7, 8, 9, 10, 11, 12, 13, 14, 15, 16, 17, 18, 19, 20,$
$21, 22, 23, 24, 25, 26, 27, 28, 29, 30, 31, 32, 33, 34, 35, 36, 37, 38, 39, 40,$
$41, 42, 43, 44, 45, 46, 47, 48, 49, 50, 51, 52, 53, 54, 55, 56, 57, 58, 59, 60,$
$61, 62, 63, 64, 65, 66, 67, 68, 69, 70, 71, 72, 73, 74, 75, 76, 77, 78, 79, 80,$
$81, 82, 83, 84, 85, 86, 87, 88, 89, 90, 91, 92, 93, 94, 95, 96, 97, 98, 99, 100]$

computes a list of the first 100 integers and names the result `t1`. We then define $f(x) = x^2$ and use `map` to square each number in `t1`.

```
> f:=x->x^2:
> map(f,t1);
```

$[1, 4, 9, 16, 25, 36, 49, 64, 81, 100, 121, 144, 169, 196, 225, 256, 289, 324,$
$361, 400, 441, 484, 529, 576, 625, 676, 729, 784, 841, 900, 961, 1024, 1089,$
$1156, 1225, 1296, 1369, 1444, 1521, 1600, 1681, 1764, 1849, 1936, 2025, 2116,$
$2209, 2304, 2401, 2500, 2601, 2704, 2809, 2916, 3025, 3136, 3249, 3364, 3481,$
$3600, 3721, 3844, 3969, 4096, 4225, 4356, 4489, 4624, 4761, 4900, 5041, 5184,$

5329, 5476, 5625, 5776, 5929, 6084, 6241, 6400, 6561, 6724, 6889, 7056, 7225,
7396, 7569, 7744, 7921, 8100, 8281, 8464, 8649, 8836, 9025, 9216, 9409, 9604,
9801, 10000]

The same result is accomplished by applying the function that squares its argument to `t1`.

```
> map(x->x^2,t1);
```

[1, 4, 9, 16, 25, 36, 49, 64, 81, 100, 121, 144, 169, 196, 225, 256, 289, 324,
361, 400, 441, 484, 529, 576, 625, 676, 729, 784, 841, 900, 961, 1024, 1089,
1156, 1225, 1296, 1369, 1444, 1521, 1600, 1681, 1764, 1849, 1936, 2025, 2116,
2209, 2304, 2401, 2500, 2601, 2704, 2809, 2916, 3025, 3136, 3249, 3364, 3481,
3600, 3721, 3844, 3969, 4096, 4225, 4356, 4489, 4624, 4761, 4900, 5041, 5184,
5329, 5476, 5625, 5776, 5929, 6084, 6241, 6400, 6561, 6724, 6889, 7056, 7225,
7396, 7569, 7744, 7921, 8100, 8281, 8464, 8649, 8836, 9025, 9216, 9409, 9604,
9801, 10000]

On the other hand, entering

```
> t1:=[seq(seq([a,b],a=1..5),b=1..5)];
```

$t1 := [[1,1],[2,1],[3,1],[4,1],[5,1],[1,2],[2,2],[3,2],[4,2],[5,2],$
$[1,3],[2,3],[3,3],[4,3],[5,3],[1,4],[2,4],[3,4],[4,4],[5,4],$
$[1,5],[2,5],[3,5],[4,5],[5,5]]$

is a list of 25 ordered pairs. f is a function of one variable. Given an ordered pair $\mathbf{v} = (x, y)$, $f(\mathbf{v})$ returns the ordered triple $((x, y), x^2 + y^2)$.

```
> f:=v->[[v[1],v[2]],v[1]^2+v[2]^2]:
> map(f,t1);
```

$[[[1,1],2],[[2,1],5],[[3,1],10],[[4,1],17],[[5,1],26],[[1,2],5],$
$[[2,2],8],[[3,2],13],[[4,2],20],[[5,2],29],[[1,3],10],[[2,3],13],$
$[[3,3],18],[[4,3],25],[[5,3],34],[[1,4],17],[[2,4],20],[[3,4],25],$
$[[4,4],32],[[5,4],41],[[1,5],26],[[2,5],29],[[3,5],34],[[4,5],41],$
$[[5,5],50]]$

EXAMPLE 4.2.3: Make a table of the values of the trigonometric functions $y = \sin x$, $y = \cos x$, and $y = \tan x$ for the principal angles.

SOLUTION: We first construct a set of the principal angles, which is accomplished by defining setone to be the set consisting of $n\pi/4$ for $n = 0, 1, \ldots, 8$ and settwo to be the set consisting of $n\pi/6$ for $n = 0, 1, \ldots, 12$. The principal angles are obtained by taking the union of setone and settwo. `'union'(setone,settwo)` joins the sets setone and settwo, removes repeated elements.

```
> setone:={seq(n*Pi/4,n=0..8)};
> settwo:={seq(n*Pi/6,n=0..12)};
> setthree:='union'(setone,settwo);
```

$$setone := \{0, \pi, 5/4\pi, 3/2\pi, 1/4\pi, 1/2\pi, 2\pi, 3/4\pi, 7/4\pi\}$$

$$settwo := \{0, \pi, 3/2\pi, 1/2\pi, 2\pi, 1/6\pi, 1/3\pi, 2/3\pi, 5/6\pi, 7/6\pi, 4/3\pi, 5/3\pi, \frac{11}{6}\pi\}$$

$$setthree := \{0, \pi, 5/4\pi, 3/2\pi, 1/4\pi, 1/2\pi, 2\pi, 3/4\pi, 7/4\pi, 1/6\pi, 1/3\pi, 2/3\pi, 5/6\pi, 7/6\pi, 4/3\pi, 5/3\pi, \frac{11}{6}\pi\}$$

```
> prin_vals:=convert(setthree,list);
```

$$prin\_vals := [0, \pi, 5/4\pi, 3/2\pi, 1/4\pi, 1/2\pi, 2\pi, 3/4\pi, 7/4\pi, 1/6\pi, 1/3\pi, 2/3\pi, 5/6\pi, 7/6\pi, 4/3\pi, 5/3\pi, \frac{11}{6}\pi]$$

Next, we define $f(x)$ to be the function that returns the ordered quadruple $(x, \sin x, \cos x, \tan x)$ and compute the value of $f(x)$ for each number in prin_vals with map naming the resulting table prin_vals.

```
> g:=(x,y)->is(x<y):
> prin_vals:=sort(prin_vals,g);
```

$$prin\_vals := [0, 1/6\pi, 1/4\pi, 1/3\pi, 1/2\pi, 2/3\pi, 3/4\pi, 5/6\pi, \pi, 7/6\pi, 5/4\pi, 4/3\pi, 3/2\pi, 5/3\pi, 7/4\pi, \frac{11}{6}\pi, 2\pi]$$

Finally, we use map followed by array to display s_and_c in row-and-column form

Remember that the result of using array is not a list so cannot be manipulated like a list.

```
> f:=x->[x,sin(x),cos(x)]:
> s_and_c:=array(map(f,prin_vals));
```

$$\begin{bmatrix} 0 & 0 & 1 \\ 1/6\pi & 1/2 & 1/2\sqrt{3} \\ 1/4\pi & 1/2\sqrt{2} & 1/2\sqrt{2} \\ 1/3\pi & 1/2\sqrt{3} & 1/2 \\ 1/2\pi & 1 & 0 \\ 2/3\pi & 1/2\sqrt{3} & -1/2 \\ 3/4\pi & 1/2\sqrt{2} & -1/2\sqrt{2} \\ 5/6\pi & 1/2 & -1/2\sqrt{3} \\ \pi & 0 & -1 \\ 7/6\pi & -1/2 & -1/2\sqrt{3} \\ 5/4\pi & -1/2\sqrt{2} & -1/2\sqrt{2} \\ 4/3\pi & -1/2\sqrt{3} & -1/2 \\ 3/2\pi & -1 & 0 \\ 5/3\pi & -1/2\sqrt{3} & 1/2 \\ 7/4\pi & -1/2\sqrt{2} & 1/2\sqrt{2} \\ \frac{11}{6}\pi & -1/2 & 1/2\sqrt{3} \\ 2\pi & 0 & 1 \end{bmatrix}$$

Remark. The result of using array is not a list (or table) and calculations on it using commands like map cannot be performed. array helps you see results in a more readable format. To avoid confusion, do not assign the results of using array any name: adopting this convention avoids any possible manipulation of array objects.

■

object:=name assigns the object object the name name.

We can use `map` and `seq` with any lists, including lists of functions and/or other lists.

Lists of functions are graphed with `plot`:

```
plot(listoffunctions,x=a..b)
```

graphs the list of functions of x, `listoffunctions`, for $a \leq x \leq b$.

EXAMPLE 4.2.4 (Bessel Functions): The **Bessel functions of the first kind**, $J_n(x)$, are nonsingular solutions of $x^2y'' + xy' + (x^2 - n^2)\,y = 0$. `BesselJ(n,x)` returns $J_n(x)$. Graph $J_n(x)$ for $n = 0,1,2,\ldots,8$.

SOLUTION: In `t1`, we use `seq` and `BesselJ` to create a list of $J_n(x)$ for $n = 0, 1, 2, \ldots, 8$. We then use `plot` to graph each function in `t1`, which are displayed in Figure 4-10.

```
> t1:=[seq(BesselJ(n,x),n=0..8)]:
> plot(t1,x=0..25);
```

A different effect is achieved by graphing each function separately. To do so, in `A` we plot each function using `map` and then display each plot using a `for` loop in Figure 4-11.

```
> with(plots):
> A:=map(plot,t1,x=0..25,color=black):
> for i from 1 to 9 do A[i]end do;
```

■

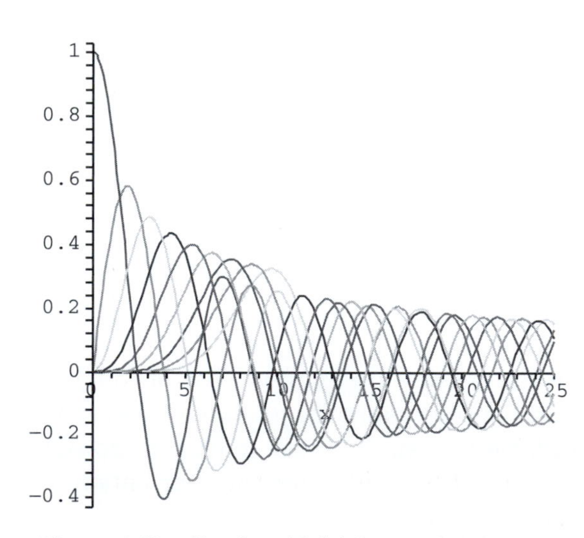

Figure 4-10 Graphs of $J_n(x)$ for $n = 0,1,2,\ldots,8$

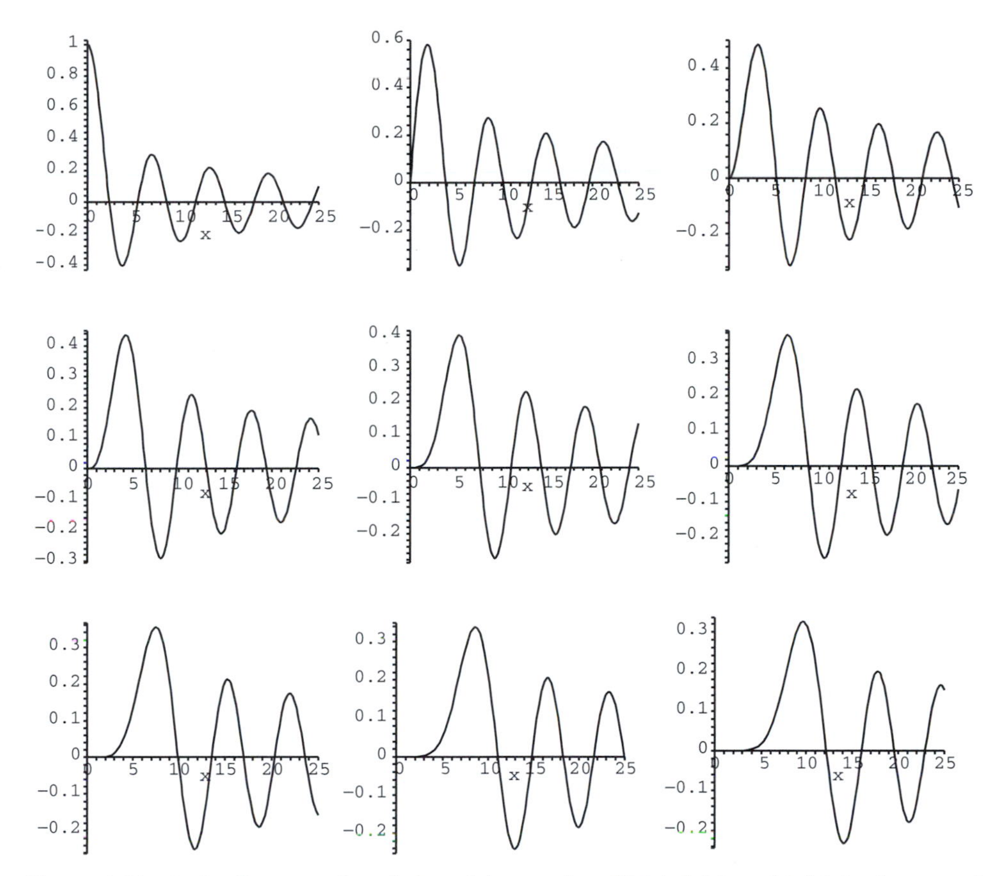

Figure 4-11 In the first row, from left to right, graphs of $J_0(x)$, $J_1(x)$, and $J_2(x)$; in the second row, from left to right, graphs of $J_3(x)$, $J_4(x)$, and $J_5(x)$; in the third row, from left to right, graphs of $J_6(x)$, $J_7(x)$, and $J_8(x)$

EXAMPLE 4.2.5 (Dynamical Systems): Let $f_c(x) = x^2 + c$ and consider the dynamical system given by $x_0 = 0$ and $x_{n+1} = f_c(x_n)$. Generate a bifurcation diagram of f_c.

Compare the approach used here with the approach used in Example 4.1.7.

SOLUTION: First, recall that `(f@@n)(x)` computes the repeated composition $f^n(x)$. Then, in terms of a composition,

$$x_{n+1} = f_c(x_n) = f_c{}^n(0).$$

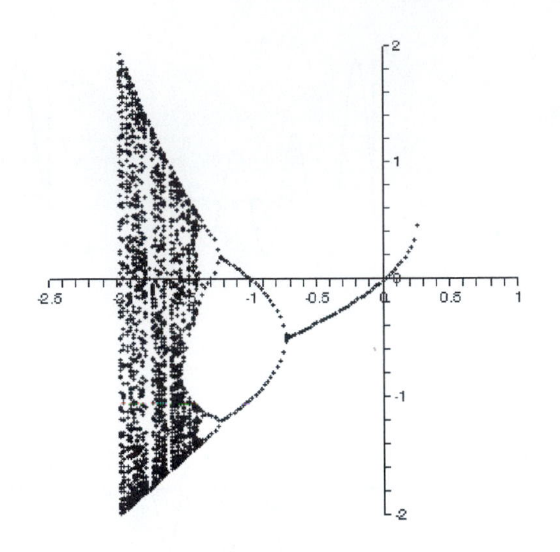

Figure 4-12 Bifurcation diagram of f_c

We will compute $f_c{}^n(0)$ for various values of c and "large" values of n so we begin by defining `cvals` to be a list of 150 equally spaced values of c between -2.5 and 1.

```
> cvals:=[seq(-2.5+3.5/149*i,i=0..149)]:
```

We then define $f_c(x) = x^2 + c$. For a given value of c, `f(c)` is a function of one variable, x, while the form `f:=(c,x)->...` results in a function of two variables.

```
> f:=c->proc(x) x^2+c end proc:
```

To iterate f_c for various values of c, we define h. For a given value of c, $h(c)$ returns the list of points $\left\{\left(c, f_c{}^{100}(0)\right), \left(c, f_c{}^{101}(0)\right), \ldots, \left(c, f_c{}^{200}(0)\right)\right\}$.

```
> h:=c->seq([c,(f(c)@@n)(0)],n=100..200):
```

We then use `map` to apply h to the list `cvals`.

```
> t1:=map(h,cvals):
```

The resulting set of points is plotted with `plot` in Figure 4-12.

```
> plot(t1,style=point,symbol=point,color=black,
   view=[-2.5..1,-2..2]);
```

■

4.2.1 More on Graphing Lists

If you do not include the option `style=point` when plotting a list of points with `plot`, successive points are connected with line segments.

EXAMPLE 4.2.6: Table 4-1 shows the percentage of the United States labor force that belonged to unions during certain years. Graph the data represented in the table.

SOLUTION: We begin by entering the data represented in the table as `dataunion`:

```
> dataunion:=[[30,11.6],[35,13.2],[40,26.9],
> [45,35.5],[50,31.5],
> [55,33.2],[60,31.4],[65,28.4],[70,27.3],
> [75,25.5],[80,21.9],
> [85,18.0],[90,16.1]]:
```

The x-coordinate of each point corresponds to the year, where x is the number of years past 1900, and the y-coordinate of each point corresponds to the percentage of the United States labor force that belonged to unions in the given year. We then use `plot` to graph the set of points represented in `dataunion` in `p1`, `p2` (illustrating the `style=point`

Table 4-1 Union membership as a percentage of the labor force

Year	Union Membership as a Percentage of the Labor Force
1930	11.6
1935	13.2
1940	26.9
1945	35.5
1950	31.5
1955	33.2
1960	31.4
1965	28.4
1970	27.3
1975	25.5
1980	21.9
1985	18.0
1990	16.1

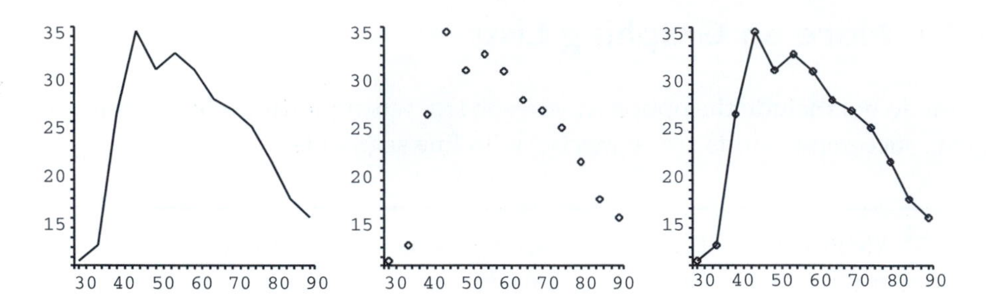

Figure 4-13 Union membership as a percentage of the labor force

symbol, and symbolsize option), and finally show the two together. All three plots are shown side-by-side in Figure 4-13 using display.

Remember that display is contained in the plots package.

```
> with(plots):
> p1:=plot(dataunion,color=black):
> display(p1);

> p2:=plot(dataunion,style=point,symbol=diamond,
> color=black,symbolsize=20):
> display(p2);
> display(p1,p2);
```

To achieve even greater control over plots of sets of points, plot each point separately and then show all the plots together with display.

For example, in t1, we plot each point separately. Successive points will appear in lighter shades of gray. On the other hand, in t2, we plot the points together; successive points are connected with thickened gray line segments.

```
> t1:=[seq(plot([dataunion[i]],style=point,
   symbol=circle,
> symbolsize=30,color=COLOR(RGB,i/13,i/13,i/13)),
   i=1..13)]:
> t2:=plot(dataunion,thickness=5,color=gray):
```

The results are shown together with display in Figure 4-14.

```
> display(t1,t2);
```

■

The select function is used to extract elements of a list that satisfy specific criteria:

```
select(list,criteria)
```

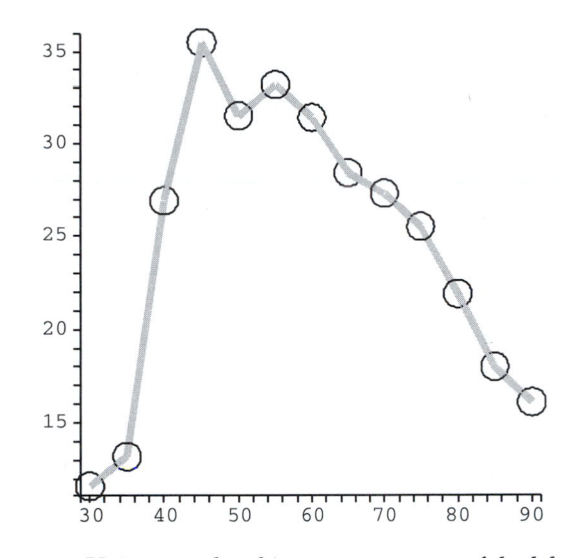

Figure 4-14 Union membership as a percentage of the labor force

returns the elements of `list` for which `criteria`, a Boolean valued function, is true. To define Boolean valued functions, use `evalb`. For example, the function

```
f:=x->evalb(x<y)
```

returns true if $x < y$ and false otherwise. Similarly, `f:=x->evalb(x=y)` returns true if $x = y$ and false otherwise.

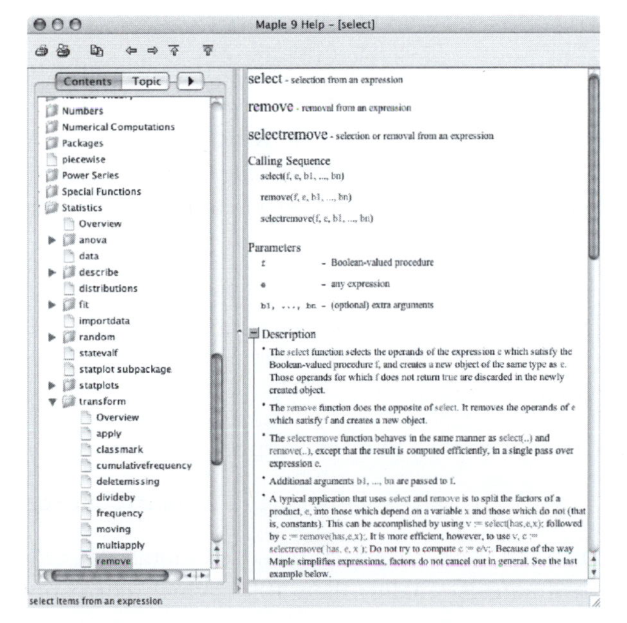

With the speed of today's computers and the power of Maple, it is relatively easy now to carry out many calculations that required supercomputers and sophisticated programming experience just a few years ago.

EXAMPLE 4.2.7 (Julia Sets): Plot Julia sets for $f(z) = \lambda \cos z$ if $\lambda = .66i$ and $\lambda = .665i$.

SOLUTION: The sets are visualized by plotting the points (a, b) for which $|f^n(a + bi)|$ is *not* large in magnitude so we begin by forming our complex grid. Using `seq`, we define `complexpts` to be a list of 22,500 points of the form $a + bi$ for 150 equally spaced real values of a between 0 and 8 and 150 equally spaced real values of b between -4 and 4 and then $f(z) = .66i \cos z$.

`nops(list)` returns the number of elements of `list`.

```
> avals:=seq(8.0*i/149,i=0..149):
> bvals:=seq(-4.0+8.0*i/149,i=0..149):
```

```
> complexpts:=[seq(seq(a+b*I,a=avals),b=bvals)]:
> nops(complexpts);
```

22500

```
> f:=proc(z) option remember;
  evalf(0.66*I*cos(z)) end proc:
```

For a given value of $c = a + bi$, $q(c)$ returns the ordered triple consisting of the real part of c, the imaginary part of c, and $|f^{25}(c)|$ unless $|f^n(c)| > 10^{10}$ for some $n < 25$ in which case the ordered triple $(\text{Re}(c), \text{Im}(c), 10^{10})$ is returned. We terminate the procedure when $|f^n(c)|$ is "large" (in this case, greater than 10^{10}) to avoid exceeding numerical precision.

```
> q:= proc(c) local i;
> for i from 1 to 24 while evalb(abs((f@@i)(c))<10^10)
> do [Re(c),Im(c),min(abs((f@@(i+1))(c)),10^10)] end do
> end proc:
```

We then use `seq` to apply q to `complexpts`.

```
> t1:=[seq(q(complexpts[i]),i=1..nops(complexpts))]:
```

In `t2`, we use `select` to select those elements of `t1` for which the third coordinate is smaller than 10^{10}, which corresponds to the

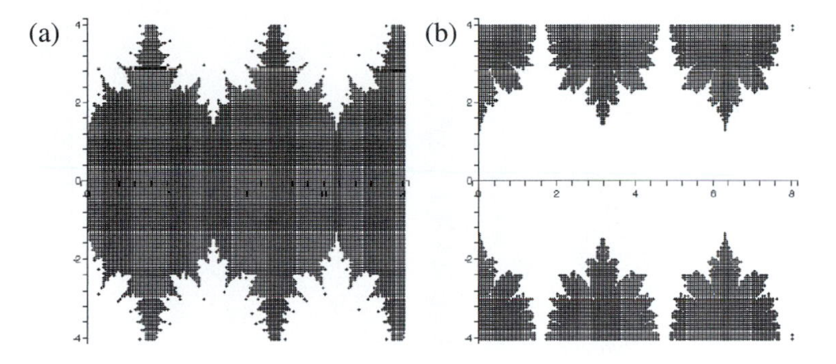

Figure 4-15 Julia set for $0.66i \cos z$

ordered triples $(a, b, |f^n(a+bi)|)$ for which $|f^n(a+bi)|$ *is not* large in magnitude.

```
> t2:=select(x->evalb(x[3]<10^10),t1):
```

```
> nops(t2);
```

$$14916$$

The first two coordinates of each ordered triple in `t2` are then obtained using `map` in `t3`. This list of points is plotted with `plot` in Figure 4-15(a).

```
> t3:=map(x->[x[1],x[2]],t2):
```

```
> plot(t3,style=point,symbol=point,color=black);
```

The inversion of Figure 4-15(a), Figure 4-15(b), is obtained by selecting those elements of `t1` for which the third coordinate is equal to 10^{10}, which corresponds to the ordered triples $(a, b, |f^n(a+bi)|)$ for which $|f^n(a+bi)|$ *is* large in magnitude.

```
> t4:=select(x->evalb(x[3]=10^10),t1):
> t5:=map(x->[x[1],x[2]],t4):
> plot(t5,style=point,symbol=point,color=black);
```

For $\lambda = 0.665i$ we use a different grid and a larger number of sample points (Figure 4-16).

```
> avals:=seq(-1+2.0*i/149,i=0..149):
> bvals:=seq(1+6.0*i/249,i=0..249):
```

```
> complexpts:=[seq(seq(a+b*I,a=avals),b=bvals)]:
```

```
> f:=proc(z) option remember;
> evalf(0.665*I*cos(z)) end proc:
```

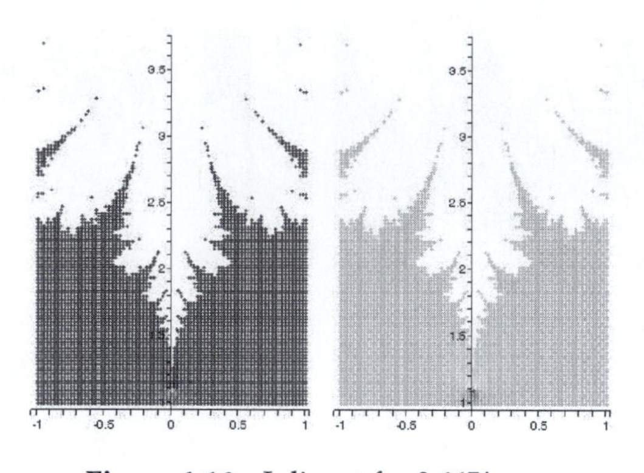

Figure 4-16 Julia set for $0.665i \cos z$

We increase the number of iterations as well.

```
> q:= proc(c) local i;
> for i from 1 to 49 while evalb(abs((f@@i)(c))<10^10)
> do [Re(c),Im(c),min(abs((f@@(i+1))(c)),10^10)] end do
> end proc:
```

```
> t1:=[seq(q(complexpts[i]),i=1..nops(complexpts))]:
```

```
> t2:=select(x->evalb(x[3]<10^10),t1):
```

In `t4`, we compute each iteration's distance from the origin. `sort` is used to find the minimum and maximum distance. We use these values to help us color-code the color of each point in the subsequent plot.

```
> t4:=sort([seq(t2[i,3],i=1..nops(t2))]):
```

```
> sm:=t4[1];
> lg:=t4[nops(t4)];
```

$$sm := 0.7985886788$$

$$lg := 2.779851981$$

To see detail, we plot each point separately and display the plots together with `display`. In Figure 4-16, the shading of the point (a, b) is assigned according to the distance of $f^{50}(a + bi)$ from the origin, calculated in `t4`. The color black indicates a distance of zero from the origin; as the distance increases, the shading of the point becomes lighter.

```
> lgdisplay1:=seq(plot([[t2[i,1],t2[i,2]]],
> style=point,symbol=point,
> color=COLOR(RGB,(t2[i,3]-sm)/(lg-sm),
  (t2[i,3]-sm)/(lg-sm),
```

```
> (t2[i,3]-sm)/(lg-sm))),i=1..nops(t2)):
> lgdisplay2:=seq(plot([[t2[i,1],t2[i,2]]],
> style=point,symbol=point,
> color=COLOR(RGB,1-(t2[i,3]-sm)/(lg-sm),
   1-(t2[i,3]-sm)/(lg-sm),
> 1-(t2[i,3]-sm)/(lg-sm))),i=1..nops(t2)):
> with(plots):
> display(lgdisplay1,scaling=constrained);
> display(lgdisplay2,scaling=constrained);
```

■

4.3 Mathematics of Finance

The use of lists and tables is quite useful in economic applications that deal with interest rates, annuities, and amortization. Maple is, therefore, of great use in these types of problems through its ability to show the results of problems in tabular form. Also, if a change is made to the problem, Maple can easily recompute the results.

In addition to defining functions yourself, you can also take advantage of the many finance-related functions defined in the `finance` package.

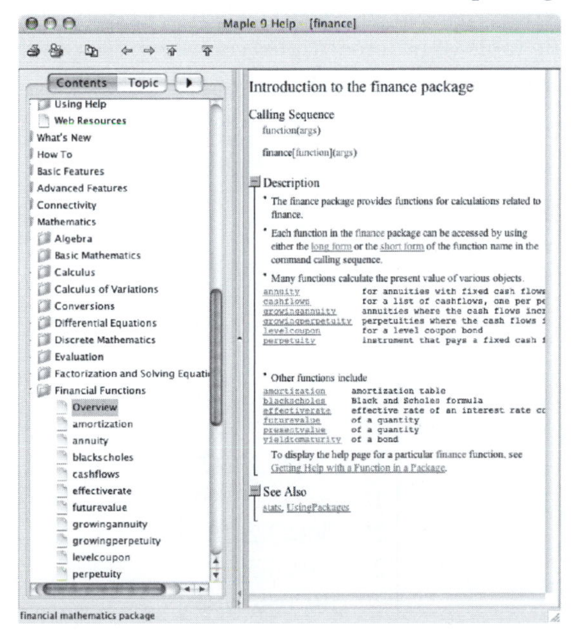

4.3.1 Compound Interest

A common problem in economics is the determination of the amount of interest earned from an investment. If P dollars are invested for t years at an annual interest rate of $r\%$ compounded m times per year, the **compound amount**, $A(t)$, at time t is given by

$$A(t) = P\left(1 + \frac{r}{m}\right)^{mt}.$$

If P dollars are invested for t years at an annual interest rate of $r\%$ compounded continuously, the compound amount, $A(t)$, at time t is given by $A(t) = Pe^{mt}$.

EXAMPLE 4.3.1: Suppose \$12,500 is invested at an annual rate of 7% compounded daily. How much money has accumulated and how much interest has been earned at the end of each 5-year period for $t = 0, 5, 10, 15, 20, 25, 30$? How much money has accumulated if interest is compounded continuously instead of daily?

SOLUTION: We define `ac(t)` to give the total value of the investment at the end of t years and `Interest(t)` to yield the total amount of interest earned at the end of t years. Then `seq` and `array` are used to produce the table of ordered triples corresponding to the year, total value of the investment, and total interest earned.

```
> ac:='ac':interest:='interest':
> ac:=t->12500*(1+.07/365)^(365*t):
> Interest:=t->ac(t)-12500:
> Time:=seq(5*n,n=0..6);
> array([seq([t,ac(t)],t=Time)]);
```

$$Time := 0, 5, 10, 15, 20, 25, 30$$

$$\begin{bmatrix} 0 & 12500 \\ 5 & 17737.75488 \\ 10 & 25170.23585 \\ 15 & 35717.07790 \\ 20 & 50683.26182 \\ 25 & 71920.58199 \\ 30 & 102056.7723 \end{bmatrix}$$

Next, we compute the value of the investment if interest is compounded continuously.

```
> array([seq([t,ac(t),Interest(t)],t=Time)]);
```

$$\begin{bmatrix} 0 & 12500 & 0 \\ 5 & 17737.75488 & 5237.75488 \\ 10 & 25170.23585 & 12670.23585 \\ 15 & 35717.07790 & 23217.07790 \\ 20 & 50683.26182 & 38183.26182 \\ 25 & 71920.58199 & 59420.58199 \\ 30 & 102056.7723 & 89556.7723 \end{bmatrix}$$

■

The problem can be redefined for arbitrary values of t, P, r, and n as follows.

```
> ac:='ac':Interest:='Interest':
> results:='results':
> ac:=(t,P,r,n)->P*(1+r/n)^(n*t):
> Interest:=(t,P,r,n)->ac(t,P,r,n)-P:
> results:=proc(Time,P,r,n)
> local k,vals,t;
> vals:=seq(
> Time[1]+k*Time[3],
> k=0..(Time[2]-Time[1])/Time[3]);
> array(
> [seq([t,ac(t,P,r,n),Interest(t,P,r,n)],
> t=vals)])
> end:
```

Hence, any problem of this type can be worked using these functions.

EXAMPLE 4.3.2: Suppose $10,000 is invested at an interest rate of 12% compounded daily. Create a table consisting of the total value of the investment and the interest earned at the end of 0, 5, 10, 15, 20, and 25 years. What is the total value and interest earned on an investment of $15,000 invested at an interest rate of 15% compounded daily at the end of 0, 10, 20, and 30 years?

SOLUTION: In this case, we use the function `results` defined above. Here, `t0=0`, `t1=25`, `m=5`, `P=10000`, `r=0.12`, and `n=365`.

```
> results([0,25,5],10000,.12,365);
```

$$\begin{bmatrix} 0 & 15000.0 & 0.0 \\ 10 & 67204.64830 & 52204.64830 \\ 20 & 301097.6502 & 286097.6502 \\ 30 & 1349010.779 & 1334010.77 \end{bmatrix}$$

If the conditions are changed to `t0=0`, `t1=30`, `m=10`, `P=15000`, `r=0.15`, and `n=365`, the desired table can be quickly calculated.

```
> results([0,30,10],15000,.15,365);
```

$$\begin{bmatrix} 0 & 15000.0 & 0.0 \\ 10 & 67204.64830 & 52204.64830 \\ 20 & 301097.6502 & 286097.6502 \\ 30 & 1349010.779 & 1334010.779 \end{bmatrix}$$

■

4.3.2 Future Value

If R dollars are deposited at the end of each period for n periods in an annuity that earns interest at a rate of $j\%$ per period, the **future value** of the annuity is

$$S_{\text{future}} = R\frac{(1+j)^n - 1}{j}.$$

EXAMPLE 4.3.3: Define a function `future` that calculates the future value of an annuity. Compute the future value of an annuity where \$250 is deposited at the end of each month for 60 months at a rate of 7% per year. Make a table of the future values of the annuity where \$150 is deposited at the end of each month for $12t$ months at a rate of 8% per year for $t = 1, 5, 9, 13, \ldots, 21, 25$.

SOLUTION: After defining `future`, we use `future` to calculate that the future value of an annuity where \$250 is deposited at the end of each month for 60 months at a rate of 7% per year is \$17,898.22.

```
> future:=(r,j,n)->r*(((1+j)^n-1)/j):
```

```
> future(250,.07/12,5*12);
```

$$17898.22420$$

For the second problem, we use `seq` and `future` to compute the future values of the annuity where \$150 is deposited at the end of each month for $12t$ months at a rate of 8% per year for $t = 1, 5, 9, 13, \ldots, 21, 25$. The first column in the following table corresponds to the time (in years) and the second column corresponds to the future value of the annuity.

```
> tocompute:=seq(1+4*i,i=0..(25-1)/4);
```

$$tocompute := 1, 5, 9, 13, 17, 21, 25$$

```
> array([seq([t,future(150,.08/12,12*t)],
> t=tocompute)]);
```

$$\begin{bmatrix} 1 & 1867.488997 \\ 5 & 11021.52910 \\ 9 & 23614.43196 \\ 13 & 40938.06180 \\ 17 & 64769.59246 \\ 21 & 97553.82186 \\ 25 & 142653.9756 \end{bmatrix}$$

■

4.3.3 Annuity Due

If R dollars are deposited at the beginning of each period for n periods with an interest rate of j% per period, the **annuity due** is

$$S_{\text{due}} = R\left[\frac{(1+j)^{n+1}-1}{j} - 1\right].$$

EXAMPLE 4.3.4: Define a function `due` that computes the annuity due. Use `due` to (a) compute the annuity due of \$500 deposited at the beginning of each month at an annual rate of 12% compounded monthly for 3 years; and (b) calculate the annuity due of \$100k deposited at the beginning of each month at an annual rate of 9% compounded monthly for 10 years for $k = 1, 2, 3, \ldots, 10$.

SOLUTION: (a) In the same manner as the previous example, we first define due and then use due to compute the annuity due of $500 deposited at the beginning of each month at an annual rate of 12% compounded monthly for 3 years.

```
> due:=(r,j,n)->r*(((1+j)^(n+1)-1)/j)-r:
> due(500,.12/12,3*12);
```

$$21753.82355$$

(b) We then use seq and due to calculate the annuity due of $100k deposited at the beginning of each month at an annual rate of 9% compounded monthly for 10 years for $k = 1, 2, 3, \ldots, 10$. The first column corresponds to the amount deposited each month at an annual rate of 9% compounded monthly and the second column corresponds to the value of the annuity.

```
> array([seq([100*k,due(100*k,.09/12,10*12)],
> k=1..10)]);
```

$$\begin{bmatrix} 100 & 19496.56341 \\ 200 & 38993.12683 \\ 300 & 58489.69024 \\ 400 & 77986.25365 \\ 500 & 97482.81707 \\ 600 & 116979.3805 \\ 700 & 136475.9439 \\ 800 & 155972.5073 \\ 900 & 175469.0707 \\ 1000 & 194965.6341 \end{bmatrix}$$

■

EXAMPLE 4.3.5: Compare the annuity due on a $100k monthly investment at an annual rate of 8% compounded monthly for $t = 5, 10, 15, 20$ and $k = 1, 2, 3, 4, 5$.

SOLUTION: We use seq and due to calculate due[100 k, 0.08/12,t 12], corresponding to the annuity due of $100k deposited monthly at an annual rate of 8% compounded monthly for t years, for $k = 1, 2, 3, 4$, and $t = 5, 10, 15$, and 20. Notice that the rows correspond to the annuity due on a \$100, \$200, \$300, \$400, and \$500 monthly investment for 5, 10, 15, and 20 years, respectively. For example, the annuity due on \$300 deposited monthly at an annual rate of 8% compounded monthly for 15 years is \$104,504.

```
> times:=seq(5*i,i=1..4):
> array([seq([seq(due(100*k,.08/12,t*12),t=times)],
> k=1..5)]);
```

$$\begin{bmatrix} 7396.670645 & 18416.56889 & 34834.51730 & 59294.72777 \\ 14793.34129 & 36833.13778 & 69669.03460 & 118589.4555 \\ 22190.01193 & 55249.70667 & 104503.5519 & 177884.1832 \\ 29586.68258 & 73666.27556 & 139338.0692 & 237178.9110 \\ 36983.35322 & 92082.84445 & 174172.5865 & 296473.6388 \end{bmatrix}$$

■

4.3.4 Present Value

Another type of problem deals with determining the amount of money that must be invested in order to insure a particular return on the investment over a certain period of time. The **present value**, P, of an annuity of n payments of R dollars each at the end of consecutive interest periods with interest compounded at a rate of j% per period is

$$P = R\frac{1-(1+j)^{-n}}{j}.$$

EXAMPLE 4.3.6: Define a function `present` to compute the present value of an annuity. (a) Find the amount of money that would have to be invested at $7\frac{1}{2}$% compounded annually to provide an ordinary annuity income of \$45,000 per year for 40 years; and (b) find the amount of money that would have to be invested at 8% compounded annually to provide an ordinary annuity income of \$20,000 + \$5000k per year for 35 years for $k = 0, 1, 2, 3, 4$, and 5 years.

SOLUTION: In the same manner as in the previous examples, we first define the function `present` which calculates the present value of an annuity. (a) We then use `present` to calculate the amount of money that would have to be invested at $7\frac{1}{2}\%$ compounded annually to provide an ordinary annuity income of $45,000 per year for 40 years.

```
> r:='r':j:='j':n:='n':
> present:=(r,j,n)->r*((1-(1+j)^(-n))/j):
> present(45000,.075,40);
```

$$566748.3899$$

(b) Also, we use `seq` to find the amount of money that would have to be invested at 8% compounded annually to provide an ordinary annuity income of $\$20,000 + \$5000k$ per year for 35 years for $k = 0, 1, 2, 3, 4$, and 5. In the table, the first column corresponds to the annuity income and the second column corresponds to the present value of the annuity.

```
> array([seq(
> [20000+5000*k,present(20000+5000*k,.08,35)],
> k=0..5)]);
```

$$\begin{bmatrix} 20000 & 233091.3644 \\ 25000 & 291364.2054 \\ 30000 & 349637.0465 \\ 35000 & 407909.8876 \\ 40000 & 466182.7286 \\ 45000 & 524455.5698 \end{bmatrix}$$

■

4.3.5 Deferred Annuities

The present value of a **deferred annuity** of R dollars per period for n periods deferred for k periods with an interest rate of j per period is

$$P_{\text{def}} = R\left[\frac{1-(1+j)^{-(n+k)}}{j} - \frac{1-(1+j)^{-k}}{j}\right].$$

EXAMPLE 4.3.7: Define a function `def(r,n,k,j)` to compute the value of a deferred annuity where r equals the amount of the deferred annuity, n equals the number of years in which the annuity is received, k equals the number of years in which the lump sum investment is made, and j equals the rate of interest. Use `def` to compute the lump sum that would have to be invested for 30 years at a rate of 15% compounded annually to provide an ordinary annuity income of $35,000 per year for 35 years. How much money would have to be invested at the ages of 25, 35, 45, 55, and 65 at a rate of $8\frac{1}{2}\%$ compounded annually to provide an ordinary annuity income of $30,000 per year for 40 years beginning at age 65?

SOLUTION: As in the previous examples, we first define `def` and then use `def` to compute the lump sum that would have to be invested for 30 years at a rate of 15% compounded annually to provide an ordinary annuity income of $35,000 per year for 35 years. The function `def` computes the present value of a deferred annuity where r equals the amount of the deferred annuity, n equals the number of years in which the annuity is received, k equals the number of years in which the lump sum investment is made, and j equals the rate of interest is defined.

```
> def:=(r,n,k,j)->
> r*((1-(1+j)^(-(n+k)))/j-(1-(1+j)^(-k))/j):
> def(35000,35,30,.15);
```

$$3497.584370$$

To answer the second question, we note that the number of years the annuity is deferred is equal to 65 (the age at retirement) minus the age at which the money is initially invested and then use `seq` and `def` to compute the amount of money that would have to be invested at the ages of 25, 35, 45, 55, and 65 at a rate of $8\frac{1}{2}\%$ compounded annually to provide an ordinary annuity income of $30,000 per year for 40 years beginning at age 65. Note that the first column corresponds to the current age of the individual, the second column corresponds to the number of years from retirement, and the third column corresponds to the present value of the annuity.

```
> k_vals:=seq(25+10*k,k=0..4):
> array([seq([k,65-k,def(30000,40,65-k,.085)],
> k=k_vals)]);
```

$$\begin{bmatrix} 25 & 40 & 12988.76520 \\ 35 & 30 & 29367.38340 \\ 45 & 20 & 66399.16809 \\ 55 & 10 & 150127.4196 \\ 65 & 0 & 339435.6102 \end{bmatrix}$$

■

4.3.6 Amortization

A loan is **amortized** if both the principal and interest are paid by a sequence of equal periodic payments. A loan of P dollars at interest rate j per period may be amortized in n equal periodic payments of R dollars made at the end of each period, where

$$R = \frac{Pj}{1-(1+j)^{-n}}.$$

The function `amort(p,j,n)` defined next determines the monthly payment needed to amortize a loan of p dollars with an interest rate of j compounded monthly over n months. A second function, `totintpaid(p,j,n)`, calculates the total amount of interest paid to amortize a loan of p dollars with an interest rate of j% compounded monthly over n months.

```
> amort:=(p,j,n)->p*j/(1-(1+j)^(-n)):
> totintpaid:=(p,j,n)->n*amort(p,j,n)-p:
```

EXAMPLE 4.3.8: What is the monthly payment necessary to amortize a loan of $75,000 with an interest rate of 9.5% compounded monthly over 20 years?

SOLUTION: The first calculation uses `amort` to determine the necessary monthly payment to amortize the loan. The second calculation determines the total amount paid on a loan of $75,000 at a rate of 9.5% compounded monthly over 20 years while the third shows how much of this amount was paid towards the interest.

```
> amort(75000,0.095/12,20*12);
```

$$699.0983810$$

```
> 240*amort(75000,0.095/12,20*12);
```

$$167783.6114$$

```
> totintpaid(75000,0.095/12,20*12);
```

$$92783.6114$$

■

EXAMPLE 4.3.9: What is the monthly payment necessary to amortize a loan of \$80,000 at an annual rate of $j\%$ in 20 years for $j = 8, 8.5, 9, 9.5, 10$, and 10.5?

SOLUTION: We use `amort` to calculate the necessary monthly payments. The first column corresponds to the annual interest rate and the second column corresponds to the monthly payment.

```
> jvals:=[.08,.085,.09,.095,.10,.105]:
> array([seq([j,amort(80000.,j/12,240)],j=jvals)]);
```

$$\begin{bmatrix} 0.08 & 669.1520417 \\ 0.085 & 694.2585990 \\ 0.09 & 719.7807647 \\ 0.095 & 745.7049399 \\ 0.10 & 772.0173256 \\ 0.105 & 798.7039096 \end{bmatrix}$$

■

In many cases, the amount paid towards the principal of the loan and the total amount that remains to be paid after a certain payment need to be computed. This is easily accomplished with the functions `unpaidbalance` and `curprinpaid` defined using the function `amort(p,j,n)` that was previously defined.

```
> unpaidbalance:=(p,j,n,m)->present(amort(p,j,n),j,n-m):
> unpaidbalance(p,j,n,m);
```

$$\frac{p\left(1-(1+j)^{-n+m}\right)}{1-(1+j)^{-n}}$$

```
> curprinpaid:=(p,j,n,m)->p-unpaidbalance(p,j,n,m):
> curprinpaid(p,j,n,m);
```

$$p-\frac{p\left(1-(1+j)^{-n+m}\right)}{1-(1+j)^{-n}}$$

EXAMPLE 4.3.10: What is the unpaid balance of the principal at the end of the fifth year of a loan of \$60,000 with an annual interest rate of 8% scheduled to be amortized with monthly payments over a period of 10 years? What is the total interest paid immediately after the 60th payment?

SOLUTION: We use the functions `unpaidbalance` and `curprinpaid`, defined above, to calculate that of the original $60,000 loan, $24,097.90 has been paid at the end of 5 years; $35,902.10 is still owed on the loan.

```
> unpaidbalance(60000,0.08/12,120,60);
```

$$35902.12153$$

```
> curprinpaid(60000,0.08/12,120,60);
```

$$24097.87847$$

■

Maple can also be used to determine the total amount of interest paid on a loan using the following function

```
> curintpaid:=(p,j,n,m)->m*amort(p,j,n)-curprinpaid(p,j,n,m):
> curintpaid(p,j,n,m);
```

$$\frac{mpj}{1-(1+j)^{-n}}-p+\frac{p\left(1-(1+j)^{-n+m}\right)}{1-(1+j)^{-n}}$$

where `curintpaid(p,j,n,m)` computes the interest paid on a loan of $\$p$ amortized at a rate of j per period over n periods immediately after the mth payment.

EXAMPLE 4.3.11: What is the total interest paid on a loan of $60,000 with an interest rate of 8% compounded monthly amortized over a period of 10 years (120 months) immediately after the 60th payment?

SOLUTION: Using `curintpaid`, we see that the total interest paid is $19,580.10.

```
> curintpaid(60000,0.08/12,120,60);
```

$$19580.05407$$

■

Using the functions defined above, amortization tables can be created that show a breakdown of the payments made on a loan.

EXAMPLE 4.3.12: What is the monthly payment necessary to amortize a loan of \$45,000 with an interest rate of 7% compounded monthly over a period of 15 years (180 months)? What is the total principal and interest paid after 0, 3, 6, 9, 12, and 15 years?

SOLUTION: We first use `amort` to calculate the monthly payment necessary to amortize the loan.

```
> amort(45000,0.07/12,15*12);
```

$$404.4727349$$

Next, we use `seq`, `curprinpaid`, and `curintpaid` to determine the interest and principal paid at the end of 0, 3, 6, 9, 12, and 15 years.

```
> tvals:=seq(3*t,t=0..5):
> array([seq([t,curprinpaid(45000,0.07/12,15*12,12*t),
   curintpaid(45000,0.07/12,15*12,12*t)],t=tvals)]);
```

$$\begin{bmatrix} 0 & 0.0 & 0.0 \\ 3 & 5668.98524 & 8892.03322 \\ 6 & 12658.42214 & 16463.61477 \\ 9 & 21275.87760 & 22407.17777 \\ 12 & 31900.55882 & 26343.51501 \\ 15 & 45000.0 & 27805.09228 \end{bmatrix}$$

Note that the first column represents the number of years, the second column represents the principal paid, and the third column represents the interest paid. Thus, at the end of 12 years, \$31,900.60 of the principal has been paid and \$26,343.50 has been paid in interest.

■

Because `curintpaid(p,j,n,y)` computes the interest paid on a loan of $\$p$ amortized at a rate of j per period over n periods immediately after the yth payment, and `curintpaid(p,j,n,y-12)` computes the interest paid on a loan of $\$p$ amortized at a rate of j per period over n periods immediately after the $(y-12)$th payment,

```
curintpaid(p,j,n,y)-curintpaid(p,j,n,y-12)
```

yields the amount of interest paid on a loan of $\$p$ amortized at a rate of j per period over n periods between the $(y-12)$th and yth payment. Consequently, the interest paid and the amount of principal paid over a year can also be computed.

EXAMPLE 4.3.13: Suppose that a loan of $45,000 with interest rate of 7% compounded monthly is amortized over a period of 15 years (180 months)? What is the principal and interest paid during each of the first 5 years of the loan?

SOLUTION: We begin by defining the functions `annualintpaid` and `annualprinpaid` that calculate the interest and principal paid during the yth year on a loan of $\$p$ amortized at a rate of j per period over n periods.

```
> annualintpaid:=(p,j,n,y)->curintpaid(p,j,n,y)-
  curintpaid(p,j,n,y-12):
> annualprinpaid:=(p,j,n,y)->curprinpaid(p,j,n,y)-
> curprinpaid(p,j,n,y-12):
```

We then use these functions along with `seq` to calculate the principal and interest paid during the first 5 years of the loan. Note that the first column represents the number of years the loan has been held, the second column represents the interest paid on the loan during the year, and the third column represents the amount of the principal that has been paid.

```
> array([seq([t,annualintpaid(45000,0.07/12,
  15*12,12*t),
> annualprinpaid(45000,0.07/12,15*12,12*t)],
  t=1..5)]);
```

$$\begin{bmatrix} 1 & 3094.263699 & 1759.40912 \\ 2 & 2967.075879 & 1886.59694 \\ 3 & 2830.693642 & 2022.97918 \\ 4 & 2684.45231 & 2169.22051 \\ 5 & 2527.63920 & 2326.03361 \end{bmatrix}$$

For example, we see that during the third year of the loan, $2830.69 was paid in interest and $2022.98 was paid on the principal.
■

4.3.7 More on Financial Planning

We can use many of the functions defined above to help make decisions about financial planning.

EXAMPLE 4.3.14: Suppose a retiree has \$1,200,000. If she can invest this sum at 7%, compounded annually, what level payment can she withdraw annually for a period of 40 years?

SOLUTION: The answer to the question is the same as the monthly payment necessary to amortize a loan of \$1,200,000 at a rate of 7% compounded annually over a period of 40 years. Thus, we use `amort` to see that she can withdraw \$90,011 annually for 40 years.

```
> amort(1200000,0.07,40);
```

$$90010.96665$$

■

EXAMPLE 4.3.15: Suppose an investor begins investing at a rate of d dollars per year at an annual rate of $j\%$. Each year the investor increases the amount invested by $i\%$. How much has the investor accumulated after m years?

SOLUTION: The following table illustrates the amount invested each year and the value of the annual investment after m years.

Year	Rate of Increase	Annual Interest	Amount Invested	Value after m Years
0		$j\%$	d	$(1+j\%)^m d$
1	$i\%$	$j\%$	$(1+i\%)d$	$(1+i\%)(1+j\%)^{m-1}d$
2	$i\%$	$j\%$	$(1+i\%)^2 d$	$(1+i\%)^2(1+j\%)^{m-2}d$
3	$i\%$	$j\%$	$(1+i\%)^3 d$	$(1+i\%)^3(1+j\%)^{m-3}d$
k	$i\%$	$j\%$	$(1+i\%)^k d$	$(1+i\%)^k(1+j\%)^{m-k}d$
m	$i\%$	$j\%$	$(1+i\%)^m d$	$(1+i\%)^m d$

It follows that the total value of the amount invested for the first k years after m years is given by:

Year	Total Investment
0	$(1+j\%)^m d$
1	$(1+j\%)^m d + (1+i\%)(1+j\%)^{m-1} d$
2	$(1+j\%)^m d + (1+i\%)(1+j\%)^{m-1} d + (1+i\%)^2(1+j\%)^{m-2} d$
3	$\sum_{n=0}^{3}(1+i\%)^n(1+j\%)^{m-n} d$
k	$\sum_{n=0}^{k}(1+i\%)^n(1+j\%)^{m-n} d$
m	$\sum_{n=0}^{m}(1+i\%)^n(1+j\%)^{m-n} d$

The command sum can be used to find a closed form of the sums $\sum_{n=0}^{k}(1+i\%)^n(1+j\%)^{m-n} d$ and $\sum_{n=0}^{m}(1+i\%)^n(1+j\%)^{m-n} d$. We use sum to find the sum $\sum_{n=0}^{k}(1+i\%)^n(1+j\%)^{m-n} d$ and name the result closedone.

```
> closedone:=sum((1+i)^n*(1+j)^(m-n)*d,n=0..k);
```

$$closedone := -(1+j)^m d\left(\frac{1+i}{1+j}\right)^{k+1}(1+j)(-i+j)^{-1} + \frac{(1+j)^m d(1+j)}{-i+j}$$

In the same way, sum is used to find a closed form of $\sum_{n=0}^{m}(1+i\%)^n(1+j\%)^{m-n} d$, naming the result closedtwo.

```
> closedtwo:=sum((1+i)^n*(1+j)^(m-n)*d,n=0..m);
```

$$closedtwo := -(1+j)^m d\left(\frac{1+i}{1+j}\right)^{m+1}(1+j)(-i+j)^{-1} + \frac{(1+j)^m d(1+j)}{-i+j}$$

These results are used to define the functions investment(d,i,j,k,m) and investmenttot(d,i,j,m) that return the value of the investment after k and m years, respectively.

```
> investment:=(d,i,j,k,m)->-(1+j)^m*d*((1+i)/
> (1+j))^(k+1)*(1+j)/(-i+j)+(1+j)^m*d*(1+j)/(-i+j):
> investmenttot:=(d,i,j,m)->-(1+j)^m*d*((1+i)/
> (1+j))^(m+1)*(1+j)/(-i+j)+(1+j)^m*d*(1+j)/(-i+j):
```

Finally, investment and investmenttot are used to illustrate various financial scenarios. In the first example, investment is used to compute the value after 25 years of investing \$6500 the first year and then increasing the amount invested 5% per year for 5, 10, 15, 20, and 25 years assuming a 15% rate of interest on the amount invested. In the second example, `investmenttot` is used to compute the value after 25 years of investing $6500 the first year and then increasing the amount invested 5% per year for 25 years assuming various rates of interest.

```
> tvals:=seq(5*t,t=1..5):
> array([seq([t,investment(6500,0.05,0.15,t,25)],
   t=tvals)]);
```

$$\begin{bmatrix} 5 & 1035064.556 \\ 10 & 1556077.818 \\ 15 & 1886680.271 \\ 20 & 2096459.926 \\ 25 & 2229572.983 \end{bmatrix}$$

```
> ivals:=seq(0.08+0.02*i,i=0..6):
> array([seq([i,investmenttot(6500,0.05,i,25)],
   i=ivals)]);
```

$$\begin{bmatrix} 0.08 & 832147.4477 \\ 0.10 & 1087125.500 \\ 0.12 & 1437837.092 \\ 0.14 & 1921899.153 \\ 0.16 & 2591635.686 \\ 0.18 & 3519665.382 \\ 0.20 & 4806524.115 \end{bmatrix}$$

■

Another interesting investment problem is discussed in the following example. In this case, Maple is useful in solving a recurrence equation that occurs in the problem. The command

```
rsolve({equations},a[n])
```

attempts to solve the recurrence equations `equations` for the variable `a(n)` with no dependence on `a(j)`, $j \leq n-1$.

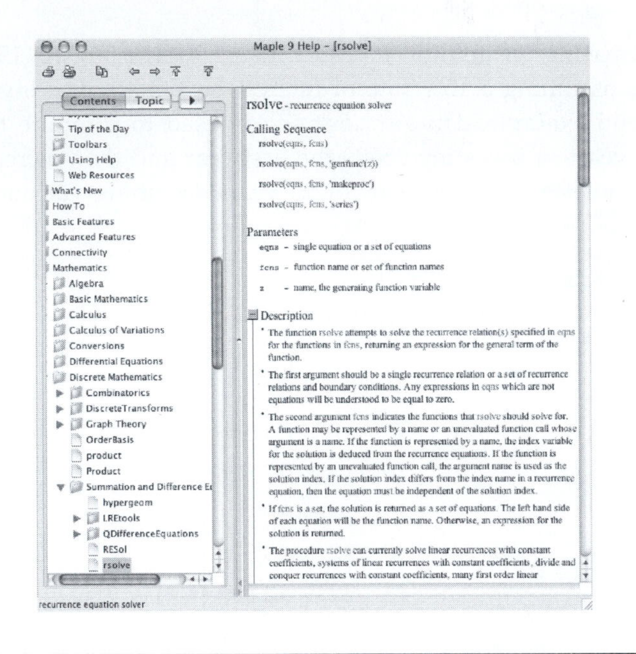

EXAMPLE 4.3.16: I am 50 years old and I have \$500,000 that I can invest at a rate of 7% annually. Furthermore, I wish to receive a payment of \$50,000 the first year. Future annual payments should include cost-of-living adjustments at a rate of 3% annually. Is \$500,000 enough to guarantee this amount of annual income if I live to be 80 years old?

SOLUTION: Instead of directly solving the above problem, let's solve a more general problem. Let a denote the amount invested and p the first-year payment. Let a_n denote the balance of the principal at the end of year n. Then, the amount of the nth payment, the interest earned on the principal, the decrease in principal, and the principal balance at the end of year n is shown in the table for various values of n. Observe that if $(1+j)^{n-1} > (1+j)a_{n-1}$, then the procedure terminates and the amount received in year n is $(1+j)a_{n-1}$.

Year	Amount	Interest	From Principal	Principal Balance
1	p	ia	$p - ia$	$a_1 = (1+i)a - p$
2	$(1+j)p$	ia_1	$(1+j)p - ia_1$	$a_2 = (1+i)a_1 - (1+j)p$
3	$(1+j)^2p$	ia_2	$(1+j)^2p - ia_2$	$a_3 = (1+i)a_2 - (1+j)^2p$
4	$(1+j)^3p$	ia_3	$(1+j)^3p - ia_3$	$a_4 = (1+i)a_3 - (1+j)^3p$
n	$(1+j)^{n-1}p$	ia_{n-1}	$(1+j)^{n-1}p - ia_{n-1}$	$a_n = (1+i)a_{n-1} - (1+j)^{n-1}p$

The recurrence equation

$$a_n = (1+i)a_{n-1} - (1+j)^{n-1}p$$

is solved for a_n with no dependence on a_{n-1}. After clearing several definitions of variable names, we use `rsolve` to solve the recurrence equation given above where the initial balance is represented by `amount`. Hence, a_n is given by the expression found in `bigstep`.

```
> eq1:=a(1)=(1+i)*amount-p:
> eq2:=a(n)=(1+i)*a(n-1)-(1+j)^(n-1)*p:
> bigstep:=rsolve(eq1,eq2,a(n));
```

$$bigstep := -\frac{(-amount - amount\, i + p)\,(1+i)^n}{1+i} + \frac{p\,(1+j)^n}{-j+i} - \frac{p\,(1+j)\,(1+i)^n}{(1+i)\,(-j+i)}$$

We then define `am(n,amount,i,p,j)` to be the explicit solution found in `bigstep`, which corresponds to the balance of the principal of a dollars invested under the above conditions at the end of the nth year.

```
> am:=(n,amount,i,p,j)->-(-amount-amount*i+p)*
   (1+i)^n/(1+i)+p*(1+j)^n/(-j+i)-
> p*(1+j)*(1+i)^n/((1+i)*(-j+i)):
```

To answer the question, we first define `annuitytable` in the following. For given a, i, p, j, and m, `annuitytable(a,i,p,j,m)` returns an ordered triple corresponding to the year, amount of income received in that year, and principal balance at the end of the year for m years.

```
> annuitytable:=(a,i,p,j,m)->array([seq([k,(1+j)^
   (k-1)*p,am(k,a,i,p,j)],k=1..m)]):
```

Then we compute `annuitytable(500000,.07,50000,.03,15)`. In this case, we see that the desired level of income is only guaranteed for 13 years, which corresponds to an age of 67, because the principal balance is negative after 13 years.

```
> annuitytable(500000,.07,50000,.03,15);
```

$$\begin{bmatrix} 1 & 50000.0 & 485000.0 \\ 2 & 51500.0 & 467450.0 \\ 3 & 53045.0 & 447126.500 \\ 4 & 54636.35000 & 423789.005 \\ 5 & 56275.44050 & 397178.794 \\ 6 & 57963.70370 & 367017.606 \\ 7 & 59702.61485 & 333006.225 \\ 8 & 61493.69325 & 294822.966 \\ 9 & 63338.50405 & 252122.071 \\ 10 & 65238.65920 & 204531.956 \\ 11 & 67195.81895 & 151653.375 \\ 12 & 69211.69355 & 93057.418 \\ 13 & 71288.04435 & 28283.392 \\ 14 & 73426.68565 & -43163.457 \\ 15 & 75629.48625 & -121814.386 \end{bmatrix}$$

We can also investigate other problems. For example, a 30-year mortgage of \$80,000 with an annual interest rate of 8.125% requires monthly payments of approximately \$600 (\$7200 annually) to amortize the loan in 30 years. However, using `annuitytable`, we see that if the amount of the payments is increased by 3% each year, the 30-year mortgage is amortized in 17 years. In the following result, the first column corresponds to the year of the loan, the second column to the annual payment, and the third column to the principal balance.

```
> annuitytable(80000,.08125,7200,.03,18);
```

$$\begin{bmatrix} 1 & 7200.0 & 79300.0 \\ 2 & 7416.0 & 78327.1251 \\ 3 & 7638.4800 & 77052.7240 \\ 4 & 7867.634400 & 75445.6234 \\ 5 & 8103.663432 & 73471.9167 \\ 6 & 8346.773333 & 71094.7367 \\ 7 & 8597.176538 & 68274.0073 \\ 8 & 8855.091828 & 64966.1789 \\ 9 & 9120.744583 & 61123.9363 \\ 10 & 9394.366925 & 56695.8890 \\ 11 & 9676.197929 & 51626.2322 \\ 12 & 9966.483871 & 45854.3798 \\ 13 & 10265.47839 & 39314.5697 \\ 14 & 10573.44273 & 31935.4357 \\ 15 & 10890.64602 & 23639.5438 \\ 16 & 11217.36540 & 14342.8915 \\ 17 & 11553.88636 & 3954.3648 \\ 18 & 11900.50295 & -7624.8458 \end{bmatrix}$$

■

Instead of defining our own functions, we could also have taken advantage of functions contained in the `finance` package. For example, the command `growingannuity(p,i,j,n)` returns the present value of an annuity of n periods invested at a rate of i per period with initial payment p. The payments increase at a rate j per period. Thus, entering

```
> with(finance):
> growingannuity(50000,.07,.03,30);
```

$$851421.9086$$

shows that to receive the desired income, I must invest $851,422. On the other hand, using `fsolve` we see that

```
> fsolve(growingannuity(x,0.07,0.03,30)=500000,x);
```

$$29362.64589$$

if I invest my $500,000 at 7% annually, I can receive an initial payment of $29,363 with subsequent 3% annual increases for 30 years. If, on the other hand, I wish to receive $50,000 my first year and guarantee annual increases of 3% annually forever, `growingperpetuity`, which is also contained in the `finance` package,

```
> growingperpetuity(50000,0.07,0.03);
```

$$1250000.0$$

shows us that I must initially invest $1,250,000. On the other hand, my $500,000 investment is enough to guarantee a first-year income of $20,000 with subsequent annual increases of 3% per year forever.

```
> fsolve(growingperpetuity(x,0.07,0.03)=500000);
```

$$20000.0$$

We can also investigate certain other problems. For example using `annuity`, which is also contained in the `finance` package, we see that a 30-year mortgage of $80,000 at $8\frac{1}{8}$% requires an annual payment of $7190 or approximately $600 per month.

```
> fsolve(annuity(x,0.08125,30)=80000);
```

$$7190.169059$$

On the other hand, using `growingannuity`, we see that if the amount of the payments is increased by 3% each year, the 30-year mortgage is amortized in 17 years!

```
> fsolve(growingannuity(7200,0.08125,0.03,k)=80000);
```

$$17.35372050$$

4.4 Other Applications

We now present several other applications that we find interesting and require the manipulation of lists. The examples also illustrate (and combine) many of the skills that were demonstrated in the earlier chapters.

4.4.1 Approximating Lists with Functions

Another interesting application of lists is that of curve-fitting.

Given a set of data points, we frequently want to approximate the data with a particular function. The command

```
fit[leastsquare[[x,y]],function,unknown parameters]([xcoords],
   [ycoords])
```

fits the list of data points `[xcoords],[ycoords]` using the function, `function`, containing parameters `unknown parameters` to be determined by the method of least-squares. The unknown parameters must appear linearly. If `function` and `unknown parameters` are not specified, a linear fit is found. Note that `fit` is contained in the `stats` package so is loaded by entering `with(stats)` before being used.

Recall from Sections 4.1 and 4.2, that when we graph lists of points with `plot`, the lists are in the form

$$(x_1, y_1), (x_2, y_2), \ldots, (x_n, y_n).$$

However, when we use `fit` to find an approximating function, the lists are of the form

$$(x_1, x_2, \ldots, x_n), (y_1, y_2, \ldots, y_n),$$

as indicated above. The following example illustrates how to use `seq` to transform a list from the form $(x_1, x_2, \ldots, x_n), (y_1, y_2, \ldots, y_n)$ to the form $(x_1, y_1), (x_2, y_2), \ldots, (x_n, y_n)$.

EXAMPLE 4.4.1: Define `datalist` to be the list of numbers consisting of 1.14479, 1.5767, 2.68572, 2.5199, 3.58019, 3.84176, 4.09957, 5.09166, 5.98085, 6.49449, and 6.12113. (a) Find a quadratic approximation of the points in `datalist`. (b) Find a fourth-degree polynomial approximation of the points in `datalist`.

SOLUTION: (a) After loading the `stats` package, the approximating function obtained via the least-squares method with `fit` is plotted along with the data points in Figure 4-17. Notice that many of the data points are not very close to the approximating function.

```
> with(stats);
```

[*anova, describe, fit, importdata, random, statevalf, statplots, transform*]

```
> datalist:=[[1,1.14479],[2,1.5767],[3,2.68572],
   [4,2.5199],[5,3.58019],[6,3.84176],
> [7,4.09957],[8,5.09166],[9,5.98085],
   [10,6.49449],[11,6.12113]]:
```

Next, we transform `datalist` from a list of the form $(x_1, x_2, \ldots, x_n)$, $(y_1, y_2, \ldots, y_n)$ to the form (x_1, y_1), (x_2, y_2), ..., (x_n, y_n) with `seq`.

`nops` returns the number of elements in a list.

```
> datalist2:=[[seq(datalist[i,1],
   i=1..nops(datalist))],
> [seq(datalist[i,2],i=1..nops(datalist))]]:
```

We then use `fit` to find the linear least-squares function that approximates the data.

```
> y:='y':
> fit1:=fit[leastsquare[[x,y]]](datalist2);
```

$$fit1 := y = 0.6432790909 + 0.5463740909\, x$$

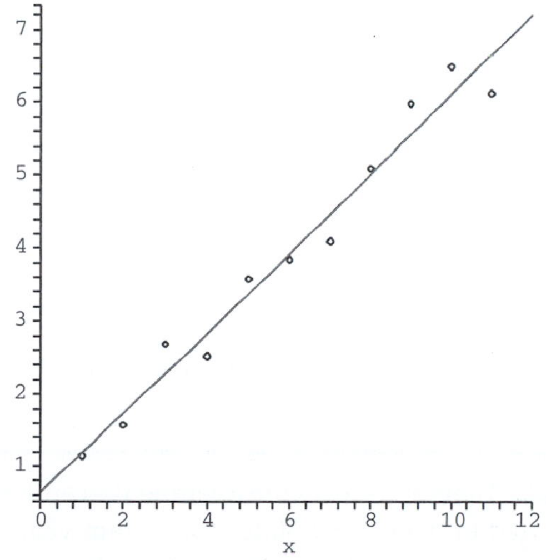

Figure 4-17 The graph of a quadratic fit shown with the data points

Note that the same results would have been obtained with the command

```
fit[leastsquare[[x,y],y=a*x+b,{a,b}]](datalist2).
```

We then use `assign` to name y the result obtained in `fit1`. The approximating function obtained via the least-squares method with `fit` is plotted along with the data points in Figure 4-17. Notice that many of the data points are not very close to the approximating function.

```
> assign(fit1):
with(plots):
p1:=plot(y,x=0..12):
p2:=plot(datalist,style=POINT):
display(p1,p2);
```

(b) A better approximation is obtained using a polynomial of higher degree (4).

```
> y:='y':
fit2:=fit[leastsquare[[x,y],y=a*x^4+b*x^3+c*x^2+d*x+e,
a,b,c,d,e]](datalist2);
```

$$fit2 := y = -0.003109847999\,x^4 + 0.07092011267\,x^3 - 0.5322814690\,x^2 + 2.027437282\,x - 0.5413294697$$

To check its accuracy, the second approximation is graphed simultaneously with the data points in Figure 4-18.

```
> assign(fit2):
p3:=plot(y,x=0..12):
display(p2,p3);
```

■

Next, consider a list of data points made up of ordered pairs, where we illustrate the use of `interp`: `interp(xcoords,ycoords,x)` fits the list of data points `data` with an $n-1$ degree polynomial in the variable x.

EXAMPLE 4.4.2: Table 4-2 shows the average percentage of petroleum products imported to the United States for certain years. (a) Graph the points corresponding to the data in the table and connect the consecutive points with line segments. (b) Use `interp` to find a function that

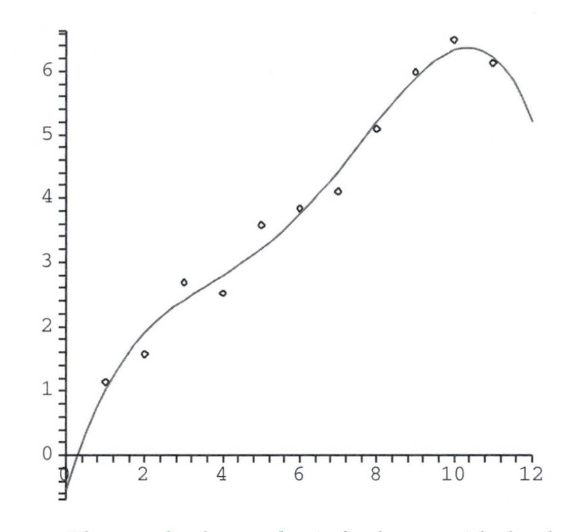

Figure 4-18 The graph of a quadratic fit shown with the data points

Table 4-2 Petroleum products imported to the United States for certain years

Year	Percent	Year	Percent
1973	34.8105	1983	28.3107
1974	35.381	1984	29.9822
1975	35.8167	1985	27.2542
1976	40.6048	1986	33.407
1977	47.0132	1987	35.4875
1978	42.4577	1988	38.1126
1979	43.1319	1989	41.57
1980	37.3182	1990	42.1533
1981	33.6343	1991	39.5108
1982	28.0988		

approximates the data in the table. (c) Find a fourth-degree polynomial approximation of the data in the table. (d) Find a trigonometric approximation of the data in the table.

SOLUTION: (a) We begin by defining `dataset` to be the set of ordered pairs represented in the table: the x-coordinate of each point represents the number of years past 1900 and the y-coordinate represents the percentage of petroleum products imported to the United States.

```
> dataset:=[[73,35],[74,35],[75,36],[76,41],[77,47],
   [78,42],[79,43],[80,37],[81,34],[82,28],[83,28],
   [84,30],[85,27],[86,33],[87,35],[88,38],[89,42],
   [90,42],[91,40]]:
```

Next, we transform `dataset` from a list of the form $(x_1, x_2, \ldots, x_n)$, $(y_1, y_2, \ldots, y_n)$ to the form (x_1, y_1), (x_2, y_2), ..., (x_n, y_n) with `seq`.

```
> dataset2:=[[seq(dataset[i,1],i=1..nops(dataset))],
   [seq(dataset[i,2],
> i=1..nops(dataset))]]:
```

(b) Then, `interp` is used to find a polynomial approximation of the data in the table.

```
> y:='y':
> fit1:=interp(dataset2[1],dataset2[2],x);
```

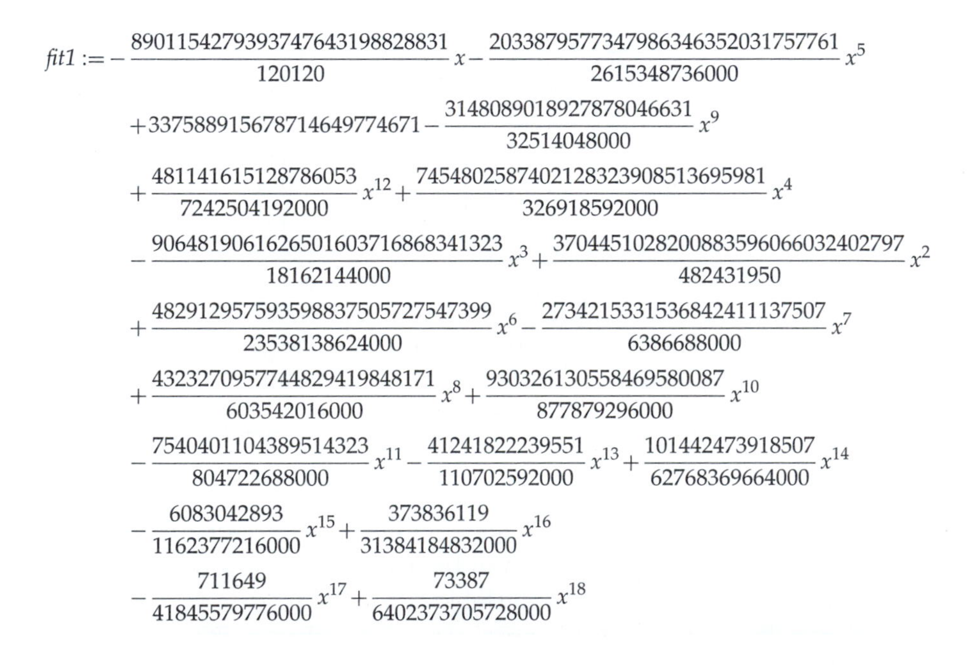

$$\begin{aligned} fit1 := &-\frac{890115427939374764319882883 1}{120120}x - \frac{2033879577347986346352031757761}{2615348736000}x^5 \\ &+3375889156787146497746 71 - \frac{314808901892787804663 1}{32514048000}x^9 \\ &+\frac{4811416151287860 53}{7242504192000}x^{12} + \frac{7454802587402128323908513695981}{326918592000}x^4 \\ &-\frac{9064819061626501603716868341323}{18162144000}x^3 + \frac{3704451028200883596066032402797}{482431950}x^2 \\ &+\frac{4829129575935988375057275 47399}{23538138624000}x^6 - \frac{2734215331536842411137507}{6386688000}x^7 \\ &+\frac{43232709577448294198481 71}{603542016000}x^8 + \frac{930326130558469580087}{877879296000}x^{10} \\ &-\frac{7540401104389514323}{804722688000}x^{11} - \frac{41241822239551}{110702592000}x^{13} + \frac{101442473918507}{62768369664000}x^{14} \\ &-\frac{6083042893}{1162377216000}x^{15} + \frac{373836119}{31384184832000}x^{16} \\ &-\frac{711649}{41845579776000}x^{17} + \frac{73387}{6402373705728000}x^{18} \end{aligned}$$

We then graph `fit1` along with the data in the table for the years corresponding to 1973 to 1991 in Figure 4-19. Although the interpolating polynomial agrees with the data exactly, the interpolating polynomial oscillates wildly.

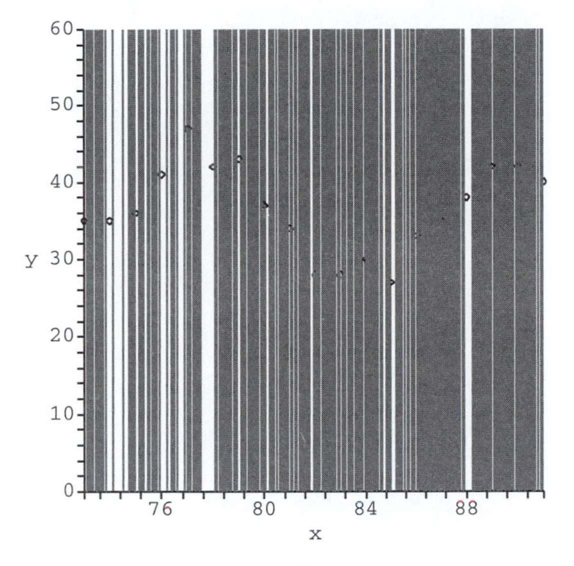

Figure 4-19 Even though interpolating polynomials agree with the data exactly, they may have extreme oscillations, even for relatively small data sets

```
> with(plots):
> p1:=plot(dataset,style=POINT,color=BLACK):
> p2:=plot(fit1,x=73..91,y=0..60):
> display(p2,p1);
```

In fact, it may be difficult to believe that the interpolating polynomial agrees with the data exactly so we use `seq` and `subs` to substitute the x-coordinates into the polynomial to confirm that it does agree exactly.

```
> seq(subs(x=t,fit1),t=dataset2[1]);
```

$$35, 35, 36, 41, 47, 42, 43, 37, 34, 28, 28, 30, 27, 33, 35, 38, 42, 42, 40$$

(c) To find a polynomial that approximates the data but does not oscillate wildly, we use `fit`. Again, we graph the fit and display the graph of the fit and the data simultaneously. In this case, the fit does not identically agree with the data but does not oscillate wildly, as illustrated in Figure 4-20.

```
> with(stats):
> y:='y':
> fit2:=fit[leastsquare[[x,y],
   y=a*x^4+b*x^3+c*x^2+d*x+e,a,b,c,d,e]](dataset2);
```

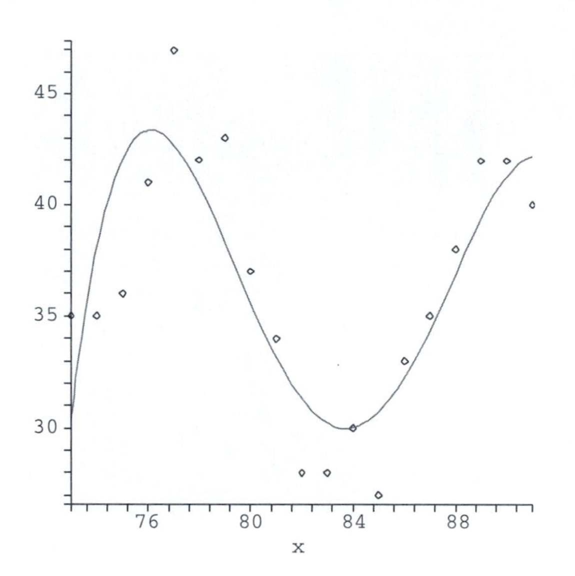

Figure 4-20 Even though the fit does not agree with the data exactly, the oscillations seen in Figure 4-19 do not occur

$$fit2 := y = -\frac{2659}{653752}x^4 + \frac{4002953}{2941884}x^3 - \frac{333893095}{1961256}x^2 + \frac{27775296511}{2941884}x - \frac{122751569}{627}$$

```
> assign(fit2):
> p3:=plot(y,x=73..91):
> display(p1,p3);
```

(d) In addition to curve-fitting with polynomials, Maple can also fit the data with trigonometric functions. In this case, we use `fit` to find an approximation of the data of the form $p = c_1 + c_2 \sin x + c_3 \sin(x/2) + c_4 \cos x + c_5 \cos(x/2)$. As in the previous two cases, we graph the fit and display the graph of the fit and the data simultaneously; the results are shown in Figure 4-21.

See texts like Abell, Braselton, and Rafter's *Statistics with Maple* [3] for a more sophisticated discussion of curve-fitting and related statistical applications.

```
> y:='y':
> fit3:=fit[leastsquare[[x,y],y=a+b*sin(x)
   +c*sin(x/2)+d*cos(x)+e*cos(x/2),
> a,b,c,d,e]](evalf(dataset2));
```

$$fit3 := y = 35.36378125 + 0.1147371447 \sin(x) + 6.159409180 \sin(1/2x) - 0.8594797306 \cos(x) + 4.267714121 \cos(1/2x)$$

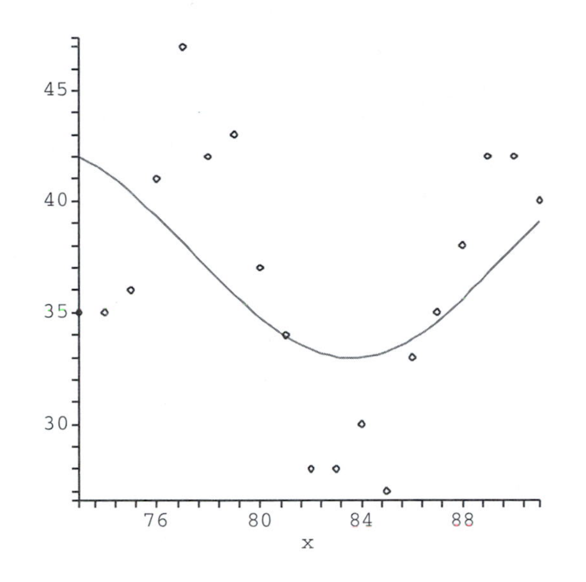

Figure 4-21 You can use `fit` to approximate data by a variety of functions

```
> y:='y':
> fit3:=fit[leastsquare[[x,y],y=a+b*sin(x/4)+
> c*cos(x/4),a,b,c]](evalf(dataset2));
```

$$fit3 := y = 37.77201950 - 4.270766858 \sin(1/4\,x) + 2.216757902 \cos(1/4\,x)$$

```
> with(plots):
> p1:=plot(dataset,style=POINT,color=BLACK):
> assign(fit3):
> p4:=plot(y,x=73..91):
> display(p1,p4);
```

■

4.4.2 Introduction to Fourier Series

Many problems in applied mathematics are solved through the use of Fourier series. Maple assists in the computation of these series in several ways. Suppose that $y = f(x)$ is defined on $-p < x < p$. Then the Fourier series for $f(x)$ is

$$\frac{1}{2}a_0 + \sum_{n=1}^{\infty}\left(a_n \cos\frac{n\pi x}{p} + b_n \sin\frac{n\pi x}{p}\right) \tag{4.1}$$

where

$$a_0 = \frac{1}{p}\int_{-p}^{p} f(x)\,dx$$

$$a_n = \frac{1}{p}\int_{-p}^{p} f(x)\cos\frac{n\pi x}{p}\,dx \quad n = 1, 2 \ldots \tag{4.2}$$

$$b_n = \frac{1}{p}\int_{-p}^{p} f(x)\sin\frac{n\pi x}{p}\,dx \quad n = 1, 2 \ldots$$

The n**th term of the Fourier series** (4.2) is

$$a_n \cos\frac{n\pi x}{p} + b_n \sin\frac{n\pi x}{p}. \tag{4.3}$$

The k**th partial sum of the Fourier series** (4.2) is

$$\frac{1}{2}a_0 + \sum_{n=1}^{k}\left(a_n \cos\frac{n\pi x}{p} + b_n \sin\frac{n\pi x}{p}\right). \tag{4.4}$$

It is a well-known theorem that if $y = f(x)$ is a periodic function with period $2p$ and $f'(x)$ is continuous on $[-p, p]$ except at finitely many points, then at each point x the Fourier series for $f(x)$ converges and

$$\frac{1}{2}a_0 + \sum_{n=1}^{\infty}\left(a_n \cos\frac{n\pi x}{p} + b_n \sin\frac{n\pi x}{p}\right) = \frac{1}{2}\left(\lim_{z\to x^+} f(z) + \lim_{z\to x^-} f(z)\right).$$

In fact, if the series $\sum_{n=1}^{\infty}(|a_n| + |b_n|)$ converges, then the Fourier series converges uniformly on $(-\infty, \infty)$.

EXAMPLE 4.4.3: Let $f(x) = \begin{cases} -x, & -1 \le x < 0 \\ 1, & 0 \le x < 1 \\ f(x-2), & x \ge 1 \end{cases}$. Compute and graph the first few partial sums of the Fourier series for $f(x)$.

SOLUTION: We begin by clearing all prior definitions of f. We then define the piecewise function $f(x)$ and graph $f(x)$ on the interval $[-1, 5]$ in Figure 4-22. Note that `elif` is used to avoid repeated use of `fi` and means "else if."

```
> f:='f':
> f:=proc(x) if x>=0 and x<1 then 1
```

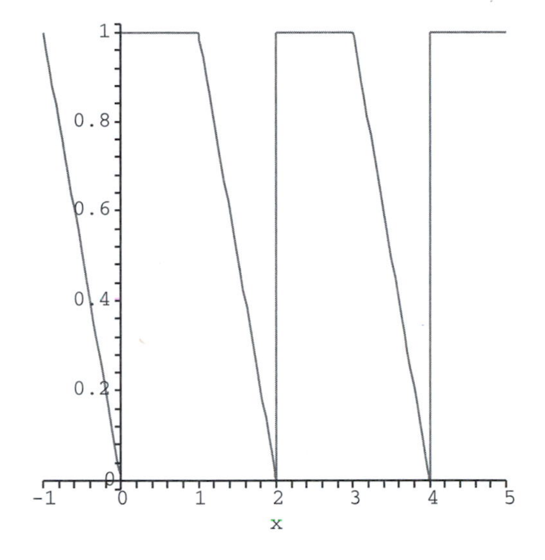

Figure 4-22 Plot of a few periods of $f(x)$

```
> elif x<0 and x>=-1 then -x
> elif x>=1 then f(x-2) fi end:
```

In the `plot` command, the function f and the variable x are contained in single quotation marks (') so that a delayed evaluation takes place. Of course, since f is defined as a procedure, operator notation may also be used to graph f.

```
> plot('f(x)','x'=-1..5,numpoints=150);
```

The Fourier series coefficients are computed with the integral formulas in (4.2). Executing the following commands defines `a[0]` to be an approximation of the integral $a_0 = \frac{1}{p}\int_{-p}^{p} f(x)\,dx$, `a[n]` to be an approximation of the integral $a_n = \frac{1}{p}\int_{-p}^{p} f(x)\cos\frac{n\pi x}{p}\,dx$, and `b[n]` to be an approximation of the integral $b_n = \frac{1}{p}\int_{-p}^{p} f(x)\sin\frac{n\pi x}{p}\,dx$.

```
> f1:=x->-x:
> f2:=x->1:
> a:='a':
> a:=table():
> a[0]:=evalf(1/2*(int(f1(x),x=-1..0)+
> int(f2(x),x=0..1))):
```

We use a `for` loop to determine a_1 through a_{12}. As with a_0 integration is performed over the two intervals $[-1,0]$ and $[0,1]$. Notice that the variable x is contained in single quotation marks so that the integrals are not evaluated until given a value of i.

```
> j:='j':
> for j from 1 to 12 do
> a[j]:=evalf(Int(f1(x)*cos(j*Pi*x),'x'=-1..0)+
> f2(x)*Int(cos(j*Pi*x),'x'=0..1)) od:
```

A similar loop is then used to compute the coefficients b_1 through b_{12}.

```
> b:='b':
> b:=table():
> j:='j':
> for j from 1 to 12 do
> b[j]:=evalf(Int(f1(x)*sin(j*Pi*x),'x'=-1..0)+
> f2(x)*Int(sin(j*Pi*x),'x'=0..1)) od:
```

We now display the coefficients computed above. The elements in the second column of the array represent the a_i's and the third column represents the b_i's.

```
> array([seq([i,a[i],b[i]],i=1..12)]);
```

$$\begin{bmatrix} 1 & -0.2026423673 & 0.3183098862 \\ 2 & -2.053102225 \times 10^{-15} & 0.1591549431 \\ 3 & -0.02251581859 & 0.1061032954 \\ 4 & 4.683922638 \times 10^{-15} & 0.07957747155 \\ 5 & -0.008105694691 & 0.06366197726 \\ 6 & -6.404955766 \times 10^{-15} & 0.05305164770 \\ 7 & -0.004135558516 & 0.04547284089 \\ 8 & -3.839413640 \times 10^{-15} & 0.03978873577 \\ 9 & -0.002501757621 & 0.03536776513 \\ 10 & -2.235182450 \times 10^{-15} & 0.03183098862 \\ 11 & -0.001674730308 & 0.02893726238 \\ 12 & -9.399385301 \times 10^{-16} & 0.02652582385 \end{bmatrix}$$

After the coefficients are calculated, the nth partial sum of the Fourier series is obtained with `sum`. The kth term of the Fourier series, $a_k \cos(k\pi x) + b_k \sin(k\pi x)$, is defined in `kterm`. Hence, the nth partial sum of the series is given by

$$a_0 + \sum_{k=1}^{n} [a_k \cos(k\pi x) + b_k \sin(k\pi x)] = \texttt{a[0]} + \sum_{k=1}^{n} \texttt{fs[k,x]},$$

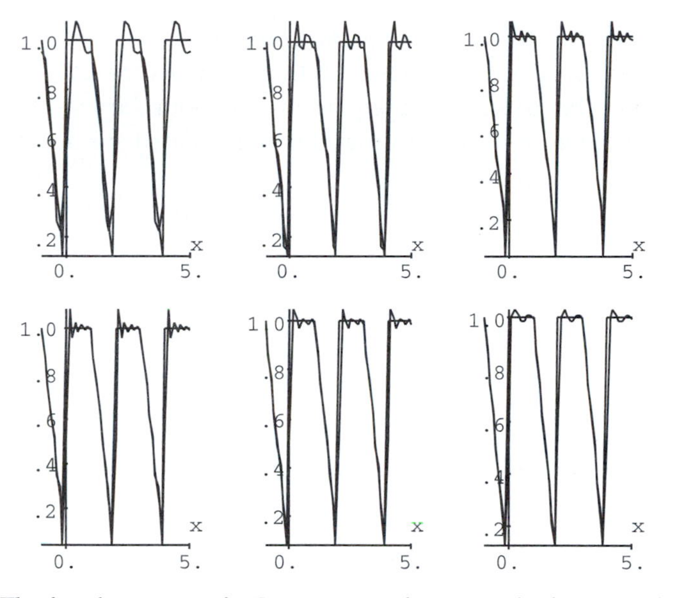

Figure 4-23 The first few terms of a Fourier series for a periodic function plotted with the function

which is defined in `fapprox` using `sum`. We illustrate the use of `fapprox` by finding `fapprox(2)`.

```
> i:='i':k:='k':
> kterm:=k->a[k]*cos(k*Pi*x)+b[k]*sin(k*Pi*x):
> fapprox:=n->a[0]+sum(kterm(i),i=1..n):
> fapprox(2);
```

$$0.7500000000 - 0.2026423673 \cos(\pi x) + 0.3183098862 \sin(\pi x)$$

$$- 2.053102225 \times 10^{-15} \cos(2\pi x) + 0.1591549431 \sin(2\pi x)$$

To see how the Fourier series approximates the periodic function, we plot the function simultaneously with the Fourier approximation for $n = 2$ and $n = 5$. The results are displayed together using `animate` and `display` in Figure 4-23.

```
> with(plots):
> A:=animate('f(x)',fapprox(n),x=-1..5,n=2..12,
> frames=6,color=black,tickmarks=[2,4]):
> display(A);
```

■

Application: The One-Dimensional Heat Equation

A typical problem in applied mathematics that involves the use of Fourier series is that of the **one-dimensional heat equation**. The boundary value problem that describes the temperature in a uniform rod with insulated surface is

$$\begin{aligned} &k\frac{\partial^2 u}{\partial x^2} = \frac{\partial u}{\partial t}, \ 0 < x < a, \ t > 0, \\ &u(0,t) = T_0, \ t > 0, \\ &u(a,t) = T_a, \ t > 0, \text{ and} \\ &u(x,0) = f(x), \ 0 < x < a. \end{aligned} \tag{4.5}$$

In this case, the rod has "fixed end temperatures" at $x = 0$ and $x = a$. $f(x)$ is the initial temperature distribution. The solution to the problem is

$$u(x,t) = \underbrace{T_0 + \frac{1}{a}(T_a - T_0)\,x}_{v(x)} + \sum_{n=1}^{\infty} b_n \sin(\lambda_n x)\, e^{-\lambda_n{}^2 kt}, \tag{4.6}$$

where

$$\lambda_n = n\pi/a \qquad \text{and} \qquad b_n = \frac{2}{a}\int_0^a (f(x) - v(x)) \sin\frac{n\pi x}{a}\,dx,$$

and is obtained through separation of variable techniques. The coefficient b_n in the solution (4.6) is the Fourier series coefficient b_n of the function $f(x) - v(x)$, where $v(x)$ is the **steady-state temperature**.

EXAMPLE 4.4.4: Solve $\begin{cases} \dfrac{\partial^2 u}{\partial x^2} = \dfrac{\partial u}{\partial t}, & 0 < x < 1, \ t > 0 \\ u(0,t) = 10, \quad u(1,t) = 10, \ t > 0 \\ u(x,0) = 10 + 20\sin^2 \pi x \end{cases}$

SOLUTION: In this case, $a = 1$ and $k = 1$. The fixed end temperatures are $T_0 = T_a = 10$, and the initial heat distribution is $f(x) = 10 + 20\sin^2 \pi x$. The steady-state temperature $v(x) = 10$ and $f(x)$ functions are defined. Also, the steady-state temperature, $v(x)$, and the eigenvalue are defined. Finally, `Int` and `evalf` are used to define a function that will be used to calculate the coefficients of the solution.

```
> f:='f':
> f:=x->10+20*sin(Pi*x)^2:
```

```
> v:=x->10:
> lambda:=n->n*Pi:
> b:=proc(n) option remember;
> evalf(2*Int((f(x)-v(x))*sin(n*Pi*x),x=0..1))
> end:
> b(1);
```

$$16.97652726$$

Notice that `b` is defined using the `remember` option so that Maple "remembers" the values of `b(n)` computed and thus avoids recomputing previously computed values.

Let $S_m = b_m \sin(\lambda_m x)\, e^{-\lambda_m{}^2 t}$. Then, the desired solution, $u(x,t)$, is given by

$$u(x,t) = v(x) + \sum_{m=1}^{\infty} S_m.$$

Let $u(x,t,n) = v(x) + \sum_{m=1}^{n} S_m$. Notice that $u(x,t,n) = u(x,t,n-1) + S_n$. Consequently, approximations of the solution to the heat equation are obtained recursively taking advantage of Maple's ability to compute recursively. The solution is first defined for $n = 1$ by `u(1)`. Subsequent partial sums, `u(n)`, are obtained by adding the nth term of the series, S_n, to `u(n-1)`.

```
> u:='u':
> u:=proc(n) option remember;
> u(n-1)+b(n)*sin(lambda(n)*x)*
> exp(-lambda(n)^2*t)
> end:
> u(1):=v(x)+b(1)*sin(lambda(1)*x)*
> exp(-lambda(1)^2*t):

> u(2);
```

$$10 + 16.97652726\, \sin(\pi\, x)\, e^{-\pi^2 t} - 6.240298228 \times 10^{-16}\, \sin(2\pi\, x)\, e^{-4\pi^2 t}$$

```
> u(8);
```

$$10+16.97652726 \sin(\pi\, x) e^{-\pi^2 t} - 6.240298228 \times 10^{-16} \sin(2\pi\, x) e^{-4\pi^2 t}$$
$$-3.395305452 \sin(3\pi\, x) e^{-9\pi^2 t} + 1.982867836 \times 10^{-15} \sin(4\pi\, x) e^{-16\pi^2 t}$$
$$-0.4850436360 \sin(5\pi\, x) e^{-25\pi^2 t} + 2.114252388 \times 10^{-15} \sin(6\pi\, x) e^{-36\pi^2 t}$$
$$-0.1616812120 \sin(7\pi\, x) e^{-49\pi^2 t} + 2.761834600 \times 10^{-15} \sin(8\pi\, x) e^{-64\pi^2 t}$$

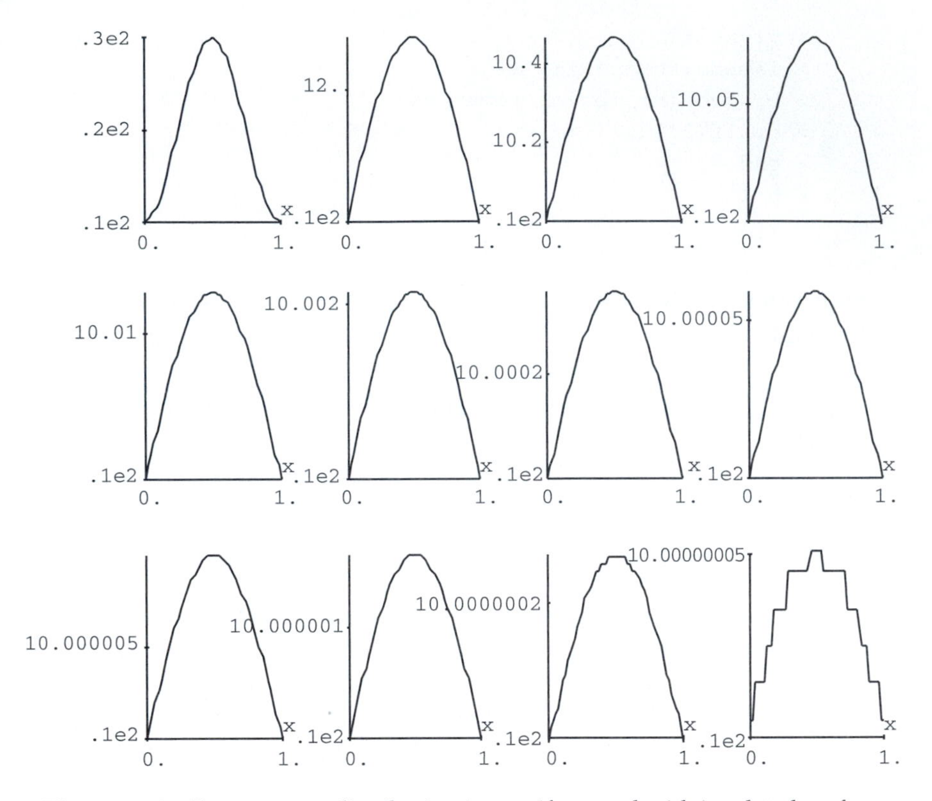

Figure 4-24 Temperature distribution in a uniform rod with insulated surface

By defining the solution in this manner a table can be created that includes the partial sums of the solution. In the following we use `animate` to graph the first, fourth, and seventh partial sums of the solution to the problem (Figure 4-24).

```
> j:='j':
> with(plots):
> A:=animate(subs(t=j,u(8)),x=0..1,j=0..2,frames=12,
    tickmarks=[2,2],color=BLACK,view=[0..1,0..30]):
> display(A);
> animate(subs(t=j,u(8)),x=0..1,j=0..2,frames=30,
    tickmarks=[2,2],color=BLACK);
```

■

Fourier series and generalized Fourier series arise in too many applications to list. Examples using them illustrate Maple's power to manipulate lists, symbolics, and graphics.

Application: The Wave Equation on a Circular Plate

For a classic approach to the subject see Graff's *Wave Motion in Elastic Solids* [10].

The vibrations of a circular plate satisfy the equation

$$D\,\nabla^4\, w(r,\theta,t) + \rho h \frac{\partial^2 w(r,\theta,t)}{\partial t^2} = q(r,\theta,t), \tag{4.7}$$

where $\nabla^4 w = \nabla^2\, \nabla^2\, w$ and ∇^2 is the **Laplacian in polar coordinates**, which is defined by

$$\nabla^2 = \frac{1}{r}\frac{\partial}{\partial r}\left(r\frac{\partial}{\partial r}\right) + \frac{1}{r^2}\frac{\partial^2}{\partial \theta^2} = \frac{\partial^2}{\partial r^2} + \frac{1}{r}\frac{\partial}{\partial r} + \frac{1}{r^2}\frac{\partial^2}{\partial \theta^2}.$$

Assuming no forcing so that $q(r,\theta,t) = 0$ and $w(r,\theta,t) = W(r,\theta)e^{-i\omega t}$, (4.7) can be written as

$$\nabla^4 W(r,\theta) - \beta^4 W(r,\theta) = 0, \qquad \beta^4 = \omega^2 \rho h / D. \tag{4.8}$$

For a clamped plate, the boundary conditions are $W(a,\theta) = \partial W(a,\theta)/\partial r = 0$ and after *much work* (see [10]) the **normal modes** are found to be

$$W_{nm}(r,\theta) = \left[J_n\left(\beta_{nm} r\right) - \frac{J_n\left(\beta_{nm} a\right)}{I_n\left(\beta_{nm} a\right)} I_n\left(\beta_{nm} r\right)\right] \begin{pmatrix} \sin n\theta \\ \cos n\theta \end{pmatrix}. \tag{4.9}$$

In (4.9), $\beta_{nm} = \lambda_{nm}/a$ where λ_{nm} is the mth solution of

$$I_n(x) J_n{}'(x) - J_n(x) I_n{}'(x) = 0, \tag{4.10}$$

where $J_n(x)$ is the Bessel function of the first kind of order n and $I_n(x)$ is the **modified Bessel function of the first kind** of order n, related to $J_n(x)$ by $i^n I_n(x) = J_n(ix)$.

See Example 4.2.4.

The Maple command `BesselI(n,x)` returns $I_n(x)$.

EXAMPLE 4.4.5: Graph the first few normal modes of the clamped circular plate.

SOLUTION: We must determine the value of λ_{nm} for several values of n and m so we begin by defining `eqn(n)(x)` to be $I_n(x)J_n{}'(x) - J_n(x)I_n{}'(x)$. The mth solution of (4.10) corresponds to the mth zero of the graph of `eqn(n)(x)` so we graph `eqn(n)(x)` for $n = 0, 1, 2,$ and 3 with `plot` and show the results in Figure 4-25.

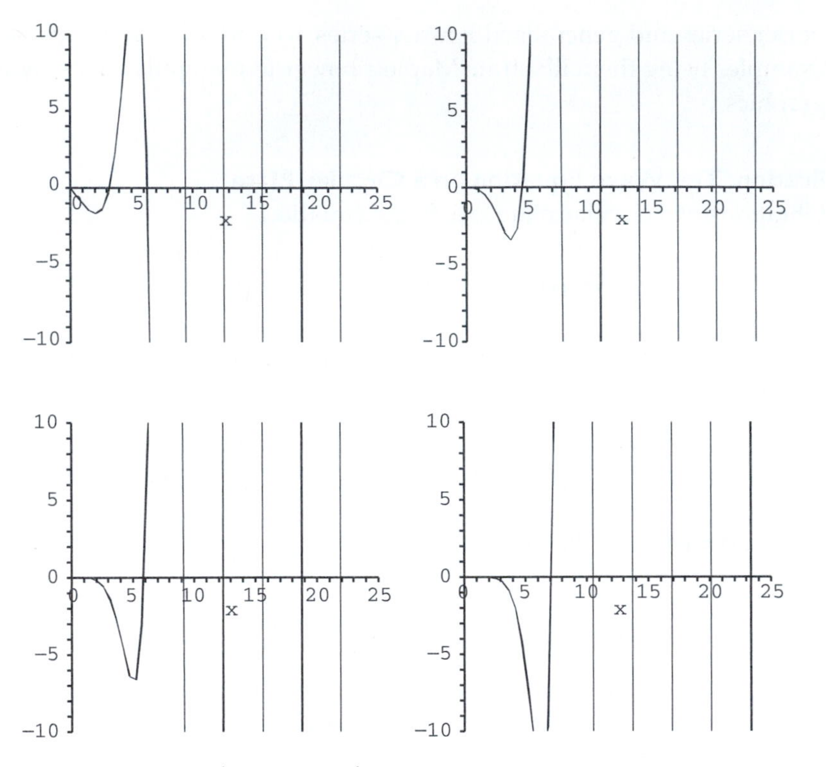

Figure 4-25 Plot of $I_n(x)J_n{}'(x) - J_n(x)I_n{}'(x)$ for $n = 0$ and 1 in the first row; $n = 2$ and 3 in the second row

```
> eqn:=n->proc(x) BesselI(n,x)*diff(BesselJ(n,x),x)-
> BesselJ(n,x)*diff(BesselI(n,x),x) end proc:

> plot(eqn(0)(x),x=0..25,view=[0..25,-10..10],
   color=black);
> plot(eqn(1)(x),x=0..25,view=[0..25,-10..10],
   color=black);
> plot(eqn(2)(x),x=0..25,view=[0..25,-10..10],
   color=black);
> plot(eqn(3)(x),x=0..25,view=[0..25,-10..10],
   color=black);
```

To determine λ_{nm} we use `fsolve`. Recall that to use `fsolve` to search for solutions of `equation` on an interval (a, b) specify the interval: `fsolve(equation,x,a..b)`. For example,

```
> lambda01:=fsolve(eqn(0)(x)=0,x,3.0..3.5);
```

$$lambda01 := 3.196220617$$

approximates λ_{01}, the first solution of (4.10) if $n = 0$. Thus,

We use the graphs in Figure 4-25 to obtain initial approximations of each solution.

```
> lambda0s:=[seq(fsolve(eqn(0)(x)=0,x,ints),
> ints=[3.0..3.2,6.0..6.5,9.0..9.5,
        12.25..13,15..16])];
```

$$lambda0s := [3.196220617, 6.306437048, 9.439499138, 12.57713064, 15.71643853]$$

approximates the first five solutions of (4.10) if $n = 0$ and then returns the specific value of each solution. We use the same steps to approximate the first five solutions of (4.10) if $n = 1, 2$, and 3.

```
> lambda1s:=[seq(fsolve(eqn(1)(x)=0,x,ints),
> ints=[4.0..5,7.0..8.5,10.0..12,14..15,17..18])];
> lambda2s:=[seq(fsolve(eqn(2)(x)=0,x,ints),
> ints=[5.0..6,9.0..10,12.0..13,15..16,18..19])];
> lambda3s:=[seq(fsolve(eqn(3)(x)=0,x,ints),
> ints=[7.0..8,10.0..11,13.0..14,16.5..17.5,20..21])];
```

$$lambda1s := [4.610899879, 7.799273801, 10.95806719, 14.10862781, 17.25572701]$$

$$lambda2s := [5.905678235, 9.196882600, 12.40222097, 15.57949149, 18.74395810]$$

$$lambda3s := [7.143531024, 10.53666987, 13.79506359, 17.00529018, 20.19231303]$$

All four lists are combined together in λs.

```
> lambdas:=[lambda0s,lambda1s,lambda2s,lambda3s];
```

$$lambdas := [[3.196220617, 6.306437048, 9.439499138, 12.57713064, 15.71643853], [4.610899879, 7.799273801, 10.95806719, 14.10862781, 17.25572701], [5.905678235, 9.196882600, 12.40222097, 15.57949149, 18.74395810], [7.143531024, 10.53666987, 13.79506359, 17.00529018, 20.19231303]]$$

For $n = 0, 1, 2$, and 3 and $m = 1, 2, 3, 4$, and 5, λ_{nm} is the mth part of the $(n+1)$st part of λs.

Observe that the value of a does not affect the shape of the graphs of the normal modes so we use $a = 1$ and then define β_{nm}.

```
> a:=1:
> beta:=(n,m)->lambdas[n+1,m]/a:
```

```
> beta(3,4);
```

$$17.00529018$$

ws is defined to be the sine part of (4.9)

```
> ws:=(n,m)->proc(r,theta)
> (BesselJ(n,beta(n,m)*r)-
> BesselJ(n,beta(n,m)*a)/BesselI(n,beta(n,m)*a)*
  BesselI(n,beta(n,m)*r))*
> sin(n*theta)
> end proc:
```

and wc to be the cosine part.

```
> wc:=(n,m)->proc(r,theta)
> (BesselJ(n,beta(n,m)*r)-
> BesselJ(n,beta(n,m)*a)/BesselI(n,beta(n,m)*a)*
  BesselI(n,beta(n,m)*r))*
> cos(n*theta)
> end proc:
```

We use `plot3d` to plot ws and wc. For example,

```
> plot3d([r*cos(theta),r*sin(theta),
  ws(3,4)(r,theta)],r=0..1,
> theta=0..2*Pi,scaling=constrained,
  view=[-1..1,-1..1,-0.5..0.5],
> grid=[40,40]);
```

graphs the sine part of $W_{34}(r,\theta)$ shown in Figure 4-26. We use `seq` together with `plot3d` followed by `display` to graph the sine part of $W_{nm}(r,\theta)$ for $n = 0$, 1, 2, and 3 and $m = 1$, 2, 3, and 4 shown in Figure 4-27.

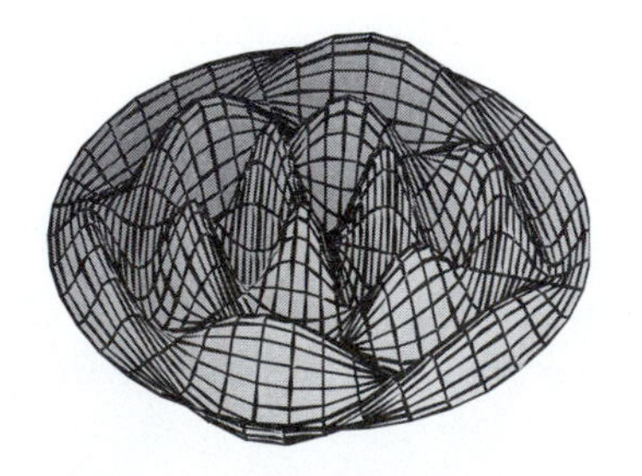

Figure 4-26 The sine part of $W_{34}(r,\theta)$

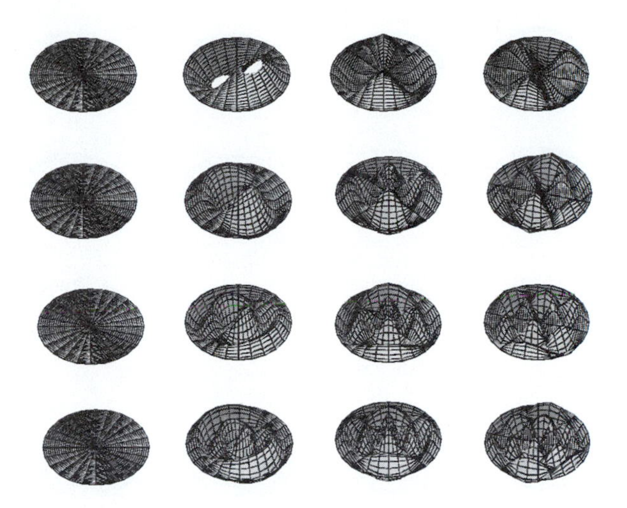

Figure 4-27 The sine part of $W_{nm}(r,\theta)$: $n = 0$ in row 1, $n = 1$ in row 2, $n = 2$ in row 3, and $n = 3$ in row 4 ($m = 1$ to 4 from left to right in each row)

```
> nandmvals:=[seq(seq([n,m],n=0..3),m=1..4)];
```

$$nandmvals := [[0,1],[1,1],[2,1],[3,1],[0,2],[1,2],[2,2],[3,2],\\ [0,3],[1,3],[2,3],[3,3],[0,4],[1,4],[2,4],[3,4]]$$

```
> wsplot:=v->plot3d([r*cos(theta),r*sin(theta),
   ws(v[1],v[2])(r,theta)],
> r=0..1,theta=0..2*Pi,scaling=constrained,
> view=[-1..1,-1..1,-0.5..0.5]):

> for i from 1 to nops(nandmvals) do
   wsplot(nandmvals[i]) end do;
```

Identical steps are followed to graph the cosine part shown in Figure 4-28.

```
> wcplot:=v->plot3d([r*cos(theta),r*sin(theta),
   wc(v[1],v[2])(r,theta)],r=0..1,theta=0..2*Pi,
   scaling=constrained,view=[-1..1,-1..1,-0.5..0.5]):
> for i from 1 to nops(nandmvals) do
   wcplot(nandmvals[i]) end do;
```

■

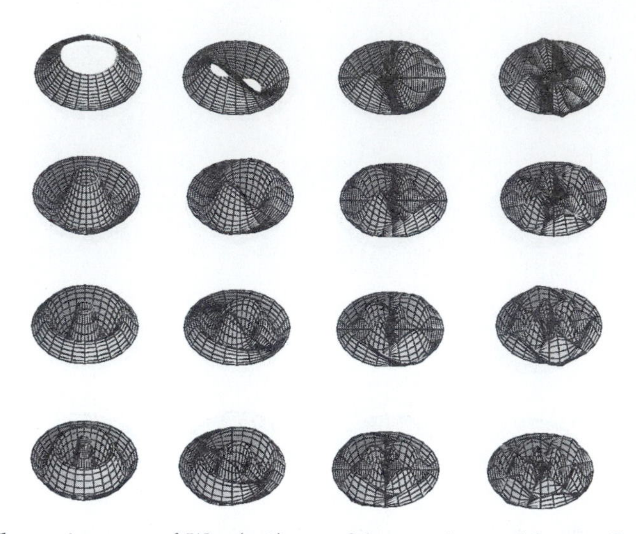

Figure 4-28 The cosine part of $W_{nm}(r,\theta)$: $n=0$ in row 1, $n=1$ in row 2, $n=2$ in row 3, and $n=3$ in row 4 ($m=1$ to 4 from left to right in each row)

4.4.3 The Mandelbrot Set and Julia Sets

See references like Barnsley's *Fractals Everywhere* [4], or Devaney and Keen's *Chaos and Fractals* [6], for detailed discussions regarding many of the topics briefly described in this section.

$f_c(x)=x^2+c$ is the special case of $p=2$ for $f_{p,c}(x)=x^p+c$.

Compare the approach here with the approach used in Example 4.2.5.

In Examples 4.1.7, 4.2.5, and 4.2.7 we illustrated several techniques for plotting bifurcation diagrams and Julia sets.

Let $f_c(x)=x^2+c$. In Example 4.2.5, we generated the c-values when plotting the bifurcation diagram of f_c. Depending upon how you think, some approaches may be easier to understand than others. With the exception of very serious calculations, the differences in the time needed to carry out the computations may be minimal so we encourage you to follow the approach that you understand. Learn new techniques as needed.

EXAMPLE 4.4.6 (Dynamical Systems): For example, entering

```
> f:='f':
> f:=c->proc(x) evalf(x^2+c) end proc;
```

$$f := c \longmapsto \textbf{proc}(x) \quad \mathit{evalf}(x^2c)\ \textbf{end proc;}$$

defines $f_c(x)=x^2+c$ so

```
> (f(-1)@@3)(x);
```

$$\left(\left(x^2-1.0\right)^2-1.0\right)^2-1.0$$

computes $f_{-1}{}^3(x)$ and

```
> seq((f(1/4)@@n)(0),n=101..200);
```

0.4906925007, 0.4907791302, 0.4908641546, 0.4909476183, 0.4910295639, 0.4911100326, 0.4911890641, 0.4912666967, 0.4913429673, 0.4914179115, 0.4914915637, 0.4915639572, 0.4916351240, 0.4917050952, 0.4917739006, 0.4918415693, 0.4919081293, 0.4919736077, 0.4920380307, 0.4921014237, 0.4921638112, 0.4922252171, 0.4922856643, 0.4923451753, 0.4924037716, 0.4924614743, 0.4925183037, 0.4925742795, 0.4926294208, 0.4926837462, 0.4927372738, 0.4927900210, 0.4928420048, 0.4928932417, 0.4929437477, 0.4929935384, 0.4930426289, 0.4930910339, 0.4931387677, 0.4931858442, 0.4932322769, 0.4932780790, 0.4933232632, 0.4933678420, 0.4934118275, 0.4934552315, 0.4934980655, 0.4935403407, 0.4935820679, 0.4936232578, 0.4936639206, 0.4937040665, 0.4937437053, 0.4937828465, 0.4938214995, 0.4938596734, 0.4938973770, 0.4939346190, 0.4939714078, 0.4940077517, 0.4940436587, 0.4940791367, 0.4941141933, 0.4941488360, 0.4941830721, 0.4942169088, 0.4942503529, 0.4942834113, 0.4943160907, 0.4943483975, 0.4943803381, 0.4944119187, 0.4944431454, 0.4944740240, 0.4945045604, 0.4945347603, 0.4945646291, 0.4945941724, 0.4946233954, 0.4946523033, 0.4946809012, 0.4947091940, 0.4947371866, 0.4947648838, 0.4947922902, 0.4948194104, 0.4948462489, 0.4948728100, 0.4948990981, 0.4949251173, 0.4949508717, 0.4949763654, 0.4950016023, 0.4950265863, 0.4950513211, 0.4950758105, 0.4951000581, 0.4951240675, 0.4951478422, 0.4951713856

returns a list of $f_{1/4}{}^n(0)$ for $n = 101, 102, \ldots, 200$. Thus,

```
> cvals:=[seq(-2+9/(4*299)*i,i=0..299)]:
> nops(cvals);
```

300

```
> lgtable:=[seq(seq([c,(f(c)@@n)(0)],c=cvals),
   n=101..200)]:
> lgtable[1..10];
```

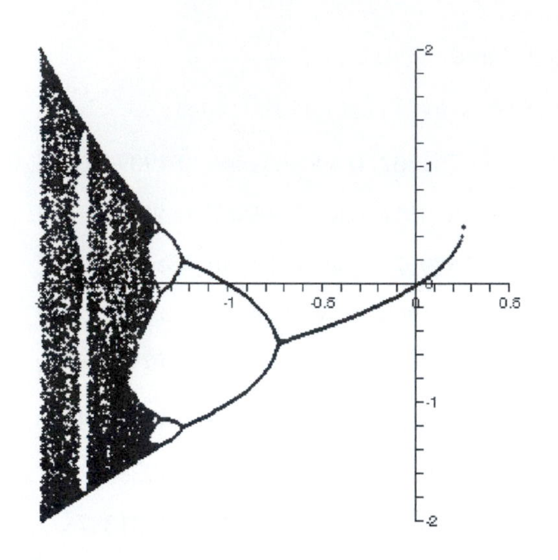

Figure 4-29 Another bifurcation diagram for f_c

$$\left[[-2,2.0],\left[-\frac{2383}{1196},-1.408377461\right],\left[-\frac{1187}{598},1.572643643\right],\right.$$
$$\left[-\frac{2365}{1196},-1.405546006\right],\left[-\frac{589}{299},-1.638767890\right],\left[-\frac{2347}{1196},1.817025091\right],$$
$$\left[-\frac{1169}{598},0.798494327\right],\left[-\frac{2329}{1196},0.120263477\right],\left[-\frac{580}{299},1.682973122\right],$$
$$\left.\left[-\frac{2311}{1196},-1.053082612\right]\right]$$

returns a list of lists of ${f_c}^n(0)$ for $n = 101, 102, \ldots, 200$ for 300 equally spaced values of c between -2 and 1. The list `lgtable` is plotted with `plot`. See Figure 4-29 and compare this result to the result obtained in Example 4.2.5.

```
> plot(lgtable,style=point,symbol=point,color=black,
> view=[-2..0.5,-2..2]);
```

For a given complex number c the **Julia set**, J_c, of $f_c(x) = x^2 + c$ is the set of complex numbers, $z = a + bi$, a, b real, for which the sequence z, $f_c(z) = z^2 + c$, $f_c\left(f_c(z)\right) = \left(z^2 + c\right)^2 + c, \ldots, {f_c}^n(z), \ldots$, does *not* tend to ∞ as $n \to \infty$:

We use the notation $f^n(x)$ to represent the composition $\underbrace{(f \circ f \circ \cdots \circ f)}_{n}(x)$.

$$J_c = \left\{z \in \mathbf{C} | z,\ z^2 + c\ \left(z^2 + c\right)^2 + c, \ldots \not\to \infty\right\}.$$

Using a dynamical system, setting $z = z_0$ and computing $z_{n+1} = f_c(z_n)$ for large n can help us determine if z is an element of J_c. In terms of a composition, computing ${f_c}^n(z)$ for large n can help us determine if z is an element of J_c.

EXAMPLE 4.4.7 (Julia Sets): Plot the Julia set of $f_c(x) = x^2 + c$ if $c = -0.122561 + 0.744862i$.

You do not need to redefine $f_c(x)$ if you have already defined it during your current Maple session.

SOLUTION: After defining $f_c(x) = x^2 + c$, we use `seq` together with the repeated composition operator `@@` to compute ordered triples of the form

$$\left(x, y, \left| {f_{-0.122561+0.744862i}}^{200}(x+iy)\right|\right)$$

for 150 equally spaced values of x between $-3/2$ and $3/2$ and 150 equally spaced values of y between $-3/2$ and $3/2$.

```
> f:=c->proc(x) evalf(x^2+c) end proc:
> xvals:=[seq(-3./2+3/149*i,i=0..149)]:
> yvals:=[seq(-3./2+3/149*i,i=0..149)]:
> g1:=[seq(seq([x,y,abs((f(-0.122561+0.744862*I)
   @@200)(x+I*y))],x=xvals),y=yvals)];
```

We select those elements of `g1` for which the third coordinate is less than ∞ with `select`, then extract a list of the first two coordinates, (x, y), from the elements of `g2` with `map`, and plot the resulting list of points in Figure 4-30 using `plot`.

```
> g2:=select(x->evalb(x[3]<Float(infinity)),g1);
> g2b:=map(x->[x[1],x[2]],g2);
> plot(g2b,style=point,color=black,symbol=point);
```

■

Of course, one can consider functions other than $f_c(x) = x^2 + c$ as well as rearrange the order in which we carry out the computations.

EXAMPLE 4.4.8 (Julia Sets): Plot the Julia set for $f(z) = .36e^z$.

SOLUTION: For this example, we begin by forming our complex grid first in `avals` and `bvals` using `seq`. We will use these results to form a list of numbers of the form $a + bi$ for 150 equally spaced values of a

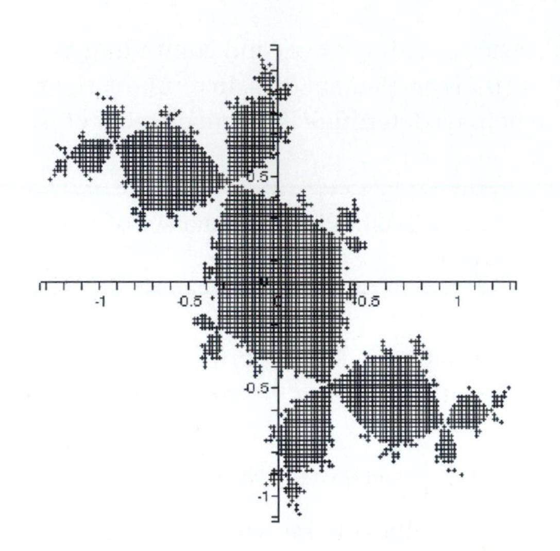

Figure 4-30 Filled Julia set for f_C

between 0 and 5 and 150 equally spaced values of b between -2.5 and 2.5.

```
> avals:=[seq(5./149*i,i=0..149)]:
> bvals:=[seq(-2.5+5/149*i,i=0..149)]:
```

After defining $f(z)$, we use `seq` to compute the ordered triple $\left(a, b, \left| f^{200}(a+bi)\right|\right)$ for each a in `avals` and each b in `bvals`.

```
> f:=z->evalf(0.36*exp(z)):
> t1:=[seq(seq([a,b,abs((f@@200)(a+b*I))],
   a=avals),b=bvals)]:
```

We then use `select` to extract those elements of `t1` for which the third coordinate is less than (complex) ∞ in `t2`. The first two coordinates of each point in `t2` are obtained in `t2b` with `map`. The resulting list of ordered pairs is plotted with `plot` and shown in Figure 4-31.

```
> t2:=select(x->evalb(x[3]<Float(infinity)),t1):
> t2b:=map(x->[x[1],x[2]],t2):
> plot(t2b,style=point,color=black,symbol=point);
```

■

You have even greater control over your graphics if you use `select` to extract points meeting specified criteria and then plot the results in different colors and/or gray levels.

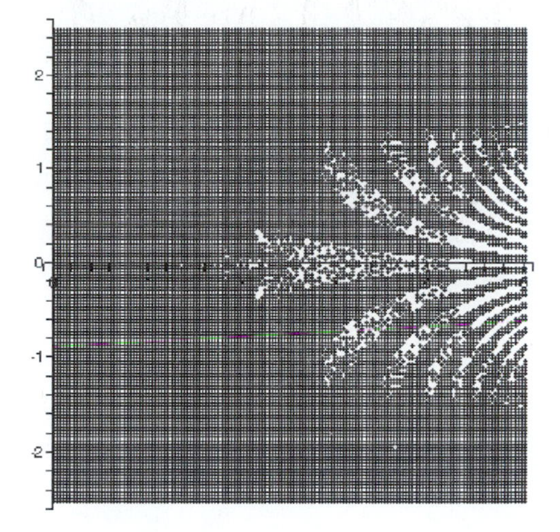

Figure 4-31 The Julia set of $f(z) = .36e^z$: the black points (a, b) are the points for which $f^{200}(a + bi)$ is finite

---

EXAMPLE 4.4.9 (Julia Sets): Plot the Julia set for $f_c(z) = z^2 - cz$ if $c = 0.737369 + 0.67549i$.

SOLUTION: We proceed as in Example 4.4.7.

```
> f:=c->proc(x) evalf(x^2-c*x) end proc:

> pts:=[seq(seq([-1.2+2.95/199*i,-0.7+2.1/199*j],
   i=0..199),j=0..199)]:

> g1:=[seq([v[1],v[2],abs((f(0.737369+0.67549*I)
   @@100)(v[1]+v[2]*I))],
> v=pts)]:

> g2:=select(v->evalb(v[3]<Float(infinity)),g1):
> nops(g2);
```

$$11213$$

After selecting the points with third coordinate less than `Float(infinity)`, we plot the remaining ones according to their distance from the origin. The effects of using various shades of gray are show in Figure 4-32.

```
> g2a:=select(v->evalb(v[3]<0.125),g1):
> g3a:=map(x->[x[1],x[2]],g2a):
```

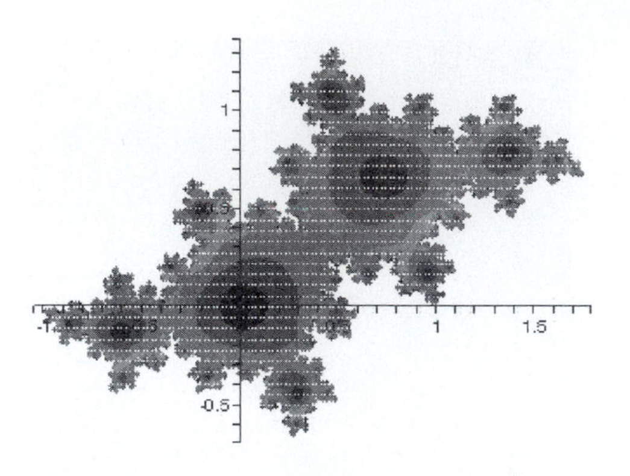

Figure 4-32 The lightest points (a, b) are the ones for which $\left| f_{0.737369+0.67549i}{}^{100}(z) \right|$ is the largest

```
> g2b:=select(v->evalb(v[3]>0.125 and v[3]<0.25),g1):
> g3b:=map(x->[x[1],x[2]],g2b):
> g2c:=select(v->evalb(v[3]>0.25 and v[3]<0.375),g1):
> g3c:=map(x->[x[1],x[2]],g2c):
> g2d:=select(v->evalb(v[3]>0.375 and v[3]<0.5),g1):
> g3d:=map(x->[x[1],x[2]],g2d):
> with(plots):
> p1:=plot(g3a,style=point,symbol=point,
   color=COLOR(RGB,0,0,0)):
> p2:=plot(g3b,style=point,symbol=point,
   color=COLOR(RGB,.15,.15,.15)):
> p3:=plot(g3c,style=point,symbol=point,
   color=COLOR(RGB,.30,.30,.30)):
> p4:=plot(g3d,style=point,symbol=point,
   color=COLOR(RGB,.45,.45,.45)):
> display(p1,p2,p3,p4,scaling=constrained);
```

■

EXAMPLE 4.4.10 (The Ikeda Map): The **Ikeda map** is defined by

$$\mathbf{F}(x, y) = \langle \gamma + \beta \left(x \cos \tau - y \sin \tau \right), \beta \left(x \sin \tau + y \cos \tau \right) \rangle, \qquad (4.11)$$

where $\tau = \mu - \alpha / \left(1 + x^2 + y^2\right)$. If $\beta = .9$, $\mu = .4$, and $\alpha = 4.0$, plot the *basins of attaction* for F if $\gamma = .92$ and $\gamma = 1.0$.

SOLUTION: The *basins of attraction* for F are the set of points (x, y) for which $\|\mathbf{F}^n(x,y)\| \nrightarrow \infty$ as $n \to \infty$.

After defining `f(γ)([x,y])` to be (4.11) and then $\beta = .9$, $\mu = .4$, and $\alpha = 4.0$, we use `seq` to define `pts` to be the list of 22,500 ordered pairs (x, y) for 150 equally spaced values of x between -2.3 and 1.3 and 150 equally spaced values of y between -2.8 and .8.

```
> f:=gamma->proc(v) evalf(
> [gamma+beta*(v[1]*cos(mu-alpha/(1+v[1]^2+v[2]^2))-
> v[2]*sin(mu-alpha/(1+v[1]^2+v[2]^2))),
> beta*(v[1]*sin(mu-alpha/(1+v[1]^2+v[2]^2))+
> v[2]*cos(mu-alpha/(1+v[1]^2+v[2]^2)))]) end proc:

> beta:=0.9:mu:=0.4:alpha:=4.0:

> xvals:=[seq(-2.3+3.6/149*i,i=0..149)]:
> yvals:=[seq(-2.8+3.6/149*i,i=0..149)]:
> pts:=[seq(seq([x,y],x=xvals),y=yvals)]:

> nops(pts);
```

$$22500$$

In `l1`, we use `seq` to compute $\left(x, y, \mathbf{F}_{.92}{}^{25}(x,y)\right)$ for each (x, y) in `pts`. In `l2`, we convert the norm of the third component of each element of `l1`.

```
> l1:=[seq([v[1],v[2],(f(.92)@@25)(v)],v=pts)]:

> g:=v->[v[1],v[2],sqrt(v[3,1]^2+v[3,2]^2)]:
> l2:=map(g,l1):

> maxl2:=max(seq(l2[i,3],i=1..nops(l2)));
```

$$maxl2 := 5.829963954$$

and see that the maximum norm is approximately 5.83 with `max`.

We then select and plot those points with norm less than 1

```
> t1:=select(x->evalb(x[3]<1),l2):
> nops(t1);
> t1b:=map(x->[x[1],x[2]],t1):
```

$$8505$$

and those points with norm between 1 and 2

```
> t2:=select(x->evalb(1 < x[3] and x[3] < 2),l2):
> nops(t2);
```

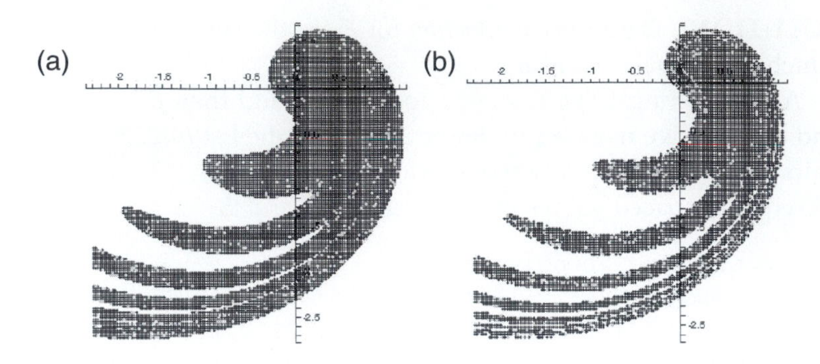

Figure 4-33 Basins of attaction for **F** if (a) $\gamma = .92$ and (b) $\gamma = 1.0$

```
> t2b:=map(x->[x[1],x[2]],t2):
```

407

and plot the results. The resulting plots are displayed together in Figure 4-33(a).

```
> p1:=plot(t1b,style=point,symbol=point,color=black):
> p2:=plot(t2b,style=point,symbol=point,color=gray):
> with(plots):
> display(p1,p2);
```

For $\gamma = 1.0$, we proceed in the same way (Figure 4-33(b)).

```
> l1b:=[seq([v[1],v[2],(f(1.0)@@25)(v)],v=pts)]:
> l2b:=map(g,l1b):
> t1b:=select(x->evalb(x[3]<1),l2b):
> t1bb:=map(x->[x[1],x[2]],t1b):
> t2b:=select(x->evalb(1 < x[3] and x[3] < 2),l2b):
> t2bb:=map(x->[x[1],x[2]],t2b):
> p1b:=plot(t1bb,style=point,symbol=point,
   color=black):
> p2b:=plot(t2bb,style=point,symbol=point,
   color=gray):
> display(p1b,p2b);
```

■

The **Mandelbrot set**, M, is the set of complex numbers, $z = a + bi$, a, b real, for which the sequence z, $f_z(z) = z^2 + z$, $f_z(f_z(z)) = (z^2 + z)^2 + z, \ldots, f_z{}^n(z), \ldots,$ does *not* tend to ∞ as $n \to \infty$:

$$M = \left\{ z \in \mathbf{C} | z,\ z^2 + z\ \left(z^2 + z\right)^2 + z, \ldots \nrightarrow \infty \right\}.$$

Using a dynamical system, setting $z = z_0$ and computing $z_{n+1} = f_{z_0}(z_n)$ for large n can help us determine if z is an element of M. In terms of a composition, computing $f_z{}^n(z)$ for large n can help us determine if z is an element of M.

EXAMPLE 4.4.11 (Mandelbrot Set): Plot the Mandelbrot set.

SOLUTION: We proceed as in Example 4.4.7 except that instead of iterating $f_c(z)$ for fixed c we iterate $f_z(z)$.

```
> f:=c->proc(x) evalf(x^2+c) end proc:
> pts:=[seq(seq([-1.5+5/(2*149)*i,-1.0+2/149*j],
   i=0..149),j=0..149)]:
> nops(pts);
```

22500

```
> g1:=[seq([v[1],v[2],abs((f(v[1]+v[2]*I)
   @@100)(v[1]+v[2]*I))],v=pts)]:
```

The following gives us the image on the left in Figure 4-34.

```
> g2:=select(v->evalb(v[3]<Float(infinity)),g1):
> g3:=map(x->[x[1],x[2]],g2):
```

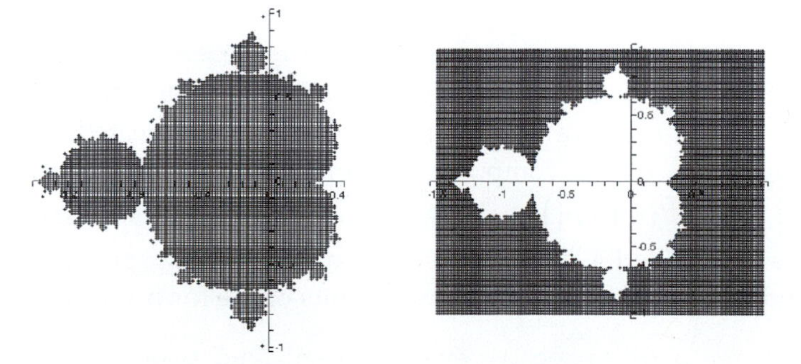

Figure 4-34 Two different views of the Mandelbrot set: on the left, the black points (a, b) are the points for which $f_{a+bi}{}^{100}(a + bi)$ is finite; on the right, the black points (a, b) are the ones for which $f_{a+bi}{}^{200}(a + bi)$ is not finite

```
> plot(g3,style=point,symbol=point,color=black,
   scaling=constrained);
```

To invert the image, we use the following to obtain the result on the right in Figure 4-34.

```
> g2b:=select(v->evalb(v[3]=Float(infinity)),g1):
> nops(g2b);
```

15512

```
> g3b:=map(x->[x[1],x[2]],g2b):
> plot(g3b,style=point,symbol=point,color=black,
   scaling=constrained);
```

■

In Example 4.4.11, the Mandelbrot set is obtained (or, more precisely, approximated) by repeatedly composing $f_z(z)$ for a grid of z-values and then deleting those for which the values exceed machine precision, `Float(infinity)`.

We can generalize by considering exponents other than 2 by letting $f_{p,c} = x^p + c$. The **generalized Mandelbrot set**, M_p, is the set of complex numbers, $z = a + bi$, a, b real, for which the sequence z, $f_{p,z}(z) = z^p + z$, $f_{p,z}\left(f_{p,z}(z)\right) = (z^p + z)^p + z, \ldots,$ $f_{p,z}{}^n(z), \ldots,$ does *not* tend to ∞ as $n \to \infty$:

$$M_p = \left\{z \in \mathbf{C} | z,\, z^p + z\, \left(z^p + z\right)^p + z,\, \ldots \nrightarrow \infty\right\}.$$

Using a dynamical system, setting $z = z_0$ and computing $z_{n+1} = f_p\,(z_n)$ for large n can help us determine if z is an element of M_p. In terms of a composition, computing $f_p{}^n(z)$ for large n can help us determine if z is an element of M_p.

EXAMPLE 4.4.12 (Generalized Mandelbrot Set): After defining $f_{p,c}(x) = x^p + c$, we use `seq`, `abs`, and the repeated composition operator, `@@`, to compute a list of ordered triples of the form $\left(x, y, \left| f_{p,x+iy}{}^{100}(x + iy) \right| \right)$ for p-values from 1.625 to 2.625 spaced by equal values of 1/3 and 125 values of $x(y)$ values equally spaced between −2 and 2, resulting in 15,625 sample points of the form $x + iy$.

```
> f:=(p,c)->proc(x) evalf(x^p+c) end proc:
> pts:=[seq(seq([-2.0+4/124*i,-2.0+4/124*j],
   i=0..124),j=0..124)]:
> pvals:=[seq(1.625+i/3,i=0..3)]:
```

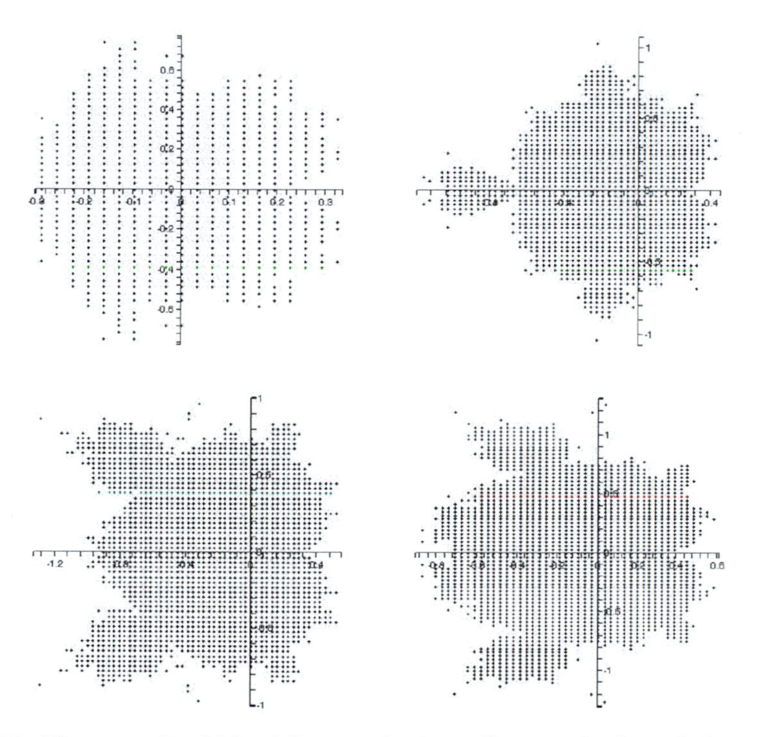

Figure 4-35 The generalized Mandelbrot set for 4 equally spaced values of p between 1.625 and 2.625

```
> nops(pts);
```

$$15625$$

```
> h:=proc(p,v) local i;
> for i from 1 to 24 while evalb(abs((f(p,v[1]+v[2]
   *I)@@i)(v[1]+v[2]*I))<10^10)
> do min(abs((f(p,v[1]+v[2]*I)@@(i+1))(v[1]+v[2]*I)),
   10^10) end do
> end proc:
```

```
> g1:=[seq([seq([v[1],v[2],h(p,v)],v=pts)],p=pvals)]:
```

Next, we select those points for which the third coordinate is less than 10^{10} with `seq` and `select`, ordered pairs of the first two coordinates are obtained in `g3`. The resulting lists of points are plotted with `plot` in Figure 4-35.

```
> g2:=[seq(select(x->evalb(x[3]<10^10),g1[i]),
   i=1..4)]:
```

```
> g3:=[seq(map(x->[x[1],x[2]],g2[i]),i=1..4)]:
> for i from 1 to 4 do plot(g3[i],style=point,
   symbol=point,color=black) end do;
```

Throughout these examples, we have typically computed the iteration $f^n(z)$ for "large" n, like values of n between 25 and 100. To indicate why we have selected those values of n, we revisit the Mandelbrot set plotted in Example 4.4.11.

EXAMPLE 4.4.13 (Mandelbrot Set): We proceed in essentially the same way as in the previous examples. After defining $f_c(x) = x^2 + c$,

```
> f:=c->proc(x) evalf(x^2+c) end proc:
> pts:=[seq(seq([-1.5+5/(2*149)*i,-1.0+2/149*j],
   i=0..149),j=0..149)]:
> nops(pts);
```

$$22500$$

we use `seq` to create a nested list. For each $n = 5, 10, 15, 25, 50$, and 100, a nested list is formed for 150 equally spaced values of y between -1 and 1 and then 150 equally spaced values of x between -1.5 and 1. At the bottom level of each nested list, the elements are of the form $\left(x, y, \left| f_{x+iy}{}^n(x+iy) \right|\right)$.

```
> g1:=[seq(
> [seq([v[1],v[2],abs((f(v[1]+v[2]*I)@@k)
   (v[1]+v[2]*I))],
> v=pts)],k=[5,10,15,25,50,100])]:
> nops(g1);
```

$$6$$

We then select those points for which the third coordinate, $\left| f_{2,x+iy}{}^n(x+iy) \right|$, is less than `Float(infinity)`,

```
> g2:=[seq(select(v->evalb(v[3]<Float(infinity)),
   g1[i]),i=1..6)]:
```

extract (x, y) from the remaining ordered triples

```
> g3:=[seq(map(x->[x[1],x[2]],g2[i]),i=1..6)]:
```

and graph the resulting sets of points using `plot` in Figure 4-36. As shown in Figure 4-36, we see that Maple's numerical precision provides decent plots are obtained when $n = 50$ or $n = 100$.

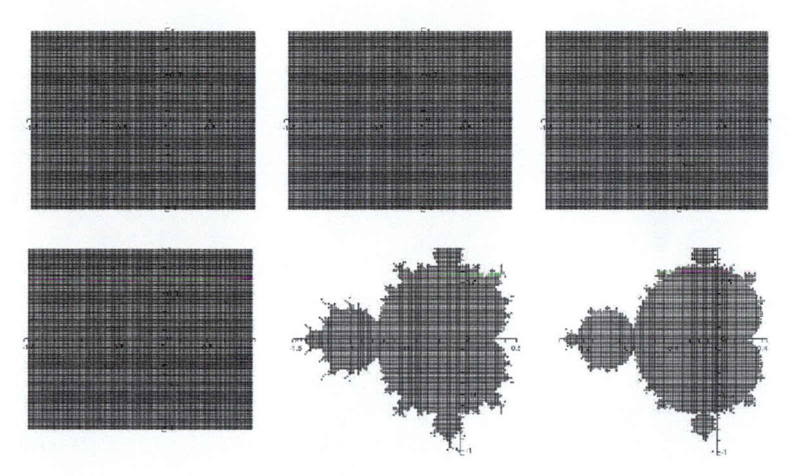

Figure 4-36 Without shading the points, the effects of iteration are difficult to see until the number of iterations is "large"

```
> for i from 1 to 6
> do plot(g3[i],style=point,symbol=point,color=black,
   scaling=constrained) end do;
```

If instead, we shade each point (x, y) according to $\left| f_{x+iy}{}^{n}(x+iy) \right|$ detail emerges quickly as shown in Figure 4-37.

```
> g2a:=[seq(select(v->evalb(v[3]<.125),g1[i]),
   i=1..6)]:
> g2b:=[seq(select(v->evalb(v[3]>.125 and v[3]<.25),
   g1[i]),i=1..6)]:
> g2c:=[seq(select(v->evalb(v[3]>.25 and v[3]<.375),
   g1[i]),i=1..6)]:
> g2d:=[seq(select(v->evalb(v[3]>.375 and v[3]<.5),
   g1[i]),i=1..6)]:

> g3a:=[seq(map(x->[x[1],x[2]],g2a[i]),i=1..6)]:
> g3b:=[seq(map(x->[x[1],x[2]],g2b[i]),i=1..6)]:
> g3c:=[seq(map(x->[x[1],x[2]],g2c[i]),i=1..6)]:
> g3d:=[seq(map(x->[x[1],x[2]],g2d[i]),i=1..6)]:

> with(plots):
> p1a:=plot(g3a[1],style=point,symbol=point,
   color=COLOR(RGB,0,0,0),
> scaling=constrained):
> p1b:=plot(g3b[1],style=point,symbol=point,
   color=COLOR(RGB,.2,.2,.2),
> scaling=constrained):
```

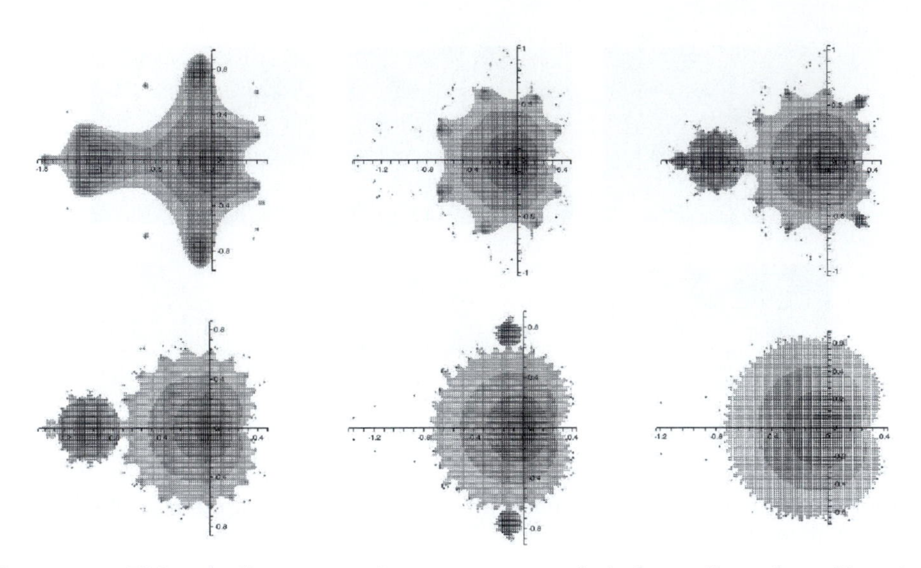

Figure 4-37 Using shading, we see that we can use a relatively small number of iterations to visualize the Mandelbrot set

```
> p1c:=plot(g3c[1],style=point,symbol=point,
   color=COLOR(RGB,.4,.4,.4),
> scaling=constrained):
> p1d:=plot(g3d[1],style=point,symbol=point,
   color=COLOR(RGB,.6,.6,.6),
> scaling=constrained):
> display(p1a,p1b,p1c,p1d);

> p2a:=plot(g3a[2],style=point,symbol=point,
   color=COLOR(RGB,0,0,0),
> scaling=constrained):
> p2b:=plot(g3b[2],style=point,symbol=point,
   color=COLOR(RGB,.2,.2,.2),
> scaling=constrained):
> p2c:=plot(g3c[2],style=point,symbol=point,
   color=COLOR(RGB,.4,.4,.4),
> scaling=constrained):
> p2d:=plot(g3d[2],style=point,symbol=point,
   color=COLOR(RGB,.6,.6,.6),
> scaling=constrained):
> display(p2a,p2b,p2c,p2d);
> p3a:=plot(g3a[3],style=point,symbol=point,
   color=COLOR(RGB,0,0,0),
> scaling=constrained):
```

```
> p3b:=plot(g3b[3],style=point,symbol=point,
   color=COLOR(RGB,.2,.2,.2),
> scaling=constrained):
> p3c:=plot(g3c[3],style=point,symbol=point,
   color=COLOR(RGB,.4,.4,.4),
> scaling=constrained):
> p3d:=plot(g3d[3],style=point,symbol=point,
   color=COLOR(RGB,.6,.6,.6),
> scaling=constrained):
> display(p3a,p3b,p3c,p3d);

> p4a:=plot(g3a[4],style=point,symbol=point,
   color=COLOR(RGB,0,0,0),
> scaling=constrained):
> p4b:=plot(g3b[4],style=point,symbol=point,
   color=COLOR(RGB,.2,.2,.2),
> scaling=constrained):
> p4c:=plot(g3c[4],style=point,symbol=point,
   color=COLOR(RGB,.4,.4,.4),
> scaling=constrained):
> p4d:=plot(g3d[4],style=point,symbol=point,
   color=COLOR(RGB,.6,.6,.6),
> scaling=constrained):
> display(p4a,p4b,p4c,p4d);

> p5a:=plot(g3a[5],style=point,symbol=point,
   color=COLOR(RGB,0,0,0),
> scaling=constrained):
> p5b:=plot(g3b[5],style=point,symbol=point,
   color=COLOR(RGB,.2,.2,.2),
> scaling=constrained):
> p5c:=plot(g3c[5],style=point,symbol=point,
   color=COLOR(RGB,.4,.4,.4),
> scaling=constrained):
> p5d:=plot(g3d[5],style=point,symbol=point,
   color=COLOR(RGB,.6,.6,.6),
> scaling=constrained):
> display(p5a,p5b,p5c,p5d);

> p6a:=plot(g3a[6],style=point,symbol=point,
   color=COLOR(RGB,0,0,0),
> scaling=constrained):
> p6b:=plot(g3b[6],style=point,symbol=point,
   color=COLOR(RGB,.2,.2,.2),
> scaling=constrained):
```

```
> p6c:=plot(g3c[6],style=point,symbol=point,
   color=COLOR(RGB,.4,.4,.4),
> scaling=constrained):
> p6d:=plot(g3d[6],style=point,symbol=point,
   color=COLOR(RGB,.6,.6,.6),
> scaling=constrained):
> display(p6a,p6b,p6c,p6d);
```

Thus, these figures indicate that for examples like the ones illustrated here similar results could have been accomplished using far smaller values of n than $n = 100$ or $n = 200$. With fast machines, the difference in the time needed to perform the calculations is minimal; $n = 100$ and $n = 200$ appear to be "safe" large values of n for well-studied examples like these.

Not even 10 years ago calculations like these required the use of a supercomputer and sophisticated computer programming. Now, they are accessible to virtually anyone working on a relatively new machine with just a few lines of Maple code. Quite amazing!

Matrices and Vectors: Topics from Linear Algebra and Vector Calculus

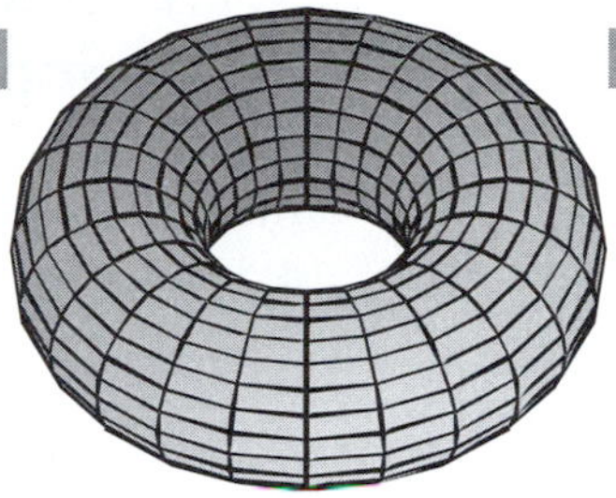

5

Chapter 5 discusses operations on matrices and vectors, including topics from linear algebra, linear programming, and vector calculus.

The `LinearAlgebra` and `linalg` packages have extensive and sophisticated capabilities with nested lists, matrices, and vectors. A brief overview of the capabilities of each package are described on the main `LinearAgebra` and `linalg` help windows, which are obtained by entering `?LinearAlgebra` and `?linalg`, respectively.

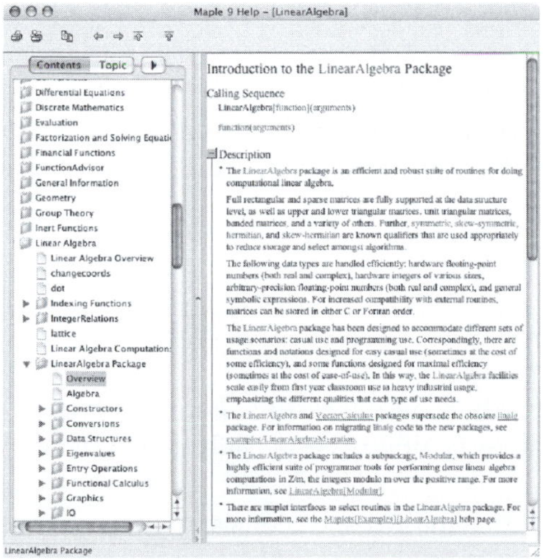

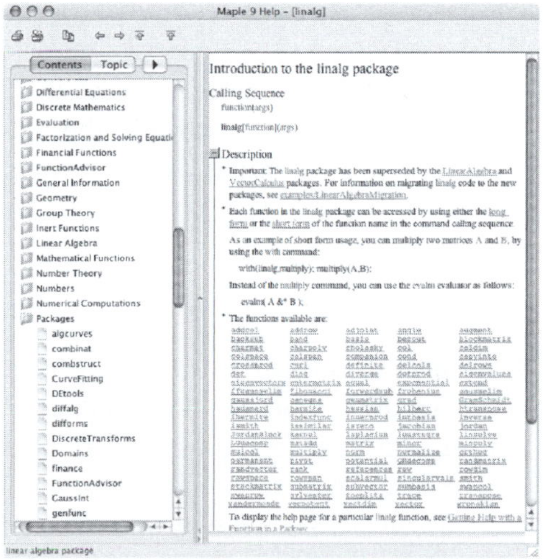

5.1 Nested Lists: Introduction to Matrices, Vectors, and Matrix Operations

5.1.1 Defining Nested Lists, Matrices, and Vectors

In Maple, a **matrix** is a list of lists where each list represents a row of the matrix or a data structure of type `Matrix`. Thus, the $m \times n$ matrix

$$\mathbf{A} = \begin{pmatrix} a_{11} & a_{12} & a_{13} & \cdots & a_{1n} \\ a_{21} & a_{22} & a_{23} & \cdots & a_{2n} \\ a_{31} & a_{32} & a_{33} & \cdots & a_{3n} \\ \vdots & \vdots & \vdots & & \vdots \\ a_{m1} & a_{m2} & a_{m3} & \cdots & a_{mn} \end{pmatrix}$$

can be entered with

```
A:=array(1..n,1..m),
```

if each entry is not assigned a value immediately, or with either

```
A:=array(1..m,1..m,list_of_rows) or A:=array(list_of_rows),
```

where `list_of_rows` is a list of lists in which the ith list in `list_of_rows` corresponds to the entries in the ith row of the matrix **A**, if the entries of **A** are assigned a value immediately. In this case, the resulting data structure is essentially a nested list.

Alternatively,

```
A:=Matrix(1..m,1..m,list_of_rows) or A:=Matrix(list_of_rows),
```

results in a data structure of type `Matrix`.

For the beginner, either method will produce decent results. For the advanced user, the `Matrix` data structure offers more flexibility, especially when dealing with large matrices.

Regardless of the way you choose to define matrices, be consistent throughout your calculations as the two data types are not interchangeable.

In both situations, use `evalm` to evaluate expressions involving matrices.

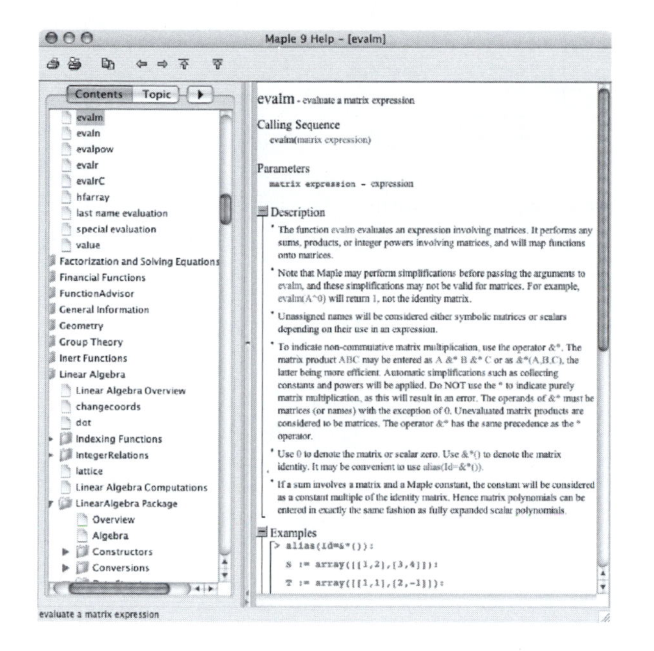

For example, to use Maple to define **A** to be the matrix $\mathbf{A} = \begin{pmatrix} a_{11} & a_{12} \\ a_{21} & a_{22} \end{pmatrix}$ enter the command

```
> A:=array([[a[1,1],a[1,2]],[a[2,1],a[2,2]]]);
```

$$A := \begin{bmatrix} a_{1,1} & a_{1,2} \\ a_{2,1} & a_{2,2} \end{bmatrix}$$

```
> whattype(A);
```

$$symbol$$

The resulting data structure is a nested list.

Entering

```
> evalm(A^2);
```

$$\begin{bmatrix} {a_{1,1}}^2 + a_{1,2}a_{2,1} & a_{1,1}a_{1,2} + a_{1,2}a_{2,2} \\ a_{2,1}a_{1,1} + a_{2,2}a_{2,1} & a_{1,2}a_{2,1} + {a_{2,2}}^2 \end{bmatrix}$$

computes $\mathbf{A}^2$ and, after the `linalg` package has been loading,

```
> with(linalg):
> inverse(A);
```

$$\begin{bmatrix} \frac{a_{2,2}}{a_{1,1}a_{2,2} - a_{1,2}a_{2,1}} & -\frac{a_{1,2}}{a_{1,1}a_{2,2} - a_{1,2}a_{2,1}} \\ -\frac{a_{2,1}}{a_{1,1}a_{2,2} - a_{1,2}a_{2,1}} & \frac{a_{1,1}}{a_{1,1}a_{2,2} - a_{1,2}a_{2,1}} \end{bmatrix}$$

returns the inverse of **A**, $\mathbf{A}^{-1}$, assuming the inverse of **A** exists.

Alternatively, you can construct matrices so that the resulting data type is of type `Matrix` with `Matrix`. Thus,

```
> B:=Matrix(2,2,[[b[1,1],b[1,2]],[b[2,1],b[2,2]]]);
```

$$B := \begin{bmatrix} b_{1,1} & b_{1,2} \\ b_{2,1} & b_{2,2} \end{bmatrix}$$

defines $\mathbf{B} = \begin{pmatrix} b_{11} & b_{12} \\ b_{21} & b_{22} \end{pmatrix}$, which is a data set of type `Matrix`.

```
> whattype(B);
```

symbol

Use `evalm` to evaluate computations involving matrices. So,

```
> evalm(B^2);
```

$$\begin{bmatrix} {b_{1,1}}^2 + b_{1,2}b_{2,1} & b_{1,1}b_{1,2} + b_{1,2}b_{2,2} \\ b_{2,1}b_{1,1} + b_{2,2}b_{2,1} & b_{1,2}b_{2,1} + {b_{2,2}}^2 \end{bmatrix}$$

computes $\mathbf{B}^2$. And, since we have already loaded the `linalg` package,

```
> inverse(B);
```

$$\begin{bmatrix} \frac{b_{2,2}}{b_{1,1}b_{2,2} - b_{1,2}b_{2,1}} & -\frac{b_{1,2}}{b_{1,1}b_{2,2} - b_{1,2}b_{2,1}} \\ -\frac{b_{2,1}}{b_{1,1}b_{2,2} - b_{1,2}b_{2,1}} & \frac{b_{1,1}}{b_{1,1}b_{2,2} - b_{1,2}b_{2,1}} \end{bmatrix}$$

computes the inverse of **B**, assuming the inverse of **B** exists. However, when dealing with `Matrix` objects, the `LinearAlgebra` package is preferred. After loading the `LinearAlgebra` package, we see that `MatrixInverse` finds the inverse of **B** (a `Matrix` object) but not the inverse of **A** (a nested list).

```
> with(LinearAlgebra):
```

```
> MatrixInverse(B);
```

$$\begin{bmatrix} \dfrac{b_{2,2}}{b_{1,1}b_{2,2}-b_{1,2}b_{2,1}} & -\dfrac{b_{1,2}}{b_{1,1}b_{2,2}-b_{1,2}b_{2,1}} \\ -\dfrac{b_{2,1}}{b_{1,1}b_{2,2}-b_{1,2}b_{2,1}} & \dfrac{b_{1,1}}{b_{1,1}b_{2,2}-b_{1,2}b_{2,1}} \end{bmatrix}$$

```
> MatrixInverse(A);

Error, (in MatrixInverse) invalid input: MatrixInverse expects
its 1st argument, M, to be of type {Matrix, list} but received A
```

Objects of type `Matrix` can also be constructed using the < . . . > shortcut. Thus,

```
> B:=<<b[1,1] | b[1,2]>,<b[2,1] | b[2,2]>>;
```

$$B := \begin{bmatrix} b_{1,1} & b_{1,2} \\ b_{2,1} & b_{2,2} \end{bmatrix}$$

```
> MatrixPower(B,2);
```

$$\begin{bmatrix} {b_{1,1}}^2 + b_{1,2}b_{2,1} & b_{1,1}b_{1,2} + b_{1,2}b_{2,2} \\ b_{2,1}b_{1,1} + b_{2,2}b_{2,1} & b_{1,2}b_{2,1} + {b_{2,2}}^2 \end{bmatrix}$$

returns the same results as those obtained previously.

Arrays and matrices do not need to be square. Thus,

```
> C:=
> <<c[1,1] | c[1,2] | c[1,3] | c[1,4]>,
  <c[2,1]|c[2,2]|c[2,3]|c[2,4]>>:
```

$$C := \begin{bmatrix} c_{1,1} & c_{1,2} & c_{1,3} & c_{1,4} \\ c_{2,1} & c_{2,2} & c_{2,3} & c_{2,4} \end{bmatrix}$$

and

```
> Matrix(2,4,symbol=c);
```

$$\begin{bmatrix} c_{1,1} & c_{1,2} & c_{1,3} & c_{1,4} \\ c_{2,1} & c_{2,2} & c_{2,3} & c_{2,4} \end{bmatrix}$$

both return the 2×4 matrix $\mathbf{C} = \begin{pmatrix} c_{11} & c_{12} & c_{13} & c_{14} \\ c_{21} & c_{22} & c_{23} & c_{24} \end{pmatrix}$. Alternatively, using `Matrix` in a slightly different manner, we see that

```
> C:=Matrix(2,4,c):
```

$$C := \begin{bmatrix} c(1,1) & c(1,2) & c(1,3) & c(1,4) \\ c(2,1) & c(2,2) & c(2,3) & c(2,4) \end{bmatrix}$$

returns an equivalent result.

More generally the command

```
array([seq([seq(f(i,j),j=1..m)],i=1..n)])
```

yields the nested list (or array) corresponding to the $n \times m$ matrix

$$\begin{pmatrix} f(1,1) & f(1,2) & \cdots & f(1,m) \\ f(2,1) & f(2,2) & \cdots & f(2,m) \\ \vdots & \vdots & \vdots & \vdots \\ f(n,1) & f(n,2) & \cdots & f(n,m) \end{pmatrix}$$

while the command

```
Matrix(n,m,f)
```

returns the $n \times m$ matrix $\begin{pmatrix} f(1,1) & f(1,2) & \cdots & f(1,m) \\ f(2,1) & f(2,2) & \cdots & f(2,m) \\ \vdots & \vdots & \vdots & \vdots \\ f(n,1) & f(n,2) & \cdots & f(n,m) \end{pmatrix}$.

EXAMPLE 5.1.1: Define **E** to be the 3×4 matrix (c_{ij}), where c_{ij}, the entry in the ith row and jth column of **E**, is the value of $\cos\left(j^2 - i^2\right)\sin\left(i^2 - j^2\right)$.

SOLUTION: We define `e(i,j)` to be the value of $\cos\left(j^2 - i^2\right)\sin\left(i^2 - j^2\right)$ and then use `Matrix` to compute the 3×4 matrix `E`.

```
> e:=(i,j)->cos(j^2-i^2)*sin(i^2-j^2):
> E:=Matrix(3,4,e);
```

$$E := \begin{bmatrix} 0 & -\cos(3)\sin(3) & -\cos(8)\sin(8) & -\cos(15)\sin(15) \\ \cos(3)\sin(3) & 0 & -\cos(5)\sin(5) & -\cos(12)\sin(12) \\ \cos(8)\sin(8) & \cos(5)\sin(5) & 0 & -\cos(7)\sin(7) \end{bmatrix}$$

Use `evalf` to approximate each entry.

```
> evalf(E);
```

$$\begin{bmatrix} 0.0 & 0.1397077491 & 0.1439516583 & 0.4940158121 \\ -0.1397077491 & 0.0 & 0.2720105555 & 0.4527891810 \\ -0.1439516583 & -0.2720105555 & 0.0 & -0.4953036778 \end{bmatrix}$$

On the other hand, using `array` and `seq`, we define `altE` to be an equivalent array.

```
> altE:=array([seq([seq(e(i,j),j=1..4)],i=1..3)]);
```

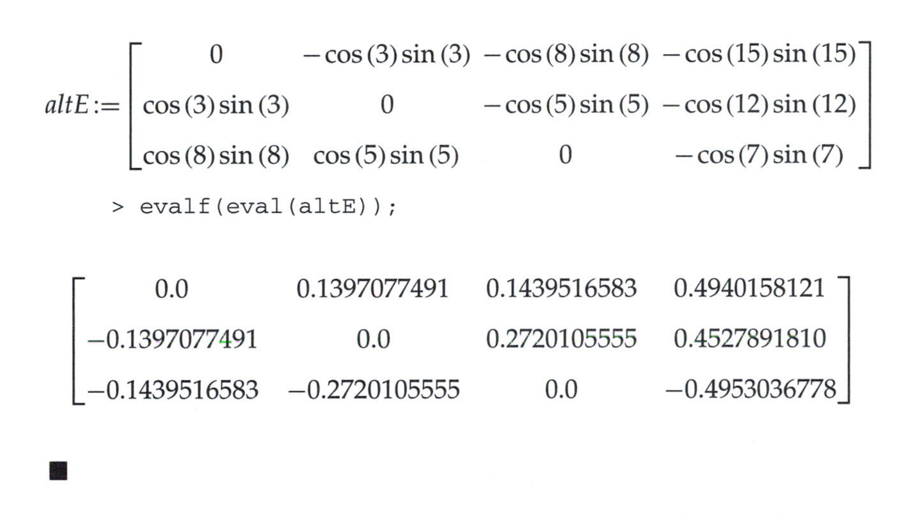

$$altE := \begin{bmatrix} 0 & -\cos(3)\sin(3) & -\cos(8)\sin(8) & -\cos(15)\sin(15) \\ \cos(3)\sin(3) & 0 & -\cos(5)\sin(5) & -\cos(12)\sin(12) \\ \cos(8)\sin(8) & \cos(5)\sin(5) & 0 & -\cos(7)\sin(7) \end{bmatrix}$$

```
> evalf(eval(altE));
```

$$\begin{bmatrix} 0.0 & 0.1397077491 & 0.1439516583 & 0.4940158121 \\ -0.1397077491 & 0.0 & 0.2720105555 & 0.4527891810 \\ -0.1439516583 & -0.2720105555 & 0.0 & -0.4953036778 \end{bmatrix}$$

■

EXAMPLE 5.1.2: Define the matrix $\mathbf{I}_3 = \begin{pmatrix} 1 & 0 & 0 \\ 0 & 1 & 0 \\ 0 & 0 & 1 \end{pmatrix}$.

SOLUTION: The matrix $\mathbf{I}_3$ is the 3×3 **identity matrix**. Generally, the $n \times n$ matrix with 1's on the diagonal and 0's elsewhere is the $n \times n$ identity matrix. The command `IdentityMatrix(n)`, which is contained in the `LinearAlgebra` package, returns the $n \times n$ identity matrix. Thus,

```
> with(LinearAlgebra):
```

```
> IdentityMatrix(3);
```

$$\begin{bmatrix} 1 & 0 & 0 \\ 0 & 1 & 0 \\ 0 & 0 & 1 \end{bmatrix}$$

returns $\mathbf{I}_3$.

■

In Maple, a **vector** can be viewed as a list of numbers and, thus, entered in the same manner as lists. For example, to use Maple to define the row vector `vectorv` to be $\begin{pmatrix} v_1 & v_2 & v_3 \end{pmatrix}$ enter

```
> vvec:=array([v[1],v[2],v[3]]);
```

$$vvec := [v_1, v_2, v_3]$$

Similarly, to define the column vector `vectorv` to be $\begin{pmatrix} v_1 \\ v_2 \\ v_3 \end{pmatrix}$ enter

```
> vvec:=array([[v[1]],[v[2]],[v[3]]]);
```

$$vvec := \begin{bmatrix} v_1 \\ v_2 \\ v_3 \end{bmatrix}$$

In these two cases, the end result is a list, which can be manipulated with commands contained in the `linalg` package but not with commands contained in the `LinearAlgebra` package.

Alternatively, you can use `Vector` to create a Maple object of type `Vector` that can be manipulated by commands contained in the `LinearAlgebra` and `linalg` packages. By default, `Vector` returns a column vector. Entering

```
> vvec:=Vector([v[1],v[2],v[3]]);
```

$$vvec := \begin{bmatrix} v_1 \\ v_2 \\ v_3 \end{bmatrix}$$

returns an object of type `Vector`; the result corresponds to the vector $\begin{pmatrix} v_1 \\ v_2 \\ v_3 \end{pmatrix}$.

For a row vector, specify `row` in the `Vector` command:

```
> vvec:=Vector[row]([v[1],v[2],v[3]]);
```

$$vvec := \begin{bmatrix} v_1 & v_2 & v_3 \end{bmatrix}$$

EXAMPLE 5.1.3: Define the vector $\mathbf{w} = \begin{pmatrix} -4 \\ -5 \\ 2 \end{pmatrix}$, `vectorv` to be the vector $\begin{pmatrix} v_1 & v_2 & v_3 & v_4 \end{pmatrix}$ and `zerovec` to be the vector $\begin{pmatrix} 0 & 0 & 0 & 0 & 0 \end{pmatrix}$.

SOLUTION: To define **w**, we enter

```
> w:=Vector([-4,-5,-2]);
```

$$w := \begin{bmatrix} -4 \\ -5 \\ -2 \end{bmatrix}$$

To define `vectorv`, we use `Vector[row]`.

```
> vvec:=Vector[row](4,symbol=v);
```

$$vvec := \begin{bmatrix} v_1 & v_2 & v_3 & v_4 \end{bmatrix}$$

To define `zerovec`, we use `ZeroVector`.

```
> with(LinearAlgebra):
> zerovec:=ZeroVector(5);
```

$$zerovec := \begin{bmatrix} 0 \\ 0 \\ 0 \\ 0 \\ 0 \end{bmatrix}$$

For a row vector, specify `row` in brackets.

```
> zerovec:=ZeroVector[row](5):
```

$$zerovec := \begin{bmatrix} 0 & 0 & 0 & 0 & 0 \end{bmatrix}$$

■

5.1.2 Extracting Elements of Matrices

Elements of matrices and vectors are extracted in the same way as they are for nested lists. Once you have defined the matrix **A**, `A[i,j]` returns the element in the ith row and jth column; `A[m..n,p..q]` returns the submatrix (a_{ij}), $i = m \dots n$, $j = p \dots q$. To illustrate, we define the 5×7 matrix **A** with `Matrix`.

```
> A:=Matrix(5,7,symbol=a);
```

$$A := \begin{bmatrix} a_{1,1} & a_{1,2} & a_{1,3} & a_{1,4} & a_{1,5} & a_{1,6} & a_{1,7} \\ a_{2,1} & a_{2,2} & a_{2,3} & a_{2,4} & a_{2,5} & a_{2,6} & a_{2,7} \\ a_{3,1} & a_{3,2} & a_{3,3} & a_{3,4} & a_{3,5} & a_{3,6} & a_{3,7} \\ a_{4,1} & a_{4,2} & a_{4,3} & a_{4,4} & a_{4,5} & a_{4,6} & a_{4,7} \\ a_{5,1} & a_{5,2} & a_{5,3} & a_{5,4} & a_{5,5} & a_{5,6} & a_{5,7} \end{bmatrix}$$

Entering

```
> A[3,4];
```

$$a_{3,4}$$

returns a_{34}, the entry in the third row and fourth column of **A**. On the other hand,

```
> A[2..5,4..6];
```

$$\begin{bmatrix} a_{2,4} & a_{2,5} & a_{2,6} \\ a_{3,4} & a_{3,5} & a_{3,6} \\ a_{4,4} & a_{4,5} & a_{4,6} \\ a_{5,4} & a_{5,5} & a_{5,6} \end{bmatrix}$$

returns the submatrix (a_{ij}), $i = 2, 3, 4, 5$, $j = 4, 5, 6$ of **A** while

```
> A[[1,2,3],[4,5,6]];
```

$$\begin{bmatrix} a_{1,4} & a_{1,5} & a_{1,6} \\ a_{2,4} & a_{2,5} & a_{2,6} \\ a_{3,4} & a_{3,5} & a_{3,6} \end{bmatrix}$$

returns the submatrix (a_{ij}), $i = 1, 2, 3$, $j = 4, 5, 6$ of **A**.

Rows and columns of a `Matrix` object can be extracted using the `Row` and `Column` commands contained in the `LinearAlgebra` package. After loading the `LinearAlgebra` package, we use `Row` and `Column`

```
> with(LinearAlgebra):
> Row(A,2);
```

```
> Column(A,1);
```

$$\begin{bmatrix} a_{2,1} & a_{2,2} & a_{2,3} & a_{2,4} & a_{2,5} & a_{2,6} & a_{2,7} \end{bmatrix}$$

$$\begin{bmatrix} a_{1,1} \\ a_{2,1} \\ a_{3,1} \\ a_{4,1} \\ a_{5,1} \end{bmatrix}$$

to extract the second row and first column of **A**. Similarly,

```
> [Row(A,3..5)];
```

$$\left[\begin{bmatrix} a_{3,1} & a_{3,2} & a_{3,3} & a_{3,4} & a_{3,5} & a_{3,6} & a_{3,7} \end{bmatrix}, \begin{bmatrix} a_{4,1} & a_{4,2} & a_{4,3} & a_{4,4} & a_{4,5} & a_{4,6} & a_{4,7} \end{bmatrix}, \begin{bmatrix} a_{5,1} & a_{5,2} & a_{5,3} & a_{5,4} & a_{5,5} & a_{5,6} & a_{5,7} \end{bmatrix}\right]$$

returns a list of rows 3 to 5 of **A** while

```
> [Column(A,[1,3,5])];
```

$$\left[\begin{bmatrix} a_{1,1} \\ a_{2,1} \\ a_{3,1} \\ a_{4,1} \\ a_{5,1} \end{bmatrix}, \begin{bmatrix} a_{1,3} \\ a_{2,3} \\ a_{3,3} \\ a_{4,3} \\ a_{5,3} \end{bmatrix}, \begin{bmatrix} a_{1,5} \\ a_{2,5} \\ a_{3,5} \\ a_{4,5} \\ a_{5,5} \end{bmatrix}\right]$$

returns a list consisting of **A**'s columns 1, 3, and 5.

EXAMPLE 5.1.4: Define `mb` to be the matrix $\begin{pmatrix} 10 & -6 & -9 \\ 6 & -5 & -7 \\ -10 & 9 & 12 \end{pmatrix}$. (a) Extract the third row of `mb`; (b) extract the element in the first row and third column of `mb`.

SOLUTION: We begin by defining `mb`. `mb[i,j]` yields the (unique) number in the *i*th row and *j*th column of `mb`.

```
> mb:=<<10|-6|-9>,<6|-5|-7>,<-10|9|12>>;
```

$$mb := \begin{bmatrix} 10 & -6 & -9 \\ 6 & -5 & -7 \\ -10 & 9 & 12 \end{bmatrix}$$

```
> mb[1,3];
```

$$-9$$

After loading the `LinearAlgebra` package, you can use `Row` and `Column` to extract rows and columns of a matrix. Thus,

```
> with(LinearAlgebra):
> Row(mb,1);
```

$$\begin{bmatrix} 10 & -6 & -9 \end{bmatrix}$$

returns the first row of `mb` and

```
> Column(mb,2);
```

$$\begin{bmatrix} -6 \\ -5 \\ 9 \end{bmatrix}$$

returns the second column of `mb`.

■

5.1.3 Basic Computations with Matrices

Maple performs all of the usual operations on matrices. Matrix addition ($\mathbf{A} + \mathbf{B}$), scalar multiplication ($k\mathbf{A}$), matrix multiplication (when defined) ($\mathbf{AB}$), and combinations of these operations are all possible. The **transpose** of $\mathbf{A}$, $\mathbf{A}^t$, is obtained by interchanging the rows and columns of $\mathbf{A}$ and is computed with the command `Transpose(A)`. If $\mathbf{A}$ is a square matrix, the determinant of $\mathbf{A}$ is obtained with `Determinant(A)`. Both `Transpose` and `Determinant` are contained in the `LinearAlgebra` package.

If $\mathbf{A}$ and $\mathbf{B}$ are $n \times n$ matrices satisfying $\mathbf{AB} = \mathbf{BA} = \mathbf{I}$, where $\mathbf{I}$ is the $n \times n$ matrix with 1's on the diagonal and 0's elsewhere (the $n \times n$ identity matrix), $\mathbf{B}$ is called the **inverse** of $\mathbf{A}$ and is denoted by $\mathbf{A}^{-1}$. If the inverse of a matrix $\mathbf{A}$ exists, the inverse is found with `MatrixInverse(A)`. As with `Transpose` and `Determinant`, `MatrixInverse` is contained in the `LinearAlgebra` package. Thus, assuming that $\begin{pmatrix} a & b \\ c & d \end{pmatrix}$ has an inverse $(ad - bc \neq 0)$, the inverse is

```
> with(LinearAlgebra):
> MatrixInverse(<<a|b>,<c|d>>);
```

$$\begin{bmatrix} \dfrac{d}{ad-bc} & -\dfrac{b}{ad-bc} \\ -\dfrac{c}{ad-bc} & \dfrac{a}{ad-bc} \end{bmatrix}$$

EXAMPLE 5.1.5: Let $A = \begin{pmatrix} 3 & -4 & 5 \\ 8 & 0 & -3 \\ 5 & 2 & 1 \end{pmatrix}$ and $B = \begin{pmatrix} 10 & -6 & -9 \\ 6 & -5 & -7 \\ -10 & 9 & 12 \end{pmatrix}$. Compute (a) $\mathbf{A} + \mathbf{B}$; (b) $\mathbf{B} - 4\mathbf{A}$; (c) the inverse of $\mathbf{AB}$; (d) the transpose of $(\mathbf{A} - 2\mathbf{B})\,\mathbf{B}$; and (e) $\det \mathbf{A} = |\mathbf{A}|$.

SOLUTION: We enter A (corresponding to $\mathbf{A}$) and B (corresponding to $\mathbf{B}$) as matrices. We suppress the output by ending each command with a colon.

```
> with(LinearAlgebra):
> A:=<<3|-4|5>,<8|0|-3>,<5|2|1>>:
> B:=<<10|-6|-9>,<6|-5|-7>,<-10|9|12>>:
```

Entering

```
> A+B;
```

$$\begin{bmatrix} 13 & -10 & -4 \\ 14 & -5 & -10 \\ -5 & 11 & 13 \end{bmatrix}$$

adds matrix A to B and expresses the result in traditional matrix form. Entering

```
> B-4*A;
```

$$\begin{bmatrix} -2 & 10 & -29 \\ -26 & -5 & 5 \\ -30 & 1 & 8 \end{bmatrix}$$

subtracts 4 times matrix A from B and expresses the result in traditional matrix form.

Use `evalm` together with the noncommutative multiplication operator, `&*`, or `Multiply`, which is contained in the `LinearAlgebra` package, to compute matrix products when they are defined. Thus, both

```
> evalm(A&*B);
```

$$\begin{bmatrix} -2 & 10 & -29 \\ -26 & -5 & 5 \\ -30 & 1 & 8 \end{bmatrix}$$

and

```
> Multiply(A,B);
```

$$\begin{bmatrix} -2 & 10 & -29 \\ -26 & -5 & 5 \\ -30 & 1 & 8 \end{bmatrix}$$

compute the matrix product **AB** and

```
> MatrixInverse(Multiply(A,B));
```

$$\begin{bmatrix} \frac{59}{380} & \frac{53}{190} & -\frac{167}{380} \\ -\frac{223}{570} & -\frac{92}{95} & \frac{979}{570} \\ \frac{49}{114} & \frac{18}{19} & -\frac{187}{114} \end{bmatrix}$$

computes the inverse of the matrix product **AB**. Similarly, entering

```
> Transpose(Multiply(A-2*B,B));
```

$$\begin{bmatrix} -352 & -90 & 384 \\ 269 & 73 & -277 \\ 373 & 98 & -389 \end{bmatrix}$$

computes the transpose of $(\mathbf{A} - 2\mathbf{B})\,\mathbf{B}$ and entering

```
> Determinant(A);
```

$$190$$

computes the determinant of A.

■

EXAMPLE 5.1.6: Compute **AB** and **BA** if $\mathbf{A} = \begin{pmatrix} -1 & -5 & -5 & -4 \\ -3 & 5 & 3 & -2 \\ -4 & 4 & 2 & -3 \end{pmatrix}$ and $\mathbf{B} = \begin{pmatrix} 1 & -2 \\ -4 & 3 \\ 4 & -4 \\ -5 & -3 \end{pmatrix}$.

SOLUTION: Because **A** is a 3×4 matrix and **B** is a 4×2 matrix, **AB** is defined and is a 3×2 matrix. We define `A` and `B` with the following commands.

Remember that you do not need to reload the `LinearAlgebra` package if you have already loaded it during your *current* Maple session.

```
> with(LinearAlgebra):
> A:=Matrix([[-1,-5,-5,-4],[-3,5,3,-2],[-4,4,2,-3]]):
> B:=Matrix([[1,-2],[-4,3],[4,-4],[-5,-3]]):
```

We then compute the product using both `evalm` together with the noncommutative multiplication operator, `&*`, and `Multiply`.

```
> evalm(A&*B);
```

$$\begin{bmatrix} 19 & 19 \\ -1 & 15 \\ 3 & 21 \end{bmatrix}$$

```
> Multiply(A,B);
```

$$\begin{bmatrix} 19 & 19 \\ -1 & 15 \\ 3 & 21 \end{bmatrix}$$

However, the matrix product **BA** is not defined and Maple produces error messages when we attempt to compute it. (The symbol <> means "not equal to.")

```
> Multiply(B,A);

Error, (in MatrixMatrixMultiply) first matrix column
dimension (2) <> second matrix row dimension (3)
```

■

You can use either ^ or `MatrixPower`, which is contained in the `LinearAlgebra` package, to compute powers of matrices.

EXAMPLE 5.1.7: Compute $\mathbf{B}^3$ if $\mathbf{B} = \begin{pmatrix} -2 & 3 & 4 & 0 \\ -2 & 0 & 1 & 3 \\ -1 & 4 & -6 & 5 \\ 4 & 8 & 11 & -4 \end{pmatrix}$.

SOLUTION: After defining **B**, we compute $\mathbf{B}^3$. The same result is obtained by entering `MatrixPower(B,3)`.

```
> B:=Matrix([[-2,3,4,0],[-2,0,1,3],[-1,4,-6,5],
   [4,8,11,-4]]):
> B^3;
```

$$\begin{bmatrix} 137 & 98 & 479 & -231 \\ -121 & 65 & -109 & 189 \\ -309 & 120 & -871 & 646 \\ 520 & 263 & 1381 & -738 \end{bmatrix}$$

```
> MatrixPower(B,3);
```

$$\begin{bmatrix} 137 & 98 & 479 & -231 \\ -121 & 65 & -109 & 189 \\ -309 & 120 & -871 & 646 \\ 520 & 263 & 1381 & -738 \end{bmatrix}$$

■

If $|\mathbf{A}| \neq 0$, the inverse of **A** can be computed using the formula

$$\mathbf{A}^{-1} = \frac{1}{|\mathbf{A}|}\mathbf{A}^a, \tag{5.1}$$

The **cofactor matrix**, $\mathbf{A}^c$, of **A** is the matrix obtained by replacing each element of **A** by its cofactor.

where $\mathbf{A}^a$ is the *transpose of the cofactor matrix*.

If **A** has an inverse, reducing the matrix $(\mathbf{A}|\mathbf{I})$ to reduced row echelon form results in $(\mathbf{I}|\mathbf{A}^{-1})$. This method is often easier to implement than computing (5.1).

EXAMPLE 5.1.8: Calculate $\mathbf{A}^{-1}$ if $\mathbf{A} = \begin{pmatrix} 2 & -2 & 1 \\ 0 & -2 & 2 \\ -2 & -1 & -1 \end{pmatrix}$.

SOLUTION: After defining $\mathbf{A}$ and $\mathbf{I} = \begin{pmatrix} 1 & 0 & 0 \\ 0 & 1 & 0 \\ 0 & 0 & 1 \end{pmatrix}$, we compute $|\mathbf{A}| = 12$ with `Determinant`, so $\mathbf{A}^{-1}$ exists.

```
> with(LinearAlgebra):
> A:=Matrix([[2,-2,1],[0,-2,2],[-2,-1,-1]]):
> i3:=IdentityMatrix(3):
> Determinant(A);
```

$$12$$

We then form the matrix $(\mathbf{A}|\mathbf{I})$

```
> ai3:=<A | i3>;
```

$$\begin{bmatrix} 2 & -2 & 1 & 1 & 0 & 0 \\ 0 & -2 & 2 & 0 & 1 & 0 \\ -2 & -1 & -1 & 0 & 0 & 1 \end{bmatrix}$$

and then use `ReducedRowEchelonForm` to reduce $(\mathbf{A}|\mathbf{I})$ to row echelon form.

`ReducedRowEchelonForm(A)` reduces $\mathbf{A}$ to **reduced row echelon form.** `ReducedRowEchelonForm` is contained in the `LinearAlgebra` package.

```
> ReducedRowEchelonForm(ai3);
```

$$\begin{bmatrix} 1 & 0 & 0 & 1/3 & -1/4 & -1/6 \\ 0 & 1 & 0 & -1/3 & 0 & -1/3 \\ 0 & 0 & 1 & -1/3 & 1/2 & -1/3 \end{bmatrix}$$

The result indicates that $\mathbf{A}^{-1} = \begin{pmatrix} 1/3 & -1/4 & -1/6 \\ -1/3 & 0 & -1/3 \\ -1/3 & 1/2 & -1/3 \end{pmatrix}$. We check this result with `MatrixInverse`, which is contained in the `LinearAlgebra` package.

```
> MatrixInverse(A);
```

$$\begin{bmatrix} 1/3 & -1/4 & -1/6 \\ -1/3 & 0 & -1/3 \\ -1/3 & 1/2 & -1/3 \end{bmatrix}$$

■

5.1.4 Basic Computations with Vectors

Basic Operations on Vectors

Computations with vectors are performed in the same way as computations with matrices.

EXAMPLE 5.1.9: Let $\mathbf{v} = \begin{pmatrix} 0 \\ 5 \\ 1 \\ 2 \end{pmatrix}$ and $\mathbf{w} = \begin{pmatrix} 3 \\ 0 \\ 4 \\ -2 \end{pmatrix}$. (a) Calculate $\mathbf{v} - 2\mathbf{w}$ and $\mathbf{v} \cdot \mathbf{w}$. (b) Find a unit vector with same direction as $\mathbf{v}$ and a unit vector with the same direction as $\mathbf{w}$.

Remember that you do not need to reload a package if you have already loaded it during your *current* Maple session.

SOLUTION: We begin by defining $\mathbf{v}$ and $\mathbf{w}$ and then compute $\mathbf{v} - 2\mathbf{w}$ and $\mathbf{v} \cdot \mathbf{w}$.

```
> with(LinearAlgebra):
> v:=Vector([0,5,1,2]):
> w:=Vector([3,0,4,-2]):
> v-2*w;
```

$$\begin{bmatrix} -6 \\ 5 \\ -7 \\ 6 \end{bmatrix}$$

```
> v.w;
```

$$0$$

The **norm** of the vector $\mathbf{v} = \begin{pmatrix} v_1 \\ v_2 \\ \vdots \\ v_n \end{pmatrix}$ is

$$\|\mathbf{v}\| = \sqrt{{v_1}^2 + {v_2}^2 + \cdots + {v_n}^2} = \sqrt{\mathbf{v} \cdot \mathbf{v}}.$$

If k is a positive scalar, the direction of $k\mathbf{v}$ is the same as the direction of $\mathbf{v}$. Thus, if $\mathbf{v}$ is a nonzero vector, the vector $\frac{1}{\|\mathbf{v}\|}\mathbf{v}$ has the same direction as $\mathbf{v}$ and because $\left\| \frac{1}{\|\mathbf{v}\|}\mathbf{v} \right\| = \frac{1}{\|\mathbf{v}\|}\|\mathbf{v}\| = 1$, $\frac{1}{\|\mathbf{v}\|}\mathbf{v}$ is a unit vector. The command `Norm(v,Frobenius)` computes $\|\mathbf{v}\|$. (`Norm` is contained in the

`LinearAlgebra` package.) We then compute $\frac{1}{\|\mathbf{v}\|}\mathbf{v}$, calling the result `uv`, and $\frac{1}{\|\mathbf{w}\|}\mathbf{w}$. The results correspond to unit vectors with the same direction as $\mathbf{v}$ and $\mathbf{w}$, respectively.

```
> uv:=v/Norm(v,Frobenius);
```

$$\begin{bmatrix} 0 \\ 1/6\sqrt{30} \\ 1/30\sqrt{30} \\ 1/15\sqrt{30} \end{bmatrix}$$

```
> Norm(uv,Frobenius);
```

$$1$$

```
> w/Norm(w,Frobenius);
```

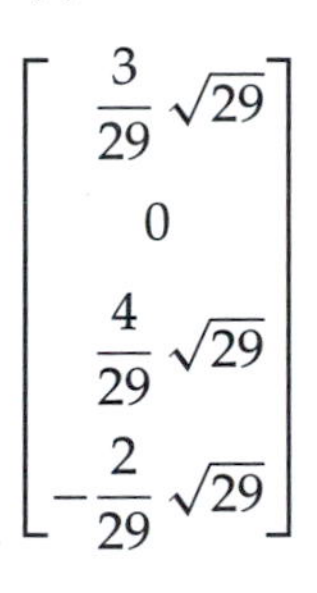

■

Basic Operations on Vectors in 3-Space

Vector calculus is discussed in Section 5.5.

We review the elementary properties of vectors in 3-space. Let

$$\mathbf{u} = \langle u_1, u_2, u_3 \rangle = u_1\mathbf{i} + u_2\mathbf{j} + u_3\mathbf{k}$$

and

$$\mathbf{v} = \langle v_1, v_2, v_3 \rangle = v_1\mathbf{i} + v_2\mathbf{j} + v_3\mathbf{k}$$

be vectors in space.

In space, the **standard unit vectors** are $\mathbf{i} = \langle 1,0,0 \rangle$, $\mathbf{j} = \langle 0,1,0 \rangle$, and $\mathbf{k} = \langle 0,0,1 \rangle$. With the exception of the cross product, the vector operations discussed here are performed in the same way for vectors in the plane as they are in space. In the plane, the **standard unit vectors** are $\mathbf{i} = \langle 1,0 \rangle$ and $\mathbf{j} = \langle 0,1 \rangle$.

1. $\mathbf{u}$ and $\mathbf{v}$ are **equal** if and only if their components are equal:

$$\mathbf{u} = \mathbf{v} \Leftrightarrow u_1 = v_1, u_2 = v_2, \text{ and } u_3 = v_3.$$

2. The **length** (or **norm**) of $\mathbf{u}$ is

$$\|\mathbf{u}\| = \sqrt{{u_1}^2 + {u_2}^2 + {u_3}^2}.$$

3. If c is a scalar (number),

$$c\mathbf{u} = \langle cu_1, cu_2, cu_3 \rangle .$$

4. The **sum** of **u** and **v** is defined to be the vector

$$\mathbf{u} + \mathbf{v} = \langle u_1 + v_1, u_2 + v_2, u_3 + v_3 \rangle .$$

A **unit vector** is a vector with length 1.

5. If $\mathbf{u} \neq \mathbf{0}$, a unit vector with the same direction as **u** is

$$\frac{1}{\|\mathbf{u}\|}\mathbf{u} = \frac{1}{\sqrt{u_1{}^2 + u_2{}^2 + u_3{}^2}} \langle u_1, u_2, u_3 \rangle .$$

6. **u** and **v** are **parallel** if there is a scalar c so that $\mathbf{u} = c\mathbf{v}$.
7. The **dot product** of **u** and **v** is

$$\mathbf{u} \cdot \mathbf{v} = u_1 v_1 + u_2 v_2 + u_3 v_3 .$$

If θ is the angle between **u** and **v**,

$$\cos\theta = \frac{\mathbf{u} \cdot \mathbf{v}}{\|\mathbf{u}\| \, \|\mathbf{v}\|} .$$

Consequently, **u** and **v** are orthogonal if $\mathbf{u} \cdot \mathbf{v} = 0$.
8. The **cross product** of **u** and **v** is

$$\begin{aligned} \mathbf{u} \times \mathbf{v} &= \begin{vmatrix} \mathbf{i} & \mathbf{j} & \mathbf{k} \\ u_1 & u_2 & u_3 \\ v_1 & v_2 & v_3 \end{vmatrix} \\ &= (u_2 v_3 - u_3 v_2)\,\mathbf{i} - (u_1 v_3 - u_3 v_1)\,\mathbf{j} + (u_1 v_2 - u_2 v_1)\,\mathbf{k}. \end{aligned}$$

You should verify that $\mathbf{u} \cdot (\mathbf{u} \times \mathbf{v}) = 0$ and $\mathbf{v} \cdot (\mathbf{u} \times \mathbf{v}) = 0$. Hence, $\mathbf{u} \times \mathbf{v}$ is orthogonal to both **u** and **v**.

Topics from linear algebra (including determinants) are discussed in more detail in the next sections. For now, we illustrate several of the basic operations listed above. In Maple, many other vector calculations take advantage of functions contained in the `VectorCalculus` package. Use Maple's **Help** facility to obtain general help regarding the `VectorCalculus` package.

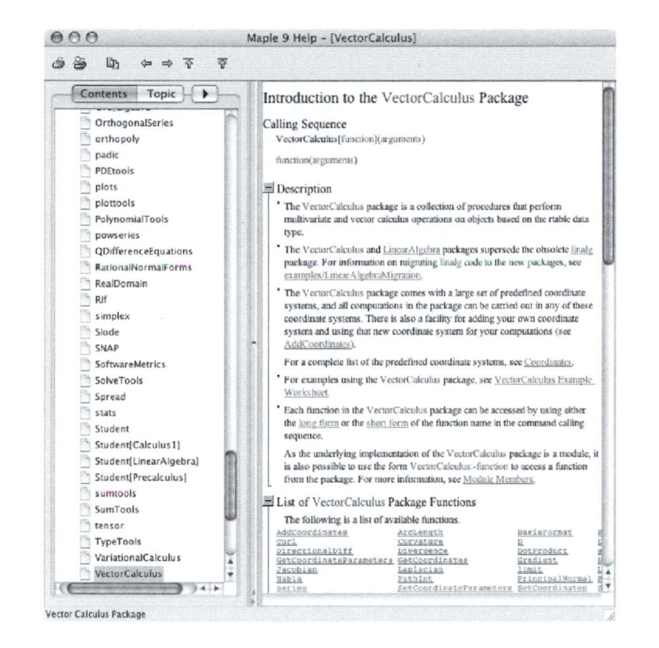

EXAMPLE 5.1.10: Let $\mathbf{u} = \langle 3,4,1 \rangle$ and $\mathbf{v} = \langle -4,3,-2 \rangle$. Calculate (a) $\mathbf{u} \cdot \mathbf{v}$, (b) $\mathbf{u} \times \mathbf{v}$, (c) $\|\mathbf{u}\|$, and (d) $\|\mathbf{v}\|$. (e) Find the angle between $\mathbf{u}$ and $\mathbf{v}$. (f) Find unit vectors with the same direction as $\mathbf{u}$, $\mathbf{v}$, and $\mathbf{u} \times \mathbf{v}$.

SOLUTION: We define $\mathbf{u} = \langle 3,4,1 \rangle$ and $\mathbf{v} = \langle -4,3,-2 \rangle$ and then illustrate the use of `DotProduct` and `CrossProduct`, which are contained in the `LinearAlgebra` package, to calculate (a)–(d).

Remark. Generally, `u.v` returns the same result as `DotProduct(u,v)`.

```
> with(LinearAlgebra):
> u:=Vector([3,4,1]):
> v:=Vector([-4,3,-2]):
```

```
> udv:=u.v;
```

$$udv := -2$$

```
> udv:=DotProduct(u,v);
```

$$udv := -2$$

```
> ucv:=CrossProduct(u,v);
```

$$\begin{bmatrix} -11 \\ 2 \\ 25 \end{bmatrix}$$

Both `sqrt(v.v)` and `Norm(v,Frobenius)` return the norm of $\mathbf{v}$.

```
> nv:=sqrt(v.v);
```

$$\sqrt{29}$$

```
> nv:=Norm(v,Frobenius);
```

$$\sqrt{29}$$

```
> nv:=Norm(u,Frobenius);
```

$$\sqrt{26}$$

(e) We use the formula $\theta = \cos^{-1}\left(\frac{\mathbf{u}\cdot\mathbf{v}}{\|\mathbf{u}\|\,\|\mathbf{v}\|}\right)$ to find the angle θ between $\mathbf{u}$ and $\mathbf{v}$.

```
> ev:=arccos(u.v/(nu*nv));
```

$$\pi - \arccos\left(\frac{1}{377}\sqrt{26}\sqrt{29}\right)$$

```
> evalf(ev);
```

$$1.643696585$$

(f) Unit vectors with the same direction as $\mathbf{u}$, $\mathbf{v}$, and $\mathbf{u} \times \mathbf{v}$ are found next. You can use the formula or `Normalize(v,Frobenius)` to find a unit vector with the same direction as $\mathbf{v}$. (`Normalize` is contained in the `LinearAlgebra` package.)

```
> normv:=v/Norm(v,Frobenius);
```

$$\begin{bmatrix} -\frac{4}{29}\sqrt{29} \\ \frac{3}{29}\sqrt{29} \\ -\frac{2}{29}\sqrt{29} \end{bmatrix}$$

```
> normv:=Normalize(v,Frobenius);
```

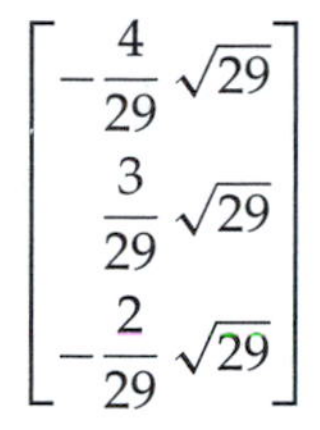

$$\begin{bmatrix} -\frac{4}{29}\sqrt{29} \\ \frac{3}{29}\sqrt{29} \\ -\frac{2}{29}\sqrt{29} \end{bmatrix}$$

```
> normu:=Normalize(u,Frobenius);
```

$$\begin{bmatrix} \frac{3}{26}\sqrt{26} \\ 2/13\sqrt{26} \\ 1/26\sqrt{26} \end{bmatrix}$$

```
> nuucrossuv:=Normalize(CrossProduct(normu,normv));
```

$$\begin{bmatrix} -\frac{11}{25} \\ \frac{2}{25} \\ 1 \end{bmatrix}$$

We can graphically confirm that these three vectors are orthogonal by graphing all three vectors with the `arrow` function, which is contained in the `plottools` package.

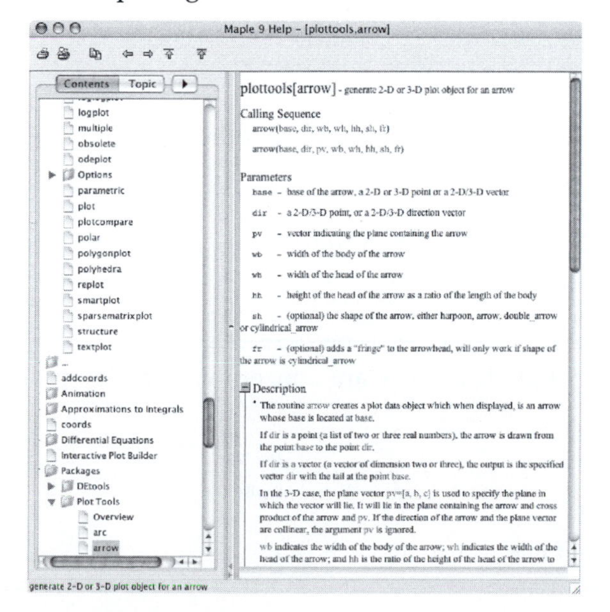

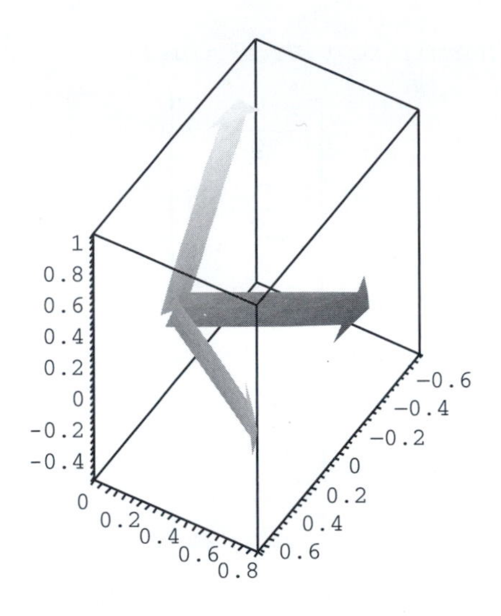

Figure 5-1 Orthogonal vectors

We show the vectors in Figure 5-1.

```
> with(plots):
> with(plottools):
> l1:=arrow(Vector([0,0,0]),normu,.2,.4,.1):
> l2:=arrow(Vector([0,0,0]),normv,.2,.4,.1):
> l3:=arrow(Vector([0,0,0]),
   nuucrossuv,.2,.4,.1):

> display(l1,l2,l3);
```

In the plot, the vectors do appear to be orthogonal as expected.

■

With the exception of the cross product, the calculations described above can also be performed on vectors in the plane.

EXAMPLE 5.1.11: If $\mathbf{u}$ and $\mathbf{v}$ are nonzero vectors, the **projection** of $\mathbf{u}$ onto $\mathbf{v}$ is

$$\text{proj}_{\mathbf{v}}\mathbf{u} = \frac{\mathbf{u} \cdot \mathbf{v}}{\|\mathbf{v}\|^2}\mathbf{v}.$$

Find $\text{proj}_{\mathbf{v}}\mathbf{u}$ if $\mathbf{u} = \langle -1, 4 \rangle$ and $\mathbf{v} = \langle 2, 6 \rangle$.

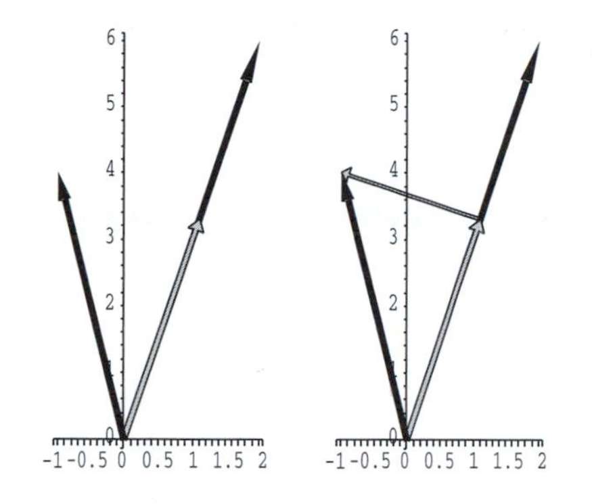

Figure 5-2 Projection of a vector

SOLUTION: We define $\mathbf{u} = \langle -1, 4\rangle$ and $\mathbf{v} = \langle 2, 6\rangle$ and then compute $\text{proj}_{\mathbf{v}}\mathbf{u}$.

```
> u:=Vector([-1,4]):
> v:=Vector([2,6]):
> projvu:=(u.v)*v/(v.v);
```

$$projvu := Vector_{column}\left(\left[\left[\frac{11}{10}\right], \left[\frac{33}{10}\right]\right], [\text{"}x\text{"}, \text{"}y\text{"}]\right)$$

Finally, we graph $\mathbf{u}$, $\mathbf{v}$, and $\text{proj}_{\mathbf{v}}\mathbf{u}$ together using `arrow` and `display` in Figure 5-2.

```
> with(plots):
> with(plottools):
> p1:=arrow(Vector([0,0]),u,.1,.2,.1,color=black):
> p2:=arrow(Vector([0,0]),v,.1,.2,.1,color=black):
> p3:=arrow(Vector([0,0]),projvu,.05,.25,.05,
   color=gray):
> display(p1,p3,p2,scaling=constrained);
```

```
> with(plots):
> with(plottools):
> p1:=arrow(Vector([0,0]),u,.1,.2,.1,color=black):
> p2:=arrow(Vector([0,0]),v,.1,.2,.1,color=black):
> p3:=arrow(Vector([0,0]),projvu,.05,.25,.05,
   color=gray):
> p4:=arrow(projvu,u-projvu,.05,.25,.05,color=gray):
> display(p1,p3,p2,p4,scaling=constrained);
```

In the graph, notice that $\mathbf{u} = \text{proj}_{\mathbf{v}}\mathbf{u} + (\mathbf{u} - \text{proj}_{\mathbf{v}}\mathbf{u})$ and the vector $\mathbf{u} - \text{proj}_{\mathbf{v}}\mathbf{u}$ is perpendicular to $\mathbf{v}$.

■

5.2 Linear Systems of Equations

Maple offers several commands for solving systems of linear equations, however, that do not depend on the computation of the inverse of **A**. The command

```
solve({eqn1,eqn2,...,eqnm},{var1,var2,...,varn})
```

solves an $m \times n$ system of linear equations (m equations and n unknown variables). Note that both the equations as well as the variables are entered as sets. If one wishes to solve for all variables that appear in a system, the command `solve({eqn1,eqn2,...eqnn})` attempts to solve `eqn1, eqn2, ..., eqnn` for all variables that appear in them.

Generally, `solve(set_of_equations,set_of_variables)` will solve most linear systems of equations you encounter. In some situations, you may wish to see detail and construct solutions to a system using particular methods.

The following commands that are contained in the `LinearAlgebra` package are especially helpful in solving linear systems of equations.

1. `Column(A,i)` returns the ith column of **A**.
2. `LinearSolve(A,b)` solves the matrix equations $\mathbf{Ax} = \mathbf{b}$ for **x**.
3. `MatrixInverse(A)` finds the inverse of the matrix **A**, if it exists.
4. `ReducedRowEchelonForm(A)` reduces **A** to reduced row echelon form.
5. `RowOperation(A,i,k)` returns the matrix obtained by multiplying row i of matrix **A** by k.
6. `RowOperation(A,[j,i],k)` returns the matrix obtained by multiplying row i of matrix **A** by k and adding the result to row j of matrix **A**.

In the following examples, we load the `LinearAlgebra` package first so that each example can be completed independently of the others. Remember that you do not need to reload a package if you have already loaded it during your *current* Maple session.

5.2.1 Calculating Solutions of Linear Systems of Equations

To solve the system of linear equations $\mathbf{Ax} = \mathbf{b}$, where **A** is the coefficient matrix, **b** is the known vector, and **x** is the unknown vector, we often proceed as follows: if $\mathbf{A}^{-1}$ exists, then $\mathbf{AA}^{-1}\mathbf{x} = \mathbf{A}^{-1}\mathbf{b}$ so $\mathbf{x} = \mathbf{A}^{-1}\mathbf{b}$.

EXAMPLE 5.2.1: Solve the matrix equation $\begin{pmatrix} 3 & 0 & 2 \\ -3 & 2 & 2 \\ 2 & -3 & 3 \end{pmatrix}\begin{pmatrix} x \\ y \\ z \end{pmatrix} = \begin{pmatrix} 3 \\ -1 \\ 4 \end{pmatrix}$.

SOLUTION: The solution is given by $\begin{pmatrix} x \\ y \\ z \end{pmatrix} = \begin{pmatrix} 3 & 0 & 2 \\ -3 & 2 & 2 \\ 2 & -3 & 3 \end{pmatrix}^{-1}\begin{pmatrix} 3 \\ -1 \\ 4 \end{pmatrix}$.

We proceed by defining `A` and `b` and then using `MatrixInverse` to calculate `MatrixInverse(A).b` naming the resulting output `Vector([x,y,z])`.

```
> with(LinearAlgebra):
> A:=Matrix([[3,0,2],[-3,2,2],[2,-3,3]]):
> b:=Vector([3,-1,4]):
> Vector([x,y,z]):=MatrixInverse(A).b;
```

$$\begin{bmatrix} \frac{13}{23} \\ -\frac{7}{23} \\ \frac{15}{23} \end{bmatrix}$$

We verify that the result is the desired solution by calculating `A &* Vector([x,y,z])`. Because the result of this procedure is $\begin{pmatrix} 3 \\ -1 \\ 4 \end{pmatrix}$, we conclude that the solution to the system is $\begin{pmatrix} x \\ y \\ z \end{pmatrix} = \begin{pmatrix} 13/23 \\ -7/23 \\ 15/23 \end{pmatrix}$.

```
> evalm(A &* Vector([x,y,z]));
```

$$[3, -1, 4]$$

Instead of using the noncommutative multiplication operator, `&*`, we could also have used `Multiply`, which is contained in the `LinearAlgebra` package, to perform the verification.

```
> Multiply(A, Vector([x,y,z]));
```

$$\begin{bmatrix} 3 \\ -1 \\ 4 \end{bmatrix}$$

We note that this matrix equation is equivalent to the system of equations

$$3x + 2z = 3$$
$$-3x + 2y + 2z = -1,$$
$$2x - 3y + 3z = 4$$

which we are able to solve with `solve`.

```
> solve(3*x+2*z=3,-3*x+2*y+2*z=-1,2*x-3*y+3*z=4);
```

$$\left\{z = \frac{15}{23}, y = -\frac{7}{23}, x = \frac{13}{23}\right\}$$

■

In addition to using `solve` to solve a system of linear equations, the command

`LinearSolve(A,b)`

calculates the solution vector $\mathbf{x}$ of the system $\mathbf{Ax} = \mathbf{b}$. `LinearSolve` generally solves a system more quickly than does `solve`.

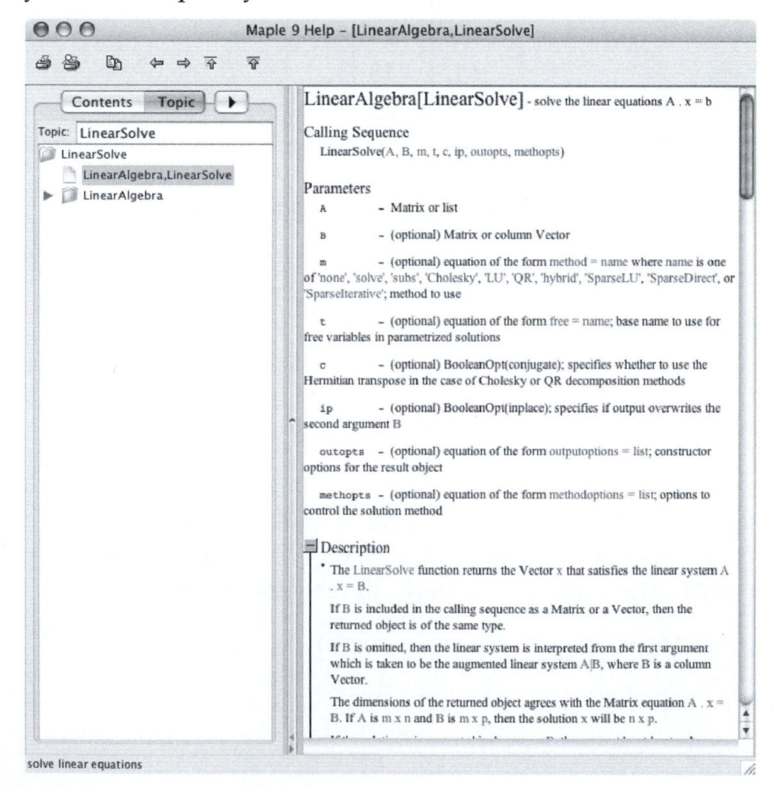

EXAMPLE 5.2.2: Solve the system $\begin{cases} x - 2y + z = -4 \\ 3x + 2y - z = 8 \\ -x + 3y + 5z = 0 \end{cases}$ for x, y, and z.

SOLUTION: In this case, entering

```
> with(LinearAlgebra):
> solve(x-2*y+z=-4,3*x+2*y-z=8,-x+3*y+5*z=0);
```

$$\{z = -1,\ y = 2,\ x = 1\}$$

solves the system for x, y, and z.

Another way to solve systems of equations is based on the matrix form of the system of equations, $\mathbf{Ax} = \mathbf{b}$. This system of equations is equivalent to the matrix equation

$$\begin{pmatrix} 1 & -2 & 1 \\ 3 & 2 & -1 \\ -1 & 3 & 5 \end{pmatrix} \begin{pmatrix} x \\ y \\ z \end{pmatrix} = \begin{pmatrix} -4 \\ 8 \\ 0 \end{pmatrix}.$$

The matrix of coefficients for this example is entered as `A` along with the vector of right-hand side values `b`. After defining the vector of variables, `vectorx`, the system $\mathbf{Ax} = \mathbf{b}$ is solved explicitly with the command `LinearSolve`.

```
> A:=Matrix([[1,-2,1],[3,2,-1],[-1,3,5]]):
> b:=Vector([-4,8,0]):
> xvec:=LinearSolve(A,b);
```

$$xvec := \begin{bmatrix} 1 \\ 2 \\ -1 \end{bmatrix}$$

We verify the solution by computing $\mathbf{Ax}$.

```
> A.xvec;
```

$$\begin{bmatrix} -4 \\ 8 \\ 0 \end{bmatrix}$$

■

EXAMPLE 5.2.3: Solve the system $\begin{cases} 2x - 4y + z = -1 \\ 3x + y - 2z = 3 \\ -5x + y - 2z = 4 \end{cases}$. Verify that the result returned satisfies the system.

SOLUTION: To solve the system using `solve`, we define `eqs` and then use `solve` to solve the set of equations `eqs`. The resulting output is named `sols`.

```
> with(LinearAlgebra):
> eqs:=2*x-4*y+z=-1,3*x+y-2*z=3,-5*x+y-2*z=4:
> sols:=solve(eqs);
```

$$sols := \left\{y = -\frac{15}{56}, z = -\frac{51}{28}, x = -1/8\right\}$$

To verify that the result given in `sols` is the desired solution, we replace each occurrence of x, y, and z in `eqs` by the values found in `sols` using `subs`. Because the result indicates each of the three equations is satisfied, we conclude that the values given in `sols` are the components of the desired solution.

```
> subs(sols,eqs);
```

$$\{-1 = -1, 3 = 3, 4 = 4\}$$

To solve the system using `LinearSolve`, we note that the system is equivalent to the matrix equation $\begin{pmatrix} 2 & -4 & 1 \\ 3 & 1 & -2 \\ -5 & 1 & -2 \end{pmatrix} \begin{pmatrix} x \\ y \\ z \end{pmatrix} = \begin{pmatrix} -1 \\ 3 \\ 4 \end{pmatrix}$, define `A` and `b`, and use `LinearSolve` to solve this matrix equation.

```
> A:=Matrix([[2,-4,1],[3,1,-2],[-5,1,-2]]):
> b:=Vector([-1,3,4]):
> solvector:=LinearSolve(A,b);
```

$$\begin{bmatrix} -1/8 \\ -\frac{15}{56} \\ -\frac{51}{28} \end{bmatrix}$$

To verify that the results are correct, we compute `A.solvector`. Because the result is $\begin{pmatrix} -1 \\ 3 \\ 4 \end{pmatrix}$, we conclude that the solution to the system is $\begin{pmatrix} x \\ y \\ z \end{pmatrix} = \begin{pmatrix} -1/8 \\ -15/36 \\ -51/28 \end{pmatrix}$.

```
> A.solvector;
```

$$\begin{bmatrix} -1 \\ 3 \\ 4 \end{bmatrix}$$

■

EXAMPLE 5.2.4: Solve the system of equations

$$\begin{cases} 4x_1 + 5x_2 - 5x_3 - 8x_4 - 2x_5 = 5 \\ 7x_1 + 2x_2 - 10x_3 - x_4 - 6x_5 = -4 \\ 6x_1 + 2x_2 + 10x_3 - 10x_4 + 7x_5 = -7 \\ -8x_1 - x_2 - 4x_3 + 3x_5 = 5 \\ 8x_1 - 7x_2 - 3x_3 + 10x_4 + 5x_5 = 7 \end{cases}.$$

SOLUTION: We solve the system in two ways. First, we use `solve` to solve the system.

```
> solve(4*x[1]+5*x[2]-5*x[3]-8*x[4]-2*x[5]
   =5,7*x[1]+2*x[2]-10*x[3]-x[4]-6*x[5]=-4,
> 6*x[1]+2*x[2]+10*x[3]-10*x[4]+7*x[5]
   =-7,-8*x[1]-x[2]-4*x[3]+3*x[5]=5,
> 8*x[1]-7*x[2]-3*x[3]+10*x[4]+5*x[5]=7);
```

$$\left\{x_3 = -\frac{7457}{9939}, x_1 = \frac{1245}{6626}, x_5 = \frac{49327}{9939}, x_4 = \frac{38523}{6626}, x_2 = \frac{113174}{9939}\right\}$$

We also use `LinearSolve` after defining `A` and `b`. As expected, in each case the results are the same.

```
> with(LinearAlgebra):
> A:=Matrix([[4,5,-5,-8,-2],[7,2,-10,-1,-6],
> [6,2,10,-10,7],[-8,-1,-4,0,3],[8,-7,-3,10,5]]):
```

```
> b:=Vector([5,-4,-7,5,7]):
> LinearSolve(A,b);
```

$$\begin{bmatrix} \frac{1245}{6626} \\ \frac{113174}{9939} \\ -\frac{7457}{9939} \\ \frac{38523}{6626} \\ \frac{49327}{9939} \end{bmatrix}$$

■

5.2.2 Gauss-Jordan Elimination

Given the matrix equation $\mathbf{Ax} = \mathbf{b}$, where

$$\mathbf{A} = \begin{pmatrix} a_{11} & a_{12} & \cdots & a_{1n} \\ a_{21} & a_{22} & \cdots & a_{2n} \\ \vdots & \vdots & \ddots & \vdots \\ a_{m1} & a_{m2} & \cdots & a_{mn} \end{pmatrix}, \quad \mathbf{x} = \begin{pmatrix} x_1 \\ x_2 \\ \vdots \\ x_n \end{pmatrix}, \quad \text{and} \quad \mathbf{b} = \begin{pmatrix} b_1 \\ b_2 \\ \vdots \\ b_m \end{pmatrix},$$

the $m \times n$ matrix $\mathbf{A}$ is called the **coefficient matrix** for the matrix equation $\mathbf{Ax} = \mathbf{b}$ and the $m \times (n+1)$ matrix

$$\begin{pmatrix} a_{11} & a_{12} & \cdots & a_{1n} & b_1 \\ a_{21} & a_{22} & \cdots & a_{2n} & b_2 \\ \vdots & \vdots & \ddots & \vdots & \vdots \\ a_{m1} & a_{m2} & \cdots & a_{mn} & b_m \end{pmatrix}$$

is called the **augmented** (or **associated**) **matrix** for the matrix equation. We may enter the augmented matrix associated with a linear system of equations directly: once you have defined **A** as a matrix and **b** as a vector, the augmented matrix is formed with the command `<A | b>`.

EXAMPLE 5.2.5: Solve the system $\begin{cases} -2x + y - 2x = 4 \\ 2x - 4y - 2z = -4 \\ x - 4y - 2z = 3 \end{cases}$ using Gauss-Jordan elimination.

SOLUTION: The system is equivalent to the matrix equation

$$\begin{pmatrix} -2 & 1 & -2 \\ 2 & -4 & -2 \\ 1 & -4 & -2 \end{pmatrix} \begin{pmatrix} x \\ y \\ z \end{pmatrix} = \begin{pmatrix} 4 \\ -4 \\ 3 \end{pmatrix}.$$

The augmented matrix associated with this system is

$$\begin{pmatrix} -2 & 1 & -2 & 4 \\ 2 & -4 & -2 & -4 \\ 1 & -4 & -2 & 3 \end{pmatrix}$$

which we construct using the command `<A | b>`. We proceed by loading the `LinearAlgebra` package, defining `A` and `b`, and then constructing the augmented matrix which we name `augm`.

```
> with(LinearAlgebra):
> A:=Matrix([[-2,1,-2],[2,-4,-2],[1,-4,-2]]):
> b:=Vector([4,-4,3]):
> augm:=<A | b >;
```

$$augm := \begin{bmatrix} -2 & 1 & -2 & 4 \\ 2 & -4 & -2 & -4 \\ 1 & -4 & -2 & 3 \end{bmatrix}$$

We calculate the solution by row-reducing `augm` using the built-in command `ReducedRowEchelonForm`. Generally, `ReducedRowEchelonForm(A)` reduces **A** to **reduced row echelon form.**

```
> rrefAb:=ReducedRowEchelonForm(augm);
```

$$rrefAb := \begin{bmatrix} 1 & 0 & 0 & -7 \\ 0 & 1 & 0 & -4 \\ 0 & 0 & 1 & 3 \end{bmatrix}$$

From this result, we see that the solution is

$$\begin{pmatrix} x \\ y \\ z \end{pmatrix} = \begin{pmatrix} -7 \\ -4 \\ 3 \end{pmatrix},$$

which we extract from `rrefAb` with `Column`.

```
> solvec:=Column(rrefAb,4);
```

$$solvec := \begin{bmatrix} -7 \\ -4 \\ 3 \end{bmatrix}$$

We verify the solution by computing `A.solvec`.

```
> A.solvec;
```

$$\begin{bmatrix} 4 \\ -4 \\ 3 \end{bmatrix}$$

■

If you wish to implement the reduction of **A** to reduced row echelon form yourself, use `RowOperation`, which is contained in the `LinearAlgebra` package.

`ColumnOperation` works in the same way as `RowOperation` but performs the corresponding column operations on a matrix.

EXAMPLE 5.2.6: Solve

$$\begin{aligned} -3x + 2y - 2z &= -10 \\ 3x - y + 2z &= 7 \\ 2x - y + z &= 6. \end{aligned}$$

SOLUTION: The associated matrix is $\mathbf{A} = \begin{pmatrix} -3 & 2 & -2 & -10 \\ 3 & -1 & 2 & 7 \\ 2 & -1 & 1 & 6 \end{pmatrix}$, defined in `A`.

```
> with(LinearAlgebra):
> A:=Matrix([[-3,2,-2,-10],[3,-1,2,7],[2,-1,1,6]]):
```

We eliminate methodically. First, we multiply row 1 by $-1/3$ so that the first entry in the first column is 1.

```
> A:=RowOperation(A,1,-1/3);
```

$$A := \begin{bmatrix} 1 & -2/3 & 2/3 & 10/3 \\ 3 & -1 & 2 & 7 \\ 2 & -1 & 1 & 6 \end{bmatrix}$$

We now eliminate below. First, we multiply row 1 by -3 and add it to row 2 and then we multiply row 1 by -2 and add it to row 3.

```
> A:=RowOperation(A,[2,1],-3);
```

$$A := \begin{bmatrix} 1 & -2/3 & 2/3 & 10/3 \\ 0 & 1 & 0 & -3 \\ 2 & -1 & 1 & 6 \end{bmatrix}$$

```
> A:=RowOperation(A,[3,1],-2);
```

$$A := \begin{bmatrix} 1 & -2/3 & 2/3 & 10/3 \\ 0 & 1 & 0 & -3 \\ 0 & 1/3 & -1/3 & -2/3 \end{bmatrix}$$

Observe that the first nonzero entry in the second row is 1. We eliminate below this entry by adding $-1/3$ times row 2 to row 3.

```
> A:=RowOperation(A,[3,2],-1/3);
```

$$A := \begin{bmatrix} 1 & -2/3 & 2/3 & 10/3 \\ 0 & 1 & 0 & -3 \\ 0 & 0 & -1/3 & 1/3 \end{bmatrix}$$

We multiply the third row by -3 so that the first nonzero entry is 1.

```
> A:=RowOperation(A,3,-3);
```

$$A := \begin{bmatrix} 1 & -2/3 & 2/3 & 10/3 \\ 0 & 1 & 0 & -3 \\ 0 & 0 & 1 & -1 \end{bmatrix}$$

This matrix is equivalent to the system

$$\begin{aligned} x - \frac{2}{3}y + \frac{2}{3}z &= \frac{10}{3} \\ y &= -3 \\ z &= -1, \end{aligned}$$

which shows us that the solution is $x = 2, y = -3, z = -1$.

Working backwards with `BackwardSubstitute` confirms this.

```
> BackwardSubstitute(A);
```

$$\begin{bmatrix} 2 \\ -3 \\ -1 \end{bmatrix}$$

We confirm the result directly with `solve`.

```
> solve(-3*x+2*y-2*z=-10,3*x-y+2*z=7,2*x-y+z=6);
```

$$\{z = -1, x = 2, y = -3\}$$

■

EXAMPLE 5.2.7: Solve

$$-3x_1 + 2x_2 + 5x_3 = -12$$
$$3x_1 - x_2 - 4x_3 = 9$$
$$2x_1 - x_2 - 3x_3 = 7.$$

SOLUTION: The associated matrix is $\mathbf{A} = \begin{pmatrix} -3 & 2 & 5 & -12 \\ 3 & -1 & -4 & 9 \\ 2 & -1 & -3 & 7 \end{pmatrix}$, which is reduced to row echelon form with `ReducedRowEchelonForm`.

```
> with(LinearAlgebra):
> A:=Matrix([[-3,2,5,-12],[3,-1,-4,9],[2,-1,-3,7]]):
> rrefA:=ReducedRowEchelonForm(A);
```

$$rrefA := \begin{bmatrix} 1 & 0 & -1 & 2 \\ 0 & 1 & 1 & -3 \\ 0 & 0 & 0 & 0 \end{bmatrix}$$

The result means that the original system is equivalent to

$$\begin{aligned} x_1 - x_3 &= 2 \\ x_2 + x_3 &= -3 \end{aligned} \quad \text{or} \quad \begin{aligned} x_1 &= 2 + x_3 \\ x_2 &= -3 - x_3 \end{aligned}$$

so x_3 is *free*. That is, for any real number t, a solution to the system is

$$\begin{pmatrix} x_1 \\ x_2 \\ x_3 \end{pmatrix} = \begin{pmatrix} 2+t \\ -3-t \\ t \end{pmatrix} = \begin{pmatrix} 2 \\ -3 \\ 0 \end{pmatrix} + t \begin{pmatrix} 1 \\ -1 \\ 1 \end{pmatrix}.$$

The system has infinitely many solutions.

Equivalent results are obtained with `solve`.

```
> solve(-3*x[1]+2*x[2]+5*x[3]=-12,3*x[1]-x[2]-4*x[3]
  =9,2*x[1]-x[2]-3*x[3]=7);
```

$$\{x_3 = x_1 - 2, x_2 = -x_1 - 1, x_1 = x_1\}$$

```
> solve(-3*x[1]+2*x[2]+5*x[3]=-12,3*x[1]-x[2]-4*x[3]
  =9,2*x[1]-x[2]-3*x[3]=7,x[1],x[2]);
```

$$\{x_1 = 2 + x_3, x_2 = -3 - x_3\}$$

■

EXAMPLE 5.2.8: Solve

$$\begin{aligned}-3x_1 + 2x_2 + 5x_3 &= -14\\ 3x_1 - x_2 - 4x_3 &= 11\\ 2x_1 - x_2 - 3x_3 &= 8.\end{aligned}$$

SOLUTION: The associated matrix is $\mathbf{A} = \begin{pmatrix} -3 & 2 & 5 & -14 \\ 3 & -1 & -4 & 11 \\ 2 & -1 & -3 & 8 \end{pmatrix}$, which is reduced to row echelon form with `ReducedRowEchelonForm`.

```
> with(LinearAlgebra):
> A:=Matrix([[-3,2,5,-14],[3,-1,-4,11],[2,-1,-3,8]]):
> ReducedRowEchelonForm(A);
```

$$\begin{bmatrix} 1 & 0 & -1 & 0 \\ 0 & 1 & 1 & 0 \\ 0 & 0 & 0 & 1 \end{bmatrix}$$

The result shows that the original system is equivalent to

$$\begin{aligned}x_1 - x_3 &= 0\\ x_2 + x_3 &= 0\\ 0 &= 1.\end{aligned}$$

Of course, 0 is not equal to 1: the last equation is false. The system has no solutions.

We check the calculation with `solve`. In this case, Maple returns nothing, which indicates that `solve` cannot find any solutions to the system.

```
> solve(-3*x[1]+2*x[2]+5*x[3]=-14,
> 3*x[1]-x[2]-4*x[3]=11,2*x[1]-x[2]-3*x[3]=8);
```

Generally, if Maple returns nothing, the result means either that there is no solution or that Maple cannot solve the problem. In such a situation, we must always check using another method.

■

EXAMPLE 5.2.9: The **nullspace** of $\mathbf{A}$ is the set of solutions to the system of equations $\mathbf{Ax} = \mathbf{0}$. Find the nullspace of $\mathbf{A} = \begin{pmatrix} 3 & 2 & 1 & 1 & -2 \\ 3 & 3 & 1 & 2 & -1 \\ 2 & 2 & 1 & 1 & -1 \\ -1 & -1 & 0 & -1 & 0 \\ 5 & 4 & 2 & 2 & -3 \end{pmatrix}$.

SOLUTION: Observe that row-reducing $(\mathbf{A}|\mathbf{0})$ is equivalent to row-reducing $\mathbf{A}$. After defining $\mathbf{A}$, we use `ReducedRowEchelonForm` to row 4 reduce $\mathbf{A}$.

```
> with(LinearAlgebra):
> A:=Matrix([[3,2,1,1,-2],[3,3,1,2,-1],[2,2,1,1,-1],
> [-1,-1,0,-1,0],[5,4,2,2,-3]]):
> ReducedRowEchelonForm(A);
```

$$\begin{bmatrix} 1 & 0 & 0 & 0 & -1 \\ 0 & 1 & 0 & 1 & 1 \\ 0 & 0 & 1 & -1 & -1 \\ 0 & 0 & 0 & 0 & 0 \\ 0 & 0 & 0 & 0 & 0 \end{bmatrix}$$

The result indicates that the solutions of $\mathbf{Ax} = \mathbf{0}$ are

$$\mathbf{x} = \begin{pmatrix} x_1 \\ x_2 \\ x_3 \\ x_4 \\ x_5 \end{pmatrix} = \begin{pmatrix} t \\ -s-t \\ s+t \\ s \\ t \end{pmatrix} = s \begin{pmatrix} 0 \\ -1 \\ 1 \\ 1 \\ 0 \end{pmatrix} + t \begin{pmatrix} 1 \\ -1 \\ 1 \\ 0 \\ 1 \end{pmatrix},$$

where s and t are any real numbers. The dimension of the nullspace, the **nullity**, is 2; a basis for the nullspace is

$$\left\{ \begin{pmatrix} 0 \\ -1 \\ 1 \\ 1 \\ 0 \end{pmatrix}, \begin{pmatrix} 1 \\ -1 \\ 1 \\ 0 \\ 1 \end{pmatrix} \right\}.$$

You can use the command `NullSpace[A]`, which is contained in the `LinearAlgebra` package, to find a basis of the nullspace of a matrix $\mathbf{A}$ directly.

```
> NullSpace(A);
```

$$\left\{ \begin{bmatrix} 0 \\ -1 \\ 1 \\ 1 \\ 0 \end{bmatrix}, \begin{bmatrix} 1 \\ -1 \\ 1 \\ 0 \\ 1 \end{bmatrix} \right\}$$

■

5.3 Selected Topics from Linear Algebra

5.3.1 Fundamental Subspaces Associated with Matrices

Let $\mathbf{A} = (a_{ij})$ be an $n \times m$ matrix with entry a_{ij} in the ith row and jth column. The **row space** of **A**, row(**A**), is the spanning set of the rows of **A**; the **column space** of **A**, col(**A**), is the spanning set of the columns of **A**. If **A** is any matrix, then the dimension of the column space of **A** is equal to the dimension of the row space of **A**. The dimension of the row space (column space) of a matrix **A** is called the **rank** of **A**. The **nullspace** of **A** is the set of solutions to the system of equations $\mathbf{Ax} = \mathbf{0}$. The nullspace of **A** is a subspace and its dimension is called the **nullity** of **A**. The rank of **A** is equal to the number of nonzero rows in the row echelon form of **A**, the nullity of **A** is equal to the number of zero rows in the row echelon form of **A**. Thus, if **A** is a square matrix, the sum of the rank of **A** and the nullity of **A** is equal to the number of rows (columns) of **A**.

1. `NullSpace(A)` returns a list of vectors which form a basis for the nullspace (or kernel) of the matrix **A**.
2. `ColumnSpace(A)` returns a list of vectors which form a basis for the column space of the matrix **A**.
3. `ReducedRowEchelonForm(A)` yields the reduced row echelon form of the matrix **A**.

`NullSpace`, `ColumnSpace`, and `ReducedRowEchelonForm` are contained in the `LinearAlgebra` package.

EXAMPLE 5.3.1: Place the matrix

$$\mathbf{A} = \begin{pmatrix} -1 & -1 & 2 & 0 & -1 \\ -2 & 2 & 0 & 0 & -2 \\ 2 & -1 & -1 & 0 & 1 \\ -1 & -1 & 1 & 2 & 2 \\ 1 & -2 & 2 & -2 & 0 \end{pmatrix}$$

in reduced row echelon form. What is the rank of **A**? Find a basis for the nullspace of **A**.

SOLUTION: We begin by defining the matrix `A`. Then, `ReducedRowEchelonForm` is used to place `A` in reduced row echelon form.

```
> with(LinearAlgebra):
> A:=Matrix([[-1,-1,2,0,-1],[-2,2,0,0,-2],
> [2,-1,-1,0,1],[-1,-1,1,2,2],[1,-2,2,-2,0]]):
> ReducedRowEchelonForm(A);
```

$$\begin{bmatrix} 1 & 0 & 0 & -2 & 0 \\ 0 & 1 & 0 & -2 & 0 \\ 0 & 0 & 1 & -2 & 0 \\ 0 & 0 & 0 & 0 & 1 \\ 0 & 0 & 0 & 0 & 0 \end{bmatrix}$$

Because the row-reduced form of `A` contains four nonzero rows, the rank of **A** is 4 and thus the nullity is 1. We obtain a basis for the nullspace with `NullSpace`.

```
> NullSpace(A);
```

$$\left\{ \begin{bmatrix} 2 \\ 2 \\ 2 \\ 1 \\ 0 \end{bmatrix} \right\}$$

As expected, because the nullity is 1, a basis for the nullspace contains one vector.

■

EXAMPLE 5.3.2: Find a basis for the column space of

$$\mathbf{B} = \begin{pmatrix} 1 & -2 & 2 & 1 & -2 \\ 1 & 1 & 2 & -2 & -2 \\ 1 & 0 & 0 & 2 & -1 \\ 0 & 0 & 0 & -2 & 0 \\ -2 & 1 & 0 & 1 & 2 \end{pmatrix}.$$

SOLUTION: A basis for the column space of **B** is the same as a basis for the row space of the transpose of **B**. We begin by defining `B` and then using `Transpose` to compute the transpose of `B`, naming the resulting output `Bt`.

```
> with(LinearAlgebra):
> B:=Matrix([[1,-2,2,1,-2],[1,1,2,-2,-2],[1,0,0,2,-1],
> [0,0,0,-2,0],[-2,1,0,1,2]]):
```

```
> Bt:=Transpose(B);
```

$$Bt := \begin{bmatrix} 1 & 1 & 1 & 0 & -2 \\ -2 & 1 & 0 & 0 & 1 \\ 2 & 2 & 0 & 0 & 0 \\ 1 & -2 & 2 & -2 & 1 \\ -2 & -2 & -1 & 0 & 2 \end{bmatrix}$$

Next, we use `ReducedRowEchelonForm` to row reduce `Bt` and name the result `rrBtt`. A basis for the column space consists of the first four elements of `rrBtt`. We also use `Transpose` to show that the first four elements of `rrBtt` are the same as the first four columns of the transpose of `rrBtt`. Thus, the jth column of a matrix **A** can be extracted from **A** with `Row(Transpose(A),j)`.

```
> rrBt:=ReducedRowEchelonForm(Bt):
> rrBtt:=Transpose(rrBt);
```

$$rrBtt := \begin{bmatrix} 1 & 0 & 0 & 0 & 0 \\ 0 & 1 & 0 & 0 & 0 \\ 0 & 0 & 1 & 0 & 0 \\ 0 & 0 & 0 & 1 & 0 \\ -1/3 & 1/3 & -2 & -3 & 0 \end{bmatrix}$$

More easily, a basis for the column space of **B** is found with `ColumnSpace`.

```
> ColumnSpace(B);
```

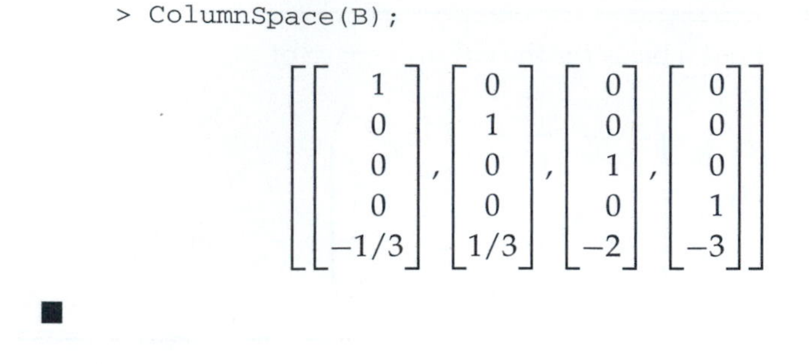

■

5.3.2 The Gram-Schmidt Process

A set of vectors $\{\mathbf{v}_1, \mathbf{v}_2, \ldots, \mathbf{v}_n\}$ is **orthonormal** means that $\|\mathbf{v}_i\| = 1$ for all values of i and $\mathbf{v}_i \cdot \mathbf{v}_j = 0$ for $i \neq j$. Given a set of linearly independent vectors $S = \{\mathbf{v}_1, \mathbf{v}_2, \ldots, \mathbf{v}_n\}$, the set of all linear combinations of the elements of S, $V = \text{span } S$, is a vector space. Note that if S is an orthonormal set and $\mathbf{u} \in \text{span } S$, then $\mathbf{u} = (\mathbf{u} \cdot \mathbf{v}_1)\, \mathbf{v}_1 + (\mathbf{u} \cdot \mathbf{v}_2)\, \mathbf{v}_2 + \cdots + (\mathbf{u} \cdot \mathbf{v}_n)\, \mathbf{v}_n$. Thus, we may easily express $\mathbf{u}$ as a linear combination of the vectors in S. Consequently, if we are given any vector space, V, it is frequently convenient to be able to find an orthonormal basis of V. We may use the **Gram-Schmidt process** to find an orthonormal basis of the vector space $V = \text{span } \{\mathbf{v}_1, \mathbf{v}_2, \ldots, \mathbf{v}_n\}$.

We summarize the algorithm of the Gram-Schmidt process so that given a set of n linearly independent vectors $S = \{\mathbf{v}_1, \mathbf{v}_1, \ldots, \mathbf{v}_n\}$, where $V = \text{span } \{\mathbf{v}_1, \mathbf{v}_2, \ldots, \mathbf{v}_n\}$, we can construct a set of orthonormal vectors $\{\mathbf{u}_1, \mathbf{u}_2, \ldots, \mathbf{u}_n\}$ so that $V = \text{span } \{\mathbf{u}_1, \mathbf{u}_2, \ldots, \mathbf{u}_n\}$.

1. Let $\mathbf{u}_1 = \dfrac{1}{\|\mathbf{v}\|}\mathbf{v}$;

2. Compute $\text{proj}_{\{\mathbf{u}_1\}}\mathbf{v}_2 = (\mathbf{u}_1 \cdot \mathbf{v}_2)\, \mathbf{u}_1$, $\mathbf{v}_2 - \text{proj}_{\{\mathbf{u}_1\}}\mathbf{v}_2$, and let

$$\mathbf{u}_2 = \frac{1}{\left\|\mathbf{v}_2 - \text{proj}_{\{\mathbf{u}_1\}}\mathbf{v}_2\right\|}\left(\mathbf{v}_2 - \text{proj}_{\{\mathbf{u}_1\}}\mathbf{v}_2\right).$$

   Then, span $\{\mathbf{u}_1, \mathbf{u}_2\} = \text{span } \{\mathbf{v}_1, \mathbf{v}_2\}$ and

$$\text{span } \{\mathbf{u}_1, \mathbf{u}_2, \mathbf{v}_3, \ldots, \mathbf{v}_n\} = \text{span } \{\mathbf{v}_1, \mathbf{v}_1, \ldots, \mathbf{v}_n\};$$

3. Generally, for $3 \leq i \leq n$, compute

$$\text{proj}_{\{\mathbf{u}_1, \mathbf{u}_2, \ldots, \mathbf{u}_n\}}\mathbf{v}_i = (\mathbf{u}_1 \cdot \mathbf{v}_i)\, \mathbf{u}_1 + (\mathbf{u}_2 \cdot \mathbf{v}_i)\, \mathbf{u}_2 + \cdots + \left(\mathbf{u}_{i-1} \cdot \mathbf{v}_i\right) \mathbf{u}_{i-1},$$

$\mathbf{v}_i - \text{proj}_{\{\mathbf{u}_1,\mathbf{u}_2,\ldots,\mathbf{u}_n\}}\mathbf{v}_i$, and let

$$\mathbf{u}_i = \frac{1}{\left\|\text{proj}_{\{\mathbf{u}_1,\mathbf{u}_2,\ldots,\mathbf{u}_n\}}\mathbf{v}_i\right\|}\left(\text{proj}_{\{\mathbf{u}_1,\mathbf{u}_2,\ldots,\mathbf{u}_n\}}\mathbf{v}_i\right).$$

Then, span $\{\mathbf{u}_1, \mathbf{u}_2, \ldots, \mathbf{u}_i\} =$ span $\{\mathbf{v}_1, \mathbf{v}_2, \ldots, \mathbf{v}_i\}$ and

$$\text{span}\left\{\mathbf{u}_1, \mathbf{u}_2, \ldots, \mathbf{u}_i, \mathbf{v}_{i+1}, \ldots, \mathbf{v}_n\right\} = \text{span}\left\{\mathbf{v}_1, \mathbf{v}_2, \mathbf{v}_3, \ldots, \mathbf{v}_n\right\};$$

and

4. Because span $\{\mathbf{u}_1, \mathbf{u}_2, \ldots, \mathbf{u}_n\} =$ span $\{\mathbf{v}_1, \mathbf{v}_2, \ldots, \mathbf{v}_n\}$ and $\{\mathbf{u}_1, \mathbf{u}_2, \ldots, \mathbf{u}_n\}$ is an orthonormal set, $\{\mathbf{u}_1, \mathbf{u}_2, \ldots, \mathbf{u}_n\}$ is an orthonormal basis of V.

The Gram-Schmidt process is well suited to computer arithmetic. Given a set of vectors, V, the `LinearAlgebra` command `GramSchmidt` returns a set of orthogonal vectors with the same span as V; including the `normalized=true` option in the `GramSchmidt` command results in an orthonormal set of vectors with the same span as V.

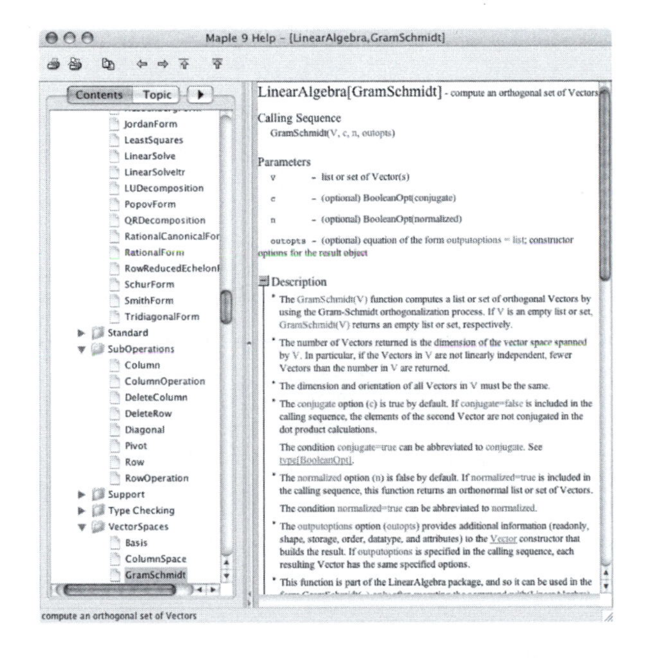

EXAMPLE 5.3.3: Use the Gram-Schmidt process to transform the basis $S = \left\{ \begin{pmatrix} -2 \\ -1 \\ -2 \end{pmatrix}, \begin{pmatrix} 0 \\ -1 \\ 2 \end{pmatrix}, \begin{pmatrix} 1 \\ 3 \\ -2 \end{pmatrix} \right\}$ of $\mathbf{R}^3$ into an orthonormal basis.

SOLUTION: We proceed by defining v1, v2, and v3 to be the vectors in the basis S and use `GramSchmidt({v1,v2,v3})` to find an orthogonal basis

```
> with(LinearAlgebra):
> v1:=Vector([-2,-1,-2]):
> v2:=Vector([0,-1,2]):
> v3:=Vector([1,3,-2]):
> GramSchmidt(v1,v2,v3);
```

$$\left\{ \begin{bmatrix} -2 \\ -1 \\ -2 \end{bmatrix}, \begin{bmatrix} -2/3 \\ -4/3 \\ 4/3 \end{bmatrix}, \begin{bmatrix} -4/9 \\ 4/9 \\ 2/9 \end{bmatrix} \right\}$$

and then `GramSchmidt({v1,v2,v3},normalized=true)` to find an orthonormal basis.

```
> GramSchmidt(v1,v2,v3,normalized=true);
```

$$\left\{ \begin{bmatrix} -2/3 \\ -1/3 \\ -2/3 \end{bmatrix}, \begin{bmatrix} -1/3 \\ -2/3 \\ 2/3 \end{bmatrix}, \begin{bmatrix} -2/3 \\ 2/3 \\ 1/3 \end{bmatrix} \right\}$$

■

EXAMPLE 5.3.4: Compute an orthonormal basis for the subspace of $\mathbf{R}^4$ spanned by the vectors $\begin{pmatrix} 2 \\ 4 \\ 4 \\ 1 \end{pmatrix}$, $\begin{pmatrix} -4 \\ 1 \\ -3 \\ 2 \end{pmatrix}$, and $\begin{pmatrix} 1 \\ 4 \\ 4 \\ -1 \end{pmatrix}$. Also, verify that the basis vectors are orthogonal and have norm 1.

SOLUTION: With `GramSchmidt`, we compute orthogonal and orthonormal basis vectors. The orthogonality of these vectors is then verified. The norm of each vector is then found to be 1.

```
> with(LinearAlgebra):
> oset1:=GramSchmidt([Vector([2,4,4,1]),
> Vector([-4,1,-3,2]),Vector([1,4,4,-1])]);
```

$$oset1 := \left[\begin{bmatrix} 2 \\ 4 \\ 4 \\ 1 \end{bmatrix}, \begin{bmatrix} -\frac{120}{37} \\ \frac{93}{37} \\ -\frac{55}{37} \\ \frac{88}{37} \end{bmatrix}, \begin{bmatrix} -\frac{449}{457} \\ \frac{268}{457} \\ \frac{156}{457} \\ -\frac{798}{457} \end{bmatrix} \right]$$

```
> oset2:=GramSchmidt([Vector([2,4,4,1]),
> Vector([-4,1,-3,2]),Vector([1,4,4,-1])],
  normalized=true);
```

$$oset2 := \left[\begin{bmatrix} \frac{2}{37}\sqrt{37} \\ \frac{4}{37}\sqrt{37} \\ \frac{4}{37}\sqrt{37} \\ 1/37\sqrt{37} \end{bmatrix}, \begin{bmatrix} -\frac{60}{16909}\sqrt{33818} \\ \frac{93}{33818}\sqrt{33818} \\ -\frac{55}{33818}\sqrt{33818} \\ \frac{44}{16909}\sqrt{33818} \end{bmatrix}, \begin{bmatrix} -\frac{449}{934565}\sqrt{934565} \\ \frac{268}{934565}\sqrt{934565} \\ \frac{156}{934565}\sqrt{934565} \\ -\frac{798}{934565}\sqrt{934565} \end{bmatrix} \right]$$

The three vectors are extracted with `oset2` with `oset2[1]`, `oset2[2]`, and `oset2[3]`.

```
> map(Norm,oset2,Frobenius);
```

$$[1, 1, 1]$$

```
> array([seq([seq(oset2[i].oset2[j],i=1..3)],j=1..3)]);
```

$$\begin{bmatrix} 1 & 0 & 0 \\ 0 & 1 & 0 \\ 0 & 0 & 1 \end{bmatrix}$$

■

5.3.3 Linear Transformations

A function $T : \mathbf{R}^n \longrightarrow \mathbf{R}^m$ is a **linear transformation** means that T satisfies the properties $T(\mathbf{u} + \mathbf{v}) = T(\mathbf{u}) + T(\mathbf{v})$ and $T(c\mathbf{u}) = cT(\mathbf{u})$ for all vectors $\mathbf{u}$ and $\mathbf{v}$ in $\mathbf{R}^n$ and all real numbers c. Let $T : \mathbf{R}^n \longrightarrow \mathbf{R}^m$ be a linear transformation and suppose $T(\mathbf{e}_1) = \mathbf{v}_1$, $T(\mathbf{e}_2) = \mathbf{v}_2, \ldots, T(\mathbf{e}_n) = \mathbf{v}_n$ where $\{\mathbf{e}_1, \mathbf{e}_2, \ldots, \mathbf{e}_n\}$ represents the standard basis of $\mathbf{R}^n$ and $\mathbf{v}_1, \mathbf{v}_2, \ldots, \mathbf{v}_n$ are (column) vectors in $\mathbf{R}^m$. The **associated matrix** of T is the $m \times n$ matrix $\mathbf{A} = \begin{pmatrix} \mathbf{v}_1 & \mathbf{v}_2 & \cdots & \mathbf{v}_n \end{pmatrix}$:

$$\text{if } \mathbf{x} = \begin{pmatrix} x_1 \\ x_2 \\ \vdots \\ x_n \end{pmatrix}, \quad T(\mathbf{x}) = T\left(\begin{pmatrix} x_1 \\ x_2 \\ \vdots \\ x_n \end{pmatrix}\right) = \mathbf{A}\mathbf{x} = \begin{pmatrix} \mathbf{v}_1 & \mathbf{v}_2 & \cdots & \mathbf{v}_n \end{pmatrix} \begin{pmatrix} x_1 \\ x_2 \\ \vdots \\ x_n \end{pmatrix}$$

Moreover, if $\mathbf{A}$ is any $m \times n$ matrix, then $\mathbf{A}$ is the associated matrix of the linear transformation defined by $T(\mathbf{x}) = \mathbf{A}\mathbf{x}$. In fact, a linear transformation T is completely determined by its action on any basis.

The **kernel** of the linear transformation T, $\ker(T)$, is the set of all vectors $\mathbf{x}$ in $\mathbf{R}^n$ such that $T(\mathbf{x}) = \mathbf{0}$: $\ker(T) = \{x \in \mathbf{R}^n | T(\mathbf{x}) = \mathbf{0}\}$. The kernel of T is a subspace of $\mathbf{R}^n$. Because $T(\mathbf{x}) = \mathbf{A}\mathbf{x}$ for all $\mathbf{x}$ in $\mathbf{R}^n$, $\ker(T) = \{x \in \mathbf{R}^n | T(\mathbf{x}) = \mathbf{0}\} = \{x \in \mathbf{R}^n | \mathbf{A}\mathbf{x} = \mathbf{0}\}$ so the kernel of T is the same as the nullspace of $\mathbf{A}$.

EXAMPLE 5.3.5: Let $T : \mathbf{R}^5 \longrightarrow \mathbf{R}^3$ be the linear transformation defined by $T(\mathbf{x}) = \begin{pmatrix} 0 & -3 & -1 & -3 & -1 \\ -3 & 3 & -3 & -3 & -1 \\ 2 & 2 & -1 & 1 & 2 \end{pmatrix} \mathbf{x}$. (a) Calculate a basis for the kernel of the linear transformation. (b) Determine which of the vectors $\begin{pmatrix} 4 \\ 2 \\ 0 \\ 0 \\ -6 \end{pmatrix}$ and $\begin{pmatrix} 1 \\ 2 \\ -1 \\ -2 \\ 3 \end{pmatrix}$ is in the kernel of T.

SOLUTION: We begin by defining `A` to be the matrix $\mathbf{A} = \begin{pmatrix} 0 & -3 & -1 & -3 & -1 \\ -3 & 3 & -3 & -3 & -1 \\ 2 & 2 & -1 & 1 & 2 \end{pmatrix}$ and then defining `t`. A basis for the kernel of T is the same as a basis for the nullspace of $\mathbf{A}$ found with `NullSpace`.

```
> with(LinearAlgebra):
> A:=Matrix([[0,-3,-1,-3,-1],[-3,3,-3,-3,-1],
    [2,2,-1,1,2]]):
> t:=x->A.x:

> na:=NullSpace(A);
```

$$na := \left\{ \begin{bmatrix} -\frac{2}{3} \\ -\frac{1}{3} \\ 0 \\ 0 \\ 1 \end{bmatrix}, \begin{bmatrix} -\frac{6}{13} \\ -\frac{8}{13} \\ -\frac{15}{13} \\ 1 \\ 0 \end{bmatrix} \right\}$$

Because $\begin{pmatrix} 4 \\ 2 \\ 0 \\ 0 \\ -6 \end{pmatrix}$ is a linear combination of the vectors that form a basis for the kernel, $\begin{pmatrix} 4 \\ 2 \\ 0 \\ 0 \\ -6 \end{pmatrix}$ is in the kernel while $\begin{pmatrix} 1 \\ 2 \\ -1 \\ -2 \\ 3 \end{pmatrix}$ is not. These results are verified more easily by evaluating t for each vector.

```
> t(na[1]);
```

$$\begin{bmatrix} 0 \\ 0 \\ 0 \end{bmatrix}$$

```
> t(Vector([4,2,0,0,-6]));
```

$$\begin{bmatrix} 0 \\ 0 \\ 0 \end{bmatrix}$$

Of course, any linear combination of the vectors in na is in the nullspace of $\mathbf{A}$.

```
> t(a*na[1]+b*na[2]);
```

$$\begin{bmatrix} 0 \\ 0 \\ 0 \end{bmatrix}$$

```
> t(Vector([1,2,-1,-2,3]));
```

$$\begin{bmatrix} -2 \\ 9 \\ 11 \end{bmatrix}$$

■

Application: Rotations

Let $\mathbf{x} = \begin{pmatrix} x_1 \\ x_2 \end{pmatrix}$ be a vector in $\mathbf{R}^2$ and θ an angle. Then, there are numbers r and ϕ given by $r = \sqrt{x_1{}^2 + x_2{}^2}$ and $\phi = \tan^{-1}(x_2/x_1)$ so that $x_1 = r\cos\phi$ and $x_2 = r\sin\phi$. When we rotate $\mathbf{x} = \begin{pmatrix} x_1 \\ x_2 \end{pmatrix} = \begin{pmatrix} r\cos\phi \\ r\sin\phi \end{pmatrix}$ through the angle θ, we obtain the vector $\mathbf{x}' = \begin{pmatrix} r\cos(\theta+\phi) \\ r\sin(\theta+\phi) \end{pmatrix}$. Using the trigonometric identities $\sin(\theta \pm \phi) = \sin\theta\cos\phi \pm \sin\phi\cos\theta$ and $\cos(\theta\pm\phi) = \cos\theta\cos\phi \mp \sin\theta\sin\phi$ we rewrite

$$\mathbf{x}' = \begin{pmatrix} r\cos(\theta+\phi) \\ r\sin(\theta+\phi) \end{pmatrix} = \begin{pmatrix} r\cos\theta\cos\phi - r\sin\theta\sin\phi \\ r\sin\theta\cos\phi + r\sin\phi\cos\theta \end{pmatrix} = \begin{pmatrix} \cos\theta & -\sin\theta \\ \sin\theta & \cos\theta \end{pmatrix} \begin{pmatrix} r\cos\phi \\ r\sin\phi \end{pmatrix}$$

$$= \begin{pmatrix} \cos\theta & -\sin\theta \\ \sin\theta & \cos\theta \end{pmatrix} \begin{pmatrix} x_1 \\ x_2 \end{pmatrix}.$$

Thus, the vector $\mathbf{x}'$ is obtained from $\mathbf{x}$ by computing $\begin{pmatrix} \cos\theta & -\sin\theta \\ \sin\theta & \cos\theta \end{pmatrix}\mathbf{x}$. Generally, if θ represents an angle, the linear transformation $T : \mathbf{R}^2 \longrightarrow \mathbf{R}^2$ defined by $T(\mathbf{x}) = \begin{pmatrix} \cos\theta & -\sin\theta \\ \sin\theta & \cos\theta \end{pmatrix}\mathbf{x}$ is called the **rotation of $\mathbf{R}^2$ through the angle θ**.

We can use the `rotate` command that is contained in the `plottools` package to rotate two- and three-dimensional graphics objects.

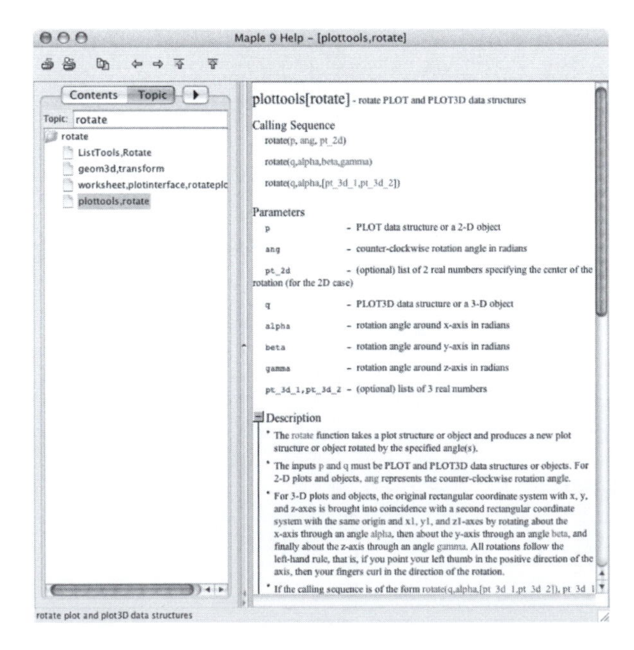

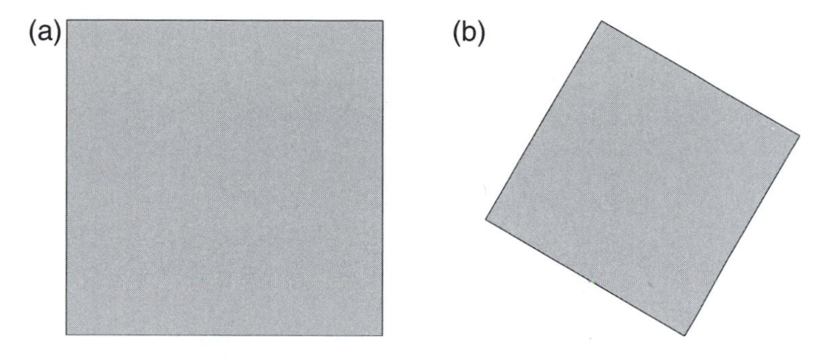

Figure 5-3 (a) A square. (b) A rotated square

As an illustration of `rotate`, we first load the `plots` and `plottools` packages

```
> with(plots):
> with(plottools):
```

and then use `rectangle` to create a gray square with corners at $(-1/2, -1/2)$ and $(1/2, 1/2)$. The resulting graphics object, named `sq`, is displayed with `display` in Figure 5-3(a).

```
> sq:=rectangle([-1/2,-1/2],[1/2,1/2],color=GRAY):
> display(sq,axes=NONE);
```

Next, we use `rotate` to rotate the square counter-clockwise $\pi/3$ radians. We use `display` to see the rotated square in Figure 5-3(b).

```
> display(rotate(sq, Pi/3),axes=NONE);
```

We can rotate the square through various angles and either animate the result or display the result as an array with `display` together with the option `insequence=true`. We begin by defining `thetavals` to be nine equally spaced numbers between 0 and $\pi/2$. Then, we use `seq` and `display` to rotate the square about each of the angles in `thetavals` and name the resulting list of graphics objects `toanimate`.

```
> thetavals:=seq(j*Pi/16,j=0..8):
```

```
> toanimate:=[seq(display(rotate(sq,theta),axes=NONE),
   theta=thetavals)]:
```

When we use `display` to display `toanimate`, all nine graphs are shown together, as illustrated in Figure 5-4.

```
> display(toanimate);
```

Figure 5-4 Rotated squares displayed together

To display `toanimate` as an array or as an animation, we use `display` together with the option `insequence=true`. Entering

```
> p:=display(toanimate,insequence=true):
> display(p);
```

displays `p` as an array of graphics in Figure 5-5, while entering

```
> display(p,insequence=true);
```

animates the graphics in `p`.

5.3.4 Eigenvalues and Eigenvectors

Let $\mathbf{A}$ be an $n \times n$ matrix. λ is an **eigenvalue** of $\mathbf{A}$ if there is a *nonzero* vector, $\mathbf{v}$, called an **eigenvector**, satisfying

$$\mathbf{Av} = \lambda\mathbf{v}. \tag{5.2}$$

We find the eigenvalues of $\mathbf{A}$ by solving the **characteristic polynomial**

$$|\mathbf{A} - \lambda\mathbf{I}| = 0 \tag{5.3}$$

Figure 5-5 Rotated squares shown as an array

for λ. Once we find the eigenvalues, the corresponding eigenvectors are found by solving

$$(\mathbf{A} - \lambda\mathbf{I})\,\mathbf{v} = \mathbf{0} \tag{5.4}$$

for $\mathbf{v}$.

If $\mathbf{A}$ is a square matrix,

```
Eigenvalues(A)
```

finds the eigenvalues of $\mathbf{A}$ and

```
Eigenvectors(A)
```

finds the eigenvalues and corresponding eigenvectors of $\mathbf{A}$.

```
CharacteristicPolynomial(A,lambda)
```

finds the characteristic polynomial of **A** as a function of `λ`. `Eigenvalues`, `Eigenvectors`, and `CharacteristicPolynomial` are contained in the `LinearAlgebra` package.

EXAMPLE 5.3.6: Find the eigenvalues and corresponding eigenvectors for each of the following matrices. (a) $\mathbf{A} = \begin{pmatrix} -3 & 2 \\ 2 & -3 \end{pmatrix}$; (b) $\mathbf{A} = \begin{pmatrix} 1 & -1 \\ 1 & 3 \end{pmatrix}$; (c) $\mathbf{A} = \begin{pmatrix} 0 & 1 & 1 \\ 1 & 0 & 1 \\ 1 & 1 & 0 \end{pmatrix}$; (d) $\mathbf{A} = \begin{pmatrix} -1/4 & 2 \\ -8 & -1/4 \end{pmatrix}$.

SOLUTION: (a) We begin by finding the eigenvalues. Solving

$$|\mathbf{A} - \lambda \mathbf{I}| = \begin{vmatrix} -3-\lambda & 2 \\ 2 & -3-\lambda \end{vmatrix} = \lambda^2 + 6\lambda + 5 = 0$$

gives us $\lambda_1 = -5$ and $\lambda_2 = -1$.

Observe that the same results are obtained using `CharacteristicPolynomial` and `Eigenvalues`.

```
> with(LinearAlgebra):
> A:=Matrix([[-3,2],[2,-3]]):
> factor(CharacteristicPolynomial(A,lambda));
```

$$(\lambda + 5)\,(\lambda + 1)$$

```
> e1:=Eigenvalues(A);
```

$$e1 := \begin{bmatrix} -1 \\ -5 \end{bmatrix}$$

We now find the corresponding eigenvectors. Let $\mathbf{v}_1 = \begin{pmatrix} x_1 \\ y_1 \end{pmatrix}$ be an eigenvector corresponding to λ_1, then

$$(\mathbf{A} - \lambda_1 \mathbf{I})\,\mathbf{v}_1 = \mathbf{0}$$

$$\left[\begin{pmatrix} -3 & 2 \\ 2 & -3 \end{pmatrix} - (-5)\begin{pmatrix} 1 & 0 \\ 0 & 1 \end{pmatrix}\right]\begin{pmatrix} x_1 \\ y_1 \end{pmatrix} = \begin{pmatrix} 0 \\ 0 \end{pmatrix}$$

$$\begin{pmatrix} 2 & 2 \\ 2 & 2 \end{pmatrix}\begin{pmatrix} x_1 \\ y_1 \end{pmatrix} = \begin{pmatrix} 0 \\ 0 \end{pmatrix},$$

which row-reduces to

$$\begin{pmatrix} 1 & 1 \\ 0 & 0 \end{pmatrix} \begin{pmatrix} x_1 \\ y_1 \end{pmatrix} = \begin{pmatrix} 0 \\ 0 \end{pmatrix}.$$

That is, $x_1 + y_1 = 0$ or $x_1 = -y_1$. Hence, for any value of $y_1 \neq 0$,

$$\mathbf{v}_1 = \begin{pmatrix} x_1 \\ y_1 \end{pmatrix} = \begin{pmatrix} -y_1 \\ y_1 \end{pmatrix} = y_1 \begin{pmatrix} -1 \\ 1 \end{pmatrix}$$

is an eigenvector corresponding to λ_1. Of course, this represents infinitely many vectors. But, they are all linearly dependent. Choosing $y_1 = 1$ yields $\mathbf{v}_1 = \begin{pmatrix} -1 \\ 1 \end{pmatrix}$. Note that you might have chosen $y_1 = -1$ and obtained $\mathbf{v}_1 = \begin{pmatrix} 1 \\ -1 \end{pmatrix}$. However, both of our results are "correct" because these vectors are linearly dependent.

Similarly, letting $\mathbf{v}_2 = \begin{pmatrix} x_2 \\ y_2 \end{pmatrix}$ be an eigenvector corresponding to λ_2 we solve $(\mathbf{A} - \lambda_2\mathbf{I})\,\mathbf{v}_1 = \mathbf{0}$:

$$\begin{pmatrix} -2 & 2 \\ 2 & -2 \end{pmatrix} \begin{pmatrix} x_2 \\ y_2 \end{pmatrix} = \begin{pmatrix} 0 \\ 0 \end{pmatrix} \quad \text{or} \quad \begin{pmatrix} 1 & -1 \\ 0 & 0 \end{pmatrix} \begin{pmatrix} x_2 \\ y_2 \end{pmatrix} = \begin{pmatrix} 0 \\ 0 \end{pmatrix}.$$

Thus, $x_2 - y_2 = 0$ or $x_2 = y_2$. Hence, for any value of $y_2 \neq 0$,

$$\mathbf{v}_2 = \begin{pmatrix} x_2 \\ y_2 \end{pmatrix} = \begin{pmatrix} y_2 \\ y_2 \end{pmatrix} = y_2 \begin{pmatrix} 1 \\ 1 \end{pmatrix}$$

is an eigenvector corresponding to λ_2. Choosing $y_2 = 1$ yields $\mathbf{v}_2 = \begin{pmatrix} 1 \\ 1 \end{pmatrix}$.

We confirm these results using `ReducedRowEchelonForm`.

```
> i2:=IdentityMatrix(2):
> ev1:=A-e1[1]*i2;
```

$$ev1 := \begin{bmatrix} -2 & 2 \\ 2 & -2 \end{bmatrix}$$

```
> ReducedRowEchelonForm(ev1);
```

$$\begin{bmatrix} 1 & -1 \\ 0 & 0 \end{bmatrix}$$

```
> ev2:=A-e1[2]*i2;
```

$$ev2 := \begin{bmatrix} 2 & 2 \\ 2 & 2 \end{bmatrix}$$

```
> ReducedRowEchelonForm(ev2);
```

$$\begin{bmatrix} 1 & 1 \\ 0 & 0 \end{bmatrix}$$

We obtain the same results using `Eigenvectors`.

```
> [Eigenvectors(A)];
```

$$\left[\begin{bmatrix} -5 \\ -1 \end{bmatrix}, \begin{bmatrix} -1 & 1 \\ 1 & 1 \end{bmatrix}\right]$$

(b) In this case, we see that $\lambda = 2$ has multiplicity 2. There is only one linearly independent eigenvector, $\mathbf{v} = \begin{pmatrix} -1 \\ 1 \end{pmatrix}$, corresponding to λ.

```
> A:=Matrix([[1,-1],[1,3]]):
> factor(CharacteristicPolynomial(A,lambda));
```

$$(\lambda - 2)^2$$

```
> [Eigenvectors(A)];
```

$$\left[\begin{bmatrix} 2 \\ 2 \end{bmatrix}, \begin{bmatrix} -1 & 0 \\ 1 & 0 \end{bmatrix}\right]$$

(c) The eigenvalue $\lambda_1 = 2$ has corresponding eigenvector $\mathbf{v}_1 = \begin{pmatrix} 1 \\ 1 \\ 1 \end{pmatrix}$.

The eigenvalue $\lambda_{2,3} = -1$ has multiplicity 2. In this case, there are two linearly independent eigenvectors corresponding to this eigenvalue: $\mathbf{v}_2 = \begin{pmatrix} -1 \\ 0 \\ 1 \end{pmatrix}$ and $\mathbf{v}_3 = \begin{pmatrix} -1 \\ 1 \\ 0 \end{pmatrix}$.

```
> A:=Matrix([[0,1,1],[1,0,1],[1,1,0]]):
> factor(CharacteristicPolynomial(A,lambda));
```

$$(\lambda - 2)\,(\lambda + 1)^2$$

```
> [Eigenvectors(A)];
```

$$\left[\begin{bmatrix} 2 \\ -1 \\ -1 \end{bmatrix}, \begin{bmatrix} 1 & -1 & -1 \\ 1 & 1 & 0 \\ 1 & 0 & 1 \end{bmatrix}\right]$$

(d) In this case, the eigenvalues $\lambda_{1,2} = -\frac{1}{4} \pm 4i$ are complex conjugates. We see that the eigenvectors $\mathbf{v}_{1,2} = \begin{pmatrix} 0 \\ 2 \end{pmatrix} \pm \begin{pmatrix} 1 \\ 0 \end{pmatrix} i$ are complex conjugates as well.

```
> A:=Matrix([[-1/4,2],[-8,-1/4]]):
> factor(CharacteristicPolynomial(A,lambda));
```

$$\lambda^2 + 1/2\,\lambda + \frac{257}{16}$$

```
> factor(CharacteristicPolynomial(A,lambda),I);
```

$$-1/16\,(-4\,\lambda - 1 + 16\,i)\,(4\,\lambda + 1 + 16\,i)$$

```
> [Eigenvectors(A)];
```

$$\left[\begin{bmatrix} -1/4 + 4\,i \\ -1/4 - 4\,i \end{bmatrix}, \begin{bmatrix} -1/2\,i & 1/2\,i \\ 1 & 1 \end{bmatrix}\right]$$

■

5.3.5 Jordan Canonical Form

Let $\mathbf{N}_k = (n_{ij}) = \begin{cases} 1, & j = i + 1 \\ 0, & \text{otherwise} \end{cases}$ represent a $k \times k$ matrix with the indicated elements. The $k \times k$ **Jordan block matrix** is given by $\mathbf{B}(\lambda) = \lambda\mathbf{I} + \mathbf{N}_k$ where λ is a constant:

$$\mathbf{N}_k = \begin{pmatrix} 0 & 1 & 0 & \cdots & 0 \\ 0 & 0 & 1 & \cdots & 0 \\ \vdots & \vdots & \vdots & & \vdots \\ 0 & 0 & 0 & \cdots & 1 \\ 0 & 0 & 0 & \cdots & 0 \end{pmatrix} \quad \text{and} \quad \mathbf{B}(\lambda) = \lambda\mathbf{I} + \mathbf{N}_k = \begin{pmatrix} \lambda & 1 & 0 & \cdots & 0 \\ 0 & \lambda & 1 & \cdots & 0 \\ \vdots & \vdots & \vdots & & \vdots \\ 0 & 0 & 0 & \cdots & 1 \\ 0 & 0 & 0 & \cdots & \lambda \end{pmatrix}.$$

Hence, $\mathbf{B}(\lambda)$ can be defined as $\mathbf{B}(\lambda) = (b_{ij}) = \begin{cases} \lambda, & i = j \\ 1, & j = i + 1 \\ 0, & \text{otherwise} \end{cases}$. A **Jordan matrix** has the form

$$\mathbf{J} = \begin{pmatrix} \mathbf{B}_1(\lambda) & 0 & \cdots & 0 \\ 0 & \mathbf{B}_2(\lambda) & \cdots & 0 \\ \vdots & \vdots & & \vdots \\ 0 & 0 & \cdots & \mathbf{B}_n(\lambda) \end{pmatrix}$$

where the entries $\mathbf{B}_j(\lambda), j = 1, 2, \ldots, n$ represent Jordan block matrices.

Suppose that $\mathbf{A}$ is an $n \times n$ matrix. Then there is an invertible $n \times n$ matrix $\mathbf{C}$ such that $\mathbf{C}^{-1}\mathbf{AC} = \mathbf{J}$ where $\mathbf{J}$ is a Jordan matrix with the eigenvalues of $\mathbf{A}$ as diagonal elements. The matrix $\mathbf{J}$ is called the **Jordan canonical form** of $\mathbf{A}$. The `LinearAlgebra` command

```
JordanForm(A)
```

returns the Jordan canonical form, $\mathbf{J}$,

```
JordanForm(A,output=Q)
```

returns the matrix $\mathbf{C}$, and

```
JordanForm(A,output=[J,Q])
```

returns a list consisting of $\mathbf{J}$ and $\mathbf{C}$.

For a given matrix $\mathbf{A}$, the unique monic polynomial q of least degree satisfying $q(\mathbf{A}) = 0$ is called the **minimal polynomial of A**. Let p denote the characteristic polynomial of $\mathbf{A}$. Because $p(\mathbf{A}) = 0$, it follows that q divides p. We can use the Jordan canonical form of a matrix to determine its characteristic and minimal polynomials. Note that the `LinearAlgebra` command

```
MinimalPolynomial(A,lambda)
```

finds the minimal polynomial of $\mathbf{A}$.

EXAMPLE 5.3.7: Find the Jordan canonical form, $\mathbf{J}_{\mathbf{A}}$, of $\mathbf{A} = \begin{pmatrix} 2 & 9 & -9 \\ 0 & 8 & -6 \\ 0 & 9 & -7 \end{pmatrix}$.

SOLUTION: After defining `A`, we use `JordanForm` to find the Jordan canonical form of `A` and name the resulting output `jA`.

```
> with(LinearAlgebra):
> A:=Matrix([[2,9,-9],[0,8,-6],[0,9,-7]]):
> jA:=JordanForm(A);
```

$$jA := \begin{bmatrix} -1 & 0 & 0 \\ 0 & 2 & 0 \\ 0 & 0 & 2 \end{bmatrix}$$

```
> jAQ:=JordanForm(A,output=Q);
```

$$jAQ := \begin{bmatrix} -3 & 3 & 0 \\ -2 & 4 & 1 \\ -3 & 4 & 1 \end{bmatrix}$$

```
> jAQ:=[JordanForm(A,output=[J,Q])];
```

$$jAQ := \left[\begin{bmatrix} -1 & 0 & 0 \\ 0 & 2 & 0 \\ 0 & 0 & 2 \end{bmatrix}, \begin{bmatrix} -3 & 3 & 0 \\ -2 & 4 & 1 \\ -3 & 4 & 1 \end{bmatrix}\right]$$

The Jordan matrix corresponds to the first element of `jAQ` extracted with `jAQ[1]`. We also verify that the matrices $\mathbf{J}$ = `jAQ[1]` and $\mathbf{C}$ = `jAQ[2]` satisfy the relationship $\mathbf{A} = \mathbf{CAC}^{-1}$.

```
> evalm(jAQ[2]&*jAQ[1&*MatrixInverse(jAQ[2]));
```

$$\begin{bmatrix} 2 & 9 & -9 \\ 0 & 8 & -6 \\ 0 & 9 & -7 \end{bmatrix}$$

From the Jordan matrix, we see that the characteristic polynomial is $(x+1)(x-2)^2$. We also use `CharacteristicPolynomial` to find the characteristic polynomial of `A` and then verify that `A` satisfies its characteristic polynomial.

```
> p:=CharacteristicPolynomial(A,lambda);
```

$$p := \lambda^3 - 3\lambda^2 + 4$$

```
> A^3-3*A^2+4*IdentityMatrix(3);
```

$$\begin{bmatrix} 0 & 0 & 0 \\ 0 & 0 & 0 \\ 0 & 0 & 0 \end{bmatrix}$$

From the Jordan form, we see that the minimal polynomial of **A** is $(x+1)(x-2)$ and confirm with `MinimalPolynomial`. We define the minimal polynomial to be `q` and then verify that `A` satisfies its minimal polynomial.

```
> q:=expand((lambda+1)*(lambda-2));
```

$$q := \lambda^2 - \lambda - 2$$

```
> MinimalPolynomial(A,lambda);
```

$$\lambda^2 - \lambda - 2$$

```
> A^2-A-2*IdentityMatrix(3);
```

$$\begin{bmatrix} 0 & 0 & 0 \\ 0 & 0 & 0 \\ 0 & 0 & 0 \end{bmatrix}$$

As expected, `q` divides `p`.

```
> simplify(p/q);
```

$$\lambda - 2$$

■

EXAMPLE 5.3.8: If $\mathbf{A} = \begin{pmatrix} 3 & 8 & 6 & -1 \\ -3 & 2 & 0 & 3 \\ 3 & -3 & -1 & -3 \\ 4 & 8 & 6 & -2 \end{pmatrix}$, find the characteristic and minimal polynomials of **A**.

SOLUTION: As in the previous example, we first define `A` and then use `JordanForm` to find the Jordan canonical form of **A**.

```
> with(LinearAlgebra):
> A:=Matrix([[3,8,6,-1],[-3,2,0,3],[3,-3,-1,-3],
   [4,8,6,-2]]):
> jAQ:=[JordanForm(A,output=[J,Q])];
```

$$\left[\begin{bmatrix} -1 & 0 & 0 & 0 \\ 0 & 2 & 1 & 0 \\ 0 & 0 & 2 & 0 \\ 0 & 0 & 0 & -1 \end{bmatrix}, \begin{bmatrix} 0 & -2 & 2 & 1 \\ -2 & 0 & -1 & -3 \\ 3 & 0 & 1 & 4 \\ 2 & -2 & 2 & 4 \end{bmatrix}\right]$$

The Jordan canonical form of **A** is the first element of `jAQ`. From this result, we see that the minimal polynomial of **A** is $(x+1)(x-2)^2$ and the characteristic polynomial is $(x+1)^2(x-2)^2$. We confirm this result with `MinimalPolynomial` and define `q` to be the minimal polynomial of **A** and then verify that `A` satisfies `q`.

```
> q:=MinimalPolynomial(A,lambda);
```

$$\lambda^3 - 3\lambda^2 + 4$$

```
> A^3-3*A^2+4*IdentityMatrix(4);
```

$$\begin{bmatrix} 0 & 0 & 0 & 0 \\ 0 & 0 & 0 & 0 \\ 0 & 0 & 0 & 0 \\ 0 & 0 & 0 & 0 \end{bmatrix}$$

The characteristic polynomial is obtained next with `Characteristic-Polynomial` and named `p`. As expected, `q` divides `p`, verified with `simplify`.

```
> p:=CharacteristicPolynomial(A,lambda);
```

$$\lambda^4 - 2\lambda^3 - 3\lambda^2 + 4\lambda + 4$$

```
> simplify(p/q);
```

$$\lambda + 1$$

■

5.3.6 The QR Method

The **conjugate transpose** (or **Hermitian adjoint matrix**) of the $m \times n$ complex matrix **A** which is denoted by $\mathbf{A}^*$ is the transpose of the complex conjugate of **A**. Symbolically, we have $\mathbf{A}^* = \left(\bar{\mathbf{A}}\right)^t$. A complex matrix **A** is **unitary** if $\mathbf{A}^* = \mathbf{A}^{-1}$. Given a matrix **A**, there is a unitary matrix **Q** and an upper triangular matrix **R** such that $\mathbf{A} = \mathbf{QR}$. The product matrix **QR** is called the **QR factorization of A**. The `LinearAlgebra` command

```
QRDecomposition(A)
```

determines the QR decomposition of the matrix **A** and returns the list `[Q,R]`, where `Q` is an orthogonal matrix, `R` is an upper triangular matrix and $\mathbf{A} =$ `QR`.

EXAMPLE 5.3.9: Find the QR factorization of the matrix $\mathbf{A} = \begin{pmatrix} 4 & -1 & 1 \\ -1 & 4 & 1 \\ 1 & 1 & 4 \end{pmatrix}$.

SOLUTION: We define A and then use QRDecomposition to find the QR decomposition of A, naming the resulting output qrm.

```
> with(LinearAlgebra):
> A:=Matrix([[4,-1,1],[-1,4,1],[1,1,4]]):
> qrm:=[QRDecomposition(A)];
```

$$qrm := \left[\begin{bmatrix} 2/3\sqrt{2} & 1/33\sqrt{22} & -1/11\sqrt{11} \\ -1/6\sqrt{2} & \frac{13}{66}\sqrt{22} & -1/11\sqrt{11} \\ 1/6\sqrt{2} & \frac{5}{66}\sqrt{22} & 3/11\sqrt{11} \end{bmatrix}, \begin{bmatrix} 3\sqrt{2} & -7/6\sqrt{2} & 7/6\sqrt{2} \\ 0 & 5/6\sqrt{22} & \frac{35}{66}\sqrt{22} \\ 0 & 0 & \frac{10}{11}\sqrt{11} \end{bmatrix}\right]$$

We verify that the results returned are the QR decomposition of **A**.

```
> evalm(qrm[1]&*qrm[2]);
```

$$\begin{bmatrix} 4 & -1 & 1 \\ -1 & 4 & 1 \\ 1 & 1 & 4 \end{bmatrix}$$

■

One of the most efficient and most widely used methods for numerically calculating the eigenvalues of a matrix is the QR method. Given a matrix **A**, then there is a Hermitian matrix **Q** and an upper triangular matrix **R** such that $\mathbf{A} = \mathbf{QR}$. If we define a sequence of matrices $\mathbf{A}_1 = \mathbf{A}$, factored as $\mathbf{A}_1 = \mathbf{Q}_1\mathbf{R}_1$; $\mathbf{A}_2 = \mathbf{R}_1\mathbf{Q}_1$, factored as $\mathbf{A}_2 = \mathbf{R}_2\mathbf{Q}_2$; $\mathbf{A}_3 = \mathbf{R}_2\mathbf{Q}_2$, factored as $\mathbf{A}_3 = \mathbf{R}_3\mathbf{Q}_3$; and in general, $\mathbf{A}_k = \mathbf{R}_{k+1}\mathbf{Q}_{k+1}$, $k = 1, 2, \ldots$ Then, the sequence $\{\mathbf{A}_n\}$ converges to a triangular matrix with the eigenvalues of **A** along the diagonal or to a nearly triangular matrix from which the eigenvalues of **A** can be calculated rather easily.

EXAMPLE 5.3.10: Consider the 3×3 matrix $\mathbf{A} = \begin{pmatrix} 4 & -1 & 1 \\ -1 & 4 & 1 \\ 1 & 1 & 4 \end{pmatrix}$.

Approximate the eigenvalues of **A** with the QR method.

SOLUTION: We define the sequence a and qr recursively. We define a and qr using proc with the remember option so that Maple "remembers" the values of a and qr computed, and thus avoids recomputing values previously computed. This is of particular advantage when computing a(n) and qr(n) for large values of n.

```
> a:='a':qr:='qr':
> a:=proc(n) option remember;
> evalm(qr(n-1)[2]&*qr(n-1)[1])
> end proc:
> qr:=proc(n) option remember;
> [QRDecomposition(Matrix(a(n)))]
> end proc:
> a(1):=A:
> qr(1):=[QRDecomposition(A)]:
```

We illustrate a(n) and qr(n) by computing qr(9) and a(10). Note that computing a(10) requires the computation of qr(9). From the results, we suspect that the eigenvalues of **A** are 5 and 2.

```
> qr(9);
```

$$\left[\begin{bmatrix} -0.99999996134530366 & -0.00000008926329305327337O3 & -0.000278045649784791985 \\ 0.000000223167013828192205 & -0.99999988403599250 & -0.000481588985849757377 \\ -0.000278045574553285468 & -0.000481589029284698676 & 0.99999984538132070 \end{bmatrix},\right.$$

$$\left.\begin{bmatrix} -4.99999955127346496 & 0.00000156171596893394247 & -0.00194631923823934379 \\ 0.0 & -4.99999864781953196 & -0.00337112366200416738 \\ 0.0 & 0.0 & 2.00000072057584699 \end{bmatrix}\right]$$

```
> a(10);
```

$$\begin{bmatrix} 4.999999899 & -0.0000001780733705 & -0.000556091565 \\ -0.0000001785087514 & 4.999999691 & -0.000963178863 \\ -0.0005560913497 & -0.0009631784058 & 2.000000412 \end{bmatrix}$$

Next, we compute a(n) for $n = 5$, 10, and 15. We obtain further evidence that the eigenvalues of **A** are 5 and 2.

```
> array([[a(5)],[a(10)],[a(15)]]);
```

$$\left[\begin{array}{c}\begin{bmatrix} 4.999017283 & -0.001701002608 & 0.0542613566 \\ -0.001701003045 & 4.997055705 & 0.0939218941 \\ 0.05426135638 & 0.09392189364 & 2.003927012 \end{bmatrix} \\ \begin{bmatrix} 4.999999899 & -0.0000001780733705 & -0.000556091565 \\ -0.0000001785087514 & 4.999999691 & -0.000963178863 \\ -0.0005560913497 & -0.0009631784058 & 2.000000412 \end{bmatrix} \\ \begin{bmatrix} 5.000000002 & 0.0000000004463944226 & 0.00000569459130 \\ 1.102914355 \times 10^{-11} & 5.000000001 & 0.00000986340624 \\ 0.000005694375901 & 0.000009862948399 & 2.000000001 \end{bmatrix}\end{array}\right]$$

We verify that the eigenvalues of **A** are indeed 5 and 2 with `Eigenvalues`.

```
> Eigenvalues(A):
```

$$\begin{bmatrix} 5.0 + 0.0\,i \\ 1.99999999999999956 + 0.0\,i \\ 5.0 + 0.0\,i \end{bmatrix}$$

■

5.4 Maxima and Minima Using Linear Programming

5.4.1 The Standard Form of a Linear Programming Problem

We call the linear programming problem of the following form the **standard form** of the linear programming problem:

$$\text{Minimize } Z = \underbrace{c_1x_1 + c_2x_2 + \cdots + c_nx_n}_{\text{function}}, \text{ subject to the restrictions}$$

$$\begin{cases} a_{11}x_1 + a_{12}x_2 + \cdots + a_{1n}x_n \geq b_1 \\ a_{21}x_1 + a_{22}x_2 + \cdots + a_{2n}x_n \geq b_2 \\ \vdots \\ a_{m1}x_1 + a_{m2}x_2 + \cdots + a_{mn}x_n \geq b_m \end{cases}, \tag{5.5}$$

and $x_1 \geq 0, x_2 \geq 0, \ldots, x_n \geq 0$.

The Maple command `minimize`, contained in the `simplex` package, solves the standard form of the linear programming problem. Similarly, the command `maximize`, also contained in the `simplex` package, solves the linear programming problem: Maximize $Z = \underbrace{c_1x_1 + c_2x_2 + \cdots + c_nx_n}_{\texttt{function}}$, subject to the restrictions

$$\begin{cases} a_{11}x_1 + a_{12}x_2 + \cdots + a_{1n}x_n \geq b_1 \\ a_{21}x_1 + a_{22}x_2 + \cdots + a_{2n}x_n \geq b_2 \\ \vdots \\ a_{m1}x_1 + a_{m2}x_2 + \cdots + a_{mn}x_n \geq b_m \end{cases},$$

and $x_1 \geq 0, x_2 \geq 0, \ldots, x_n \geq 0$.

Enter `?simplex` for basic help regarding the `simplex` package.

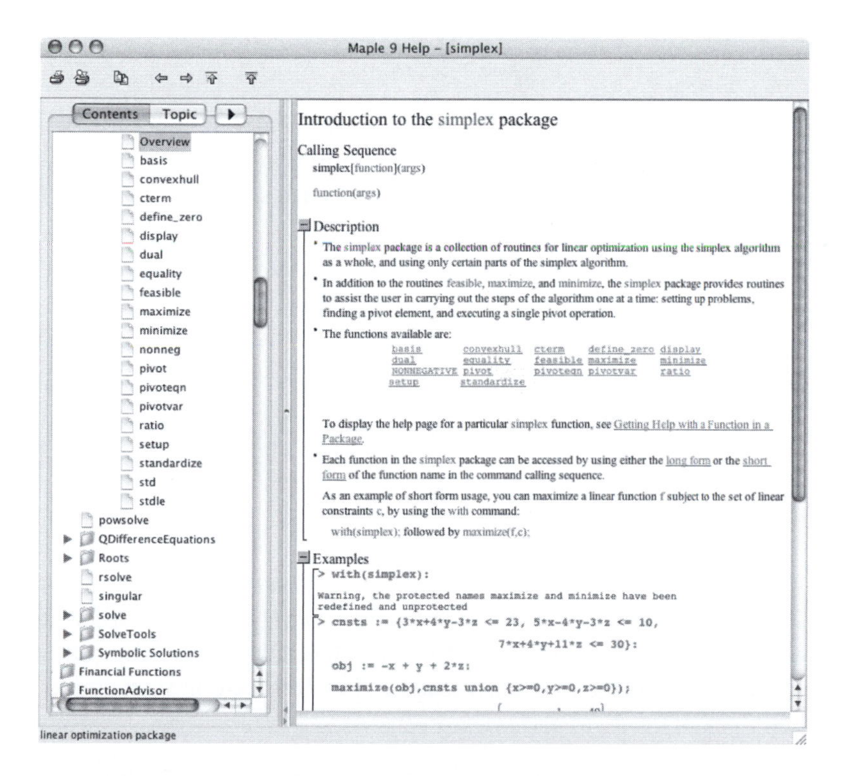

As when using commands contained in any other package, be sure to load the `simplex` package prior to using the commands `minimize` and `maximize`.

EXAMPLE 5.4.1: Maximize $Z(x_1, x_2, x_3) = 4x_1 - 3x_2 + 2x_3$ subject to the constraints $3x_1 - 5x_2 + 2x_3 \leq 60$, $x_1 - x_2 + 2x_3 \leq 10$, $x_1 + x_2 - x_3 \leq 20$, and x_1, x_2, x_3 all non-negative.

SOLUTION: In this case, we begin by clearing all prior definitions of the variables we will use, if any, and then loading the `simplex` package. After loading the `simplex` package, we define the objective function Z. In an effort to limit the amount of typing required to complete the problem, the set of inequalities is assigned the name `constraints`. The symbol "<=", obtained by typing the "<" key and then the "=" key, represents "less than or equal to" and is used in `constraints`.

```
> 'Z':x1:='x1':x2:='x2':x3:='x3':
> with(simplex):
> Z:=4*x1-3*x2+2*x3:
> constraints:=3*x1-5*x2+x3<=60,
> x1-x2+2*x3<=10,x1+x2-x3<=20:
```

The solution to the problem in which the non-negativity constraint is not considered is determined with `maximize` and named `sols_one` so that the maximum value of Z can be determined with `subs(sols_one,Z)`.

```
> sols_one:=maximize(Z,constraints);
```

$$sols\_one := \left\{x3 = -5, x1 = \frac{35}{2}, x2 = -5/2\right\}$$

```
> subs(sols_one,Z);
```

$$\frac{135}{2}$$

The non-negative constraints are indicated with `NONNEGATIVE` in the maximize command or entered explicitly along with the other constraints. We make use of the `NONNEGATIVE` setting to find the solution in `sols_two`. The maximum value of Z in this case is then found to be 45.

```
> sols_two:=maximize(Z,constraints,NONNEGATIVE);
```

$$sols\_two := \{x3 = 0, x2 = 5, x1 = 15\}$$

```
> assign(sols_two):
> Z;
```

$$45$$

■

We demonstrate the use of `minimize` in the following example.

EXAMPLE 5.4.2: Minimize $Z(x,y,z) = 4x - 3y + 2z$ subject to the constraints $3x - 5y + z \leq 60$, $x - y + 2z \leq 10$, $x + y - z \leq 20$, and x, y, and z all non-negative.

SOLUTION: We begin by loading the `simplex` package. The point at which Z has a minimum value is found with `minimize` and named `vals`. The value of $Z(x,y,z)$ at this point is then found to be -90 with `subs`, which substitutes the values of x, y, and z that were determined with `minimize` into the function $Z = Z(x,y,z)$.

```
> with(simplex):
> vals:=minimize(4*x-3*y+2*z,3*x-5*y+2*z<=60,
> x-y+2*z<=10,x+y-z<=20,NONNEGATIVE);
```

$$vals := \{y = 50, z = 30, x = 0\}$$

```
> subs(vals,4*x-3*y+2*z);
```

$$-90$$

We conclude that the minimum value is -90 and occurs if $x_1 = 0$, $x_2 = 50$, and $x_3 = 30$.

■

5.4.2 The Dual Problem

Given the standard form of the linear programming problem in (5.5), the **dual problem** is as follows: "Maximize $Y = \sum_{i=1}^{m} b_i y_y$ subject to the constraints $\sum_{i=1}^{m} a_{ij} y_i \leq c_{ij}$ for $j = 1, 2, \ldots, n$ and $y_i \geq 0$ for $i = 1, 2, \ldots, m$." Similarly, for the problem: "Maximize $Z = \sum_{j=1}^{n} c_j x_j$ subject to the constraints $\sum_{j=1}^{n} a_{ij} x_j \leq b_j$ for $i = 1, 2, \ldots, m$ and $x_j \geq 0$ for $j = 1, 2, \ldots, n$," the dual problem is as follows: "Minimize $Y = \sum_{i=1}^{m} b_i y_i$ subject to the constraints $\sum_{i=1}^{m} a_{ij} y_i \geq c_j$ for $j = 1, 2, \ldots, n$ and $y_i \geq 0$ for $i = 1, 2, \ldots, m$."

The simplex package contains the command dual(func,conlist,var), which produces the dual of the linear programming problem in standard form with objective function func and constraints conlist. The dual problem that results is given in terms of the dual variables var1, var2, ..., varn.

EXAMPLE 5.4.3: Maximize $Z = 6x + 8y$ subject to the constraints $5x + 2y \leq 20$, $x + 2y \leq 10$, $x \geq 0$, and $y \geq 0$. State the dual problem and find its solution.

SOLUTION: First, we solve the problem in its original form by using steps similar to those used in the previous example. The point at which the maximum value of the objective function occurs is determined and named max_sols. This maximum value is 45 which is found with subs.

```
> with(simplex):
> Z:=6*x+8*y:
> constraints:=5*x+2*y<=20,x+2*y<=10:
> max_sols:=maximize(Z,constraints,NONNEGATIVE);
```

$$max\_sols := \left\{y = \frac{15}{4}, x = 5/2\right\}$$

```
> subs(max_sols,Z);
```

$$45$$

In this case, we can graph the feasibility set determined by the constraints with inequal, which is contained in the plots package, as shown in Figure 5-6.

```
> with(plots):
> inequal( 5*x+2*y<=20,x+2*y<=10,x>=0,y>=0,
   x=-1..10,y=-1..10,
> optionsexcluded=(color=white,thickness=2) );
```

Because in this problem we have $c_1 = 6$, $c_2 = 8$, $b_1 = 20$, and $b_2 = 10$, the dual problem is as follows: Minimize $Z = 20y_1 + 10y_2$ subject to the constraints $5y_1 + y_2 \geq 6$, $2y_1 + 2y_2 \geq 8$, $y_1 \geq 0$, and $y_2 \geq 0$.

Since y is specified in the dual command, the two dual variables are given by y1 and y2. The resulting problem is assigned the name dual_problem. Then, the solution of the dual problem is easily found with minimize. The point at which this minimum value occurs is called

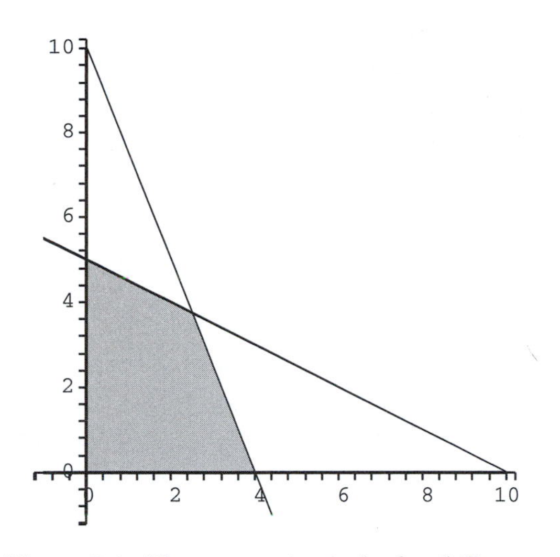

Figure 5-6 The gray region is the feasibility set

min_sols. By extracting the objective function of the dual problem with dual_problem[1], we are able to substitute the values in min_sols into this function with subs to see that we obtain the same optimal value as that found for the original problem.

```
> dual_problem:=dual(Z,constraints,y);
```

$$dual\_problem := 20\,y1 + 10\,y2,\ \{6 \leq 5\,y1 + y2, 8 \leq 2\,y1 + 2\,y2\}$$

```
> min_sols:=minimize(dual_problem,NONNEGATIVE);
```

$$min\_sols := \{y1 = 1/2, y2 = 7/2\}$$

```
> dual_problem[1];
```

$$20\,y1 + 10\,y2$$

```
> subs(min_sols,dual_problem[1]);
```

$$45$$

■

Of course, linear programming models can involve numerous variables. Consider the following: given the standard form linear programming problem in (5.5), let $\mathbf{x} = \begin{pmatrix} x_1 \\ x_2 \\ \vdots \\ x_n \end{pmatrix}$, $\mathbf{b} = \begin{pmatrix} b_1 \\ b_2 \\ \vdots \\ b_m \end{pmatrix}$, $\mathbf{c} = \begin{pmatrix} c_1 & c_2 & \cdots & c_n \end{pmatrix}$, and $\mathbf{A}$ denote the $m \times n$ matrix $\mathbf{A} = \begin{pmatrix} a_{11} & a_{12} & \cdots & a_{1n} \\ a_{21} & a_{22} & \cdots & a_{2n} \\ \vdots & \vdots & & \vdots \\ a_{m1} & a_{m2} & \cdots & a_{mn} \end{pmatrix}$. Then the standard form of the linear programming problem is equivalent to finding the vector $\mathbf{x}$ that maximizes $Z = \mathbf{c} \cdot \mathbf{x}$ subject to the restrictions $\mathbf{Ax} \geq \mathbf{b}$ and $x_1 \geq 0, x_2 \geq 0, \ldots, x_n \geq 0$. The dual problem is: "Minimize $Y = \mathbf{y} \cdot \mathbf{b}$ where $\mathbf{y} = \begin{pmatrix} y_1 & y_2 & \cdots & y_m \end{pmatrix}$ subject to the restrictions $\mathbf{yA} \leq \mathbf{c}$ (componentwise) and $y_1 \geq 0, y_2 \geq 0, \ldots, y_m \geq 0$."

The command

```
LinearProgramming[c,A,b]
```

finds the vector **x** that minimizes the quantity `Z=c.x` subject to the restrictions `A.x>=b` and `x>=0`. `LinearProgramming` does not yield the minimum value of `Z` as did `ConstrainedMin` and `ConstrainedMax` and the value must be determined from the resulting vector.

EXAMPLE 5.4.4: Maximize $Z = 5x_1 - 7x_2 + 7x_3 + 5x_4 + 6x_5$ subject to the constraints $2x_1 + 3x_2 + 3x_3 + 2x_4 + 2x_5 \geq 10$, $6x_1 + 5x_2 + 4x_3 + x_4 + 4x_5 \geq 30$, $-3x_1 - 2x_2 - 3x_3 - 4x_4 \geq -5$, $-x_1 - x_2 - x_4 \geq -10$, and $x_1 \geq 0$ for $i = 1$, 2, 3, 4, and 5. State the dual problem. What is its solution?

SOLUTION: For this problem, $\mathbf{x} = \begin{pmatrix} x_1 \\ x_2 \\ x_3 \\ x_4 \\ x_5 \end{pmatrix}$, $\mathbf{b} = \begin{pmatrix} 10 \\ 30 \\ -5 \\ -10 \end{pmatrix}$, $\mathbf{c} = \begin{pmatrix} 5 & -7 & 7 & 5 & 6 \end{pmatrix}$, and $\mathbf{A} = \begin{pmatrix} 2 & 3 & 3 & 2 & 2 \\ 6 & 5 & 4 & 1 & 4 \\ -3 & -2 & -3 & -4 & 0 \\ -1 & -1 & 0 & -1 & 0 \end{pmatrix}$. First, matrix **A** is entered and named `A` and then the vectors **c** and **b** are entered.

```
> with(linalg):
> Z:='Z':A:='A':c:='c':b:='b':x:='x':
```

```
> A:=array([[2,3,3,2,2],[6,5,4,1,4],
> [-3,-2,-3,-4,0],[-1,-1,0,-1,0],
> [0$5]]);
```

$$\begin{bmatrix} 2 & 3 & 3 & 2 & 2 \\ 6 & 5 & 4 & 1 & 4 \\ -3 & -2 & -3 & -4 & 0 \\ -1 & -1 & 0 & -1 & 0 \\ 0 & 0 & 0 & 0 & 0 \end{bmatrix}$$

```
> c:=array([[5,-7,7,5,6]]):
> b:=vector([10,30,-5,-10,0]):
> x:=vector(5):
```

The objective function Z is defined by finding the scalar (dot) product of the solution vector $\mathbf{x}$ and the vector c with `multiply`, which is found in the `linalg` package. This objective function is extracted from the output list that results with `Z[1]`.

```
> Z:=multiply(c,x);
```

$$Z := vector\left([5x_1 - 7x_2 + 7x_3 + 5x_4 + 6x_5]\right)$$

```
> Z[1];
```

$$5x_1 - 7x_2 + 7x_3 + 5x_4 + 6x_5$$

The left-hand sides of the constraints are then defined in `prod` by multiplying the coefficient matrix $\mathbf{A}$ by the solution vector $\mathbf{x}$. Notice that the output list is made up of four components that can be extracted with `prod[1]`, `prod[2]`, `prod[3]`, and `prod[4]`, respectively. In `constraints`, the set of constraints for this problem are constructed by substituting $i = 1, 2, 3, 4$ into the inequality `prod[i]>=b[i]`. Hence, a constraint is produced for each value of i.

```
> prod:=multiply(A,x);
```

$$\left[2x_1 + 3x_2 + 3x_3 + 2x_4 + 2x_5,\ 6x_1 + 5x_2 + 4x_3 + x_4 + 4x_5,\right.$$
$$\left. -3x_1 - 2x_2 - 3x_3 - 4x_4,\ -x_1 - x_2 - x_4, 0\right]$$

```
> constraints:=seq(prod[i]>=b[i],i=1..4);
```

$$constraints := \{10 \leq 2x_1 + 3x_2 + 3x_3 + 2x_4 + 2x_5,$$
$$30 \leq 6x_1 + 5x_2 + 4x_3 + x_4 + 4x_5,$$
$$-5 \leq -3x_1 - 2x_2 - 3x_3 - 4x_4,\ -10 \leq -x_1 - x_2 - x_4\}$$

The point at which the minimum value of the objective function Z[1] subject to the constraints given in constraints is found with minimize with the NONNEGATIVE setting so that the non-negativty constraints are considered by the simplex method. Of course, the simplex package must be loaded before this command is used. The solution is assigned the name vals. Finally, the components in vals are assigned to the solution vector **x** so that the minimum value of the objective function is determined by taking the scalar product of c and **x** which is equivalent to substituting the components of **x** into the objective function. The scalar product of 34/5 is found with multiply.

```
> with(simplex):
> vals:=minimize(Z[1],constraints,NONNEGATIVE);
```

$$vals := \left\{x_1 = 0,\ x_3 = 0,\ x_4 = 0,\ x_5 = \frac{35}{8},\ x_2 = 5/2\right\}$$

```
> assign(vals):
> multiply(c,x);
```

$$vector\left(\left[\frac{35}{4}\right]\right)$$

Next, we solve the corresponding dual problem. The vector **Y** is defined below that has as its components the dual variables y1, y2, y3, y4, and y5. The objective function of the dual problem is then defined by multiplying **Y** by the vector **b**. Similarly, the left-hand sides of the constraints are determined by multiplying the vector **Y** by the matrix of coefficients **A**.

```
> Y:=array([[y1,y2,y3,y4,y5]]):
> multiply(Y,b);
```

$$vector\left([10\,y1 + 30\,y2 - 5\,y3 - 10\,y4]\right)$$

```
> multiply(Y,A);
```

$$\begin{bmatrix} 2\,y1 + 6\,y2 - 3\,y3 - y4 \\ 3\,y1 + 5\,y2 - 2\,y3 - y4 \\ 3\,y1 + 4\,y2 - 3\,y3,\ 2\,y1 + y2 - 4\,y3 - y4 \\ 2\,y1 + 4\,y2 \end{bmatrix}$$

Hence, we may state the dual problem as:

Minimize $Y = 10y_1 + 30y_2 - 5y_3 - 10y_4$ subject to the constraints

$$\begin{cases} 2y_1 + 6y_2 - 3y_3 - y_4 \le 5 \\ 3y_1 + 5y_2 - 2y_3 - y_4 \le -7 \\ 3y_1 + 4y_2 - 3y_3 \le 7 \\ 2y_1 + y_2 - 4y_3 - y_4 \le 5 \\ 2y_1 + 4y_2 \le 6 \end{cases},$$

and $y_i \ge 0$ for $i = 1, 2, 3,$ and 4.

■

Application: A Transportation Problem

A certain company has two factories, F1 and F2, each producing two products, P1 and P2, that are to be shipped to three distribution centers, D1, D2, and D3. The following table illustrates the cost associated with shipping each product from the factory to the distribution center, the minimum number of each product each distribution center needs, and the maximum output of each factory. How much of each product should be shipped from each plant to each distribution center to minimize the total shipping costs?

	F1/P1	F1/P2	F2/P1	F2/P2	Minimum
D1/P1	\$0.75		\$0.80		500
D1/P2		\$0.50		\$0.40	400
D2/P1	\$1.00		\$0.90		300
D2/P2		\$0.75		\$1.20	500
D3/P1	\$0.90		\$0.85		700
D3/P2		\$0.80		\$0.95	300
Maximum Output	1000	400	800	900	

SOLUTION: Let x_1 denote the number of units of P1 shipped from F1 to D1; x_2 the number of units of P2 shipped from F1 to D1; x_3 the number of units of P1 shipped from F1 to D2; x_4 the number of units of P2 shipped from F1 to D2; x_5 the number of units of P1 shipped from F1 to D3; x_6 the number of units of P2 shipped from F1 to D3; x_7 the number of units of P1 shipped from F2 to D1; x_8 the number of units of P2 shipped from F2 to D1; x_9 the number of units of P1 shipped from F2 to D2; x_{10} the number of units of P2 shipped from F2 to D2; x_{11} the number of units of P1 shipped from F2 to D3; and x_{12} the number of units of P2 shipped from F2 to D3.

Then, it is necessary to minimize the number

$$Z = .75x_1 + .5x_2 + x_3 + .75x_4 + .9x_5 + .8x_6 + .8x_7 + .4x_8 + .9x_9 + 1.2x_{10} + .85x_{11} + .95x_{12}$$

subject to the constraints $x_1 + x_3 + x_5 \leq 1000$, $x_2 + x_4 + x_6 \leq 400$, $x_7 + x_9 + x_{11} \leq 800$, $x_8 + x_{10} + x_{12} \leq 900$, $x_1 + x_7 \geq 500$, $x_3 + x_9 \geq 500$, $x_5 + x_{11} \geq 700$, $x_2 + x_8 \geq 400$, $x_4 + x_{10} \geq 500$, $x_6 + x_{12} \geq 300$, and x_i non-negative for $i = 1, 2, \ldots, 12$.

In order to solve this linear programming problem, the objective function that computes the total cost, the 12 variables, and the set of inequalities must be entered. The coefficients of the objective function are given in the vector **c**. We will use several of the commands in the `linalg` package; we begin by loading this package. The objective function is defined by computing the dot product of the vectors **x** and **c** with `dotprod`. (Recall, that a similar computation was performed with `multiply` in a previous example.) The list of constraints are entered explicitly in `constraints`.

```
> with(linalg):
> c:=vector(
> [.75,.5,1,.75,.9,.8,.8,.4,.9,1.2,.85,.95]):
> x:=vector(12):
```

```
Warning, the names basis and pivot have been redefined
```

```
> Z:=dotprod(x,c);
```

$$Z := 0.75\,x_1 + 0.5\,x_2 + x_3 + 0.75\,x_4 + 0.9\,x_5 + 0.8\,x_6 + 0.8\,x_7 + 0.4\,x_8 + 0.9\,x_9 + 1.2\,x_{10} + 0.85\,x_{11} + 0.95\,x_{12}$$

```
> constraints:=x[1]+x[3]+x[5]<=1000,
  x[2]+x[4]+x[6]<=400,x[7]+x[9]+x[11]<=800,
  x[8]+x[10]+x[12]<=900,x[1]+x[7]>=500,
  x[3]+x[9]>=300,x[5]+x[11]>=700,
  x[2]+x[8]>=400,x[4]+x[10]>=500,
  x[6]+x[12]>=300:
```

The `simplex` package is then loaded so that the `minimize` command can be used. This is done in `min_vals`, which determines the variable values at which the minimum occurs. These values are assigned to the components of the vector **x** with `assign`. Therefore, the total number of units produced of each product at each factory is easily found by entering

```
x[1]+x[3]+x[5], x[2]+x[4]+x[6], ..., x[6]+x[12].
```

Also, the minimum value of the objective function 2115 is determined by entering Z. From these results, we see that F1 produces 700 units of P1, F1 produces 400 units of P2, F2 produces 800 units of P1, F2 produces 800 units of P2, and each distribution center receives exactly the minimum number of each product it requests.

```
> with(simplex):
  min_vals:=minimize(Z,constraints,NONNEGATIVE);
```

$$min\_vals := \{x_1 = 500, x_{12} = 300, x_3 = 0, x_2 = 0, x_8 = 400, x_9 = 300, x_6 = 0, x_7 = 0, x_{10} = 100, x_4 = 400, x_{11} = 500, x_5 = 200\}$$

```
> assign(min_vals):
> Z;
> x[1]+x[3]+x[5];
> x[2]+x[4]+x[6];
> x[7]+x[9]+x[11];
> x[3]+x[9];
> x[5]+x[11];
> x[2]+x[8];
> x[4]+x[10];
> x[6]+x[12];
```

$$2115.0$$

$$700$$

$$400$$

$$800$$

$$300$$

$$700$$

$$400$$

$$500$$

$$300$$

■

5.5 Selected Topics from Vector Calculus

Basic operations on two- and three-dimensional vectors are discussed in Section 5.1.4.

5.5.1 Vector-Valued Functions

We now turn our attention to vector-valued functions. In particular, we consider vector-valued functions of the following forms.

$$\text{Plane curves:} \quad \mathbf{r}(t) = x(t)\mathbf{i} + y(t)\mathbf{j} \tag{5.6}$$

$$\text{Space curves:} \quad \mathbf{r}(t) = x(t)\mathbf{i} + y(t)\mathbf{j} + z(t)\mathbf{k} \tag{5.7}$$

$$\text{Parametric surfaces:} \quad \mathbf{r}(s,t) = x(s,t)\mathbf{i} + y(s,t)\mathbf{j} + z(s,t)\mathbf{k} \tag{5.8}$$

$$\text{Vector fields in the plane:} \quad \mathbf{F}(x,y) = P(x,y)\mathbf{i} + Q(x,y)\mathbf{j} \tag{5.9}$$

$$\text{Vector fields in space:} \quad \mathbf{F}(x,y,z) = P(x,y,z)\mathbf{i} + Q(x,y,z)\mathbf{j} + R(x,y,z)\mathbf{k} \tag{5.10}$$

For the vector-valued functions (5.6) and (5.7), differentiation and integration is carried out term-by-term, provided that all the terms are differentiable and integrable. Suppose that C is a smooth curve defined by $\mathbf{r}(t)$, $a \le t \le b$.

1. If $\mathbf{r}'(t) \neq \mathbf{0}$, the **unit tangent vector**, $\mathbf{T}(t)$, is

$$\mathbf{T}(t) = \frac{\mathbf{r}'(t)}{\|\mathbf{r}'(t)\|}.$$

2. If $\mathbf{T}'(t) \neq \mathbf{0}$, the **principal unit normal vector**, $\mathbf{N}(t)$, is

$$\mathbf{N}(t) = \frac{\mathbf{T}'(t)}{\|\mathbf{T}'(t)\|}.$$

3. The **arc length function**, $s(t)$, is

$$s(t) = \int_a^t \|\mathbf{r}'(u)\|\, du.$$

In particular, the length of C on the interval $[a,b]$ is $\int_a^b \|\mathbf{r}'(t)\|\, dt$.

4. The **curvature**, κ, of C is

It is a good exercise to show that the curvature of a circle of radius r is $1/r$.

$$\kappa = \frac{\|\mathbf{T}'(t)\|}{\|\mathbf{r}'(t)\|} = \frac{\mathbf{a}(t)\cdot\mathbf{N}(t)}{\|\mathbf{v}(t)\|^2} = \frac{\|\mathbf{r}'(t)\times\mathbf{r}''(t)\|}{\|\mathbf{r}'(t)\|^3},$$

where $\mathbf{v}(t) = \mathbf{r}'(t)$ and $\mathbf{a}(t) = \mathbf{r}''(t)$

Use the `VectorCalculus` and `LinearAlgebra` packages to perform operations on vector-valued functions.

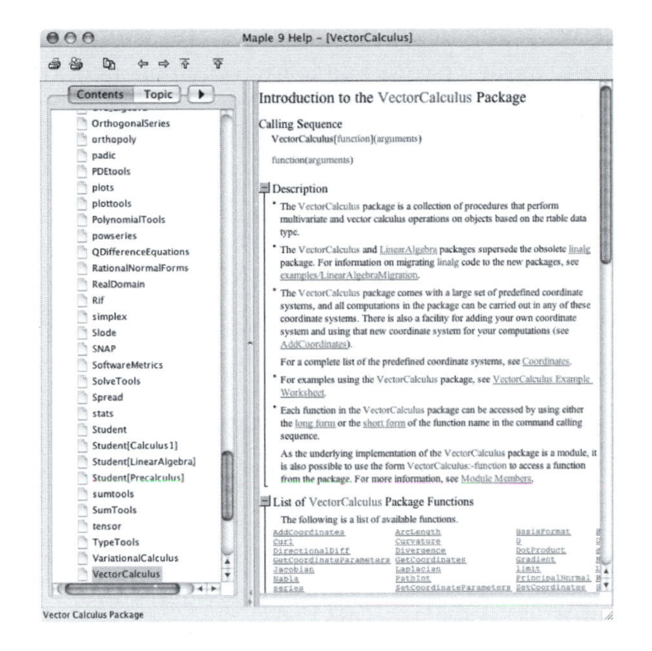

EXAMPLE 5.5.1 (Folium of Descartes): Consider the **Folium of Descartes**,

$$\mathbf{r}(t) = \frac{3at}{1+t^3}\mathbf{i} + \frac{3at^2}{1+t^3}\mathbf{j}$$

for $t \neq -1$, if $a = 1$. (a) Find $\mathbf{r}'(t)$, $\mathbf{r}''(t)$, and $\int \mathbf{r}(t)\,dt$. (b) Find $\mathbf{T}(t)$ and $\mathbf{N}(t)$. (c) Find the curvature, κ. (d) Find the length of the loop of the Folium.

SOLUTION: (a) After loading the `VectorCalculus` package and defining $\mathbf{r}(t)$,

```
> with(VectorCalculus):
> with(LinearAlgebra):
> r:=t->Vector([3*a*t/(1+t^3),3*a*t^2/(1+t^3)]):
> a:=1:
```

we compute $\mathbf{r}'(t)$, $\mathbf{r}''(t)$, and $\int \mathbf{r}(t)\,dt$ with `diff` and `int`. We name $\mathbf{r}'(t)$ `dr`, $\mathbf{r}''(t)$ `dr2`, and $\int \mathbf{r}(t)\,dt$ `ir`. Observe that when the `VectorCalculus` package is loaded, `int` is redefined, but not `integrate`. Thus, `int(r(t),t)` antidifferentiates each component of $\mathbf{r}(t)$ but `integrate(r(t),t)` does not. To use `integrate`, we must use `map` to apply `integrate` to each component of $\mathbf{r}(t)$.

```
> dr:=simplify(diff(r(t),t));
```

$$dr := \begin{bmatrix} -3\,\dfrac{2\,t^3-1}{\left(t^3+1\right)^2} \\ -3\,\dfrac{t\,\left(t^3-2\right)}{\left(t^3+1\right)^2} \end{bmatrix}$$

```
> dr2:=simplify(diff(r(t),t$2));
```

$$dr2 := \begin{bmatrix} 18\,\dfrac{t^2\,\left(t^3-2\right)}{\left(t^3+1\right)^3} \\ 6\,\dfrac{t^6-7\,t^3+1}{\left(t^3+1\right)^3} \end{bmatrix}$$

```
> ir:=map(integrate,r(t),t);
```

$$ir := \begin{bmatrix} -\ln\left(t+1\right)+1/2\,\ln\left(t^2-t+1\right)+\sqrt{3}\arctan\left(1/3\,\left(2\,t-1\right)\sqrt{3}\right) \\ \ln\left(t^3+1\right) \end{bmatrix}$$

```
> ir:=int(r(t),t);
```

$$ir := \begin{bmatrix} -\ln\left(t+1\right)+1/2\,\ln\left(t^2-t+1\right)+\sqrt{3}\arctan\left(1/3\,\left(2\,t-1\right)\sqrt{3}\right) \\ \ln\left(t^3+1\right) \end{bmatrix}$$

(b) Maple does not automatically make assumptions regarding the value of t, so it does not algebraically simplify $\|\mathbf{r}'(t)\|$ as we might typically do unless we use `radsimp`.

`radsimp(Sqrt(x^2))` returns x.

```
> simplify(sqrt(dr.dr));
```

$$3\,\sqrt{\frac{4\,t^6-4\,t^3+1+t^8-4\,t^5+4\,t^2}{\left(t^3+1\right)^4}}$$

```
> nr:=radsimp(sqrt(dr.dr));
```

$$nr := 3\,\frac{\sqrt{4\,t^6-4\,t^3+1+t^8-4\,t^5+4\,t^2}}{\left(t^3+1\right)^2}$$

We perform the same steps to compute the unit normal vector, $\mathbf{N}(t)$. In particular, note that `dutb` $= \|\mathbf{T}'(t)\|$.

```
> ut:=1/nr*dr;
```

$$\begin{bmatrix} -3\,\frac{2\,t^3-1}{nr\,(t^3+1)^2} \\ -3\,\frac{t\,(t^3-2)}{nr\,(t^3+1)^2} \end{bmatrix}$$

Alternatively, TangentVector, which is contained in the VectorCalculus package, computes $\mathbf{r}'(t)$

```
> simplify(TangentVector(r(t)));
```

$$\begin{bmatrix} -3\,\frac{2\,t^3-1}{(t^3+1)^2} \\ -3\,\frac{t\,(t^3-2)}{(t^3+1)^2} \end{bmatrix}$$

so TangentVector followed by Normalize, which is contained in the LinearAlgebra package, returns $\mathbf{T}(t)$.

```
> simplify(Normalize(TangentVector(r(t))));
```

$$\begin{bmatrix} -3\,(2\,t^3-1)\left(\max\left(3\left|\frac{2\,t^3-1}{(t^3+1)^2}\right|, 3\left|\frac{t\,(t^3-2)}{(t^3+1)^2}\right|\right)\right)^{-1}(t^3+1)^{-2} \\ -3\,t\,(t^3-2)\left(\max\left(3\left|\frac{2\,t^3-1}{(t^3+1)^2}\right|, 3\left|\frac{t\,(t^3-2)}{(t^3+1)^2}\right|\right)\right)^{-1}(t^3+1)^{-2} \end{bmatrix}$$

We perform the same steps to compute the unit normal vector, $\mathbf{N}(t)$.

```
> dut:=simplify(diff(ut,t));
```

$$dut := \begin{bmatrix} 18\,\frac{t^2\,(t^3-2)}{nr\,(t^3+1)^3} \\ 6\,\frac{t^6-7\,t^3+1}{nr\,(t^3+1)^3} \end{bmatrix}$$

```
> duta:=simplify(dut.dut);
```

$$duta := 36\,\frac{9\,t^{10}-36\,t^7+36\,t^4+t^{12}-14\,t^9+51\,t^6-14\,t^3+1}{nr^2\,(t^3+1)^6}$$

```
> dutb:=radsimp(sqrt(duta);
```

$$dutb := 6\,\frac{\sqrt{9\,t^{10}-36\,t^{7}+36\,t^{4}+t^{12}-14\,t^{9}+51\,t^{6}-14\,t^{3}+1}}{nr\,\left(t^{3}+1\right)^{3}}$$

```
> nt:=simplify(1/dutb*dut);
```

$$nt := \begin{bmatrix} 3\,\dfrac{t^{2}\,(t^{3}-2)}{\sqrt{9\,t^{10}-36\,t^{7}+36\,t^{4}+t^{12}-14\,t^{9}+51\,t^{6}-14\,t^{3}+1}} \\ \dfrac{t^{6}-7\,t^{3}+1}{\sqrt{9\,t^{10}-36\,t^{7}+36\,t^{4}+t^{12}-14\,t^{9}+51\,t^{6}-14\,t^{3}+1}} \end{bmatrix}$$

Alternatively, use `PrincipalNormal`, which is contained in the `VectorCalculus` package,

```
> radsimp(PrincipalNormal(r(t)));
```

$$\begin{bmatrix} 2\,\dfrac{t\,(t^{9}-3\,t^{3}-2)}{(4\,t^{6}-4\,t^{3}+1+t^{8}-4\,t^{5}+4\,t^{2})^{3/2}} \\ -2\,\dfrac{2\,t^{9}+3\,t^{6}-1}{(4\,t^{6}-4\,t^{3}+1+t^{8}-4\,t^{5}+4\,t^{2})^{3/2}} \end{bmatrix}$$

followed by `Normalize`.

```
> radsimp(Normalize(PrincipalNormal(r(t))))
```

$$\begin{bmatrix} 2\,t\,(t^{9}-3\,t^{3}-2)\left(4\,t^{6}-4\,t^{3}+1+t^{8}-4\,t^{5}+4\,t^{2}\right)^{-3/2} \\ \times\left(\max\left(2\left|\dfrac{2\,t^{9}+3\,t^{6}-1}{(4\,t^{6}-4\,t^{3}+1+t^{8}-4\,t^{5}+4\,t^{2})^{3/2}}\right|,\right.\right. \\ \left.\left.2\left|\dfrac{t\,(t^{9}-3\,t^{3}-2)}{(4\,t^{6}-4\,t^{3}+1+t^{8}-4\,t^{5}+4\,t^{2})^{3/2}}\right|\right)\right)^{-1} \\ -2\left(2\,t^{9}+3\,t^{6}-1\right)\left(4\,t^{6}-4\,t^{3}+1+t^{8}-4\,t^{5}+4\,t^{2}\right)^{-3/2} \\ \times\left(\max\left(2\left|\dfrac{2\,t^{9}+3\,t^{6}-1}{(4\,t^{6}-4\,t^{3}+1+t^{8}-4\,t^{5}+4\,t^{2})^{3/2}}\right|,\right.\right. \\ \left.\left.2\left|\dfrac{t\,(t^{9}-3\,t^{3}-2)}{(4\,t^{6}-4\,t^{3}+1+t^{8}-4\,t^{5}+4\,t^{2})^{3/2}}\right|\right)\right)^{-1} \end{bmatrix}$$

(c) We use the formula $\kappa = \dfrac{\|\mathbf{T}'(t)\|}{\|\mathbf{r}'(t)\|}$ to determine the curvature in `curvature`

```
> curvature:=simplify(dutb/nr);
```

$$curvature := 2/3\,\frac{\left(t^3+1\right)^4}{\left(4\,t^6-4\,t^3+1+t^8-4\,t^5+4\,t^2\right)^{3/2}}$$

and confirm the result with `Curvature`, which is contained in the `VectorCalculus` package.

```
> radsimp(Curvature(r(t)));
```

$$2/3\,\frac{\left(t^3+1\right)^4}{\left(4\,t^6-4\,t^3+1+t^8-4\,t^5+4\,t^2\right)^{3/2}}$$

We graphically illustrate the unit tangent and normal vectors at $\mathbf{r}(1) = \langle 3/2, 3/2 \rangle$. First, we compute the unit tangent and normal vectors if $t = 1$ using `subs`.

```
> ut1:=subs(t=1,ut);
```

$$ut1 := \begin{bmatrix} -3/4\,nr^{-1} \\ 3/4\,nr^{-1} \end{bmatrix}$$

```
> nt1:=subs(t=1,nt)
```

$$nt1 := \begin{bmatrix} -\frac{3}{34}\sqrt{34} \\ -\frac{5}{34}\sqrt{34} \end{bmatrix}$$

We then compute the curvature if $t = 1$ in `smallk`. The center of the osculating circle at $\mathbf{r}(1)$ is found in `x0` and `y0`.

The radius of the osculating circle is $1/\kappa$; the position vector of the center is $\mathbf{r} + \frac{1}{\kappa}\mathbf{N}$.

```
> smallk:=subs(t=1,curvature);
```

$$smallk := 8/3\,\sqrt{2}$$

```
> evalf(smallk);
```

$$3.771236166$$

```
> evalf(1/smallk);
```

$$0.2651650429$$

Note that RadiusofCurvature, which is contained in the VectorCalculus package, computes $1/\kappa$.

```
> radsimp(RadiusOfCurvature(r(t)));
```

$$3/2\,\frac{\left(4\,t^6-4\,t^3+1+t^8-4\,t^5+4\,t^2\right)^{3/2}}{\left(t^3+1\right)^4}$$

```
> subs(t=1,radsimp(RadiusOfCurvature(r(t))));
> evalf(subs(t=1,radsimp(RadiusOfCurvature(r(t)))));
```

$$3/16\,\sqrt{2}$$

$$0.2651650429$$

```
> x0:=subs(t=1,r(t)[1]-dr.dr*dr[2]/(dr[1]*dr2[2]
   -dr2[1]*dr[2]));
```

$$x0 := \frac{21}{16}$$

```
> y0:=subs(t=1,r(t)[2]-dr.dr*dr[2]/(dr[1]*dr2[2]
   -dr2[1]*dr[2]));
```

$$y0 := \frac{21}{16}$$

circle([x0, y0], r) is a two-dimensional graphics object that represents a circle of radius r centered at the point (x_0, y_0). Use display to display the graph.

We now load the plots package and graph **r**(t) with plot. The unit tangent and normal vectors at **r**(1) are graphed with arrow in p3 and p4. The osculating circle at **r**(1) is graphed with circle in c1. All four graphs are displayed together with display in Figure 5-7.

```
> with(plots):
> p1:=plot([r(t)[1],r(t)[2],t=-100..100],
> view=[-2..3,-2..3],color=black):
> p2:=circle([x0,y0],1/smallk):
> p3:=arrow(r(1),ut1):
> p4:=arrow(r(1),nt1):
> display(p1,p2,p3,p4,view=[-2..3,-2..3]);
```

(d) The loop is formed by graphing **r**(t) for $t \geq 0$. Hence, the length of the loop is given by the improper integral $\int_0^\infty \|\mathbf{r}(t)\|\,dt$, which we compute with evalf and int.

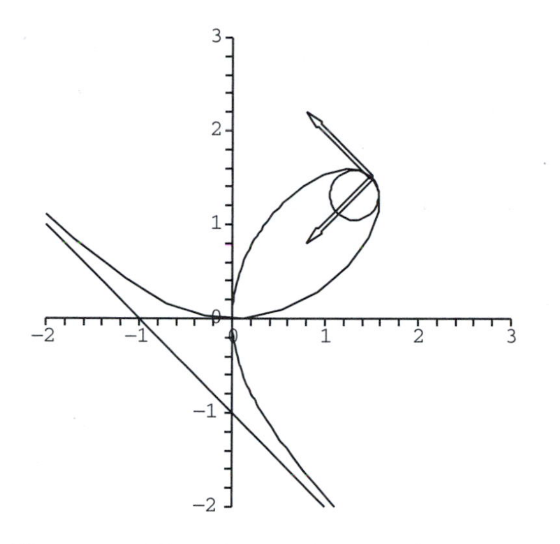

Figure 5-7 The Folium with an osculating circle

```
> evalf(int(nr,t=0..infinity));
```

$$4.917488722$$

■

Recall that the **gradient** of $z = f(x,y)$ is the vector-valued function $\nabla f(x,y) = \langle f_x(x,y), f_y(x,y)\rangle$. Similarly, we define the **gradient** of $w = f(x,y,z)$ to be

$$\nabla f(x,y,z) = \langle f_x(x,y,z), f_y(x,y,z), f_z(x,y,z)\rangle = \frac{\partial f}{\partial x}\mathbf{i} + \frac{\partial f}{\partial y}\mathbf{j} + \frac{\partial f}{\partial z}\mathbf{k}. \tag{5.11}$$

A vector field $\mathbf{F}$ is **conservative** if there is a function f, called a **potential function**, satisfying $\nabla f = \mathbf{F}$. In the special case that $\mathbf{F}(x,y) = P(x,y)\mathbf{i} + Q(x,y)\mathbf{j}$, $\mathbf{F}$ is conservative if and only if

$$\frac{\partial P}{\partial y} = \frac{\partial Q}{\partial x}.$$

The **divergence** of the vector field $\mathbf{F}(x,y,z) = P(x,y,z)\mathbf{i} + Q(x,y,z)\mathbf{j} + R(x,y,z)\mathbf{k}$ is the scalar field

$$\text{div}\mathbf{F} = \nabla \cdot \mathbf{F} = \frac{\partial P}{\partial x} + \frac{\partial Q}{\partial y} + \frac{\partial R}{\partial z}. \tag{5.12}$$

Use the command

```
F:=(x,y,z)->VectorField(<P(x,y,z),Q(x,y,z),R(x,y,z)>,
    'cartesian'[x,y,z])
```

to define the Cartesian vector field $\mathbf{F}(x,y,z) = P(x,y,z)\mathbf{i} + Q(x,y,z)\mathbf{j} + R(x,y,z)\mathbf{k}$. In general, when defining a vector field using `VectorField`, you *must* specify the coordinate system used.

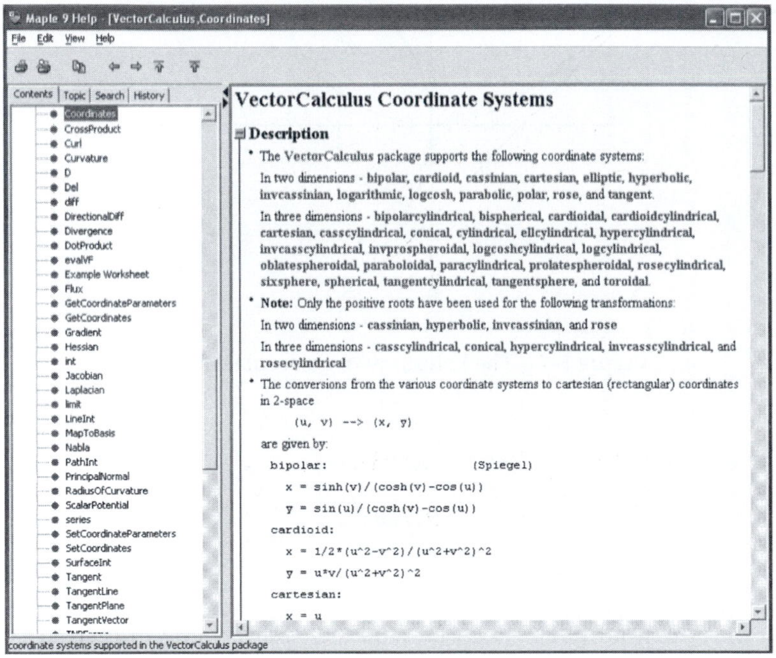

The `Divergence` command, which is contained in the `VectorCalculus` package, can be used to find the divergence of a vector field:

```
Divergence(F(x,y,z))
```

computes the divergence of $\mathbf{F}(x,y,z) = P(x,y,z)\mathbf{i} + Q(x,y,z)\mathbf{j} + R(x,y,z)\mathbf{k}$. The **laplacian** of the scalar field $w = f(x,y,z)$ is defined to be

$$\operatorname{div}(\nabla f) = \nabla \cdot (\nabla f) = \nabla^2 f = \frac{\partial^2 f}{\partial x^2} + \frac{\partial^2 f}{\partial y^2} + \frac{\partial^2 f}{\partial z^2} = \Delta f. \tag{5.13}$$

In the same way that `Divergence` computes the divergence of a vector field, `Laplacian`, which is also contained in the `VectorCalculus` package, computes the laplacian of a scalar field.

The **curl** of the vector field $\mathbf{F}(x,y,z) = P(x,y,z)\mathbf{i} + Q(x,y,z)\mathbf{j} + R(x,y,z)\mathbf{k}$ is

$$\begin{aligned}
\text{curl}\mathbf{F}(x,y,z) &= \nabla \times \mathbf{F}(x,y,z) \\
&= \begin{vmatrix} \mathbf{i} & \mathbf{j} & \mathbf{k} \\ \dfrac{\partial}{\partial x} & \dfrac{\partial}{\partial y} & \dfrac{\partial}{\partial z} \\ P(x,y,z) & Q(x,y,z) & R(x,y,z) \end{vmatrix} \\
&= \left(\frac{\partial R}{\partial y} - \frac{\partial Q}{\partial z}\right)\mathbf{i} - \left(\frac{\partial R}{\partial x} - \frac{\partial P}{\partial z}\right)\mathbf{j} + \left(\frac{\partial Q}{\partial x} - \frac{\partial P}{\partial y}\right)\mathbf{k}.
\end{aligned} \tag{5.14}$$

If $\mathbf{F}(x,y,z) = P(x,y,z)\mathbf{i} + Q(x,y,z)\mathbf{j} + R(x,y,z)\mathbf{k}$, $\mathbf{F}$ is conservative if and only if $\text{curl}\mathbf{F}(x,y,z) = \mathbf{0}$, in which case $\mathbf{F}$ is said to be **irrotational**.

EXAMPLE 5.5.2: Determine if

$$\mathbf{F}(x,y) = \left(1 - 2x^2\right) y e^{-x^2-y^2}\mathbf{i} + \left(1 - 2y^2\right) x e^{-x^2-y^2}\mathbf{j}$$

is conservative. If $\mathbf{F}$ is conservative find a potential function for $\mathbf{F}$.

SOLUTION: We define $\mathbf{F}(x,y)$ using `VectorField`. Then we use `diff` and `simplify` to see that $P_y(x,y) = Q_x(x,y)$. Hence, $\mathbf{F}$ is conservative.

```
> with(VectorCalculus):
> with(LinearAlgebra):
> F:=(x,y)->VectorField(<(1-2*x^2)*y*exp(-x^2-y^2),
> (1-2*y^2)*x*exp(-x^2-y^2)>,'cartesian'[x,y]):
> simplify(diff(F(x,y)[1],y));
```

$$e^{-x^2-y^2}\left(-2x^2 + 1 + 4y^2x^2 - 2y^2\right)$$

```
> simplify(diff(F(x,y)[2],x));
```

$$e^{-x^2-y^2}\left(-2x^2 + 1 + 4y^2x^2 - 2y^2\right)$$

We use `int` to find f satisfying $\nabla f = \mathbf{F}$.

```
> i1:=simplify(int(F(x,y)[1],x))+g(y);
```

$$i1 := yxe^{-x^2-y^2} + g(y)$$

```
> solve(diff(i1,y)=F(x,y)[2],diff(g(y),y));
```

$$0$$

Therefore, $g(y) = C$, where C is an arbitrary constant. Letting $C = 0$ gives us the potential function. We confirm this result using `ScalarPotential`, which is contained in the `VectorCalculus` package.

```
> f:=simplify(ScalarPotential(F(x,y)));
```

$$f := yxe^{-x^2-y^2}$$

Remember that the vectors **F** are perpendicular to the level curves of f. To see this, we normalize **F** in `nF` with `Normalize`.

```
> nF:=Normalize(F(x,y));
```

$$nf := \left[\begin{array}{c} \dfrac{\left(-2\,x^2+1\right)ye^{-x^2-y^2}}{\max\left(e^{-Re(x^2+y^2)}\left|\left(-2\,x^2+1\right)y\right|, e^{-Re(x^2+y^2)}\left|\left(-2\,y^2+1\right)x\right|\right)} \\ \dfrac{\left(-2\,y^2+1\right)xe^{-x^2-y^2}}{\max\left(e^{-Re(x^2+y^2)}\left|\left(-2\,x^2+1\right)y\right|, e^{-Re(x^2+y^2)}\left|\left(-2\,y^2+1\right)x\right|\right)} \end{array}\right]$$

We then graph several level curves of f in `cp` with `contourplot` and several vectors of `nF` with `fieldplot`, which is contained in the `plots` package, in `fp`. We show the graphs together with `Show` in Figure 5-8.

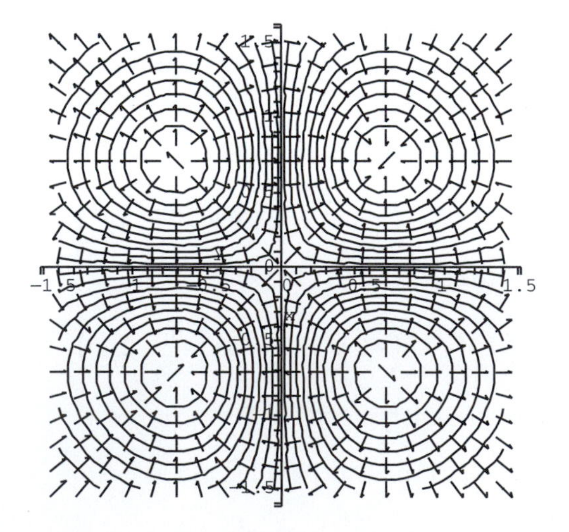

Figure 5-8 The vectors **F** are perpendicular to the level curves of f

```
> with(plots):
> fp:=fieldplot(nF,x=-3/2..3/2,y=-3/2..3/2):
> cp:=contourplot(f,x=-3/2..3/2,y=-3/2..3/2,
  color=black,contours=20):
> display(fp,cp);
```

Note that we can use `gradplot`, which is contained in the `plots` package, to graph several vectors of ∇f. However, the vectors are scaled and it can be difficult to see that the vectors are perpendicular to the level curves of f. The advantage of proceeding this way is that by graphing unit vectors, it is easier to see that the vectors are perpendicular to the level curves of f in the resulting plot.

■

EXAMPLE 5.5.3: (a) Show that

$$\mathbf{F}(x,y,z) = -10xy^2\mathbf{i} + \left(3z^3 - 10x^2y\right)\mathbf{j} + 9yz^2\mathbf{k}$$

is irrotational. (b) Find f satisfying $\nabla f = \mathbf{F}$. (c) Compute div $\mathbf{F}$ and $\nabla^2 f$.

SOLUTION: (a) After defining $\mathbf{F}(x,y,z)$, we use `Curl`, which is contained in the `VectorCalculus` package, to see that curl $\mathbf{F}(x,y,z) = \mathbf{0}$.

```
> with(VectorCalculus):
> F:=(x,y,z)->VectorField(<-10*x*y^2,3*z^3-10*x^2*y,
  9*y*z^2>,'cartesian'[x,y,z]):
> Curl(F(x,y,z));
```

$$\begin{bmatrix} 0 \\ 0 \\ 0 \end{bmatrix}$$

(b) We then use `ScalarPotential` to find $w = f(x,y,z)$ satisfying $\nabla f = \mathbf{F}$.

```
> lf:=ScalarPotential(F(x,y,z));
```

$$lf := -5\,y^2x^2 + 3\,z^3y$$

Thus, $f(x,y,z) = -5x^2y^2 + 3yz^3$. ∇f is orthogonal to the level surfaces of f. To illustrate this, we use `implicitplot3d`, which is contained in the `plots` package, to graph the level surface of $w = f(x,y,z)$

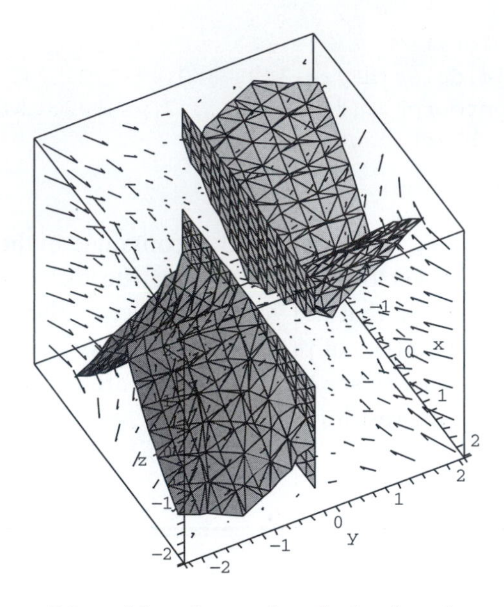

Figure 5-9 ∇f is orthogonal to the level surfaces of f

corresponding to $w = 0$ for $-2 \le x \le 2$, $-2 \le y \le 2$, and $-2 \le z \le 2$ in `pf`. We then use `gradplot3d`, which is contained in the `plots` package, to graph several vectors in the gradient field of f over the same domain in `gradf`. The two plots are shown together with `display` in Figure 5-9. In the plot, notice that the vectors appear to be perpendicular to the surface.

```
> with(plots):
> pf:=implicitplot3d(lf,x=-2..2,y=-2..2,z=-2..2,
   grid=[15,15,15]):
> pg:=gradplot3d(lf,x=-2..2,y=-2..2,z=-2..2,
   color=black):
> display(pf,pg,axes=boxed);
```

For (c), we take advantage of `Divergence` and `Laplacian`. As expected, the results are the same.

```
> Divergence(F(x,y,z));
```

$$-10\,y^2 - 10\,x^2 + 18\,yz$$

```
> Laplacian(lf,'cartesian'[x,y,z]);
```

$$-10\,y^2 - 10\,x^2 + 18\,yz$$

■

5.5.2 Line Integrals

If $\mathbf{F}$ is continuous on the smooth curve C with parametrization $\mathbf{r}(t)$, $a \le t \le b$, the **line integral** of $\mathbf{F}$ on C is

$$\int_C \mathbf{F} \cdot d\mathbf{r} = \int_a^b \mathbf{F} \cdot \mathbf{r}'(t)\, dt. \tag{5.15}$$

If you can parametrize C, you can use `LineInt`, which is contained in the `VectorCalculus` package, to compute $\int_C \mathbf{F} \cdot d\mathbf{r}$.

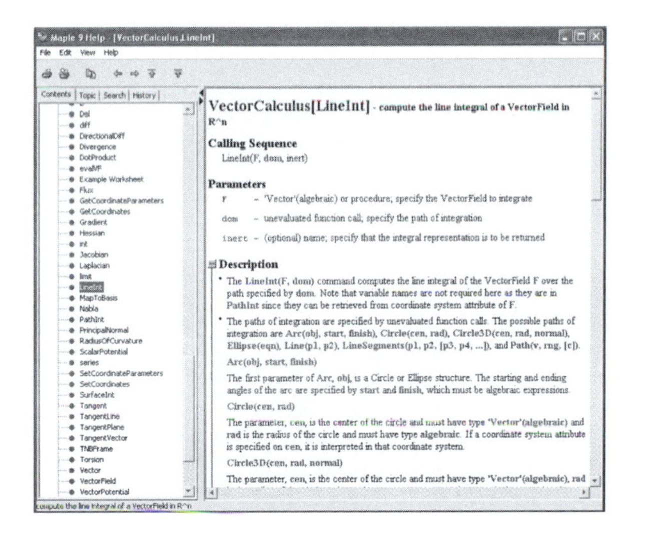

If $\mathbf{F}$ is conservative and C is piecewise smooth, line integrals can be evaluated using the *Fundamental Theorem of Line Integrals*.

Theorem 19 (Fundamental Theorem of Line Integrals): *If* $\mathbf{F}$ *is conservative and the curve* C *defined by* $\mathbf{r}(t)$, $a \le t \le b$ *is piecewise smooth,*

$$\int_C \mathbf{F} \cdot d\mathbf{r} = f\left(\mathbf{r}(b)\right) - f\left(\mathbf{r}(a)\right) \tag{5.16}$$

where $\mathbf{F} = \nabla f$.

EXAMPLE 5.5.4: Find $\int_C \mathbf{F} \cdot d\mathbf{r}$ where $\mathbf{F}(x,y) = \left(e^{-y} - ye^{-x}\right)\mathbf{i} + \left(e^{-x} - xe^{-y}\right)\mathbf{j}$ and C is defined by $\mathbf{r}(t) = \cos t\,\mathbf{i} + \ln\left(2t/\pi\right)\mathbf{j}$, $\pi/2 \le t \le 4\pi$.

SOLUTION: We see that **F** is conservative with `diff` and find that $f(x,y) = xe^{-y} + ye^{-x}$ satisfies $\nabla f = \mathbf{F}$ with `ScalarPotential`.

```
> with(VectorCalculus):
> F:=(x,y)->VectorField(<exp(-y)-y*exp(-x),
  exp(-x)-x*exp(-y)>,'cartesian'[x,y]):
> diff(F(x,y)[1],y);
> diff(F(x,y)[2],x);
```

$$-e^{-y} - e^{-x}$$

$$-e^{-y} - e^{-x}$$

```
> lf:=ScalarPotential(F(x,y));
```

$$lf := xe^{-y} + ye^{-x}$$

Hence, using (5.16),

$$\int_C \mathbf{F} \cdot d\mathbf{r} = \left(xe^{-y} + ye^{-x}\right)\Big]_{x=0,y=0}^{x=1,y=\ln 8} = \frac{3\ln 2}{e} + \frac{1}{8} \approx 0.890,$$

```
> xr:=t->cos(t):
> yr:=t->ln(2*t/Pi):
> [xr(Pi/2),yr(Pi/2)];
> [xr(4*Pi),yr(4*Pi)];
```

$$[0,0]$$

$$[1,3\ \ln(2)]$$

```
> simplify(subs([x=1,y=ln(8)],lf));
```

$$1/8 + 3\ \ln(2)\, e^{-1}$$

```
> evalf(simplify(subs([x=1,y=ln(8)],lf)));
```

$$0.8899837925$$

which we confirm with `LineInt`.

```
> LineInt(F(x,y),Path(<cos(t),ln(2*t/Pi)>,
  t=Pi/2..4*Pi));
```

$$1/8 + 3\ \ln(2)\, e^{-1}$$

■

If C is a piecewise smooth simple closed curve and $P(x,y)$ and $Q(x,y)$ have continuous partial derivatives, *Green's theorem* relates the line integral $\oint_C (P(x,y)\,dx + Q(x,y)\,dy)$ to a double integral.

Theorem 20 (Green's Theorem): *Let C be a piecewise smooth simple closed curve in the plane and R the region bounded by C. If P(x, y) and Q(x, y) have continuous partial derivatives on R,*

$$\oint_C (P(x,y)\,dx + Q(x,y)\,dy) = \iint_R \left(\frac{\partial Q}{\partial x} - \frac{\partial P}{\partial y}\right) dA. \tag{5.17}$$

EXAMPLE 5.5.5: Evaluate

$$\oint_C \left(e^{-x} - \sin y\right)\,dx + \left(\cos x - e^{-y}\right)\,dy$$

where C is the boundary of the region between $y = x^2$ and $x = y^2$.

We assume that the symbol $\oint$ means to evaluate the integral in the positive (or counter-clockwise) direction.

SOLUTION: After defining $P(x,y) = e^{-x} - \sin y$ and $Q(x,y) = \cos x - e^{-y}$, we use `plot` to determine the region R bounded by C in Figure 5-10.

```
> with(VectorCalculus):
> p:=(x,y)->exp(-x)-sin(y):
> q:=(x,y)->cos(x)-exp(y):
> plot([x^2,sqrt(x)],x=0..1.1,color=[black,gray]);
```

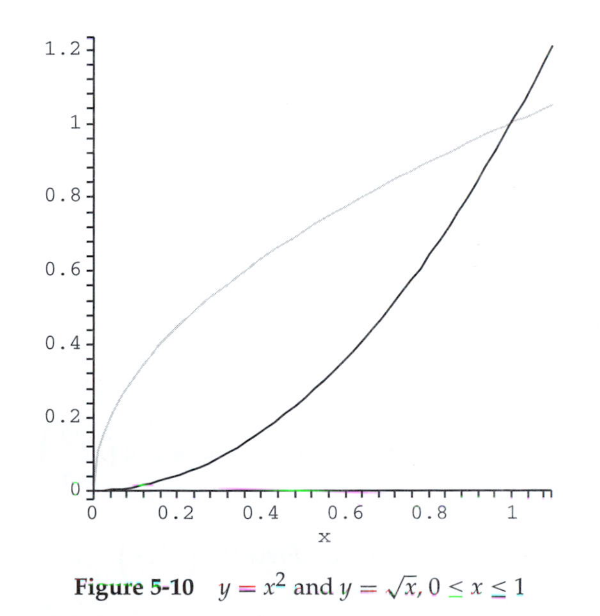

Figure 5-10 $y = x^2$ and $y = \sqrt{x}$, $0 \le x \le 1$

Using (5.17),

$$\oint_C \left(e^{-x} - \sin y\right) dx + \left(\cos x - e^{-y}\right) dy = \iint_R \left(\frac{\partial Q}{\partial x} - \frac{\partial P}{\partial y}\right) dA$$
$$= \iint_R \left(\cos y - \sin x\right) dA$$
$$= \int_0^1 \int_{x^2}^{\sqrt{x}} \left(\cos y - \sin x\right) dy\, dx,$$

```
> dqdp:=simplify(diff(q(x,y),x)-diff(p(x,y),y));
```

$$dqdp := -\sin(x) + \cos(y)$$

which we evaluate with `int`.

```
> ev:=int(int(dqdp,y=x^2..sqrt(x)),x=0..1);
```

$$ev := 4\sin(1) - 1/2\sqrt{2}\sqrt{\pi}\mathit{FresnelS}\left(\frac{\sqrt{2}}{\sqrt{\pi}}\right)$$
$$-1/2\sqrt{2}\sqrt{\pi}\mathit{FresnelC}\left(\frac{\sqrt{2}}{\sqrt{\pi}}\right) - 2$$

```
> evalf(ev);
```

$$0.151091400$$

Notice that the result is given in terms of the `FresnelS` and `FresnelC` functions, which are defined by

$$\texttt{FresnelS(x)} = \int_0^x \sin\left(\frac{\pi}{2}t^2\right) dt \quad \text{and} \quad \texttt{FresnelC(x)} = \int_0^x \cos\left(\frac{\pi}{2}t^2\right) dt.$$

A more meaningful approximation is obtained with `evalf`. We check with `LineInt`.

```
> check:=LineInt(VectorField(<p(x,y),q(x,y)>,
> 'cartesian'[x,y]),Path(<t,sqrt(t)>,t=1..0))+
> LineInt(VectorField(<p(x,y),q(x,y)>,
> 'cartesian'[x,y]),Path(<t,t^2>,t=0..1));
```

$$check := 4\sin(1) - 1/2\sqrt{2}\sqrt{\pi}\mathit{FresnelS}\left(\frac{\sqrt{2}}{\sqrt{\pi}}\right)$$
$$-1/2\sqrt{2}\sqrt{\pi}\mathit{FresnelC}\left(\frac{\sqrt{2}}{\sqrt{\pi}}\right) - 2$$

```
> evalf(check);
```

$$0.151091400$$

We conclude that

$$\int_0^1 \int_{x^2}^{\sqrt{x}} (\cos y - \sin x)\, dy\, dx \approx 0.151.$$

■

5.5.3 Surface Integrals

Let S be the graph of $z = f(x, y)$ ($y = h(x, z)$, $x = k(y, z)$) and let R_{xy} (R_{xz}, R_{yz}) be the projection of S onto the xy (xz, yz) plane. Then,

$$\iint_S g(x,y,z)\, dS = \iint_{R_{xy}} g\left(x, y, f(x,y)\right) \sqrt{\left[f_x(x,y)\right]^2 + \left[f_y(x,y)\right]^2 + 1}\, dA \tag{5.18}$$

$$= \iint_{R_{xz}} g\left(x, h(x,z), z\right) \sqrt{[h_x(x,z)]^2 + [h_z(x,z)]^2 + 1}\, dA \tag{5.19}$$

$$= \iint_{R_{yz}} g\left(k(y,z), y, z\right) \sqrt{\left[k_y(y,z)\right]^2 + \left[k_z(y,z)\right]^2 + 1}\, dA. \tag{5.20}$$

If S is defined parametrically by

$$\mathbf{r}(s,t) = x(s,t)\mathbf{i} + y(s,t)\mathbf{j} + z(s,t)\mathbf{k}, \quad (s,t) \in R$$

the formula

$$\iint_S g(x,y,z)\, dS = \iint_R g\left(\mathbf{r}(s,t)\right) \|\mathbf{r}_s \times \mathbf{r}_t\|\, dA, \tag{5.21}$$

where

$$\mathbf{r}_s = \frac{\partial x}{\partial s}\mathbf{i} + \frac{\partial y}{\partial s}\mathbf{j} + \frac{\partial z}{\partial s}\mathbf{k} \quad \text{and} \quad \mathbf{r}_t = \frac{\partial x}{\partial t}\mathbf{i} + \frac{\partial y}{\partial t}\mathbf{j} + \frac{\partial z}{\partial t}\mathbf{k},$$

is also useful.

Theorem 21 (The Divergence Theorem): *Let Q be any domain with the property that each line through any interior point of the domain cuts the boundary in exactly two points, and such that the boundary S is a piecewise smooth closed, oriented surface with unit normal $\mathbf{n}$. If $\mathbf{F}$ is a vector field that has continuous partial derivatives on Q, then*

For our purposes, a surface is **oriented** if it has two distinct sides.

$$\iiint_Q \nabla \cdot \mathbf{F}\, dV = \iiint_Q div\, \mathbf{F}\, dV = \iint_S \mathbf{F} \cdot \mathbf{n}\, dS. \tag{5.22}$$

In (5.22), $\iint_S \mathbf{F}\cdot\mathbf{n}\,dS$ is called the **outward flux** of the vector field **F** across the surface S. If S is a portion of the level curve $g(x,y)=C$ for some g, then a unit normal vector **n** may be taken to be either

$$\mathbf{n}=\frac{\nabla g}{\|\nabla g\|} \qquad \text{or} \qquad \mathbf{n}=-\frac{\nabla g}{\|\nabla g\|}.$$

If S is defined parametrically by

$$\mathbf{r}(s,t)=x(s,t)\mathbf{i}+y(s,t)\mathbf{j}+z(s,t)\mathbf{k},(s,t)\in R,$$

a unit normal vector to the surface is

$$\mathbf{n}=\frac{\mathbf{r}_s\times\mathbf{r}_t}{\|\mathbf{r}_s\times\mathbf{r}_t\|}$$

and (5.22) becomes

$$\iint_S \mathbf{F}\cdot\mathbf{n}\,dS=\iint_R \mathbf{F}\cdot(\mathbf{r}_s\times\mathbf{r}_t)\,dA.$$

EXAMPLE 5.5.6: Find the outward flux of the vector field

$$\mathbf{F}(x,y,z)=\left(xz+xyz^2\right)\mathbf{i}+\left(xy+x^2yz\right)\mathbf{j}+\left(yz+xy^2z\right)\mathbf{k}$$

through the surface of the cube cut from the first octant by the planes $x=1, y=1$, and $z=1$.

SOLUTION: By the Divergence theorem,

$$\iint_{\text{cube surface}} \mathbf{F}\cdot\mathbf{n}\,dA=\iiint_{\text{cube interior}} \nabla\cdot\mathbf{F}\,dV.$$

Hence, without the Divergence theorem, calculating the outward flux would require six separate integrals, corresponding to the six faces of the cube. After defining **F**, we compute $\nabla\cdot\mathbf{F}$ with `Divergence`.

`Divergence` is contained in the `VectorCalculus` package. You do not need to reload the `VectorCalculus` package if you have already loaded it during your *current* Maple session.

```
> with(VectorCalculus):
> F:=(x,y,z)->VectorField(<x*z+x*y*z^2,
> x*y+x^2*y*z,y*z+x*y^2*z>,'cartesian'[x,y,z]):
> divF:=Divergence(F(x,y,z));
```

$$divF := z+yz^2+x+x^2z+y+xy^2$$

The outward flux is then given by

$$\iiint_{\text{cube interior}} \nabla \cdot \mathbf{F}\, dV = \int_0^1 \int_0^1 \int_0^1 \nabla \cdot \mathbf{F}\, dz\, dy\, dx = 2,$$

which we compute with `int`.

```
> int(int(int(divF,x=0..1),y=0..1),z=0..1);
```

$$2$$

■

Theorem 22 (Stoke's Theorem): *Let S be an oriented surface with finite surface area, unit normal $\mathbf{n}$, and boundary C. Let $\mathbf{F}$ be a continuous vector field defined on S such that the components of $\mathbf{F}$ have continuous partial derivatives at each nonboundary point of S. Then,*

$$\oint_C \mathbf{F} \cdot d\mathbf{r} = \iint_S \mathit{curl}\, \mathbf{F} \cdot \mathbf{n}\, dS. \tag{5.23}$$

In other words, the surface integral of the normal component of the curl of $\mathbf{F}$ taken over S equals the line integral of the tangential component of the field taken over C. In particular, if $\mathbf{F} = P(x,y,z)\mathbf{i} + Q(x,y,z)\mathbf{j} + R(x,y,z)\mathbf{k}$, then

$$\int_C (P(x,y,z)dx + Q(x,y,z)dy + R(x,y,z)dz) = \iint_S \text{curl}\, \mathbf{F} \cdot \mathbf{n}\, dS.$$

EXAMPLE 5.5.7: Verify Stoke's theorem for the vector field

$$\mathbf{F}(x,y,z) = \left(x^2 - y\right)\mathbf{i} + \left(y^2 - z\right)\mathbf{j} + \left(x + z^2\right)\mathbf{k}$$

and S the portion of the paraboloid $z = f(x,y) = 9 - \left(x^2 + y^2\right)$, $z > 0$.

SOLUTION: After loading the `VectorCalculus` and `LinearAlgebra` packages, we define $\mathbf{F}$ and f. The curl of $\mathbf{F}$ is computed with `Curl` in `curlF`.

```
> with(VectorCalculus):
> with(LinearAlgebra):
> F:=(x,y,z)->VectorField(<x^2-y,y^2-z,x+z^2>,
   'cartesian'[x,y,z]):
> f:=(x,y)->9-(x^2+y^2):
```

```
> curlF:=Curl(F(x,y,z));
```

$$curlF := \begin{bmatrix} 1 \\ -1 \\ 1 \end{bmatrix}$$

Next, we define the function $h(x,y,z) = z - f(x,y)$. A normal vector to the surface is given by ∇h. A unit normal vector, **n**, is then given by $\mathbf{n} = \frac{\nabla h}{\| \nabla h\|}$, which is computed in un.

```
> h:=(x,y,z)->z-f(x,y):
> normtosurf:=Gradient(h(x,y,z),'cartesian'[x,y,z]);
```

$$normtosurf := \begin{bmatrix} 2x \\ 2y \\ 1 \end{bmatrix}$$

```
> un:=Normalize(normtosurf);
```

$$un := \begin{bmatrix} 2\frac{x}{\max(1,2|x|,2|y|)} \\ 2\frac{y}{\max(1,2|x|,2|y|)} \\ (\max(1,2|x|,2|y|))^{-1} \end{bmatrix}$$

```
> un:=simplify(normtosurf/sqrt(normtosurf.normtosurf));
```

$$un := \begin{bmatrix} 2\frac{x}{\sqrt{1+4x^2+4y^2}} \\ 2\frac{y}{\sqrt{1+4x^2+4y^2}} \\ \frac{1}{\sqrt{1+4x^2+4y^2}} \end{bmatrix}$$

The dot product curl $\mathbf{F} \cdot \mathbf{n}$ is computed in g.

```
> g:=simplify(curlF.un);
```

$$g := \frac{2x - 2y + 1}{\sqrt{1+4x^2+4y^2}}$$

Using the surface integral evaluation formula (5.18),

$$\iint_S \text{curl}\,\mathbf{F}\cdot\mathbf{n}\,dS = \iint_R g(x,y,f(x,y))\sqrt{[f_x(x,y)]^2+[f_y(x,y)]^2+1}\,dA$$

$$= \int_{-3}^{3}\int_{-\sqrt{9-x^2}}^{\sqrt{9-x^2}} g(x,y,f(x,y))\sqrt{[f_x(x,y)]^2+[f_y(x,y)]^2+1}\,dy\,dx$$

$$= 9\pi,$$

In this example, R, the projection of $f(x,y)$ onto the xy-plane, is the region bounded by the graph of the circle $x^2+y^2=9$.

which we compute with `integrate`.

```
> tointegrate:=subs(z=f(x,y),sqrt(diff(f(x,y),x)^2
  +diff(f(x,y),y)^2+1)*g);
```

$$tointegrate := 2x - 2y + 1$$

```
> i1:=integrate(integrate(tointegrate,
  y=-sqrt(9-x^2)..sqrt(9-x^2)),x=-3..3);
```

$$i1 := 9\pi$$

To verify Stoke's theorem, we must compute the associated line integral. Notice that the boundary of $z = f(x,y) = 9 - (x^2+y^2)$, $z = 0$, is the circle $x^2+y^2 = 9$ with parametrization $x = 3\cos t$, $y = 3\sin t$, $z = 0$, $0 \le t \le 2\pi$. This parametrization is substituted into $\mathbf{F}(x,y,z)$ and named `pvf`.

```
> pvf:=subs([x=3*cos(t),y=3*sin(t),z=0],F(x,y,z));
```

$$pvf := \begin{bmatrix} 9\,(\cos(t))^2 - 3\sin(t) \\ 9\,(\sin(t))^2 \\ 3\cos(t) \end{bmatrix}$$

To evaluate the line integral along the circle, we next define the parametrization of the circle and calculate $d\mathbf{r}$. The dot product of `pvf` and `dr` represents the integrand of the line integral.

```
> r:=t->Vector([3*cos(t),3*sin(t),0]):
> dr:=diff(r(t),t);
```

$$dr := \begin{bmatrix} -3\sin(t) \\ 3\cos(t) \\ 0 \end{bmatrix}$$

```
> tointegrate:=pvf[1]*dr[1]+pvf[2]*dr[2]+pvf[3]*dr[3]:
```

As before with x and y, we instruct Maple to assume that t is real, compute the dot product of `pvf` and `dr` and evaluate the line integral with `integrate`.

```
> integrate(tointegrate,t=0..2*Pi);
```

$$9\pi$$

As expected, the result is 9π.

■

5.5.4 A Note on Nonorientability

See "When is a surface *not* orientable?" by Braselton, Abell, and Braselton [5] for a detailed discussion regarding the examples in this section.

Suppose that S is the surface determined by

$$\mathbf{r}(s,t) = x(s,t)\mathbf{i} + y(s,t)\mathbf{j} + z(s,t)\mathbf{k}, \quad (s,t) \in R$$

and let

$$\mathbf{n} = \frac{\mathbf{r}_s \times \mathbf{r}_t}{\|\mathbf{r}_s \times \mathbf{r}_t\|} \quad \text{or} \quad \mathbf{n} = -\frac{\mathbf{r}_s \times \mathbf{r}_t}{\|\mathbf{r}_s \times \mathbf{r}_t\|}, \tag{5.24}$$

where

$$\mathbf{r}_s = \frac{\partial x}{\partial s}\mathbf{i} + \frac{\partial y}{\partial s}\mathbf{j} + \frac{\partial z}{\partial s}\mathbf{k} \quad \text{and} \quad \mathbf{r}_t = \frac{\partial x}{\partial t}\mathbf{i} + \frac{\partial y}{\partial t}\mathbf{j} + \frac{\partial z}{\partial t}\mathbf{k},$$

if $\|\mathbf{r}_s \times \mathbf{r}_t\| \neq 0$. If $\mathbf{n}$ is defined, $\mathbf{n}$ is orthogonal (or perpendicular) to S. We state three familiar definitions of *orientable*.

- S is **orientable** if S has a unit normal vector field, $\mathbf{n}$, that varies continuously between any two points (x_0, y_0, z_0) and (x_1, y_1, z_1) on S. (See [7].)
- S is **orientable** if S has a continuous unit normal vector field, $\mathbf{n}$. (See [7] and [16].)
- S is **orientable** if a unit vector $\mathbf{n}$ can be defined at every nonboundary point of S in such a way that the normal vectors vary continuously over the surface S. (See [13].)

A path is **order preserving** if our chosen orientation is preserved as we move along the path.

Thus, a surface like a torus is orientable.

Also see Example 2.3.18.

EXAMPLE 5.5.8 (The Torus): Using the standard parametrization of the torus,

$$x = (a + b\cos v)\cos u, \quad y = (a + b\cos v)\sin u, \quad z = c\sin v,$$

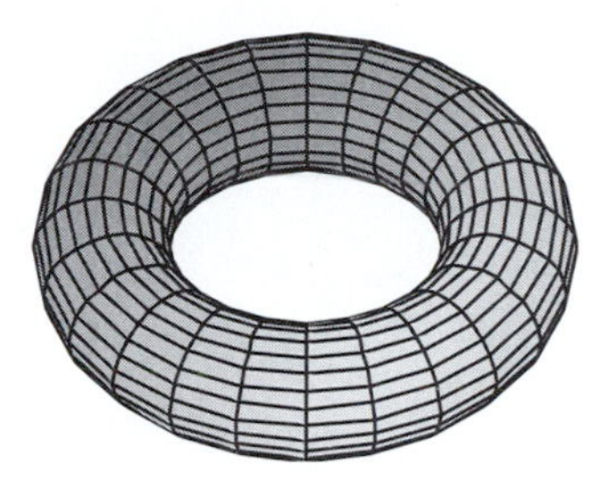

Figure 5-11 A torus

we use `plot3d` to plot the torus if $c = 3$ and $a = 1$ in Figure 5-11.

```
> with(VectorCalculus):
> with(LinearAlgebra):
> r:='r':
> c:=3:
> a:=1:
> x:=(s,t)->(c+a*cos(s))*cos(t):
> y:=(s,t)->(c+a*cos(s))*sin(t):
> z:=(s,t)->a*sin(s):
```

```
> with(plots):
> threedp1t:=plot3d([x(s,t),y(s,t),z(s,t)],
   s=-Pi..Pi,t=-Pi..Pi):
> display(threedp1t,scaling=constrained);
```

To plot a normal vector field on the torus, we compute $\frac{\partial}{\partial s}\mathbf{r}(s,t)$,

```
> r:=(s,t)->Vector([x(s,t),y(s,t),z(s,t)]):
```

```
> rs:=diff(r(s,t),s);
```

$$rs := \begin{bmatrix} -\sin(s)\cos(t) \\ -\sin(s)\sin(t) \\ \cos(s) \end{bmatrix}$$

and $\frac{\partial}{\partial t}\mathbf{r}(s,t)$.

```
> rt:=diff(r(s,t),t):
```

$$rt := \begin{bmatrix} -(3+\cos(s))\sin(t) \\ (3+\cos(s))\cos(t) \\ 0 \end{bmatrix}$$

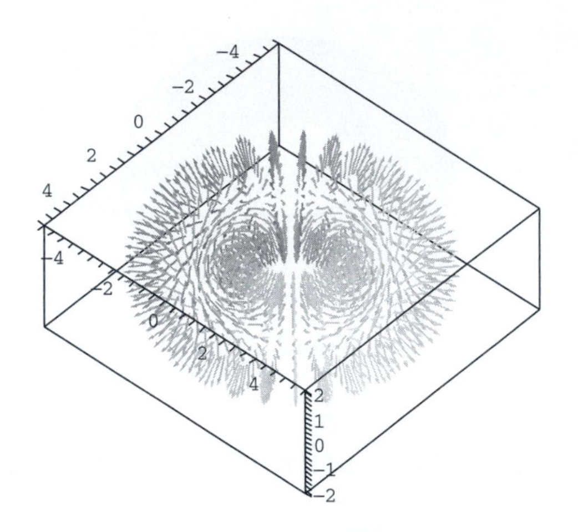

Figure 5-12 Unit normal vector field on a torus

The cross product $\frac{\partial}{\partial s}\mathbf{r}(s,t) \times \frac{\partial}{\partial t}\mathbf{r}(s,t)$ is formed in `rscrossrt`.

```
> rscrossrt:=CrossProduct(rs,rt);
```

$$rscrossrt := \begin{bmatrix} -\cos(s)(3+\cos(s))\cos(t) \\ -\cos(s)(3+\cos(s))\sin(t) \\ -\sin(s)(\cos(t))^2(3+\cos(s)) - \sin(s)(\sin(t))^2(3+\cos(s)) \end{bmatrix}$$

Using (5.24), we define `un`.

```
> simplify(sqrt(rscrossrt.rscrossrt));
```

$$csgn(3+\cos(s))(3+\cos(s))$$

```
> un:=simplify(-rscrossrt/sqrt(rscrossrt.rscrossrt));
```

$$un := \begin{bmatrix} csgn(3+\cos(s))\cos(s)\cos(t) \\ csgn(3+\cos(s))\cos(s)\sin(t) \\ \sin(s)\, csgn(3+\cos(s)) \end{bmatrix}$$

To plot the normal vector field on the torus, we take advantage of the command `arrow`, which is contained in the **plots** package (see Figure 5-12).

```
> tvals:=evalf(seq(-Pi+2*Pi/29*i,i=0..29)):
> svals:=evalf(seq(-Pi+2*Pi/29*i,i=0..29)):
> vecs:=[seq(seq(arrow(r(s0,t0),
> evalf(subs([s=s0,t=t0],un))),t0=tvals),s0=svals)]:

> display(vecs);
```

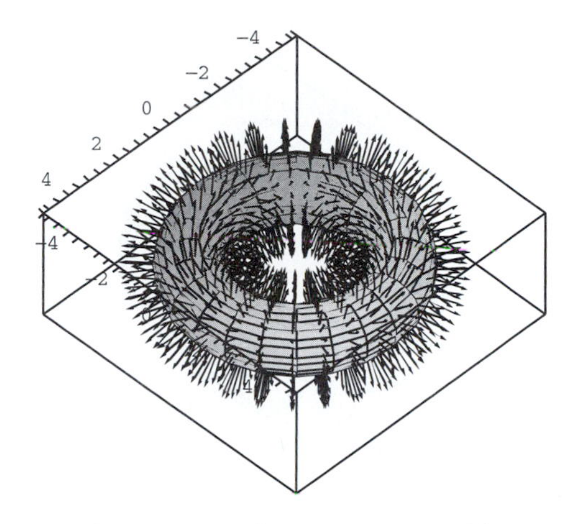

Figure 5-13 The torus is orientable

We use `display` to see the vector field on the torus together in Figure 5-13. Regardless of the viewing angle, the figure looks the same; the torus is orientable.

```
> pp2:=display(vecs,color=black):
> display(threedp1t,pp2,view=[-5..5,-5..5,-2..2],
> axes=boxed,scaling=constrained);
```

If a 2-manifold, S, has an **order reversing path** (or **not order preserving path**), S is **nonorientable** (or **not orientable**).

Determining whether a given surface S is orientable or not may be a difficult problem.

EXAMPLE 5.5.9 (The Möbius Strip): The *Möbius strip* is frequently cited as an example of a nonorientable surface with boundary: it has one side and is physically easy to construct by hand by half twisting and taping (or pasting) together the ends of a piece of paper (for example, see [5], [7], [13], and [16]). A parametrization of the Möbius strip is $\mathbf{r}(s,t) = x(s,t)\mathbf{i} + y(s,t)\mathbf{j} + z(s,t)\mathbf{k}$, $-1 \le s \le 1$, $-\pi \le t \le \pi$, where

$$x = \left[c + s\cos\left(\frac{1}{2}t\right)\right]\cos t, \quad y = \left[c + s\cos\left(\frac{1}{2}t\right)\right]\sin t, \quad \text{and}$$

$$z = s\sin\left(\frac{1}{2}t\right), \tag{5.25}$$

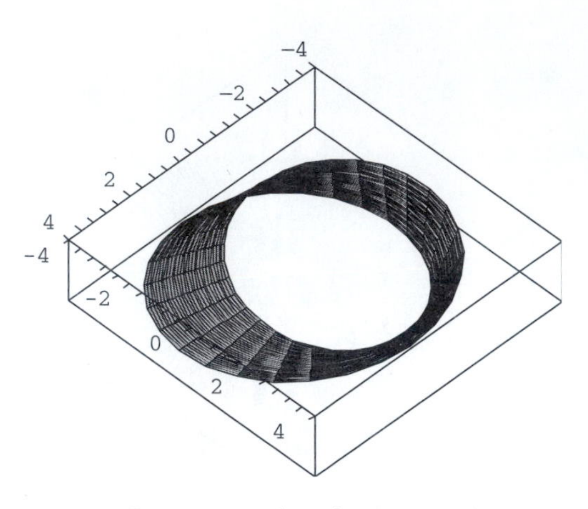

Figure 5-14 Parametric plot of equations (5.25) if $c = 3$

and we assume that $c > 1$. In Figure 5-14, we graph the Möbius strip using $c = 3$.

```
> with(VectorCalculus):
> with(plots):
> c:=3:
> x:=(s,t)->(c+s*cos(t/2))*cos(t):
> y:=(s,t)->(c+s*cos(t/2))*sin(t):
> z:=(s,t)->s*sin(t/2):
> r:=(s,t)->Vector([x(s,t),y(s,t),z(s,t)]):

> threedp1:=plot3d([x(s,t),y(s,t),z(s,t)],s=-1..1,
  t=-Pi..Pi,
> grid=[30,30],view=[-4..4,-4..4,-1..1],
  scaling=constrained,
> axes=boxed):
> display(threedp1);
```

Although it is relatively easy to see in the plot that the Möbius strip has only one side, the fact that a unit vector, **n**, normal to the Möbius strip at a point P reverses its direction as **n** moves around the strip to P is not obvious to the novice.

With Maple, we compute $\|\mathbf{r}_s \times \mathbf{r}_t\|$ and $\mathbf{n} = \dfrac{\mathbf{r}_s \times \mathbf{r}_t}{\|\mathbf{r}_s \times \mathbf{r}_t\|}$.

```
> rs:=diff(r(s,t),s);
```

$$rs := \begin{bmatrix} \cos(1/2\,t)\cos(t) \\ \cos(1/2\,t)\sin(t) \\ \sin(1/2\,t) \end{bmatrix}$$

```
> rt:=diff(r(s,t),t);
```

$$rt := \begin{bmatrix} -1/2\,s\sin(1/2\,t)\cos(t) - (s\cos(1/2\,t)+3)\sin(t) \\ -1/2\,s\sin(1/2\,t)\sin(t) + (s\cos(1/2\,t)+3)\cos(t) \\ 1/2\,s\cos(1/2\,t) \end{bmatrix}$$

```
> rscrossrt:=simplify(CrossProduct(Vector(rs),Vector(rt)));
```

$$rscrossrt := \begin{bmatrix} -\left(-3+2\,(\cos(1/2\,t))^3 s + 6\,(\cos(1/2\,t))^2 - 2\,s\cos(1/2\,t)\right)\sin(1/2\,t) \\ -3\,s\,(\cos(1/2\,t))^2 + 2\,(\cos(1/2\,t))^4 s + 6\,(\cos(1/2\,t))^3 - 6\,\cos(1/2\,t) + 1/2\,s \\ (s\cos(1/2\,t)+3)\cos(1/2\,t) \end{bmatrix}$$

```
> simplify(sqrt(rscrossrt.rscrossrt));
```

$$1/2\sqrt{4\,s^2\,(\cos(1/2\,t))^2 + 24\,s\cos(1/2\,t) + s^2 + 36}$$

```
> un:=simplify(rscrossrt/sqrt(rscrossrt.rscrossrt));
```

$$un := \begin{bmatrix} -2\,\dfrac{\left(-3+2\,(\cos(1/2\,t))^3 s + 6\,(\cos(1/2\,t))^2 - 2\,s\cos(1/2\,t)\right)\sin(1/2\,t)}{\sqrt{4\,s^2\,(\cos(1/2\,t))^2 + 24\,s\cos(1/2\,t) + s^2 + 36}} \\ \dfrac{-6\,s\,(\cos(1/2\,t))^2 + 4\,(\cos(1/2\,t))^4 s + 12\,(\cos(1/2\,t))^3 - 12\,\cos(1/2\,t) + s}{\sqrt{4\,s^2\,(\cos(1/2\,t))^2 + 24\,s\cos(1/2\,t) + s^2 + 36}} \\ 2\,\dfrac{(s\cos(1/2\,t)+3)\cos(1/2\,t)}{\sqrt{4\,s^2\,(\cos(1/2\,t))^2 + 24\,s\cos(1/2\,t) + s^2 + 36}} \end{bmatrix}$$

Consider the path C given by $\mathbf{r}(0,t)$, $-\pi \le t \le \pi$ that begins and ends at $\langle -3,0,0\rangle$. On C, $\mathbf{n}(0,t)$ is given by

```
> curvec:=simplify(subs(s=0,un));
```

$$curvec := \begin{bmatrix} -\left(-1+2\,(\cos(1/2\,t))^2\right)\sin(1/2\,t) \\ -2\,(\sin(1/2\,t))^2\cos(1/2\,t) \\ \cos(1/2\,t) \end{bmatrix}$$

At $t=-\pi$, $\mathbf{n}(0,-\pi) = \langle 1,0,0\rangle$, while at $t=\pi$, $\mathbf{n}(0,\pi) = \langle -1,0,0\rangle$.

```
> r(0,-Pi);
```

$$\begin{bmatrix} -3 \\ 0 \\ 0 \end{bmatrix}$$

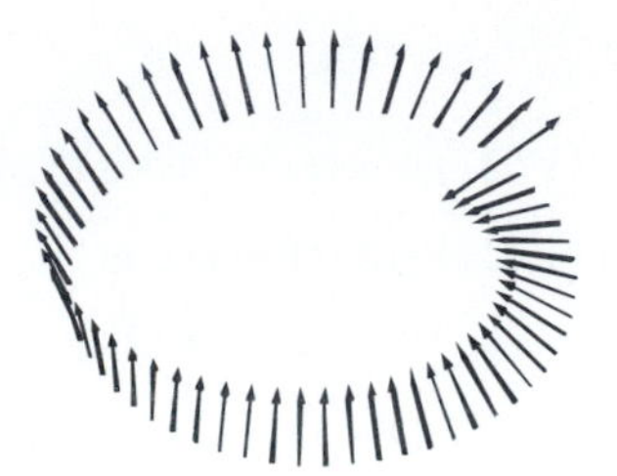

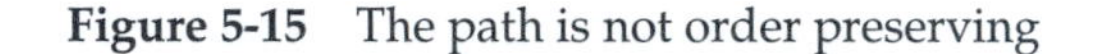

Figure 5-15 The path is not order preserving

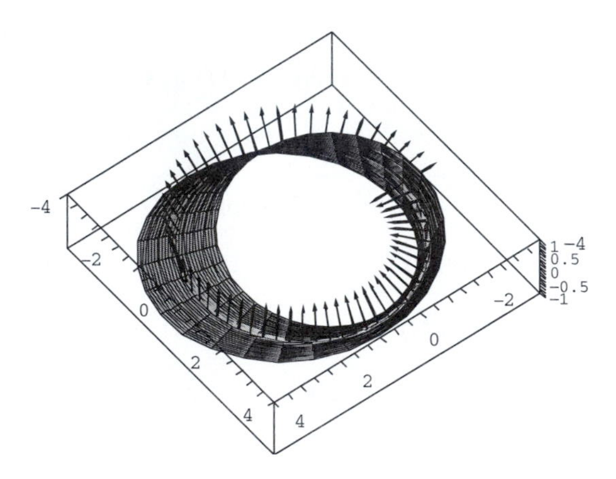

Figure 5-16 A Möbius strip with an orientation reversing path

```
> r(0,Pi);
```

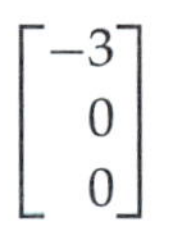

$$\begin{bmatrix} -3 \\ 0 \\ 0 \end{bmatrix}$$

As **n** moves along C from $\mathbf{r}(0, -\pi)$ to $\mathbf{r}(0, \pi)$, the orientation of **n** reverses, as shown in Figure 5-15.

```
> tvals:=evalf(seq(-Pi+2*Pi/59*i,i=0..59)):
> vecs:=[seq(arrow(r(0,t0),
   evalf(subs([s=0,t=t0],un))),t0=tvals)]:
> pp2:=display(vecs,color=black,scaling=constrained):
> display(pp2);
> display(threedp1,pp2,view=[-4..4,-4..4,-1..1],
> scaling=constrained,axes=boxed);
```

The orientation reversing path is shown on the Möbius strip in Figure 5-16. C is an orientation reversing path and we can conclude that the Möbius strip is not orientable.

EXAMPLE 5.5.10 (The Klein Bottle): The *Klein bottle* is an interesting surface with neither an inside nor an outside, which indicates to us that it is not orientable. In Figure 5-17(a) we show the "usual" *immersion* of the Klein bottle. Although the Klein bottle does not intersect itself, it is not possible to visualize it in Euclidean 3-space without it doing so. Visualizations of 2-manifolds like the Klein bottle's "usual" rendering in Euclidean 3-space are called *immersions*. The "usual" immersion of the Klein bottle has parametrization $\mathbf{r}(u,v) = x(u,v)\mathbf{i} + y(u,v)\mathbf{j} + z(u,v)\mathbf{k}$, where

$$x = \begin{cases} 6(1+\sin u)\cos u + r\cos u\cos v, \; 0 \le u \le \pi \\ 6(1+\sin u)\cos u + r\cos u\cos(v+\pi), \; \pi \le u \le 2\pi \end{cases},$$

$$y = \begin{cases} 16\sin u + r\sin u\cos v, \; 0 \le u \le \pi \\ 16\sin u, \; \pi \le u \le 2\pi \end{cases}, \tag{5.26}$$

$$z = r\sin v,$$

$$r = r\left(1 - \frac{1}{2}\cos u\right), \; 0 \le u \le 2\pi, \; 0 \le v \le 2\pi.$$

(See [11] for a non-technical discussion of immersions.)

```
> x1:='x1':x2:='x2':
> y1:='y1':y2:='y2':
> z:='z':
> r:=u->4*(1-1/2*cos(u)):
> x1:=(u,v)->6*(1+sin(u))*cos(u)+r(u)*cos(u)*cos(v):
> x2:=(u,v)->6*(1+sin(u))*cos(u)+r(u)*cos(v+Pi):
> y1:=(u,v)->16*sin(u)+r(u)*sin(u)*cos(v):
> y2:=(u,v)->16*sin(u):
> z:=(u,v)->r(u)*sin(v):

> with(plots):
> kb1a:=plot3d([x1(s,t),y1(s,t),z(s,t)],s=0..Pi,
> t=0..2*Pi,grid=[30,30],scaling=constrained):
> kb1b:=plot3d([x2(s,t),y2(s,t),z(s,t)],s=Pi..2*Pi,
> t=0..2*Pi,grid=[30,30],scaling=constrained):
> display(kb1a,kb1b,scaling=constrained);

Warning, the name changecoords has been redefined
```

Figure 5-17(b) shows the *Figure-8* immersion of the Klein bottle. Notice that it is not easy to see that the Klein bottle has neither an inside nor an outside in the figure.

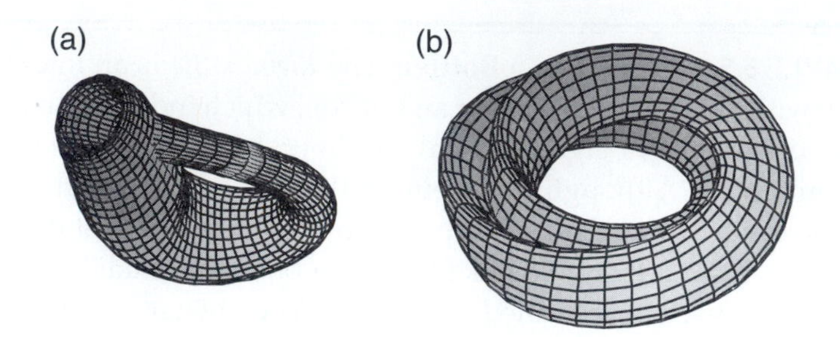

Figure 5-17 Two different immersions of the Klein bottle: (a) The "usual" immersion; (b) the Figure-8 immersion

```
> a:=3:
> x:=(u,v)->(a+cos(u/2)*sin(v)
   -sin(u/2)*sin(2*v))*cos(u):
> y:=(u,v)->(a+cos(u/2)*sin(v)
   -sin(u/2)*sin(2*v))*sin(u):
> z:=(u,v)->sin(u/2)*sin(v)+cos(u/2)*sin(2*v):
> kb2:=plot3d([x(u,v),y(u,v),z(u,v)],u=-Pi..Pi,
> v=-Pi..Pi,grid=[40,40],scaling=constrained):
> display(kb2);
```

In fact, to many readers it may not be clear whether the Klein bottle is orientable or nonorientable, especially when we compare the graph to the graphs of the Möbius strip and torus in the previous examples.

A parametrization of the Figure-8 immersion of the Klein bottle (see [17]) is $\mathbf{r}(s,t) = x(s,t)\mathbf{i} + y(s,t)\mathbf{j} + z(s,t)\mathbf{k}$, $-\pi \le s \le \pi$, $-\pi \le t \le \pi$, where

$$x = \left[c + \cos\left(\frac{1}{2}s\right)\sin t - \sin\left(\frac{1}{2}s\right)\sin 2t\right]\cos s,$$

$$y = \left[c + \cos\left(\frac{1}{2}s\right)\sin t - \sin\left(\frac{1}{2}s\right)\sin 2t\right]\sin s, \tag{5.27}$$

and

$$z = \sin\left(\frac{1}{2}s\right)\sin t + \cos\left(\frac{1}{2}s\right)\sin 2t.$$

The plot in Figure 5-17(b) uses equation (5.27) if $c = 3$.

Using (5.24), let

$$\mathbf{n} = \frac{\mathbf{r}_s \times \mathbf{r}_t}{\|\mathbf{r}_s \times \mathbf{r}_t\|}.$$

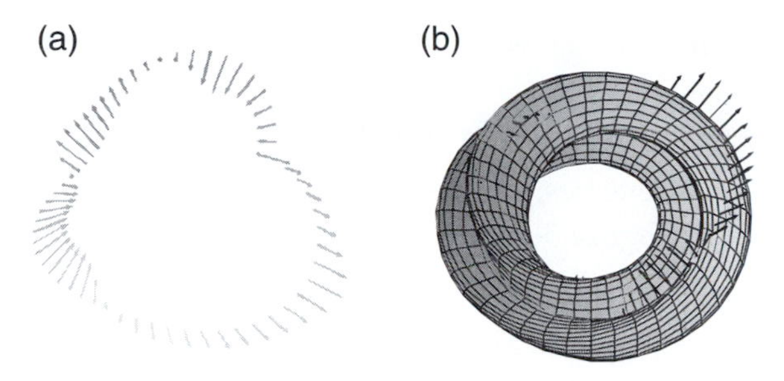

Figure 5-18 (a) An orientation reversing path. (b) The Figure-8 immersion of the Klein bottle with an orientation reversing path

Let C be the path given by

$$\mathbf{r}(t,t) = x(t,t)\mathbf{i} + y(t,t)\mathbf{j} + z(t,t)\mathbf{k}, \quad -\pi \le t \le \pi \tag{5.28}$$

that begins and ends at $\mathbf{r}(-\pi,-\pi) = \mathbf{r}(\pi,\pi) = \langle -3,0,0\rangle$ and where the components are given by (5.27). The components of $\mathbf{r}$ and $\mathbf{n}$ are computed with Maple. The final calculations are quite lengthy so we suppress the output of the last few by placing a colon (:) at the end of those commands.

```
> with(LinearAlgebra):
> with(VectorCalculus):
> r:=(s,t)->Vector([x(s,t),y(s,t),z(s,t)]):
> rs:=diff(r(s,t),s);
```

$$rs := \begin{bmatrix} \left(-1/2 \sin\left(1/2\,s\right)\sin\left(t\right) - 1/2 \cos\left(1/2\,s\right)\sin\left(2\,t\right)\right)\cos\left(s\right) \\ \quad - \left(3 + \cos\left(1/2\,s\right)\sin\left(t\right) - \sin\left(1/2\,s\right)\sin\left(2\,t\right)\right)\sin\left(s\right) \\ \left(-1/2 \sin\left(1/2\,s\right)\sin\left(t\right) - 1/2 \cos\left(1/2\,s\right)\sin\left(2\,t\right)\right)\sin\left(s\right) \\ \quad + \left(3 + \cos\left(1/2\,s\right)\sin\left(t\right) - \sin\left(1/2\,s\right)\sin\left(2\,t\right)\right)\cos\left(s\right) \\ 1/2 \cos\left(1/2\,s\right)\sin\left(t\right) - 1/2 \sin\left(1/2\,s\right)\sin\left(2\,t\right) \end{bmatrix}$$

```
> rt:=diff(r(s,t),t);
```

$$rt := \begin{bmatrix} \left(\cos\left(1/2\,s\right)\cos\left(t\right) - 2 \sin\left(1/2\,s\right)\cos\left(2\,t\right)\right)\cos\left(s\right) \\ \left(\cos\left(1/2\,s\right)\cos\left(t\right) - 2 \sin\left(1/2\,s\right)\cos\left(2\,t\right)\right)\sin\left(s\right) \\ \sin\left(1/2\,s\right)\cos\left(t\right) + 2 \cos\left(1/2\,s\right)\cos\left(2\,t\right) \end{bmatrix}$$

```
> rscrossrt:=CrossProduct(rs,rt):
> normcross:=sqrt(rscrossrt.rscrossrt):
> un:=-rscrossrt/normcross:
```

At $t = -\pi$, $\mathbf{n}(-\pi,-\pi) = \left\langle \frac{1}{\sqrt{5}}, 0, \frac{2}{\sqrt{5}} \right\rangle$, while at $t = \pi$, $\mathbf{n}(\pi,\pi) = \left\langle -\frac{1}{\sqrt{5}}, 0, -\frac{2}{\sqrt{5}} \right\rangle$ so as $\mathbf{n}$ moves along C from $\mathbf{r}(-\pi,-\pi)$ to $\mathbf{r}(\pi,\pi)$, the orientation of $\mathbf{n}$ reverses. This orientation reversing path is shown on the Klein bottle (Figure 5-18b).

```
> with(plots):
> svals:=seq(-Pi+2*Pi/59*i,i=0..59):
> vecs:=seq(arrow(r(s0,s0),evalf(subs([s=s0,t=s0],
   un))),s0=svals):
> display(vecs,scaling=constrained);

> pp2:=display(vecs,color=black):
> display(kb2,pp2,scaling=constrained);
```

Applications Related to Ordinary and Partial Differential Equations

6

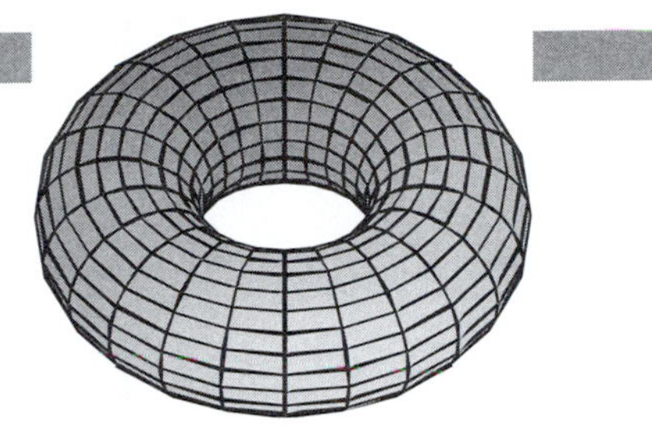

Chapter 6 discusses Maple's differential equations commands. The examples used to illustrate the various commands are similar to examples routinely done in one- and two-semester differential equations courses.

For more detailed discussions regarding Maple and differential equations see references like Abell and Braselton's *Differential Equations with Maple* [1].

6.1 First-Order Differential Equations

6.1.1 Separable Equations

Because they are solved by integrating, separable differential equations are usually the first introduced in the introductory differential equations course.

Definition 2 (Separable Differential Equation). *A differential equation of the form*

$$f(y)\,dy = g(t)\,dt \tag{6.1}$$

is called a first-order ***separable differential equation****.*

We solve separable differential equations by integrating.

Remark. The command

```
dsolve(diff(y(t),t)=f(t,y(t)),y(t))
```

attempts to solve $y' = dy/dt = f(t, y)$ for y.

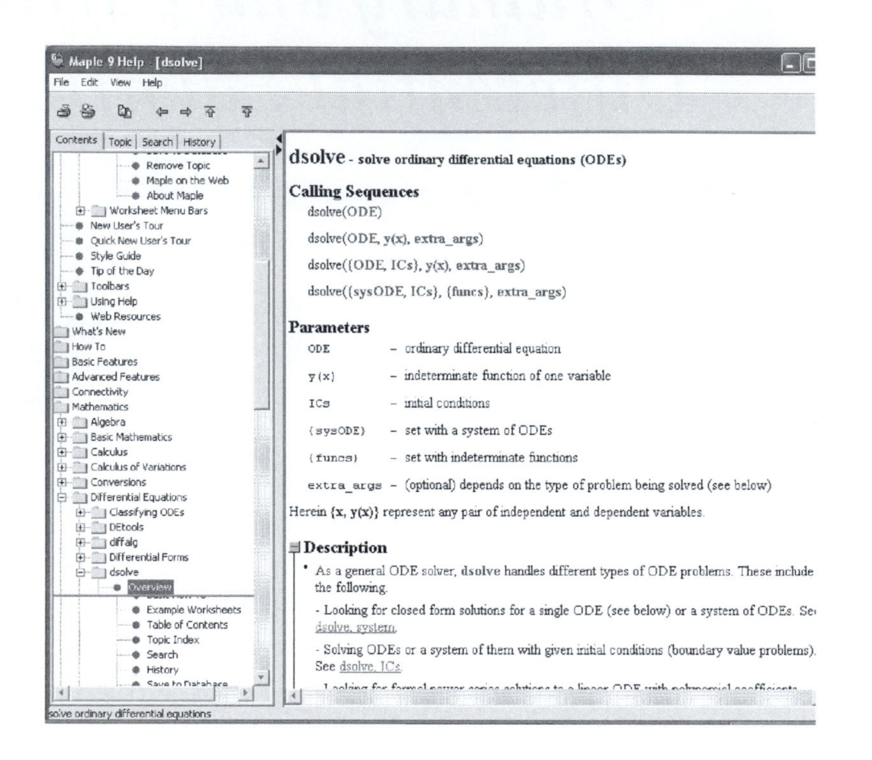

EXAMPLE 6.1.1: Solve each of the following equations: (a) $y' - y^2 \sin t = 0$; (b) $y' = \alpha y \left(1 - \frac{1}{K} y\right)$, $K, \alpha > 0$ constant.

SOLUTION: (a) The equation is separable:

$$\frac{1}{y^2} dy = \sin t \, dt$$

$$\int \frac{1}{y^2} dy = \int \sin t \, dt$$

$$-\frac{1}{y} = -\cos t + C$$

$$y = \frac{1}{\cos t + C}.$$

We check our result with `dsolve`.

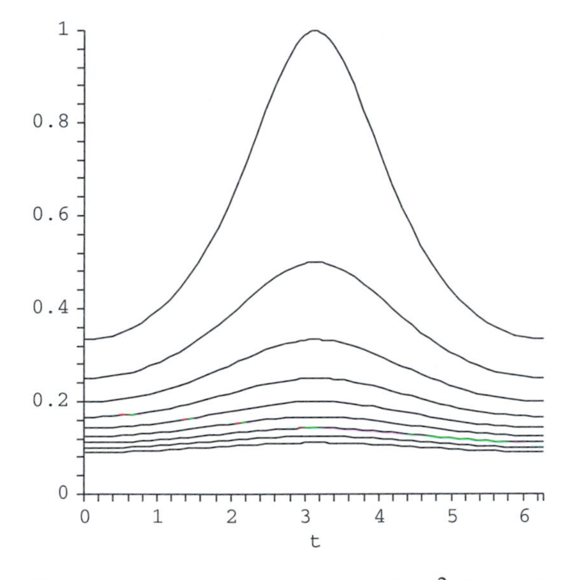

Figure 6-1 Several solutions of $y' - y^2 \sin t = 0$

```
> sola:=dsolve(diff(y(t),t)-y(t)^2*sin(t)=0,y(t));
```

$$sola := y(t) = (\cos(t) + \_C1)^{-1}$$

The formula for the solution is the right-hand side of $y(t) = \dfrac{1}{\cos t + C}$, which we obtain with `rhs`.

```
> rhs(sola);
```

$$(\cos(t) + \_C1)^{-1}$$

We then graph the solution for various values of C with `plot` in Figure 6-1.

```
> toplota:=seq(subs(_C1=i,rhs(sola)),i=2..10);
```

$$toplota := (\cos(t)+2)^{-1}, (\cos(t)+3)^{-1}, (\cos(t)+4)^{-1},$$
$$(\cos(t)+5)^{-1}, (\cos(t)+6)^{-1}, (\cos(t)+7)^{-1},$$
$$(\cos(t)+8)^{-1}, (\cos(t)+9)^{-1}, (\cos(t)+10)^{-1}$$

```
> plot([toplota],t=0..2*Pi,view=[0..2*Pi,0..1],
  color=black);
```

(b) After separating variables, we use partial fractions to integrate.

$$y' = \alpha y \left(1 - \frac{1}{K} y\right)$$

$$\frac{1}{\alpha y \left(1 - \frac{1}{K} y\right)} dy = dt$$

$$\frac{1}{\alpha}\left(\frac{1}{y} + \frac{1}{K - y}\right) = dt$$

$$\frac{1}{\alpha}\left(\ln|y| - \ln|K - y|\right) = C_1 + t$$

$$\frac{y}{K - y} = Ce^{\alpha t}$$

$$y = \frac{CKe^{\alpha t}}{Ce^{\alpha t} - 1}$$

We check the calculations with Maple. First, we use `convert` with the `parfrac` option to find the partial fraction decomposition of $\frac{1}{\alpha y \left(1 - \frac{1}{K} y\right)}$.

```
> s1:=convert(1/(alpha*y*(1-1/k*y)),parfrac,y);
```

$$s1 := y^{-1} - (-k + y)^{-1}$$

Then, we use `integrate` to check the integration.

```
> s2:=integrate(s1,y);
```

$$s2 := \ln(y) - \ln(-k + y)$$

Last, we use `solve` to solve $\frac{1}{\alpha}\left(\ln|y| - \ln|K - y|\right) = C + t$ for y.

```
> simplify(solve(s2=c+t,y));
```

$$\frac{ke^{c+t}}{-1 + e^{c+t}}$$

We can use `dsolve` to find a general solution of the equation

```
> solb:=dsolve(diff(y(t),t)=alpha*y(t)*
  (1-1/k*y(t)),y(t));
```

$$solb := y(t) = \frac{k}{1 + e^{-t}\_C1\,k}$$

as well as to find the solution that satisfies the initial condition $y(0) = y_0$.

```
> solc:=dsolve(diff(y(t),t)=alpha*y(t)*(1-1/k*y(t)),
  y(0)=y0,y(t));
```

$$solc := y(t) = k\left(1 + \frac{e^{-t}(k - y0)}{y0}\right)^{-1}$$

The equation $y' = \alpha y\left(1 - \frac{1}{K}y\right)$ is called the **Logistic equation** (or **Verhulst equation**) and is used to model the size of a population that is not allowed to grow in an unbounded manner. Assuming that $y(0) > 0$, then all solutions of the equation have the property that $\lim_{t\to\infty} y(t) = K$.

To see this, we set $\alpha = K = 1$ and use `fieldplot`, which is contained in the `plots` package, to graph the direction field associated with the equation in Figure 6-2.

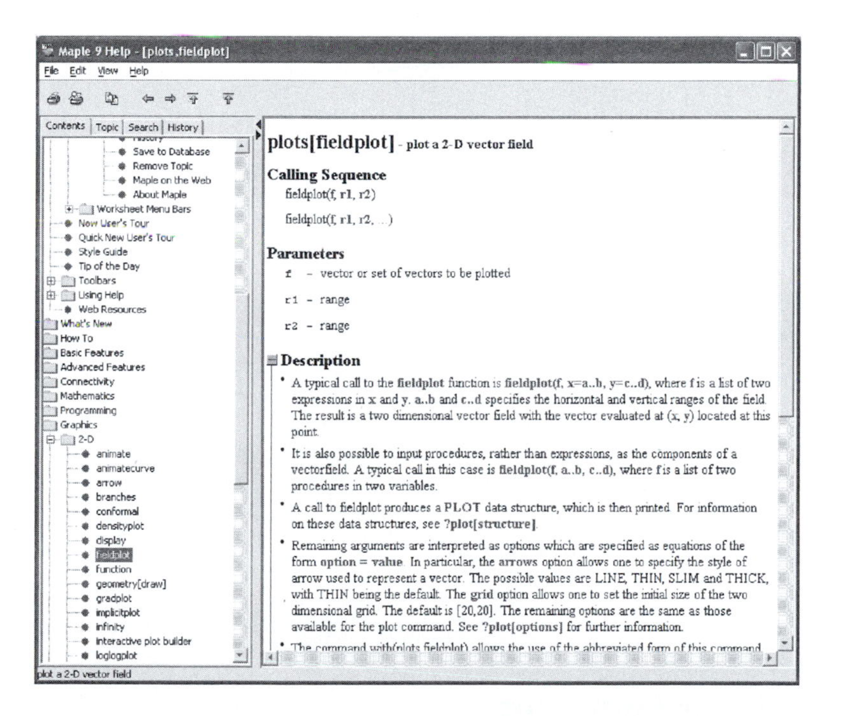


```
> with(plots):
> pvf:=fieldplot([1,y*(1-y)],t=0..5,y=0..5/2,
  scaling=constrained):
> display(pvf);
```

The property is more easily seen when we graph various solutions along with the direction field as done next in Figure 6-3.

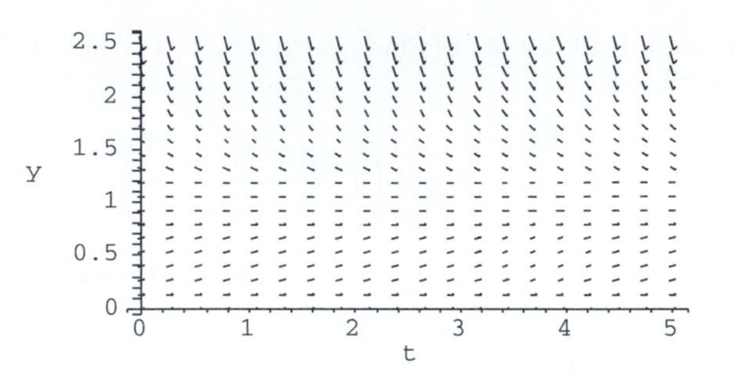

Figure 6-2 A typical direction field for the Logistic equation

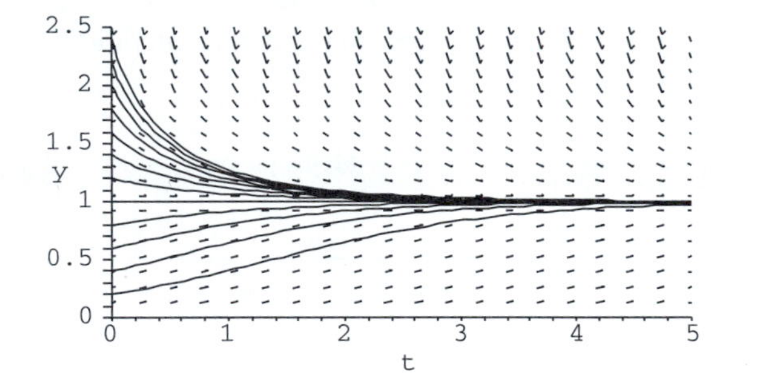

Figure 6-3 A typical direction field for the Logistic equation along with several solutions

```
> k:=1:
> alpha:=1:
> yvals:=seq(i/5,i=1..12):
> toplot:=seq(subs(y=y0,rhs(solc)),y0=yvals):
> sols:=plot([toplot],t=0..5,color=black):
> display(pvf,sols,color=black,view=[0..5,0..5/2],
   scaling=constrained);
```

■

6.1.2 Linear Equations

Definition 3 (First-Order Linear Equation). *A differential equation of the form*

$$a_1(t)\frac{dy}{dt} + a_0(t)y = f(t), \tag{6.2}$$

where $a_1(t)$ is not identically the zero function, is a first-order ***linear differential equation***.

Assuming that $a_1(t)$ is not identically the zero function, dividing (6.2) by $a_1(t)$ gives us the **standard form** of the first-order linear equation:

$$\frac{dy}{dt} + p(t)y = q(t). \tag{6.3}$$

If $q(t)$ is identically the zero function, we say that the equation is **homogeneous**. The **corresponding homogeneous equation** of (6.3) is

$$\frac{dy}{dt} + p(t)y = 0. \tag{6.4}$$

Observe that (6.4) is separable:

$$\begin{aligned}\frac{dy}{dt} + p(t)y &= 0\\ \frac{1}{y}dy &= -p(t)\,dt\\ \ln|y| &= -\int p(t)\,dt + C\\ y &= Ce^{-\int p(t)\,dt}.\end{aligned}$$

Notice that any constant multiple of a solution to a linear homogeneous equation is also a solution. Now suppose that y is any solution of (6.3) and y_p is a particular solution of (6.3). Then,

A **particular solution** is a specific solution to the equation that does not contain any arbitrary constants.

$$\begin{aligned}(y - y_p)' + p(t)(y - y_p) &= y' + p(t)y - (y_p{}' + p(t)y_p)\\ &= q(t) - q(t) = 0.\end{aligned}$$

Thus, $y - y_p$ is a solution to the corresponding homogeneous equation of (6.3). Hence,

$$\begin{aligned}y - y_p &= Ce^{-\int p(t)\,dt}\\ y &= Ce^{-\int p(t)\,dt} + y_p\\ y &= y_h + y_p,\end{aligned}$$

where $y_h = Ce^{-\int p(t)\,dt}$. That is, a general solution of (6.3) is

$$y = y_h + y_p,$$

where y_p is a particular solution to the nonhomogeneous equation and y_h is a general solution to the corresponding homogeneous equation. Thus, to solve (6.3), we need to first find a general solution to the corresponding homogeneous equation, y_h, which we can accomplish through separation of variables, and then find a particular solution, y_p, to the nonhomogeneous equation.

If y_h is a solution to the corresponding homogeneous equation of (6.3) then for any constant C, Cy_h is also a solution to the corresponding homogeneous equation. Hence, it is impossible to find a particular solution to (6.3) of this form. Instead, we search for a particular solution of the form $y_p = u(t)y_h$, where $u(t)$ is *not* a constant function. Assuming that a particular solution, y_p, to (6.3) has the form $y_p = u(t)y_h$, differentiating gives us

$$y_p{}' = u'y_h + uy_h{}'$$

and substituting into (6.3) results in

$$y_p{}' + p(t)y_p = u'y_h + uy_h{}' + p(t)uy_h = q(t).$$

y_h is a solution to the corresponding homogeneous equation so $y_h{}' + p(t)y_h = 0$.

Because $uy_h{}' + p(t)uy_h = u\left[y_h{}' + p(t)y_h\right] = u \cdot 0 = 0$, we obtain

$$\begin{aligned} u'y_h &= q(t) \\ u' &= \frac{1}{y_h}q(t) \\ u' &= e^{\int p(t)\,dt}q(t) \\ u &= \int e^{\int p(t)\,dt}q(t)\,dt \end{aligned}$$

so

$$y_p = u(t)\,y_h = Ce^{-\int p(t)\,dt}\int e^{\int p(t)\,dt}q(t)\,dt.$$

Because we can include an arbitrary constant of integration when evaluating $\int e^{\int p(t)\,dt}q(t)\,dt$, it follows that we can write a general solution of (6.3) as

$$y = e^{-\int p(t)\,dt}\int e^{\int p(t)\,dt}q(t)\,dt. \tag{6.5}$$

Alternatively, multiplying (6.3) by the **integrating factor** $\mu(t) = e^{\int p(t)\,dt}$ gives us the same result:

$$\begin{aligned} e^{\int p(t)\,dt}\frac{dy}{dt} + p(t)e^{\int p(t)\,dt}y &= q(t)e^{\int p(t)\,dt} \\ \frac{d}{dt}\left(e^{\int p(t)\,dt}y\right) &= q(t)e^{\int p(t)\,dt} \\ e^{\int p(t)\,dt}y &= \int q(t)e^{\int p(t)\,dt}\,dt \\ y &= e^{-\int p(t)\,dt}\int q(t)e^{\int p(t)\,dt}\,dt. \end{aligned}$$

Thus, first-order linear equations can always be solved, although the resulting integrals may be difficult or impossible to evaluate exactly.

Maple is able to solve the general form of the first-order equation, the initial-value problem $y' + p(t)y = q(t)$, $y(0) = y_0$,

```
> dsolve(diff(y(t),t)+p(t)*y(t)=q(t),y(t));
```

$$y(t) = \left(\int q(t)\, e^{\int p(t)dt} dt + \_C1\right) e^{\int -p(t)dt}$$

```
> dsolve(diff(y(t),t)+p(t)*y(t)=q(t),y(0)=y0,y(t));
```

$$y(t) = \left(\int_0^t q(\_z1)\, e^{\int_0^{\_z1} p(\_z1)d\_z1} d\_z1 + y0\right) e^{\int_0^t -p(\_z1)d\_z1}$$

as well as the corresponding homogeneous equation,

```
> dsolve(diff(y(t),t)+p(t)*y(t)=0,y(t));
```

$$y(t) = e^{\int -p(t)dt}\, \_C1$$

```
> dsolve(diff(y(t),t)+p(t)*y(t)=0,y(0)=y0,y(t));
```

$$y(t) = e^{\int_0^t -p(\_z1)d\_z1} y0$$

although the results contain unevaluated integrals.

EXAMPLE 6.1.2 (Exponential Growth): Let $y = y(t)$ denote the size of a population at time t. If y grows at a rate proportional to the amount present, y satisfies

$$\frac{dy}{dt} = \alpha y, \tag{6.6}$$

where α is the **growth constant**. If $y(0) = y_0$, using (6.5) results in $y = y_0 e^{\alpha t}$. We use `dsolve` to confirm this result.

```
> dsolve(diff(y(t),t)=alpha*y(t),y(0)=y0,y(t));
```

$$y(t) = y0\, e^{t}$$

$dy/dt = k(y - y_s)$ models *Newton's Law of Cooling*: the rate at which the temperature, $y(t)$, changes in a heating/cooling body is proportional to the difference between the temperature of the body and the constant temperature, y_s, of the surroundings.

EXAMPLE 6.1.3: Solve each of the following equations: (a) $dy/dt = k(y - y_s)$, $y(0) = y_0$, k and y_s constant; (b) $y' - 2ty = t$; (c) $ty' - y = 4t \cos 4t - \sin 4t$.

SOLUTION: (a) By hand, we rewrite the equation and obtain

$$\frac{dy}{dt} - ky = -ky_s.$$

A general solution of the corresponding homogeneous equation

$$\frac{dy}{dt} - ky = 0$$

is $y_h = e^{kt}$. Because k and $-ky_s$ are constants, we suppose that a particular solution of the nonhomogeneous equation, y_p, has the form $y_p = A$, where A is a constant.

This will turn out to be a lucky guess. If there is not a solution of this form, we would not find one of this form.

Assuming that $y_p = A$, we have $y_p' = 0$ and substitution into the nonhomogeneous equation gives us

$$\frac{dy_p}{dt} - ky_p = -kA = -ky_s \qquad \text{so} \qquad A = y_s.$$

Thus, a general solution is $y = y_h + y_p = Ce^{kt} + y_s$. Applying the initial condition $y(0) = y_0$ results in $y = y_s + (y_0 - y_s)e^{kt}$.

We obtain the same result with dsolve. We graph the solution satisfying $y(0) = 75$ assuming that $k = -1/2$ and $y_s = 300$ in Figure 6-4. Notice that $y(t) \to y_s$ as $t \to \infty$.

```
> k:='k':
> sola:=dsolve(diff(y(t),t)=k*(y(t)-ys),y(0)=y0,y(t));
```

$$sola := y(t) = ys + e^{kt}(-ys + y0)$$

```
> tp:=subs([k=-1/2,ys=300,y0=75],rhs(sola));
> plot(tp,t=0..10,color=black);
```

$$tp := 300 - 225\,e^{-1/2\,t}$$

(b) The equation is in standard form and we identify $p(t) = -2t$. Then, the integrating factor is $\mu(t) = e^{\int p(t)\,dt} = e^{-t^2}$. Multiplying the equation by the integrating factor, $\mu(t)$, results in

$$e^{-t^2}(y' - 2ty) = te^{-t^2} \qquad \text{or} \qquad \frac{d}{dt}\left(ye^{-t^2}\right) = te^{-t^2}.$$

Integrating gives us

$$ye^{-t^2} = -\frac{1}{2}e^{-t^2} + C \qquad \text{or} \qquad y = -\frac{1}{2} + Ce^{t^2}.$$

We confirm the result with dsolve.

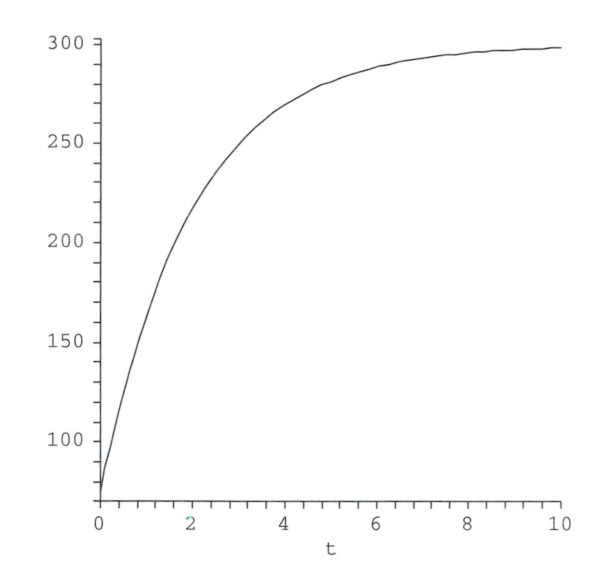

Figure 6-4 The temperature of the body approaches the temperature of its surroundings

```
> dsolve(diff(y(t),t)-2*t*y(t)=t,y(t));
```

$$y(t) = -1/2 + e^{t^2}\_C1$$

(c) In standard form, the equation is $y' - y/t = (4t\cos 4t - \sin 4t)/t$ so $p(t) = -1/t$. The integrating factor is $\mu(t) = e^{\int p(t)\,dt} = e^{-\ln t} = 1/t$ and multiplying the equation by the integrating factor and then integrating gives us

$$\begin{aligned} \frac{1}{t}\frac{dy}{dt} - \frac{1}{t^2}y &= \frac{1}{t^2}(4t\cos 4t - \sin 4t) \\ \frac{d}{dt}\left(\frac{1}{t}y\right) &= \frac{1}{t^2}(4t\cos 4t - \sin 4t) \\ \frac{1}{t}y &= \frac{\sin 4t}{t} + C \\ y &= \sin 4t + Ct, \end{aligned}$$

where we use the `integrate` function to evaluate

$$\int \frac{1}{t^2}(4t\cos 4t - \sin 4t)\,dt = \frac{\sin 4t}{t} + C.$$

```
> integrate((4*t*cos(4*t)-sin(4*t))/t^2,t);
```

$$\frac{\sin(4t)}{t}$$

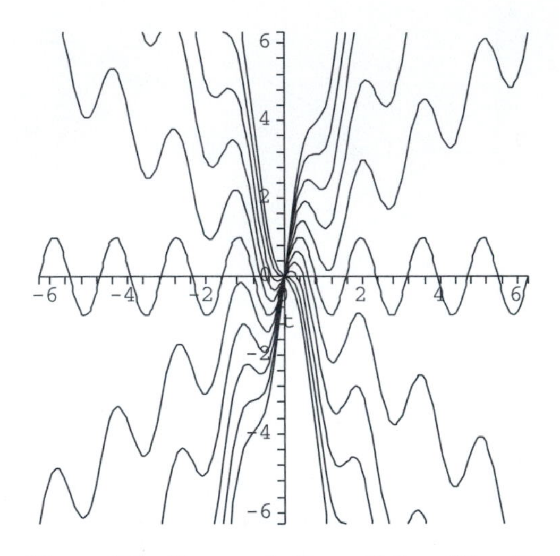

Figure 6-5 Every solution satisfies $y(0) = 0$

We confirm this result with dsolve.

```
> sol:=dsolve(diff(y(t),t)-y(t)/t=
   (4*t*cos(4*t)-sin(4*t))/t,y(t));
```

$$sol := y(t) = \sin(4t) + t\_C1$$

In the general solution, observe that *every* solution satisfies $y(0) = 0$. That is, the initial-value problem

$$\frac{dy}{dt} - \frac{1}{t}y = \frac{1}{t^2}(4t\cos 4t - \sin 4t), \qquad y(0) = 0$$

has infinitely many solutions. We see this in the plot of several solutions that is generated with plot in Figure 6-5.

```
> toplot:=seq(subs(_C1=i,rhs(sol)),i=-5..5):
> plot([toplot],t=-2*Pi..2*Pi,
   view=[-2*Pi..2*Pi,-2*Pi..2*Pi],
> color=black,scaling=constrained);
```

■

Application: Free-Falling Bodies

The motion of objects can be determined through the solution of first-order initial-value problems. We begin by explaining some of the theory that is needed to set up the differential equation that models the situation.

Newton's Second Law of Motion: The rate at which the momentum of a body changes with respect to time is equal to the resultant force acting on the body.

Because the body's momentum is defined as the product of its mass and velocity, this statement is modeled as

$$\frac{d}{dt}(mv) = F,$$

where m and v represent the body's mass and velocity, respectively, and F is the sum of the forces (the resultant force) acting on the body. Because m is constant, differentiation leads to the well-known equation

$$m\frac{dv}{dt} = F.$$

If the body is subjected only to the force due to gravity, then its velocity is determined by solving the differential equation

$$m\frac{dv}{dt} = mg \qquad \text{or} \qquad \frac{dv}{dt} = g,$$

where $g = 32\ \text{ft/s}^2$ (English system) and $g = 9.8\ \text{m/s}^2$ (international system). This differential equation is applicable only when the resistive force due to the medium (such as air resistance) is ignored. If this offsetting resistance is considered, we must discuss all of the forces acting on the object. Mathematically, we write the equation as

$$m\frac{dv}{dt} = \sum (\text{forces acting on the object})$$

where the direction of motion is taken to be the positive direction. Because air resistance acts against the object as it falls and g acts in the same direction of the motion, we state the differential equation in the form

$$m\frac{dv}{dt} = mg + (-F_R) \qquad \text{or} \qquad m\frac{dv}{dt} = mg - F_R,$$

where F_R represents this resistive force. Note that down is assumed to be the positive direction. The resistive force is typically proportional to the body's velocity, v, or the square of its velocity, v^2. Hence, the differential equation is linear or nonlinear based on the resistance of the medium taken into account.

EXAMPLE 6.1.4: An object of mass $m = 1$ is dropped from a height of 50 feet above the surface of a small pond. While the object is in the air, the force due to air resistance is v. However, when the object is in the pond, it is subjected to a buoyancy force equivalent to $6v$. Determine how much time is required for the object to reach a depth of 25 feet in the pond.

SOLUTION: This problem must be broken into two parts: an initial-value problem for the object above the pond, and an initial-value problem for the object below the surface of the pond. The initial-value problem above the pond's surface is found to be

$$\begin{cases} dv/dt = 32 - v \\ v(0) = 0 \end{cases}.$$

However, to define the initial-value problem to find the velocity of the object beneath the pond's surface, the velocity of the object when it reaches the surface must be known. Hence, the velocity of the object above the surface must be determined by solving the initial-value problem above. The equation $dv/dt = 32 - v$ is separable and solved with `dsolve` in `d1`.

```
> d1:=dsolve(diff(v(t),t)=32-v(t),v(0)=0,v(t));
```

$$d1 := v(t) = 32 - 32\,e^{-t}$$

```
> op(2,d1);
```

$$32 - 32\,e^{-t}$$

```
> rhs(d1);
```

$$32 - 32\,e^{-t}$$

In order to find the velocity when the object hits the pond's surface we must know the time at which the distance traveled by the object (or the displacement of the object) is 50. Thus, we must find the displacement function, which is done by integrating the velocity function obtaining $s(t) = 32e^{-t} + 32t - 32$.

```
> p1:=dsolve(diff(y(t),t)=op(2,d1),y(0)=0,y(t));
```

$$p1 := y(t) = 32\,e^{-t} + 32\,t - 32$$

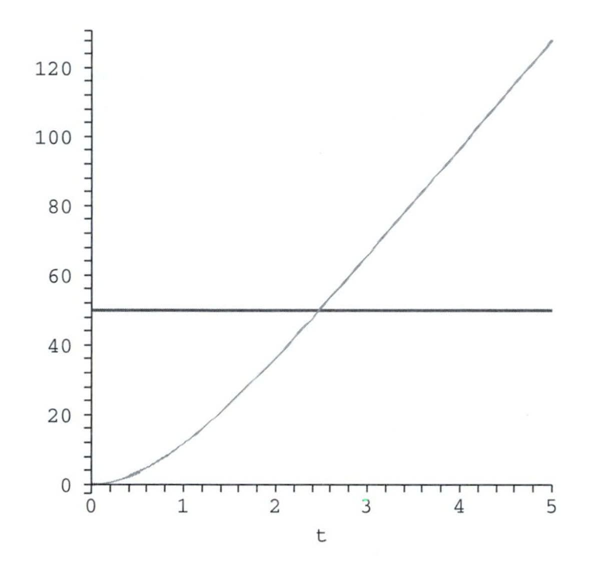

Figure 6-6 The object has traveled 50 feet when $t \approx 2.5$

The displacement function is graphed with `plot` in Figure 6-6. The value of t at which the object has traveled 50 feet is needed. This time appears to be approximately 2.5 seconds.

```
> assign(p1);
> plot(y(t),50,t=0..5);
```

A more accurate value of the time at which the object hits the surface is found using `fsolve`. In this case, we obtain $t \approx 2.47864$. The velocity at this time is then determined by substitution into the velocity function resulting in $v(2.47864) \approx 29.3166$. Note that this value is the initial velocity of the object when it hits the surface of the pond.

```
> t1:=fsolve(op(2,p1)=50,t);
```

$$t1 := 2.478643063$$

```
> v1:=evalf(subs(t=t1,op(2,d1)));
```

$$v1 := 29.31657802$$

Thus, the initial-value problem that determines the velocity of the object beneath the surface of the pond is given by

$$\begin{cases} dv/dt = 32 - 6v \\ v(0) = 29.3166 \end{cases}.$$

The solution of this initial-value problem is $v(t) = \frac{16}{3} + 23.9833e^{-t}$ and integrating to obtain the displacement function (the initial displacement is 0) we obtain $s(t) = 3.99722 - 3.99722e^{-6t} + \frac{16}{3}t$. These steps are carried out in d2 and p2.

```
> d2:=dsolve(diff(v(t),t)=32-6*v(t),
> v(0)=v1,v(t));
```

$$d2 := v(t) = 16/3 + \frac{3597486703}{150000000} e^{-6t}$$

```
> y:='y':
> p2:=dsolve(diff(y(t),t)=op(2,d2),y(0)=0,y(t));
```

$$p2 := y(t) = -\frac{3597486703}{900000000} e^{-6t} + 16/3\,t + \frac{3597486703}{900000000}$$

This displacement function is then plotted in Figure 6-7 to determine when the object is 25 feet beneath the surface of the pond. This time appears to be near 4 seconds.

```
> assign(p2):
> plot(y(t),25,t=0..5);
```

A more accurate approximation of the time at which the object is 25 feet beneath the pond's surface is obtained with fsolve. In this case, we

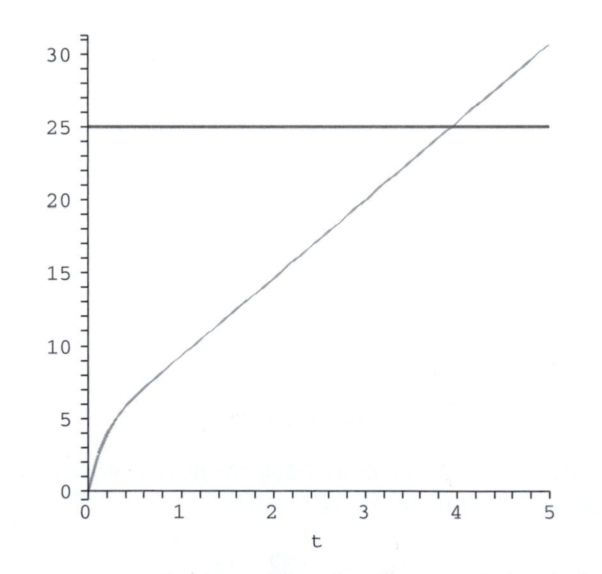

Figure 6-7 After approximately 4 seconds, the object is 25 feet below the surface of the pond

obtain $t \approx 3.93802$. Finally, the time required for the object to reach the pond's surface is added to the time needed for it to travel 25 feet beneath the surface to see that approximately 6.41667 seconds are required for the object to travel from a height of 50 feet above the pond to a depth of 25 feet below the surface.

```
> t2:=fsolve(op(2,p2)=25,t);
```

$$t2 := 3.938023604$$

```
> t1+t2;
```

$$6.416666667$$

■

6.1.3 Nonlinear Equations

Maple can solve a variety of nonlinear first-order equations that are typically encountered in the introductory differential equations course.

Use the `odeadvisor` function, which is contained in the `DEtools` package, to help you classify equations.

EXAMPLE 6.1.5: Solve each of the following equations: (a) $\left(y \cos x + 2xe^y\right)dx + \left(\sin y + x^2e^y - 1\right)dy = 0$; (b) $\left(y^2 + 2xy\right)dx - x^2dy = 0$.

SOLUTION: (a) Notice that $(\cos x + 2xe^y)\,dx + \left(\sin y + x^2e^y - 1\right)dy = 0$ can be written as $dy/dx = -\left(y\cos x + 2xe^y\right)/\left(\sin y + x^2e^y - 1\right)$.

The equation is an example of an *exact equation*. A theorem tells us that the equation

$$M(x,y)dx + N(x,y)dy = 0$$

is **exact** if and only if $\partial M/\partial y = \partial N/\partial x$.

```
> M:=(x,y)->cos(x)+2*x*exp(y):
> N:=(x,y)->sin(y)+x^2*exp(y)-1:
> diff(M(x,y),y);
```

$$2\,xe^y$$

```
> diff(N(x,y),x);
```

$$2\,xe^{y}$$

We confirm that the equation is exact with `odeadvisor`.

```
> with(DEtools):
> eq:=M(x,y(x))+N(x,y(x))*diff(y(x),x)=0:
> odeadvisor(eq);
```

$$[\_exact]$$

We solve exact equations by integrating. Let $F(x,y) = C$ satisfy $(\cos x + 2xe^y)\,dx + \left(\sin y + x^2e^y - 1\right)dy = 0$. Then,

$$F(x,y) = \int \left(\cos x + 2xe^y\right)dx = \sin x + x^2e^y + g(y),$$

where $g(y)$ is a function of y.

```
> f1:=integrate(M(x,y),x);
```

$$f1 := \sin(x) + x^2e^y$$

We next find that $g'(y) = \sin y - 1$ so $g(y) = -\cos y - y$. Hence, a general solution of the equation is

$$\sin x + x^2e^y - \cos y = C.$$

```
> f2:=diff(f1,y);
```

$$f2 := x^2e^y$$

```
> f3:=solve(f2+c=N(x,y),c);
```

$$f3 := \sin(y) - 1$$

```
> integrate(f3,y);
```

$$-\cos(y) - y$$

We confirm this result with `dsolve`. Notice that Maple cannot solve for y explicitly and returns the same implicit solution obtained by us.

```
> sol:=dsolve(eq,y(x));
```

$$sol := \sin(x) + x^2e^{y(x)} - \cos(y(x)) - y(x) + \_C1 = 0$$

Graphs of several solutions using the values of C generated in `cvals` are graphed with `contourplot` in Figure 6-8.

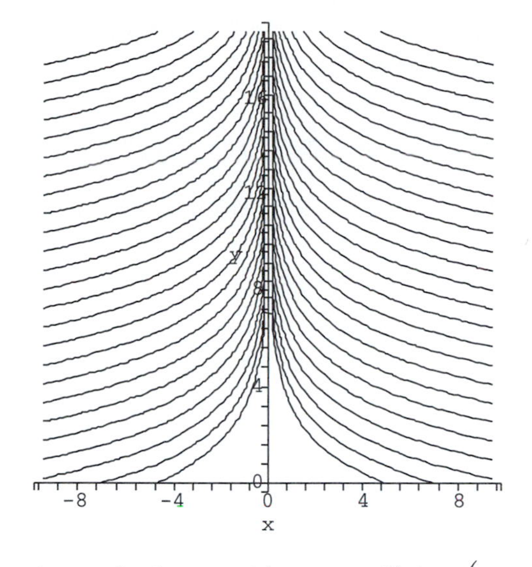

Figure 6-8 Graphs of several solutions of $(\cos x + 2xe^y)\,dx + \left(\sin y + x^2e^y - 1\right)dy = 0$

```
> sol2:=subs([_C1=0,y(x)=y],lhs(sol));
```

$$sol2 := \sin(x) + x^2e^y - \cos(y) - y$$

```
> ivals:=seq(6*Pi*i/24,i=0..24):
> cvals:=seq(subs([x=-3*Pi/2,y=i],sol2),i=ivals):
> with(plots):
> contourplot(sol2,x=-3*Pi..3*Pi,y=0..6*Pi,
   contours=[cvals],
> scaling=constrained,color=black,grid=[60,60]);
```

(b) We can write $(y^2 + 2xy)\,dx - x^2dy = 0$ as $dy/dx = (y^2 + 2xy)/x^2$. A first-order equation is **homogeneous** if it can be written in the form

$$\frac{dy}{dx} = F\left(\frac{y}{x}\right).$$

Homogeneous equations are reduced to separable equations with either the substitution $y = ux$ or $x = vy$.

In this case, we have that $dy/dx = (y/x)^2 + 2(y/x)$ so the equation is homogeneous, which we confirm with odeadvisor.

```
> eq:=(y(x)^2+2*x*y(x))-x^2*diff(y(x),x)=0:
> odeadvisor(eq);
```

$$[[homogeneous, classA], rational, Bernoulli]$$

Let $y = ux$. Then, $dy = u\,dx + x\,du$. Substituting into $(y^2 + 2xy)\,dx - x^2 dy = 0$ and separating gives us

$$\left(y^2 + 2xy\right) dx - x^2 dy = 0$$
$$\left(u^2x^2 + 2ux^2\right) dx - x^2(u\,dx + x\,du) = 0$$
$$\left(u^2 + 2u\right) dx - (u\,dx + x\,du) = 0$$
$$\left(u^2 + u\right) dx = -x\,du$$
$$\frac{1}{u\,(u+1)} du = -\frac{1}{x} dx.$$

Integrating the left- and right-hand sides of this equation with `integrate`,

```
> integrate(1/(u*(u+1)),u);
```

$$\ln(u) - \ln(u+1)$$

```
> integrate(1/x,x);
```

$$\ln(x)$$

exponentiating, resubstituting $u = y/x$, and solving for y gives us

$$\ln|u| - \ln|u+1| = -\ln|x| + C$$
$$\frac{u}{u+1} = Cx$$
$$\frac{\frac{y}{x}}{\frac{y}{x}+1} = Cx$$
$$y = \frac{Cx^2}{1 - Cx}.$$

```
> solve((y/x)/(y/x+1)=c*x,y);
```

$$-\frac{cx^2}{-1 + cx}$$

We confirm this result with `dsolve` and then graph several solutions with `plot` in Figure 6-9.

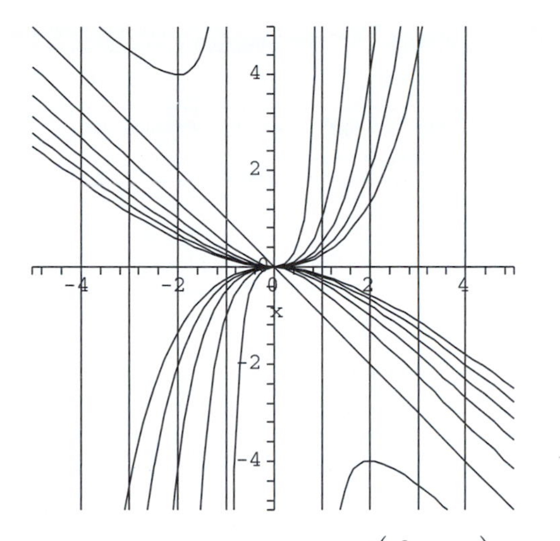

Figure 6-9 Graphs of several solutions of $\left(y^2+2xy\right)dx - x^2dy = 0$

```
> sol:=dsolve(eq,y(x));
```

$$sol := y(x) = -\frac{x^2}{x - \_C1}$$

```
> toplot:=seq(subs(_C1=i,rhs(sol)),i=-5..5):
> plot([toplot],x=-5..5,view=[-5..5,-5..5],
   scaling=constrained,color=black);
```

■

6.1.4 Numerical Methods

If numerical results are desired, use `dsolve` together with the `numeric` option:

```
dsolve({diff(y(t),t)=f(t,y(t)),y(t0)=y0},y(t),numeric)
```

attempts to generate a numerical solution of

$$\begin{cases} dy/dt = f(t,y) \\ y(t_0) = y_0 \end{cases}.$$

Use `odeplot`, which is contained in the `plots` package, to graph the numerical functions that result from using `dsolve` together with the `numeric` option.

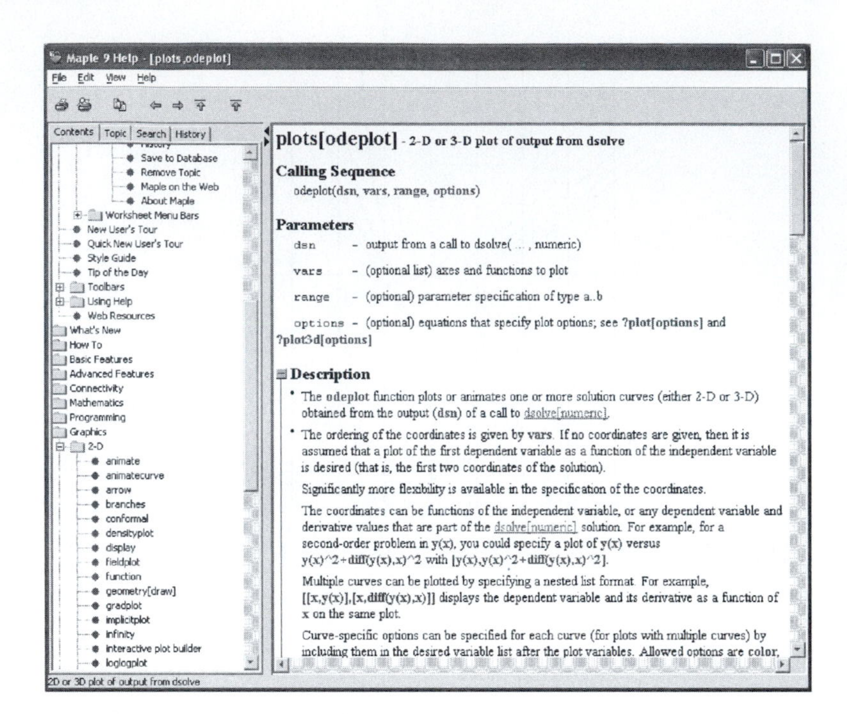

EXAMPLE 6.1.6: Consider

$$\frac{dy}{dt} = \left(t^2 - y^2\right)\sin y,\ y(0) = -1.$$

(a) Determine $y(1)$. (b) Graph $y(t)$, $-1 \le t \le 10$.

SOLUTION: We first remark that `dsolve` can neither exactly solve the differential equation $y' = \left(t^2 - y^2\right)\sin y$ nor find the solution that satisfies $y(0) = -1$.

No output means that Maple cannot solve the problem or that the problem has no solution. Generally, when Maple returns nothing, you should try other methods to determine if the problem has solutions that Maple cannot find.

```
> sol:=dsolve(diff(y(t),t)=(t^2-y(t)^2)*
   sin(y(t)),y(t));
```

$$sol :=$$

```
> sol:=dsolve(diff(y(t),t)=(t^2-y(t)^2)*sin(y(t)),
   y(0)=y0,y(t));
```

$$sol :=$$

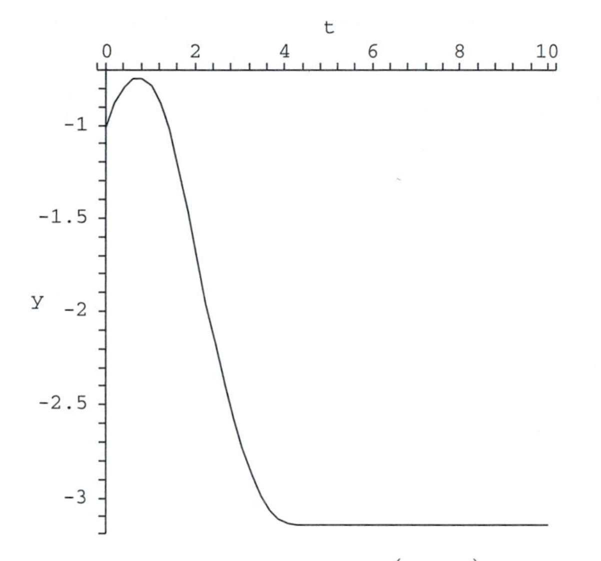

Figure 6-10 Graph of the solution to $y' = \left(t^2 - y^2\right) \sin y$, $y(0) = -1$

However, we obtain a numerical solution using `dsolve` together with the `numeric` option.

```
> sol:=dsolve(diff(y(t),t)=(t^2-y(t)^2)*sin(y(t)),
   y(0)=-1,y(t),numeric);
```

$$sol := proc(x_rkf45) \ldots endproc$$

Entering `sol(1)` evaluates the numerical solution if $t = 1$.

```
> sol(1);
```

$$[t = 1.0, y(t) = -0.7660197442785808882]$$

The result means that $y(1) \approx -.766$. We use the `odeplot` command, which is contained in the `plots` package, to graph the solution for $0 \leq t \leq 10$ in Figure 6-10.

```
> with(plots):
> odeplot(sol,[t,y(t)],t=0..10,color=black);
```

■

EXAMPLE 6.1.7 (Logistic Equation with Predation): Incorporating predation into the **logistic equation,** $y' = \alpha y\left(1 - \frac{1}{K}y\right)$, results in

$$\frac{dy}{dt} = \alpha y\left(1 - \frac{1}{K}y\right) - P(y),$$

where $P(y)$ is a function of y describing the rate of predation. A typical choice for P is $P(y) = ay^2/(b^2 + y^2)$ because $P(0) = 0$ and P is bounded above: $\lim_{t\to\infty} P(y) < \infty$.

Remark. Of course, if $\lim_{t\to\infty} y(t) = Y$, then $\lim_{t\to\infty} P(y) = aY^2/(b^2 + Y^2)$. Generally, however, $\lim_{t\to\infty} P(y) \neq a$ because $\lim_{t\to\infty} y(t) \leq K \neq \infty$, for some $K \geq 0$, in the predation situation.

If $\alpha = 1, a = 5$ and $b = 2$, graph the direction field associated with the equation as well as various solutions if (a) $K = 19$ and (b) $K = 20$.

SOLUTION: (a) We define `eqn(k)` to be

$$\frac{dy}{dt} = y\left(1 - \frac{1}{K}y\right) - \frac{5y^2}{4 + y^2}.$$

```
> with(plots):
> eqn:=k->diff(y(t),t)=y(t)*(1-1/k*y(t))
   -5*y(t)^2/(4+y(t)^2):
```

We use `fieldplot` to graph the direction field in Figure 6-11(a) and then the direction field along with the solutions that satisfy $y(0) = .5$, $y(0) = .2$, and $y(0) = 4$ in Figure 6-11(b).

```
> eqn(1);
```

$$\frac{d}{dt}y(t) = y(t)\,(1 - y(t)) - 5\,\frac{(y(t))^2}{4 + (y(t))^2}$$

```
> pvf19:=fieldplot([1,y*(1-1/19*y)-5*y^2/(4+y^2)],
   t=0..10,y=0..6,color=black):
> display(pvf19,scaling=constrained);

> n1:=dsolve(eqn(19),y(0)=0.5,y(t),numeric):
> n2:=dsolve(eqn(19),y(0)=2,y(t),numeric):
> n3:=dsolve(eqn(19),y(0)=4,y(t),numeric):
```

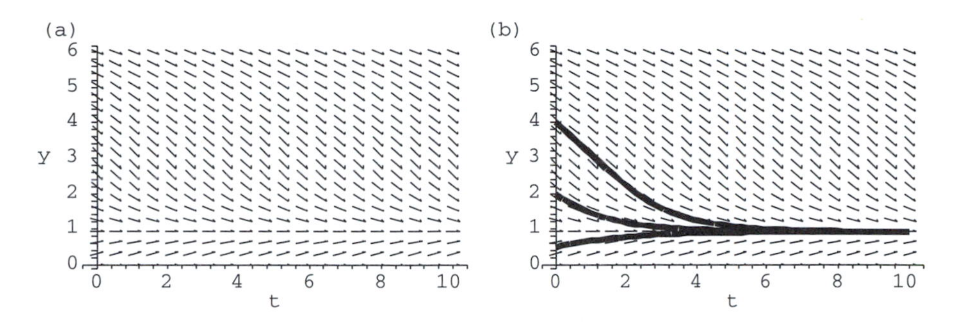

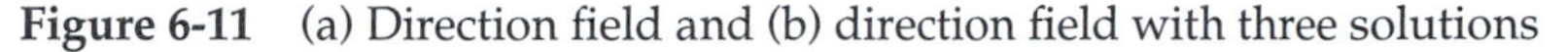
Figure 6-11 (a) Direction field and (b) direction field with three solutions

```
> sols:=map(odeplot,[n1,n2,n3],[t,y(t)],t=0..10,
  color=black,thickness=5):
> solplot:=display(sols):
> display(pvf19,solplot,scaling=constrained);
```

In the plot, notice that all nontrivial solutions appear to approach an equilibrium solution. We determine the equilibrium solution by solving $y' = 0$

```
> solve(rhs(eqn(19.))=0,y(t));
```

$$0.,0.9233508108,9.038324594-0.7858752075I,9.038324594+0.7858752075I$$

to see that it is $y \approx 0.923$.

(b) We carry out similar steps for (b). First, we graph the direction field with `fieldplot` in Figure 6-12.

```
> pvf20:=fieldplot([1,y*(1-1/20*y)-5*y^2/(4+y^2)],
  t=0..10,y=0..20,color=gray):
> display(pvf20);
```

We then use `seq` together with `dsolve` and the `numeric` option to numerically find the solution satisfying $y(0) = .5i$, for $i = 1, 2, \ldots,$ 40 and name the resulting list `numsols`. The functions contained in `numsols` are graphed with `odeplot` in `solplot`. Last, we display the direction field along with the solution graphs in `solplot` using `display` in Figure 6-13.

```
> ivals:=seq(0.5*i,i=1..40):
> numsols:=[seq(dsolve(eqn(20),y(0)=i,y(t),
  numeric),i=ivals)]:
> sols:=map(odeplot,numsols,[t,y(t)],
  t=0..10,color=black):
```

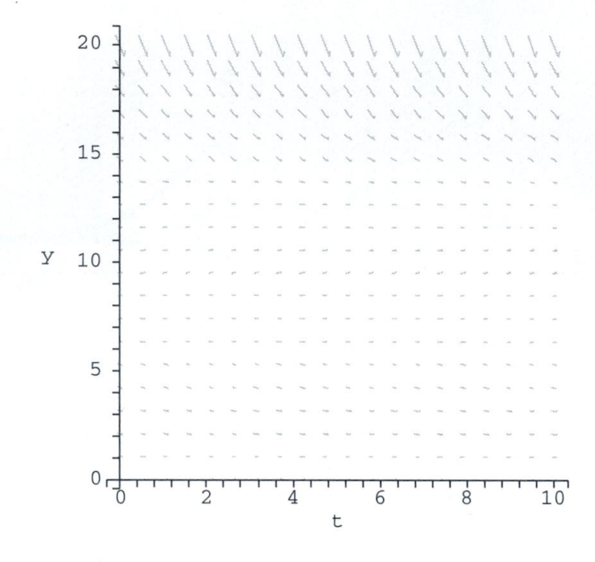

Figure 6-12 Direction field

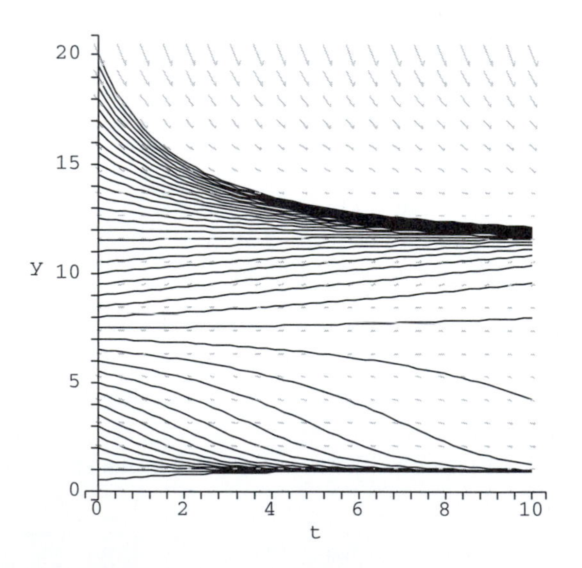

Figure 6-13 Direction field with several solutions

```
> solplot:=display(sols):
> display(pvf20,solplot);
```

Notice that there are three nontrivial equilibrium solutions that are found by solving $y' = 0$.

```
> solve(rhs(eqn(20.))=0,y(t));
```

$$0.0, 0.9267407576, 7.386450268, 11.68680897$$

In this case, $y \approx .927$ and $y \approx 11.687$ are stable while $y \approx 7.386$ is unstable.

■

6.2 Second-Order Linear Equations

We now present a concise discussion of second-order linear equations, which are extensively discussed in the introductory differential equations course.

6.2.1 Basic Theory

The **general form** of the **second-order linear equation** is

$$a_2(t)\frac{d^2y}{dt^2} + a_1(t)\frac{dy}{dt} + a_0(t)y = f(t), \tag{6.7}$$

where $a_2(t)$ is not identically the zero function.

The **standard form** of the second-order linear equation (6.7) is

$$\frac{d^2y}{dt^2} + p(t)\frac{dy}{dt} + q(t)y = f(t). \tag{6.8}$$

The **corresponding homogeneous equation** of (6.8) is

$$\frac{d^2y}{dt^2} + p(t)\frac{dy}{dt} + q(t)y = 0. \tag{6.9}$$

A **general solution** of (6.9) is $y = c_1y_1 + c_2y_2$ where

1. y_1 and y_2 are solutions of (6.9), and
2. y_1 and y_2 are *linearly independent*.

If y_1 and y_2 are solutions of (6.9), then y_1 and y_2 are **linearly independent** if and only if the **Wronskian**,

$$W\left(\{y_1, y_2\}\right) = \begin{vmatrix} y_1 & y_2 \\ y_1{}' & y_2{}' \end{vmatrix} = y_1 y_2{}' - y_1{}' y_2, \tag{6.10}$$

is not the zero function. If y_1 and y_2 are linearly independent solutions of (6.9), we call the set $S = \{y_1, y_2\}$ a **fundamental set of solutions** for (6.9).

We use the `wronskian` function, which is contained in the `linalg` package, to compute the Wronskian determinant and then use `det`, which is also contained in the `linalg` package, to compute the determinant of the result.

A particular solution, y_p, is a solution that does not contain any arbitrary constants.

Let y be a general solution of (6.8) and y_p be a particular solution of (6.8). It follows that $y - y_p$ is a solution of (6.9) so $y - y_p = y_h$ where y_h is a general solution of (6.9). Hence, $y = y_h + y_p$. That is, to solve the nonhomogeneous equation, we need a general solution, y_h, of the corresponding homogeneous equation and a particular solution, y_p, of the nonhomogeneous equation.

6.2.2 Constant Coefficients

Suppose that the coefficient functions of (6.7) are constants: $a_2(t) = a$, $a_1(t) = b$, and $a_0(t) = c$ and that $f(t)$ is identically the zero function. In this case, (6.7) becomes

$$ay'' + by' + cy = 0. \tag{6.11}$$

Now suppose that $y = e^{kt}$, k constant, is a solution of (6.11). Then, $y' = ke^{kt}$ and $y'' = k^2 e^{kt}$. Substitution into (6.11) then gives us

$$\begin{aligned} ay'' + by' + cy &= ak^2 e^{kt} + bke^{kt} + ce^{kt} \\ &= e^{kt}\left(ak^2 + bk + c\right) = 0. \end{aligned}$$

Because $e^{kt} \neq 0$, the solutions of (6.11) are determined by the solutions of

$$ak^2 + bk + c = 0, \tag{6.12}$$

called the **characteristic equation** of (6.11).

Theorem 23. *Let k_1 and k_2 be the solutions of* (6.12).

1. *If $k_1 \neq k_2$ are real and distinct, two linearly independent solutions of* (6.11) *are $y_1 = e^{k_1 t}$ and $y_2 = e^{k_2 t}$; a general solution of* (6.11) *is*

$$y = c_1 e^{k_1 t} + c_2 e^{k_2 t}.$$

2. *If* $k_1 = k_2$, *two linearly independent solutions of* (6.11) *are* $y_1 = e^{k_1 t}$ *and* $y_2 = te^{k_1 t}$; *a general solution of* (6.11) *is*

$$y = c_1 e^{k_1 t} + c_2 t e^{k_1 t}.$$

3. *If* $k_{1,2} = \alpha \pm \beta i$, $\beta \neq 0$, *two linearly independent solutions of* (6.11) *are* $y_1 = e^{\alpha t} \cos \beta t$ *and* $y_2 = e^{\alpha t} \sin \beta t$; *a general solution of* (6.11) *is*

$$y = e^{\alpha t} \left(c_1 \cos \beta t + c_2 \sin \beta t \right).$$

EXAMPLE 6.2.1: Solve each of the following equations: (a) $6y'' + y' - 2y = 0$; (b) $y'' + 2y' + y = 0$; (c) $16y'' + 8y' + 145y = 0$.

SOLUTION: (a) The characteristic equation is $6k^2 + k - 2 = (3k + 2)(2k - 1) = 0$ with solutions $k = -2/3$ and $k = 1/2$. We check with either `factor` or `solve`.

```
> factor(6*k^2+k-2);
```

$$(3\,k + 2)\,(2\,k - 1)$$

```
> solve(6*k^2+k-2=0);
```

$$1/2, -2/3$$

Then, a fundamental set of solutions is $\left\{e^{-2t/3}, e^{t/2}\right\}$ and a general solution is

$$y = c_1 e^{-2t/3} + c_2 e^{t/2}.$$

Of course, we obtain the same result with `dsolve`.

```
> dsolve(6*diff(y(t),t$2)+diff(y(t),t)-2*y(t)=0,y(t));
```

$$y\,(t) = \_C1\, e^{1/2\,t} + \_C2\, e^{-2/3\,t}$$

(b) The characteristic equation is $k^2 + 2k + 1 = (k + 1)^2 = 0$ with solution $k = -1$, which has multiplicity two, so a fundamental set of solutions is $\left\{e^{-t}, te^{-t}\right\}$ and a general solution is

$$y = c_1 e^{-t} + c_2 t e^{-t}.$$

We check the calculation in the same way as in (a).

```
> factor(k^2+2*k+1);
```

$$(k+1)^2$$

```
> solve(k^2+2*k+1=0);
```

$$-1,\ -1$$

```
> dsolve(diff(y(t),t$2)+2*diff(y(t),t)+y(t)=0,y(t));
```

$$y(t) = \_C1\, e^{-t} + \_C2\, e^{-t} t$$

(c) The characteristic equation is $16k^2 + 8k + 145 = 0$ with solutions $k_{1,2} = -\frac{1}{4} \pm 3i$ so a fundamental set of solutions is $\left\{e^{-t/4}\cos 3t, e^{-t/4}\sin 3t\right\}$ and a general solution is

$$y = e^{-t/4}\left(c_1 \cos 3t + c_2 \sin 3t\right).$$

The calculation is verified in the same way as in (a) and (b).

```
> factor(16*k^2+8*k+145,I);
```

$$(4k+1-12i)(4k+1+12i)$$

```
> solve(16*k^2+8*k+145=0);
```

$$-1/4+3i,\ -1/4-3i$$

```
> dsolve(16*diff(y(t),t$2)+8*diff(y(t),t)
  +145*y(t)=0,y(t));
```

$$y(t) = \_C1\, e^{-1/4t}\sin(3t) + \_C2\, e^{-1/4t}\cos(3t)$$

■

EXAMPLE 6.2.2: Solve

$$64\frac{d^2y}{dt^2} + 16\frac{dy}{dt} + 1025y = 0,\ y(0) = 1,\ \frac{dy}{dt}(0) = 2.$$

SOLUTION: A general solution of $64y'' + 16y' + 1025y = 0$ is $y = e^{-t/8}(c_1 \sin 4t + c_2 \cos 4t)$.

```
> gensol:=dsolve(64*diff(y(t),t$2)+16*diff(y(t ),t)
  +1025*y(t)=0,y(t));
```

$$gensol := y(t) = \_C1\, e^{-1/8\,t} \sin(4\,t) + \_C2\, e^{-1/8\,t} \cos(4\,t)$$

Applying $y(0) = 1$ shows us that $c_2 = 1$.

```
> e1:=eval(subs(t=0,rhs(gensol)));
```

$$e1 := \_C2$$

Computing y'

```
> diff(rhs(gensol),t);
```

$$-1/8\,\_C1\, e^{-1/8\,t} \sin(4\,t) + 4\,\_C1\, e^{-1/8\,t} \cos(4\,t)$$
$$-1/8\,\_C2\, e^{-1/8\,t} \cos(4\,t) - 4\,\_C2\, e^{-1/8\,t} \sin(4\,t)$$

and then $y'(0)$, shows us that $4c_1 - \frac{1}{8}c_2 = 3$.

```
> e2:=eval(subs(t=0,diff(rhs(gensol),t)));
```

$$e2 := 4\,\_C1 - 1/8\,\_C2$$

Solving for c_1 and c_2 with `solve` shows us that $c_1 = 25/32$ and $c_2 = 1$.

```
> cvals:=solve(e1=1,e2=3);
```

$$cvals := \left\{\_C2 = 1, \_C1 = \frac{25}{32}\right\}$$

Thus, $y = e^{-t/8}\left(\frac{25}{32}\sin 4t + \cos 4t\right)$, which we graph with `plot` in Figure 6-14.

```
> sol:=subs(cvals,rhs(gensol));
```

$$sol := \frac{25}{32} e^{-1/8\,t} \sin(4\,t) + e^{-1/8\,t} \cos(4\,t)$$

```
> plot(sol,t=0..8*Pi,color=black);
```

We verify the calculation with `dsolve`.

```
> dsolve(64*diff(y(t),t$2)+16*diff(y(t),t)
> +1025*y(t)=0,y(0)=1,D(y)(0)=3,y(t));
```

$$y(t) = \frac{25}{32} e^{-1/8\,t} \sin(4\,t) + e^{-1/8\,t} \cos(4\,t)$$

■

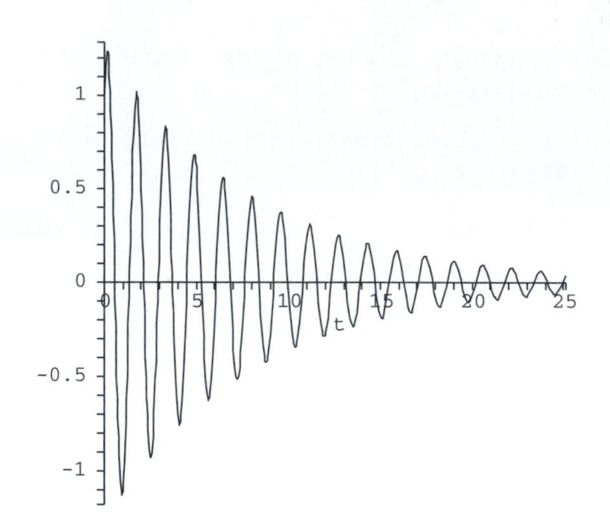

Figure 6-14 The solution to the initial-value problem tends to 0 as $t \to \infty$

Application: Harmonic Motion

Suppose that a mass is attached to an elastic spring that is suspended from a rigid support such as a ceiling. According to Hooke's law, the spring exerts a restoring force in the upward direction that is proportional to the displacement of the spring.

> **Hooke's Law:** $F = ks$, where $k > 0$ is the constant of proportionality or spring constant, and s is the displacement of the spring.

Using Hooke's Law and assuming that $x(t)$ represents the displacement of the mass from the equilibrium position at time t, we obtain the initial-value problem

$$\begin{cases} m\dfrac{d^2x}{dt^2} + kx = 0 \\ x(0) = \alpha,\ \dfrac{dx}{dt}(0) = \beta \end{cases}.$$

Note that the initial conditions give the initial displacement and velocity, respectively. This differential equation disregards all retarding forces acting on the motion of the mass and a more realistic model which takes these forces into account is needed. Studies in mechanics reveal that resistive forces due to damping are proportional to a power of the velocity of the motion. Hence, $F_R = c\,dx/dt$ or $F_R = c\,(dx/dt)^3$, where $c > 0$, are typically used to represent the damping force. Then, we have the following initial-value problem assuming that $F_R = c\,dx/dt$:

$$\begin{cases} m\dfrac{d^2x}{dt^2} + c\dfrac{dx}{dt} + kx = 0 \\ x(0) = \alpha,\ \dfrac{dx}{dt}(0) = \beta \end{cases}.$$

Problems of this type are characterized by the value of $c^2 - 4mk$ as follows:

1. $c^2 - 4mk > 0$. This situation is said to be **overdamped** because the damping coefficient c is large in comparison with the spring constant k.
2. $c^2 - 4mk = 0$. This situation is described as **critically damped** because the resulting motion is oscillatory with a slight decrease in the damping coefficient c.
3. $c^2 - 4mk > 0$. This situation is called **underdamped** because the damping coefficient c is small in comparison with the spring constant k.

EXAMPLE 6.2.3: Classify the following differential equations as overdamped, underdamped, or critically damped. Also, solve the corresponding initial-value problem using the given initial conditions and investigate the behavior of the solutions.

(a) $\dfrac{d^2x}{dt^2} + 8\dfrac{dx}{dt} + 16x = 0$ subject to $x(0) = 0$ and $\dfrac{dx}{dt}(0) = 1$;

(b) $\dfrac{d^2x}{dt^2} + 5\dfrac{dx}{dt} + 4x = 0$ subject to $x(0) = 1$ and $\dfrac{dx}{dt}(0) = 1$; and

(c) $\dfrac{d^2x}{dt^2} + \dfrac{dx}{dt} + 16x = 0$ subject to $x(0) = 0$ and $\dfrac{dx}{dt}(0) = 1$.

SOLUTION: For (a), we identify $m = 1$, $c = 8$, and $k = 16$ so that $c^2 - 4mk = 0$, which means that the differential equation $x'' + 8x' + 16x = 0$ is critically damped. After defining `DEOne`, we solve the equation subject to the initial conditions and name the resulting output `sola`. We then graph the solution shown in Figure 6-15.

```
> m:=1:c:=8:k:=16:
> c^2-4*m*k;
```

$$0$$

```
> x:='x':
> DEOne:=diff(x(t),t$2)+8*diff(x(t),t)+16*x(t)=0:
```

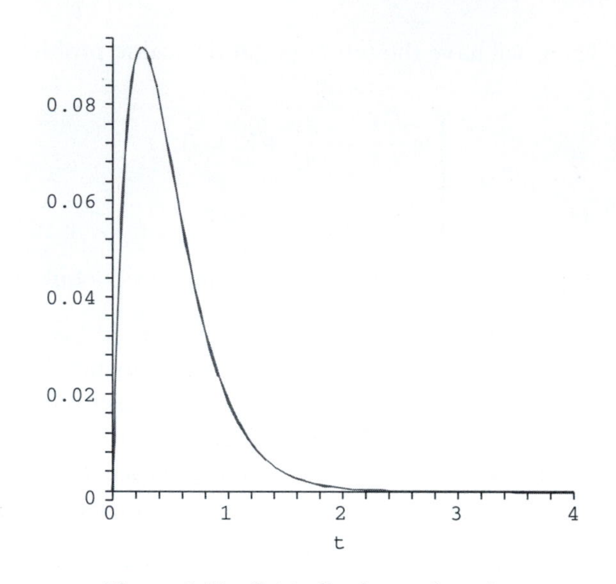

Figure 6-15 Critically damped motion

```
> sola:=dsolve(DEOne,x(0)=0,D(x)(0)=1,x(t));
```

$$sola := x(t) = e^{-4t}t$$

```
> assign(sola):
> plot(x(t),t=0..4);
```

For (b), we proceed in the same manner. We identify $m = 1$, $c = 5$, and $k = 4$ so that $c^2 - 4mk = 9$ and the equation $x'' + 5x' + 4x = 0$ is overdamped. We then define `DETwo` to be the equation and the solution to the initial-value problem obtained with `dsolve`, `solb` and then graph $x(t)$ on the interval $[0, 4]$ in Figure 6-16.

```
> m:=1:c:=5:k:=4:
> c^2-4*m*k;
```

$$9$$

```
> x:='x':
> DETwo:=diff(x(t),t$2)+5*diff(x(t),t)+4*x(t)=0:
> solb:=dsolve(DETwo,x(0)=1,D(x)(0)=-1,x(t));
> assign(solb):
> plot(x(t),t=0..6);
```

$$solb := x(t) = e^{-t}$$

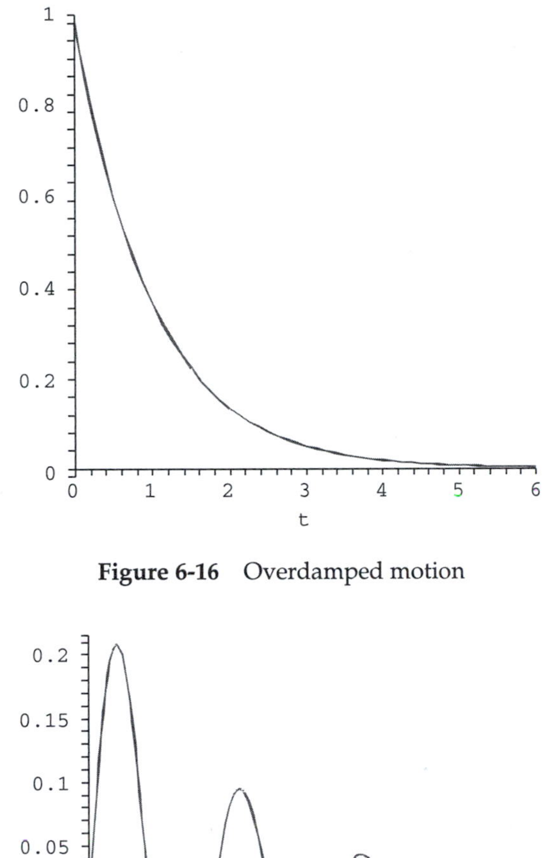

Figure 6-16 Overdamped motion

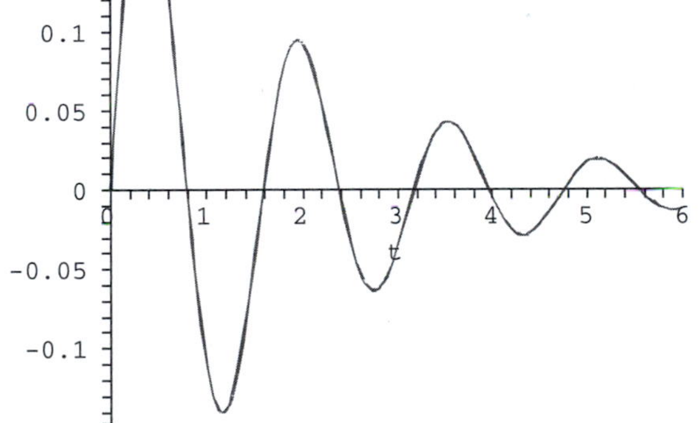

Figure 6-17 Underdamped motion

For (c), we proceed in the same manner as in (a) and (b) to show that the equation is underdamped because the value of $c^2 - 4mk$ is -63 (Figure 6-17).

```
> m:=1:c:=1:k:=16:
> c^2-4*m*k;
> x:='x':
```

```
> DEThree:=diff(x(t),t$2)+diff(x(t),t)+16*x(t)=0:
> solc:=dsolve(DEThree,x(0)=0,D(x)(0)=1,x(t));
> assign(solc):
> plot(x(t),t=0..6);
```

$$-63$$

$$solc := x(t) = 2/21\sqrt{7}e^{-1/2t}\sin\left(3/2\sqrt{7}t\right)$$

■

6.2.3 Undetermined Coefficients

If (6.7) has constant coefficients and $f(t)$ is a product of terms t^n, $e^{\alpha t}$, α constant, $\cos\beta t$, and/or $\sin\beta t$, β constant, *undetermined coefficients* can often be used to find a particular solution of (6.7). The key to implementing the method is to *judiciously* choose the correct form of y_p.

Assume that a general solution, y_h, of the corresponding homogeneous equation has been found and that each term of $f(t)$ has the form

$$t^n e^{\alpha t}\cos\beta t \qquad \text{or} \qquad t^n e^{\alpha t}\sin\beta t.$$

For *each* term of $f(t)$, write down the *associated set*

$$F = \{t^n e^{\alpha t}\cos\beta t,\ t^n e^{\alpha t}\sin\beta t,\ t^{n-1}e^{\alpha t}\cos\beta t,$$
$$t^{n-1}e^{\alpha t}\sin\beta t, \ldots, e^{\alpha t}\cos\beta t, e^{\alpha t}\sin\beta t\}.$$

If any element of F is a solution to the corresponding homogeneous equation, multiply each element of F by t^m, where m is the smallest positive integer so that none of the elements of $t^m F$ are solutions to the corresponding homogeneous equation. A particular solution will be a linear combination of the functions in all the F's.

EXAMPLE 6.2.4: Solve

$$4\frac{d^2y}{dt^2} - y = t - 2 - 5\cos t - e^{-t/2}.$$

SOLUTION: The corresponding homogeneous equation is $4y'' - y = 0$ with general solution $y_h = c_1 e^{-t/2} + c_2 e^{t/2}$.

```
> dsolve(4*diff(y(t),t$2)-y(t)=0,y(t));
```

$$y(t) = \_C1\, e^{-1/2\,t} + \_C2\, e^{1/2\,t}$$

A fundamental set of solutions for the corresponding homogeneous equation is $S = \left\{e^{-t/2}, e^{t/2}\right\}$. The associated set of functions for $t - 2$ is $F_1 = \{1, t\}$, the associated set of functions for $-5\cos t$ is $F_2 = \{\cos t, \sin t\}$, and the associated set of functions for $-e^{-t/2}$ is $F_3 = \left\{e^{-t/2}\right\}$. Note that $e^{-t/2}$ is an element of S so we multiply F_3 by t resulting in $tF_3 = \left\{te^{-t/2}\right\}$.

No element of F_1 is contained in S and no element of F_2 is contained in S.

Then, we search for a particular solution of the form

$$y_p = A + Bt + C\cos t + D\sin t + Ete^{-t/2},$$

where A, B, C, D, and E are constants to be determined.

```
> yp:=a+b*t+c*cos(t)+d*sin(t)+e*t*exp(-t/2):
```

Computing y_p' and y_p''

```
> dyp:=diff(yp,t);
```

$$dyp := b - c\sin(t) + d\cos(t) + ee^{-1/2\,t} - 1/2\,ete^{-1/2\,t}$$

```
> d2yp:=diff(yp,t$2);
```

$$d2yp := -c\cos(t) - d\sin(t) - ee^{-1/2\,t} + 1/4\,ete^{-1/2\,t}$$

and substituting into the nonhomogeneous equation results in

$$-A - Bt - 5C\cos t - 5D\sin t - 4Ee^{-t/2} = t - 2 - 5\cos t - e^{-t/2}. \tag{6.13}$$

```
> eqn:=4*diff(yp,t$2)-yp=t-2-5*cos(t)-exp(-t/2);
```

$$eqn := -5c\cos(t) - 5d\sin(t) - 4ee^{-1/2\,t} - a - bt = t - 2 - 5\cos(t) - e^{-1/2\,t}$$

Equation (6.13) is an identity: it is true for *all* values of t so the corresponding coefficients must be equal. Equating coefficients results in

$$-A = -2 \qquad -B = 1 \qquad -5C = -5 \qquad -5D = 0 \qquad -4E = -1$$

so $A = 2$, $B = -1$, $C = 1$, $D = 0$, and $E = 1/4$.

```
> cvals:=solve(-a=-2,-b=1,-5*c=-5,-5*d=0,-4*e=-1);
```

$$cvals := \{d = 0, c = 1, a = 2, b = -1, e = 1/4\}$$

Alternatively, use `solve` together with `identity`.

```
> solve(identity(eqn,t),a,b,c,d,e);
```

$$\{d = 0, c = 1, a = 2, b = -1, e = 1/4\}$$

y_p is then given by $y_p = 2 - t + \cos t + \frac{1}{4} t e^{-t/2}$

```
> subs(cvals,yp);
```

$$2 - t + \cos(t) + 1/4\, t e^{-1/2\, t}$$

and a general solution is given by

$$y = y_h + y_p = c_1 e^{-t/2} + c_e^{t/2} + 2 - t + \cos t + \frac{1}{4} t e^{-t/2}.$$

Remember that $-A - Bt - 5C\cos t - 5D\sin t - 4Ee^{-t/2} = t - 2 - 5\cos t - e^{-t/2}$ is true for *all* values of t. Evaluating for five different values of t gives us five equations that we then solve for A, B, C, D, and E, resulting in the same solutions as already obtained.

```
> sys:=seq(subs(t=i,eqn),i=0..4);
```

$$sys := -5c\cos(0) - 5d\sin(0) - 4ee^{0} - a = -2 - 5\cos(0) - e^{0},$$
$$-5c\cos(1) - 5d\sin(1) - 4ee^{-1/2} - a - b = -1 - 5\cos(1) - e^{-1/2},$$
$$-5c\cos(2) - 5d\sin(2) - 4ee^{-1} - a - 2b = -5\cos(2) - e^{-1},$$
$$-5c\cos(3) - 5d\sin(3) - 4ee^{-3/2} - a - 3b = 1 - 5\cos(3) - e^{-3/2},$$
$$-5c\cos(4) - 5d\sin(4) - 4ee^{-2} - a - 4b = 2 - 5\cos(4) - e^{-2}$$

```
> cvals:=solve(sys,a,b,c,d,e);
```

$$cvals := \{a = 2, e = 1/4, d = 0, c = 1, b = -1\}$$

Last, we check our calculations with `dsolve` and `simplify`.

```
> sol2:=dsolve(4*diff(y(t),t$2)-y(t)
    =t-2-5*cos(t)-exp(-t/2),y(t));
```

$$sol2 := y(t) = e^{1/2\, t}\_C2 + e^{-1/2\, t}\_C1$$
$$- \left((-\cos(t) - 2 + t)\, e^{1/2\, t} - 1/4\, t - 1/4\right) e^{-1/2\, t}$$

```
> simplify(sol2);
```

$$y(t) = e^{1/2t}\_C2 + e^{-1/2t}\_C1 + \cos(t) + 2 - t + 1/4\,te^{-1/2t} + 1/4\,e^{-1/2t}$$

■

EXAMPLE 6.2.5: Solve $y'' + 4y = \cos 2t$, $y(0) = 0$, $y'(0) = 0$.

SOLUTION: A general solution of the corresponding homogeneous equation is $y_h = c_1 \cos 2t + c_2 \sin 2t$. For this equation, $F = \{\cos 2t, \sin 2t\}$. Because elements of F are solutions to the corresponding homogeneous equation, we multiply each element of F by t resulting in $tF = \{t \cos 2t, t \sin 2t\}$. Therefore, we assume that a particular solution has the form

$$y_p = At \cos 2t + Bt \sin 2t,$$

where A and B are constants to be determined. Proceeding in the same manner as before, we compute y_p' and y_p''

```
> yp:=t->a*t*cos(2*t)+b*t*sin(2*t):
> diff(yp(t),t);
```

$$a\cos(2t) - 2at\sin(2t) + b\sin(2t) + 2bt\cos(2t)$$

```
> diff(yp(t),t$2);
```

$$-4a\sin(2t) - 4at\cos(2t) + 4b\cos(2t) - 4bt\sin(2t)$$

and then substitute into the nonhomogeneous equation

```
> eqn:=diff(yp(t),t$2)+4*yp(t)=cos(2*t);
```

$$eqn := -4a\sin(2t) + 4b\cos(2t) = \cos(2t)$$

Equating coefficients readily yields $A = 0$ and $B = 1/4$. Alternatively, remember that $-4A \sin 2t + 4B \cos 2t = \cos 2t$ is true for *all* values of t. Evaluating for two values of t and then solving for A and B

```
> e1:=subs(t=0,eqn):
> e2:=subs(t=Pi/4,eqn):
> cvals:=solve(e1,e2,a,b);
```

$$cvals := \{a = 0, b = 1/4\}$$

or using `solve` together with `identity`

```
> solve(identity(eqn,t),a,b);
```

$$\{a = 0, b = 1/4\}$$

gives the same result. It follows that $y_p = \frac{1}{4}t \sin 2t$ and $y = c_1 \cos 2t + c_2 \sin 2t + \frac{1}{4}t \sin 2t$.

```
> subs(cvals,yp(t));
```

$$1/4\,t \sin(2\,t)$$

```
> y:=t->c1*cos(2*t)+c2*sin(2*t)+1/4*t*sin(2*t):
```

Applying the initial conditions

```
> diff(y(t),t);
```

$$-2\,c1\, \sin(2\,t) + 2\,c2\, \cos(2\,t) + 1/4\, \sin(2\,t) + 1/2\,t \cos(2\,t)$$

```
> cvals:=solve(y(0)=0,D(y)(0)=0,c1,c2);
```

$$cvals := \{c1 = 0, c2 = 0\}$$

results in $y = \frac{1}{4}t \sin 2t$, which we graph with `plot` in Figure 6-18.

```
> subs(cvals,y(t));
```

$$1/4\,t \sin(2\,t)$$

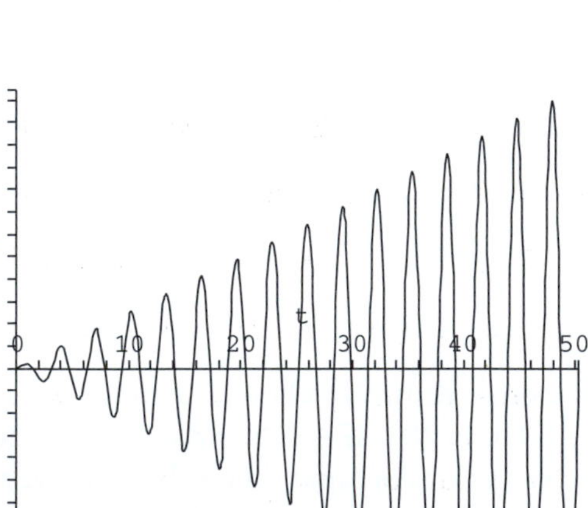

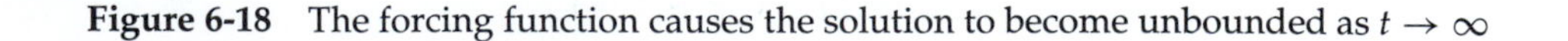

Figure 6-18 The forcing function causes the solution to become unbounded as $t \to \infty$

```
> plot(subs(cvals,y(t)),t=0..16*Pi,color=black);
```

We verify the calculation with `dsolve`.

```
> y:='y':
> dsolve(diff(y(t),t$2)+4*y(t)=cos(2*t),y(0)=0,
   D(y)(0)=0,y(t));
```

$$y(t) = 1/4\, t \sin(2\, t)$$

■

6.2.4 Variation of Parameters

Let $S = \{y_1, y_2\}$ be a fundamental set of solutions for (6.9). To solve the nonhomogeneous equation (6.8), we need to find a particular solution, y_p, of (6.8). We search for a particular solution of the form

A particular solution, y_p, is a solution that does not contain any arbitrary constants.

$$y_p = u_1(t)y_1(t) + u_2(t)y_2(t), \tag{6.14}$$

where u_1 and u_2 are functions of t. Differentiating (6.14) gives us

Observe that it is pointless to search for solutions of the form $y_p = c_1y_1 + c_2y_2$ where c_1 and c_2 are constants because for every choice of c_1 and c_2, $c_1y_1 + c_2y_2$ is a solution to the corresponding homogeneous equation.

$$y_p{}' = u_1{}'y_1 + u_1y_1{}' + u_2{}'y_2 + u_2y_2{}'.$$

Assuming that

$$y_1u_1{}' + y_2u_2{}' = 0 \tag{6.15}$$

results in $y_p{}' = u_1y_1{}' + u_2y_2{}'$. Computing the second derivative then yields

$$y_p{}'' = u_1{}'y_1{}' + u_1y_1{}'' + u_2{}'y_2{}' + u_2y_2{}''.$$

Substituting y_p, $y_p{}'$, and $y_p{}''$ into (6.8) and using the facts that

$$u_1\left(y_1{}'' + p\,y_1{}' + q\,y_1\right) = 0 \quad \text{and} \quad u_2\left(y_2{}'' + p\,y_2{}' + q\,y_2\right) = 0$$

(because y_1 and y_2 are solutions to the corresponding homogeneous equation) results in

$$\begin{aligned} \frac{d^2y_p}{dt^2} + p(t)\frac{dy_p}{dt} + q(t)y_p &= u_1{}'y_1{}' + u_1y_1{}'' + u_2{}'y_2{}' + u_2y_2{}'' \\ &\quad + p(t)\left(u_1y_1{}' + u_2y_2{}'\right) + q(t)\left(u_1y_1 + u_2y_2\right) \\ &= y_1{}'u_1{}' + y_2{}'u_2{}' = f(t). \end{aligned} \tag{6.16}$$

Observe that (6.15) and (6.16) form a system of two linear equations in the unknowns u_1' and u_2':

$$\begin{aligned} y_1 u_1' + y_2 u_2' &= 0 \\ y_1' u_1' + y_2' u_2' &= f(t). \end{aligned} \tag{6.17}$$

Applying Cramer's Rule gives us

$$u_1' = \frac{\begin{vmatrix} 0 & y_2 \\ f(t) & y_2' \end{vmatrix}}{\begin{vmatrix} y_1 & y_2 \\ y_1' & y_2' \end{vmatrix}} = -\frac{y_2(t) f(t)}{W(S)} \quad \text{and} \quad u_2' = \frac{\begin{vmatrix} y_1 & 0 \\ y_1' & f(t) \end{vmatrix}}{\begin{vmatrix} y_1 & y_2 \\ y_1' & y_2' \end{vmatrix}} = \frac{y_1(t) f(t)}{W(S)}, \tag{6.18}$$

where $W(S)$ is the Wronskian, $W(S) = \begin{vmatrix} y_1 & y_2 \\ y_1' & y_2' \end{vmatrix}$. After integrating to obtain u_1 and u_2, we form y_p and then a general solution, $y = y_h + y_p$.

If $S = \{y_1, y_2\}$ is a fundamental set of solutions for (6.9), the `DEtools` command

```
varpar([y1,y2],f(t),t)
```

solves (6.9) for y.

EXAMPLE 6.2.6: Solve $y'' + 9y = \sec 3t$, $y(0) = 0$, $y'(0) = 0$, $0 \le t < \pi/6$.

SOLUTION: The corresponding homogeneous equation is $y'' + 9y = 0$ with general solution $y_h = c_1 \cos 3t + c_2 \sin 3t$. Then, a fundamental set of solutions is $S = \{\cos 3t, \sin 3t\}$ and $W(S) = 3$, as we see using `wronskian`, `det`, and `simplify`.

```
> with(linalg):
> fs:=[cos(3*t),sin(3*t)]:
> wm:=wronskian(fs,t);
```

$$wm := \begin{bmatrix} \cos(3t) & \sin(3t) \\ -3\sin(3t) & 3\cos(3t) \end{bmatrix}$$

```
> wd:=simplify(det(wm));
```

$$wd := 3$$

We use (6.18) to find $u_1 = \frac{1}{9} \ln \cos 3t$ and $u_2 = \frac{1}{3} t$.

```
> u1:=integrate(-sin(3*t)*sec(3*t)/3,t);
```

$$u1 := 1/9 \ln(\cos(3t))$$

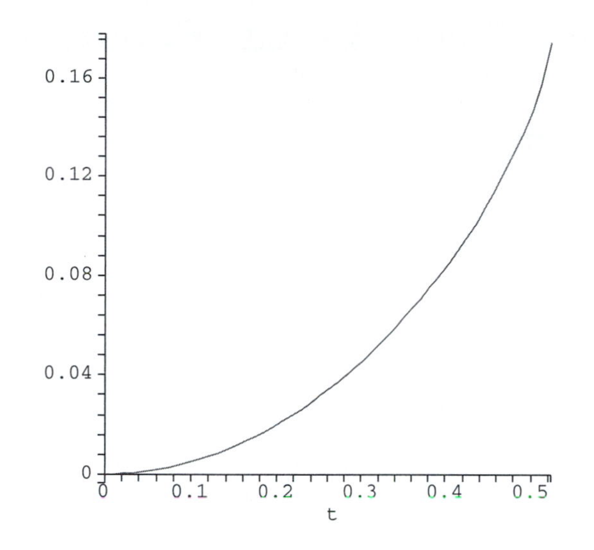

Figure 6-19 The domain of the solution is $-\pi/6 < t < \pi/6$

```
> u2:=integrate(cos(3*t)*sec(3*t)/3,t);
```

$$u2 := 1/3\,t$$

It follows that a particular solution of the nonhomogeneous equation is $y_p = \frac{1}{9}\cos 3t\ \ln\cos 3t + \frac{1}{3}t\sin 3t$ and a general solution is $y = y_h + y_p = c_1\cos 3t + c_2\sin 3t + \frac{1}{9}\cos 3t\ \ln\cos 3t + \frac{1}{3}t\sin 3t$.

```
> yp:=u1*cos(3*t)+u2*sin(3*t);
```

$$yp := 1/9\ln(\cos(3\,t))\cos(3\,t) + 1/3\,t\sin(3\,t)$$

Identical results are obtained using dsolve.

```
> dsolve(diff(y(t),t$2)+9*y(t)=sec(3*t),y(t));
```

$$y(t) = \sin(3t)\_C2 + \cos(3t)\_C1 + 1/9\ln(\cos(3t))\cos(3t) + 1/3\,t\sin(3t)$$

Applying the initial conditions gives us $c_1 = c_2 = 0$ so we conclude that the solution to the initial-value problem is $y = \frac{1}{9}\cos 3t\ \ln\cos 3t + \frac{1}{3}t\sin 3t$.

```
> sol:=dsolve(diff(y(t),t$2)+9*y(t)=sec(3*t),
   y(0)=0,D(y)(0)=0,y(t));
```

$$sol := y(t) = 1/9\ln(\cos(3\,t))\cos(3\,t) + 1/3\,t\sin(3\,t)$$

We graph the solution with plot in Figure 6-19.

```
> plot(rhs(sol),t=0..Pi/6);
```

■

6.3 Higher-Order Linear Equations

6.3.1 Basic Theory

The **standard form of the nth-order linear equation** is

$$\frac{d^n y}{dt^n} + a_{n-1}(t)\frac{d^{n-1}y}{dt^{n-1}} + \cdots + a_1(t)\frac{dy}{dt} + a_0(t)y = f(t). \tag{6.19}$$

The **corresponding homogeneous equation** of (6.19) is

$$\frac{d^n y}{dt^n} + a_{n-1}(t)\frac{d^{n-1}y}{dt^{n-1}} + \cdots + a_1(t)\frac{dy}{dt} + a_0(t)y = 0. \tag{6.20}$$

Let $y_1, y_2, \ldots, y_n$ be n solutions of (6.20). The set $S = \{y_1, y_2, \ldots, y_n\}$ is **linearly independent** if and only if the **Wronskian**,

$$W(S) = \begin{vmatrix} y_1 & y_2 & y_3 & \cdots & y_n \\ y_1' & y_2' & y_3' & \cdots & y_n' \\ y_1'' & y_2'' & y_3'' & \cdots & y_n'' \\ y_1^{(3)} & y_2^{(3)} & y_3^{(3)} & \cdots & y_n^{(3)} \\ \vdots & \vdots & \vdots & \cdots & \vdots \\ y_1^{(n-1)} & y_2^{(n-1)} & y_3^{(n-1)} & \cdots & y_n^{(n-1)} \end{vmatrix}, \tag{6.21}$$

is not identically the zero function. S is **linearly dependent** if S is not linearly independent.

If $y_1, y_2, \ldots, y_n$ are n linearly independent solutions of (6.20), we say that $S = \{y_1, y_2, \ldots, y_n\}$ is a **fundamental set** for (6.20) and a **general solution** of (6.20) is $y = c_1 y_1 + c_2 y_2 + c_3 y_3 + \cdots + c_n y_n$.

A **general solution** of (6.19) is $y = y_h + y_p$ where y_h is a general solution of the corresponding homogeneous equation and y_p is a particular solution of (6.19).

6.3.2 Constant Coefficients

If

$$\frac{d^n y}{dt^n} + a_{n-1}\frac{d^{n-1}y}{dt^{n-1}} + \cdots + a_1\frac{dy}{dt} + a_0 y = 0$$

has real constant coefficients, we assume that $y = e^{kt}$ and find that k satisfies the **characteristic equation**

$$k^n + a_{n-1}k^{n-1} + \cdots + a_1 k + a_0 = 0. \tag{6.22}$$

If a solution k of (6.22) has multiplicity m, m linearly independent solutions corresponding to k are

$$e^{kt}, te^{kt}, \ldots, t^{m-1}e^{kt}.$$

If a solution $k = \alpha + \beta i$, $\beta \neq 0$, of (6.22) has multiplicity m, $2m$ linearly independent solutions corresponding to $k = \alpha + \beta i$ (and $k = \alpha - \beta i$) are

$$e^{\alpha t}\cos\beta t,\ e^{\alpha t}\sin\beta t,\ te^{\alpha t}\cos\beta t,\ te^{\alpha t}\sin\beta t,\ \ldots, t^{m-1}e^{\alpha t}\cos\beta t,\ t^{m-1}e^{\alpha t}\sin\beta t.$$

EXAMPLE 6.3.1: Solve $12y''' - 5y'' - 6y' - y = 0$.

SOLUTION: The characteristic equation is

$$12k^3 - 5k^2 - 6k - 1 = (k-1)(3k+1)(4k+1) = 0$$

with solutions $k_1 = -1/3$, $k_2 = -1/4$ and $k_3 = 1$.

`factor(expression)` attempts to factor `expression`.

```
> y:='y':
> factor(12*k^3-5*k^2-6*k-1);
```

$$(k-1)(3k+1)(4k+1)$$

```
> solve(12*k^3-5*k^2-6*k-1=0);
```

$$1, -1/3, -1/4$$

Thus, three linearly independent solutions of the equation are $y_1 = e^{-t/3}$, $y_2 = e^{-t/4}$, and $y_3 = e^t$; a general solution is $y = c_1e^{-t/3} + c_2e^{-t/4} + c_3e^t$. We check with `dsolve`.

```
> dsolve(12*diff(y(t),t$3)-5*diff(y(t),t$2)
   -6*diff(y(t),t)-y(t)=0,y(t));
```

$$y(t) = \_C1\, e^t + \_C2\, e^{-1/3\, t} + \_C3\, e^{-1/4\, t}$$

■

EXAMPLE 6.3.2: Solve $y''' + 4y' = 0$, $y(0) = 0$, $y'(0) = 1$, $y''(0) = -1$.

SOLUTION: The characteristic equation is $k^3 + 4k = k(k^2 + 4) = 0$ with solutions $k_1 = 0$ and $k_{2,3} = \pm 2i$ that are found with `solve`.

Enter `?solve` to obtain basic help regarding the `solve` function.

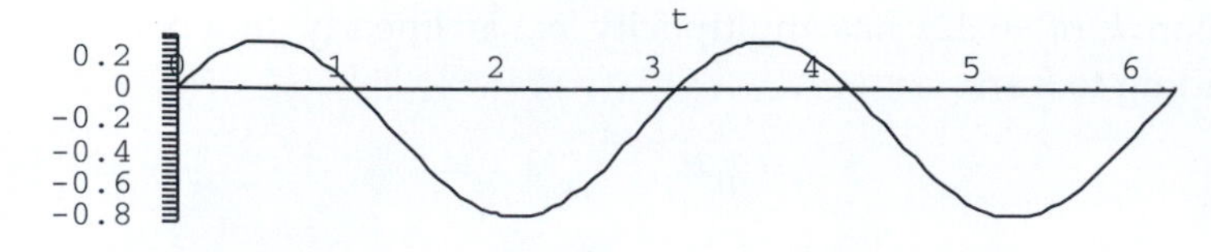

Figure 6-20 Graph of $y = -\frac{1}{4} + \frac{1}{2}\sin 2t + \frac{1}{4}\cos 2t$

```
> solve(k^3+4*k=0);
```

$$0, 2i, -2i$$

Three linearly independent solutions of the equation are $y_1 = 1$, $y_2 = \cos 2t$, and $y_3 = \sin 2t$. A general solution is $y = c_1 + c_2 \sin 2t + c_3 \cos 2t$.

```
> gensol:=dsolve(diff(y(t),t$3)+4*diff(y(t),t)=0,y(t));
```

$$gensol := y(t) = \_C1 + \_C2 \sin(2t) + \_C3 \cos(2t)$$

Application of the initial conditions shows us that $c_1 = -1/4$, $c_2 = 1/2$, and $c_3 = 1/4$ so the solution to the initial-value problem is $y = -\frac{1}{4} + \frac{1}{2}\sin 2t + \frac{1}{4}\cos 2t$. We verify the computation with `dsolve` and graph the result with `plot` in Figure 6-20.

```
> e1:=eval(subs(t=0,rhs(gensol)));
```

$$e1 := \_C1 + \_C3$$

```
> e2:=eval(subs(t=0,diff(rhs(gensol),t)));
```

$$e2 := 2\_C2$$

```
> e3:=eval(subs(t=0,diff(rhs(gensol),t$2)));
```

$$e3 := -4\_C3$$

```
> cvals:=solve(e1=0,e2=1,e3=-1);
```

$$cvals := \{\_C1 = -1/4, \_C3 = 1/4, \_C2 = 1/2\}$$

```
> partsol:=dsolve(diff(y(t),t$3)+4*diff(y(t),t)=0,
> y(0)=0,D(y)(0)=1,(D@@2)(y)(0)=-1,y(t));
```

$$partsol := y(t) = -1/4 + 1/2 \sin(2t) + 1/4 \cos(2t)$$

```
> plot(rhs(partsol),t=0..2*Pi,scaling=constrained,
   color=black);
```

■

EXAMPLE 6.3.3: Find a differential equation with general solution $y = c_1e^{-2t/3} + c_2te^{-2t/3} + c_3t^2e^{-2t/3} + c_4 \cos t + c_5 \sin t + c_6t \cos t + c_7t \sin t + c_8t^2 \cos t + c_9t^2 \sin t$.

SOLUTION: A linear homogeneous differential equation with constant coefficients that has this general solution has fundamental set of solutions

$$S = \left\{e^{-2t/3},\ te^{-2t/3},\ t^2e^{-2t/3},\ \cos t,\ \sin t,\ t\cos t,\ t\sin t,\ t^2\cos t,\ t^2\sin t\right\}$$

Hence, in the characteristic equation $k = -2/3$ has multiplicity 3 while $k = \pm i$ has multiplicity 3. The characteristic equation is

$$27\left(k+\frac{2}{3}\right)^3(k-i)^3(k+i)^3 = k^9 + 2k^8 + \frac{13}{3}k^7 + \frac{170}{27}k^6 + 7k^5 + \frac{62}{9}k^4 + 5k^3 + \frac{26}{9}k^2 + \frac{4}{3}k + \frac{8}{27},$$

where we use Maple to compute the multiplication with expand.

```
> expand(27*(k+2/3)^3*(k^2+1)^3);
```

$$27k^9+117k^7+189k^5+135k^3+54k^8+170k^6+186k^4+78k^2+36k+8$$

Thus, a differential equation obtained after dividing by 27 with the indicated general solution is

$$\frac{d^9y}{dt^9} + 2\frac{d^8y}{dt^8} + \frac{13}{3}\frac{d^7y}{dt^7} + \frac{170}{27}\frac{d^6y}{dt^6} + 7\frac{d^5y}{dt^5} + \frac{62}{9}\frac{d^4y}{dt^4} + 5\frac{d^3y}{dt^3} + \frac{26}{9}\frac{d^2y}{dt^2} + \frac{4}{3}\frac{dy}{dt} + \frac{8}{27}y = 0.$$

■

6.3.3 Undetermined Coefficients

For higher-order linear equations with constant coefficients, the method of undetermined coefficients is the same as for second-order equations discussed in Section 6.2.3, provided that the forcing function involves appropriate terms.

EXAMPLE 6.3.4: Solve

$$\frac{d^3y}{dt^3} + \frac{2}{3}\frac{d^2y}{dt^2} + \frac{145}{9}\frac{dy}{dt} = e^{-t},\ y(0) = 1,\ \frac{dy}{dt}(0) = 2,\ \frac{d^2y}{dt^2}(0) = -1.$$

SOLUTION: The corresponding homogeneous equation, $y''' + \frac{2}{3}y'' + \frac{145}{9}y' = 0$, has general solution $y_h = c_1 + (c_2 \sin 4t + c_3 \cos 4t)\, e^{-t/3}$ and a fundamental set of solutions for the corresponding homogeneous equation is $S = \left\{1, e^{-t/3}\cos 4t, e^{-t/3}\sin 4t\right\}$.

```
> dsolve(diff(y(t),t$3)+2/3*diff(y(t),t$2)
   +145/9*diff(y(t),t)=0,y(t));
```

$$y(t) = \_C1 + \_C2\, e^{-1/3\,t}\sin(4\,t) + \_C3\, e^{-1/3\,t}\cos(4\,t)$$

For e^{-t}, the associated set of functions is $F = \{e^{-t}\}$. Because no element of F is an element of S, we assume that $y_p = Ae^{-t}$, where A is a constant to be determined. After defining y_p, we compute the necessary derivatives

```
> yp:=t->a*exp(-t):
> diff(yp(t),t);
```

$$-ae^{-t}$$

```
> diff(yp(t),t$2);
```

$$ae^{-t}$$

```
> diff(yp(t),t$3);
```

$$-ae^{-t}$$

and substitute into the nonhomogeneous equation.

```
> eqn:=diff(yp(t),t$3)+2/3*diff(yp(t),t$2)
   +145/9*diff(yp(t),t)=exp(-t);
```

$$eqn := -\frac{148}{9}\,ae^{-t} = e^{-t}$$

Equating coefficients and solving for A gives us $A = -9/148$ so $y_p = -\frac{9}{148}e^{-t}$ and a general solution is $y = y_h + y_p$.

```
> solve(eqn,a);
```

$$-\frac{9}{148}$$

We verify the result with `dsolve`.

```
> gensol:=dsolve(diff(y(t),t$3)+2/3*diff(y(t),t$2)
> +145/9*diff(y(t),t)=exp(-t),y(t));
```

$$gensol := y(t) = -\frac{36}{145}\_C2\, e^{-1/3t}\cos(4t) - \frac{3}{145}\_C2\, e^{-1/3t}\sin(4t) - \frac{3}{145}e^{-1/3t}\cos(4t)\_C1 + \frac{36}{145}\_C1\, e^{-1/3t}\sin(4t) - \frac{9}{148}e^{-t} + \_C3$$

To apply the initial conditions, we compute $y(0) = 1$, $y'(0) = 2$, and $y''(0) = -1$

```
> e1:=eval(subs(t=0,rhs(gensol)))=1;
```

$$e1 := -\frac{36}{145}\_C2 - \frac{9}{148} - \frac{3}{145}\_C1 + \_C3 = 1$$

```
> e2:=eval(subs(t=0,diff(rhs(gensol),t)))=2;
```

$$e2 := \frac{9}{148} + \_C1 = 2$$

```
> e3:=eval(subs(t=0,diff(rhs(gensol),t$2)))=-1;
```

$$e3 := -\frac{9}{148} + 4\_C2 - 1/3\_C1 = -1$$

and solve for c_1, c_2, and c_3 with solve.

```
> cvals:=solve(e1,e2,e3);
```

$$cvals := \left\{\_C2 = -\frac{65}{888}, \_C1 = \frac{287}{148}, \_C3 = \frac{157}{145}\right\}$$

The solution of the initial-value problem is obtained by substituting these values into the general solution with subs.

```
> subs(cvals,rhs(gensol));
```

$$-\frac{471}{21460}e^{-1/3t}\cos(4t) + \frac{20729}{42920}e^{-1/3t}\sin(4t) - \frac{9}{148}e^{-t} + \frac{157}{145}$$

We check by using dsolve to solve the initial-value problem and graph the result with plot in Figure 6-21.

```
> sol:=dsolve(diff(y(t),t$3)+2/3*diff(y(t),t$2)
> +145/9*diff(y(t),t)=exp(-t),y(0)=-1,D(y)(0)=2,
  (D@@2)(y)(0)=-1,y(t));
```

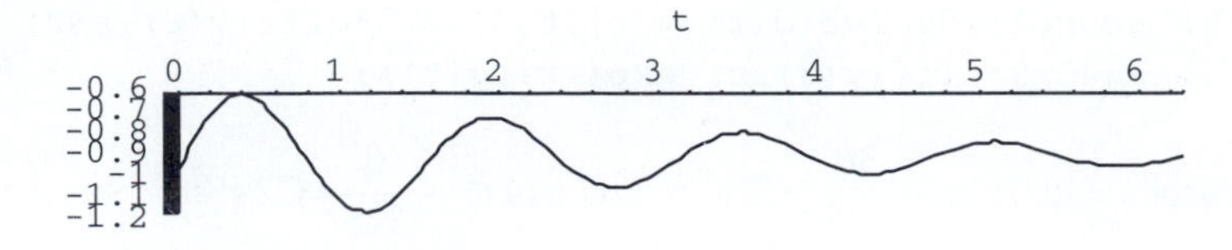

Figure 6-21 The solution of the equation that satisfies $y(0) = 1$, $y'(0) = 2$, and $y''(0) = -1$

$$sol := y(t) = -\frac{471}{21460} e^{-1/3\,t} \cos(4\,t) + \frac{20729}{42920} e^{-1/3\,t} \sin(4\,t) - \frac{9}{148} e^{-t} - \frac{133}{145}$$

```
> plot(rhs(sol),t=0..2*Pi,color=black,
   scaling=constrained);
```

■

EXAMPLE 6.3.5: Solve

$$\frac{d^8y}{dt^8} + \frac{7}{2}\frac{d^7y}{dt^7} + \frac{73}{2}\frac{d^6y}{dt^6} + \frac{229}{2}\frac{d^5y}{dt^5} + \frac{801}{2}\frac{d^4y}{dt^4} + 976\frac{d^3y}{dt^3} + 1168\frac{d^2y}{dt^2} + 640\frac{dy}{dt} + 128y = te^{-t} + \sin 4t + t.$$

SOLUTION: Solving the characteristic equation

```
> solve(k^8+7/2*k^7+73/2*k^6+229/2*k^5
   +801/2*k^4+976*k^3+1168*k^2+
> 640*k+128=0);
```

$$-1/2,\ -1,\ -1,\ -1,\ 4\,i,\ -4\,i,\ 4\,i,\ -4\,i$$

shows us that the solutions are $k_1 = -1/2$, $k_2 = -1$ with multiplicity 3, and $k_{3,4} = \pm 4i$, each with multiplicity 2. A fundamental set of solutions for the corresponding homogeneous equation is

$$S = \left\{ e^{-t/2}, e^{-t}, te^{-t}, t^2e^{-t}, \cos 4t, t\cos 4t, \sin 4t, t\sin 4t \right\}.$$

A general solution of the corresponding homogeneous equation is

$$y_h = c_1 e^{-t/2} + \left(c_2 + c_3 t + c_4 t^2\right) e^{-t} + (c_5 + c_7 t)\sin 4t + (c_6 + c_8 t)\cos 4t.$$

```
> gensol:=dsolve(diff(y(t),t$8)+7/2*diff(y(t),t$7)
> +73/2*diff(y(t),t$6)+229/2*diff(y(t),t$5)
> +801/2*diff(y(t),t$4)+976*diff(y(t),t$3)
> +1168*diff(y(t),t$2)+640*diff(y(t),t)+128*y(t)=0);
```

$$gensol := y(t) = e^{-1/2t}\_C1 + \_C2\,e^{-t} + \_C3\,e^{-t}t + \_C4\,e^{-t}t^2 + \_C5\,\sin(4t) + \_C6\,\cos(4t) + \_C7\,\sin(4t)\,t + \_C8\,\cos(4t)\,t$$

The associated set of functions for te^{-t} is $F_1 = \{e^{-t}, te^{-t}\}$. We multiply F_1 by t^n, where n is the smallest non-negative integer so that no element of t^nF_1 is an element of S: $t^3F_1 = \{t^3e^{-t}, t^4e^{-t}\}$. The associated set of functions for $\sin 4t$ is $F_2 = \{\cos 4t, \sin 4t\}$. We multiply F_2 by t^n, where n is the smallest non-negative integer so that no element of t^nF_2 is an element of S: $t^2F_2 = \{t^2\cos 4t, t^2\sin 4t\}$. The associated set of functions for t is $F_3 = \{1, t\}$. No element of F_3 is an element of S.

Thus, we search for a particular solution of the form

$$y_p = A_1t^3e^{-t} + A_2t^4e^{-t} + A_3t^2\cos 4t + A_4t^2\sin 4t + A_5 + A_6t,$$

where the A_i are constants to be determined.

After defining y_p, we compute the necessary derivatives

Remark. We have used `array` and `seq` twice for typesetting purposes. You can compute the derivatives using `array([seq([n, diff(yp(t),t\$n)], n=1..8)])`.

```
> yp:=t->a[1]*t^3*exp(-t)+a[2]*t^4*exp(-t)+
> a[3]*t^2*cos(4*t)+a[4]*t^2*sin(4*t)+a[5]+a[6]*t;
```

$$yp := t \mapsto a_1t^3e^{-t} + a_2t^4e^{-t} + a_3t^2\cos(4\,t) + a_4t^2\sin(4\,t) + a_5 + a_6t$$

```
> array([seq([n,diff(yp(t),t$n)],n=1..4)]);
```

$$\left[1, 3a_1t^2e^{-t} - a_1t^3e^{-t} + 4a_2t^3e^{-t} - a_2t^4e^{-t} + 2a_3t\cos(4t) - 4a_3t^2\sin(4t) + 2a_4t\sin(4t) + 4a_4t^2\cos(4t) + a_6\right]$$

$$\left[2, 6a_1te^{-t} - 6a_1t^2e^{-t} + a_1t^3e^{-t} + 12a_2t^2e^{-t} - 8a_2t^3e^{-t} + a_2t^4e^{-t} + 2a_3\cos(4t) - 16a_3t\sin(4t) - 16a_3t^2\cos(4t) + 2a_4\sin(4t) + 16a_4t\cos(4t) - 16a_4t^2\sin(4t)\right]$$

$$\begin{aligned}&[3, 6a_1e^{-t} - 18a_1te^{-t} + 9a_1t^2e^{-t} - a_1t^3e^{-t} + 24a_2te^{-t} - 36a_2t^2e^{-t}\\
&\quad + 12a_2t^3e^{-t} - a_2t^4e^{-t} - 24a_3\sin(4t) - 96a_3t\cos(4t) + 64a_3t^2\sin(4t)\\
&\quad + 24a_4\cos(4t) - 96a_4t\sin(4t) - 64a_4t^2\cos(4t)]\end{aligned}$$

$$\begin{aligned}&[4, 256a_3t^2\cos(4t) + 256a_4t^2\sin(4t) - 192a_3\cos(4t) + 512a_3t\sin(4t)\\
&\quad - 192a_4\sin(4t) - 512a_4t\cos(4t) - 24a_1e^{-t} + 24a_2e^{-t} + a_1t^3e^{-t}\\
&\quad + a_2t^4e^{-t} - 12a_1t^2e^{-t} - 16a_2t^3e^{-t} + 36a_1te^{-t} + 72a_2t^2e^{-t} - 96a_2te^{-t}]\end{aligned}$$

```
> array([seq([n,diff(yp(t),t$n)],n=5..8)]);
```

$$\begin{aligned}&[5, -1024\,a_3t^2\sin(4\,t) + 1024\,a_4t^2\cos(4\,t) + 60\,a_1e^{-t} + 1280\,a_3\sin(4\,t)\\
&\quad - 1280\,a_4\cos(4\,t) - 120\,a_2e^{-t} - a_1t^3e^{-t} - a_2t^4e^{-t} + 15\,a_1t^2e^{-t}\\
&\quad + 20\,a_2t^3e^{-t} + 2560\,a_3t\cos(4\,t) + 2560\,a_4t\sin(4\,t) - 60\,a_1te^{-t}\\
&\quad - 120\,a_2t^2e^{-t} + 240\,a_2te^{-t}],\end{aligned}$$

$$\begin{aligned}&[6, -4096\,a_3t^2\cos(4\,t) - 4096\,a_4t^2\sin(4\,t) + 7680\,a_3\cos(4\,t)\\
&\quad - 12288\,a_3t\sin(4\,t) + 7680\,a_4\sin(4\,t) + 12288\,a_4t\cos(4\,t) - 120\,a_1e^{-t}\\
&\quad + 360\,a_2e^{-t} + a_1t^3e^{-t} + a_2t^4e^{-t} - 18\,a_1t^2e^{-t} - 24\,a_2t^3e^{-t}\\
&\quad + 90\,a_1te^{-t} + 180\,a_2t^2e^{-t} - 480\,a_2te^{-t}],\end{aligned}$$

$$\begin{aligned}&[7, 16384\,a_3t^2\sin(4\,t) - 16384\,a_4t^2\cos(4\,t) + 210\,a_1e^{-t}\\
&\quad - 43008\,a_3\sin(4\,t) + 43008\,a_4\cos(4\,t) - 840\,a_2e^{-t} - a_1t^3e^{-t} - a_2t^4e^{-t}\\
&\quad + 21\,a_1t^2e^{-t} + 28\,a_2t^3e^{-t} - 57344\,a_3t\cos(4\,t) - 57344\,a_4t\sin(4\,t)\\
&\quad - 126\,a_1te^{-t} - 252\,a_2t^2e^{-t} + 840\,a_2te^{-t}],\end{aligned}$$

$$\begin{aligned}&[8, 65536\,a_3t^2\cos(4\,t) + 65536\,a_4t^2\sin(4\,t) - 229376\,a_3\cos(4\,t)\\
&\quad + 262144\,a_3t\sin(4\,t) - 229376\,a_4\sin(4\,t) - 262144\,a_4t\cos(4\,t)\\
&\quad - 336\,a_1e^{-t} + 1680\,a_2e^{-t} + a_1t^3e^{-t} + a_2t^4e^{-t} - 24\,a_1t^2e^{-t}\\
&\quad - 32\,a_2t^3e^{-t} + 168\,a_1te^{-t} + 336\,a_2t^2e^{-t} - 1344\,a_2te^{-t}]]\end{aligned}$$

and substitute into the nonhomogeneous equation, naming the result `eqn`. At this point we can either equate coefficients and solve for A_i or use the fact that eqn is true for *all* values of t and solve for the coefficients using solve together with identity as we do here.

```
> eqn:=simplify(diff(yp(t),t$8)+7/2*diff(yp(t),t$7)
> +73/2*diff(yp(t),t$6)+229/2*diff(yp(t),t$5)
> +801/2*diff(yp(t),t$4)+976*diff(yp(t),t$3)
> +1168*diff(yp(t),t$2)+640*diff(yp(t),t)
> +128*yp(t)=t*exp(-t)+sin(4*t)+t):
```

```
> avals:=solve(identity(eqn,t),a[1],a[2],a[3],
  a[4],a[5],a[6]);
```

$$avals := \left\{a_1 = -\frac{38}{14739}, a_3 = -\frac{107}{5109520}, a_2 = -\frac{1}{3468}, a_4 = -\frac{369}{20438080}, a_6 = \frac{1}{128}, a_5 = -\frac{5}{128}\right\}$$

y_p is obtained by substituting the values for A_i into y_p and a general solution is $y = y_h + y_p$. dsolve is able to find an exact solution, too, although dsolve does not obtain the simplified solution we obtained.

For length considerations, we have only displayed a portion of the result returned by dsolve.

```
> gensol:=dsolve(diff(y(t),t$8)+7/2*diff(y(t),t$7)
> +73/2*diff(y(t),t$6)+229/2*diff(y(t),t$5)
> +801/2*diff(y(t),t$4)+976*diff(y(t),t$3)
> +1168*diff(y(t),t$2)+640*diff(y(t),t)
> +128*y(t)=t*exp(-t)+sin(4*t)+t);
```

$$\begin{aligned} gensol := y(t) = &-\frac{256}{4225}te^{-t} + \_C8e^{-t}t^2 + \_C6t\sin(4t) + \_C7te^{-t} \\ &+ \_C5t\cos(4t) + \frac{1866128}{20757425}t + \frac{107}{20438080}t\cos(4t)\sin(8t) \\ &+ \frac{369}{81752320}t\cos(4t)\cos(8t) - \frac{2111}{1476651280}t(\sin(4t))^2 e^{-t} \\ &+ \frac{2081}{86861840}t^2(\sin(4t))^2 e^{-t} - \frac{2111}{1476651280}(\cos(4t))^2 te^{-t} \\ &+ \frac{2081}{86861840}(\cos(4t))^2 e^{-t}t^2 + \frac{369}{81752320}t\sin(4t)\sin(8t) \\ &- \frac{107}{20438080}t\sin(4t)\cos(8t) - \cdots \end{aligned}$$

■

Variation of Parameters

In the same way as with second-order equations, we assume that a particular solution of the nth-order linear equation (6.19) has the form

$y_p = u_1(t)y_1 + u_2(t)y_2 + \cdots + u_n(t)y_n$, where $S = \{y_1, y_2, \ldots, y_n\}$ is a fundamental set of solutions to the corresponding homogeneous equation (6.20). With the assumptions

$$\begin{aligned} y_p{}' &= y_1 u_1{}' + y_2 u_2{}' + \cdots + y_n u_n{}' = 0 \\ y_p{}'' &= y_1{}' u_1{}' + y_2{}' u_2{}' + \cdots + y_n{}' u_n{}' = 0 \\ &\vdots \\ y_p{}^{(n-1)} &= y_1{}^{(n-2)} u_1{}' + y_2{}^{(n-2)} u_2{}' + \cdots + y_n{}^{(n-2)} u_n{}' = 0 \end{aligned} \tag{6.23}$$

we obtain the equation

$$y_1{}^{(n-1)} u_1{}' + y_2{}^{(n-1)} u_2{}' + \cdots + y_n{}^{(n-1)} u_n{}' = f(t). \tag{6.24}$$

Equations (6.23) and (6.24) form a system of n linear equations in the unknowns $u_1{}', u_2{}', \ldots, u_n{}'$. Applying Cramer's Rule,

$$u_i{}' = \frac{W_i(S)}{W(S)}, \tag{6.25}$$

where $W(S)$ is given by (6.21) and $W_i(S)$ is the determinant of the matrix obtained by replacing the ith column of

$$\begin{pmatrix} y_1 & y_2 & \cdots & y_n \\ y_1{}' & y_2{}' & \cdots & y_n{}' \\ \vdots & \vdots & \cdots & \vdots \\ y_1{}^{(n-1)} & y_2{}^{(n-1)} & \cdots & y_n{}^{(n-1)} \end{pmatrix} \quad \text{by} \quad \begin{pmatrix} 0 \\ 0 \\ \vdots \\ f(t) \end{pmatrix}.$$

You can use the `DEtools varparam` function to implement the method of variation of parameters. After you have loaded the `DEtools` package and found and defined a fundamental set of solutions, $S = \{y_1, y_2, \ldots, y_n\}$, to the corresponding homogeneous equation, the command

```
varparam([y1(t),y2(t),...,yn(t)],f(t),t)
```

solves (6.19).

EXAMPLE 6.3.6: Solve $y^{(3)} + 4y' = \sec 2t$.

SOLUTION: A general solution of the corresponding homogeneous equation is $y_h = c_1 + c_2 \cos 2t + c_3 \sin 2t$; a fundamental set is $S = \{1, \cos 2t, \sin 2t\}$ with Wronskian $W(S) = 8$.

`wronskian` and `det` are contained in the `linalg` package.

```
> with(DEtools):
> yh:=dsolve(diff(y(t),t$3)+4*diff(y(t),t)=0,y(t));
```

$$yh := y(t) = \_C1 + \_C2 \sin(2t) + \_C3 \cos(2t)$$

```
> with(linalg):
> S:=[1,cos(2*t),sin(2*t)]:
> ws:=wronskian(S,t);
```

$$\begin{bmatrix} 1 & \cos(2t) & \sin(2t) \\ 0 & -2\sin(2t) & 2\cos(2t) \\ 0 & -4\cos(2t) & -4\sin(2t) \end{bmatrix}$$

```
> dws:=simplify(det(ws));
```

$$dws := 8$$

Using variation of parameters to find a particular solution of the nonhomogeneous equation, we let $y_1 = 1$, $y_2 = \cos 2t$, and $y_3 = \sin 2t$ and assume that a particular solution has the form $y_p = u_1 y_1 + u_2 y_2 + u_3 y_3$. Using the variation of parameters formula, we obtain

$$u_1' = \frac{1}{8}\begin{vmatrix} 0 & \cos 2t & \sin 2t \\ 0 & -2\sin 2t & 2\cos 2t \\ \sec 2t & -4\cos 2t & -4\sin 2t \end{vmatrix} = \frac{1}{4}\sec 2t \quad \text{so} \quad u_1 = \frac{1}{8}\ln|\sec 2t + \tan 2t|,$$

$$u_2' = \frac{1}{8}\begin{vmatrix} 1 & 0 & \sin 2t \\ 0 & 0 & 2\cos 2t \\ 0 & \sec 2t & -4\sin 2t \end{vmatrix} = -\frac{1}{4} \quad \text{so} \quad u_2 = -\frac{1}{4}t$$

and

$$u_3' = \frac{1}{8}\begin{vmatrix} 1 & \cos 2t & 0 \\ 0 & -2\sin 2t & 0 \\ 0 & -4\cos 2t & \sec 2t \end{vmatrix} = -\frac{1}{2}\tan 2t \quad \text{so} \quad u_3 = \frac{1}{8}\ln|\cos 2t|,$$

where we use `det`, which is contained in the `linalg` package, and `integrate` to evaluate the determinants and integrals. In the case of u_1, the output given by Maple looks different from the result we obtained by hand but using properties of logarithms ($\ln(a/b) = \ln a - \ln b$) and trigonometric identities ($\cos^2 x + \sin^2 x = 1$, $\sin 2x = 2\sin x\cos x$, $\cos^2 x - \sin^2 x = \cos 2x$, and the reciprocal identities) shows us that

$$\begin{aligned}
\frac{1}{8}\left(\ln|\cos t + \sin t| - \ln|\cos t + \sin t|\right) &= \frac{1}{8}\ln\left|\frac{\cos t + \sin t}{\cos t - \sin t}\right| \\
&= \frac{1}{8}\ln\left|\frac{\cos t + \sin t}{\cos t - \sin t}\cdot\frac{\cos t + \sin t}{\cos t + \sin t}\right| \\
&= \frac{1}{8}\ln\left|\frac{\cos^2 t + 2\cos t\sin t + \sin^2 t}{\cos^2 t - \sin^2 t}\right| \\
&= \frac{1}{8}\ln\left|\frac{1 + \sin 2t}{\cos 2t}\right| \\
&= \frac{1}{8}\ln\left|\frac{1}{\cos 2t} + \frac{\sin 2t}{\cos 2t}\right| \\
&= \frac{1}{8}\ln|\sec 2t + \tan 2t|
\end{aligned}$$

so the results obtained by hand and with Maple are the same.

```
> A:=transpose(array([[0,0,sec(2*t)],
   [cos(2*t),-2*sin(2*t),-4*cos(2*t)],
> [sin(2*t),2*cos(2*t),-4*sin(2*t)]]));
```

$$\begin{bmatrix} 0 & \cos(2t) & \sin(2t) \\ 0 & -2\sin(2t) & 2\cos(2t) \\ \sec(2t) & -4\cos(2t) & -4\sin(2t) \end{bmatrix}$$

```
> u1p:=simplify(1/8*det(transpose([[0,0,sec(2*t)],
> [cos(2*t),-2*sin(2*t),-4*cos(2*t)],
> [sin(2*t),2*cos(2*t),-4*sin(2*t)]])));
```

$$u1p := 1/4\,(\cos(2t))^{-1}$$

```
> integrate(u1p,t);
```

$$1/8\ln(\sec(2t) + \tan(2t))$$

```
> u2p:=simplify(1/8*det(transpose([[1,0,0],
   [0,0,sec(2*t)],[sin(2*t),2*cos(2*t),
   -4*sin(2*t)]])));
```

$$u2p := -1/4$$

```
> integrate(u2p,t);
```

$$-1/4\,t$$

```
> u3p:=simplify(1/8*det(transpose([[1,0,0],[cos(2*t),
   -2*sin(2*t),-4*cos(2*t)],[0,0,sec(2*t)]])));
```

$$u3p := -1/4\,\frac{\sin(2\,t)}{\cos(2\,t)}$$

```
> integrate(u3p,t);
```

$$1/8\,\ln(\cos(2\,t))$$

Thus, a particular solution of the nonhomogeneous equation is

$$y_p = \frac{1}{8}\ln|\sec 2t + \tan 2t| - \frac{1}{4}t\cos 2t + \frac{1}{8}\ln|\cos 2t|\sin 2t$$

and a general solution is $y = y_h + y_p$. We verify the calculations using `varparam` and note that `dsolve` returns an equivalent solution.

```
> varparam([1,cos(2*t),sin(2*t)],sec(2*t),t);
```

$$\_C_1 + \_C_2\cos(2\,t) + \_C_3\sin(2\,t) + 1/8\,\ln(\sec(2\,t) + \tan(2\,t))$$
$$- 1/4\,t\cos(2\,t) + 1/8\,\ln(\cos(2\,t))\sin(2\,t)$$

```
> gensol:=simplify(dsolve(diff(y(t),t$3)
   +4*diff(y(t),t)=sec(2*t),y(t)));
```

$$gensol := y(t) = -1/2\,\_C2\,\cos(2\,t) + 1/2\,\_C1\,\sin(2\,t)$$
$$+ 1/8\,\sin(2\,t) - 1/4\,t\cos(2\,t) - 1/16\,ie^{2\,it}\ln(\cos(2\,t))$$
$$+ 1/16\,ie^{2\,it} - 1/4\,i\arctan\left(e^{2\,it}\right) + 1/16\,ie^{-2\,it}\ln(\cos(2\,t))$$
$$- 1/16\,ie^{-2\,it} + \_C3$$

■

6.3.4 Laplace Transform Methods

The *method of Laplace transforms* can be useful when the forcing function is piecewise-defined or periodic.

Definition 4 (Laplace Transform and Inverse Laplace Transform). *Let $y = f(t)$ be a function defined on the interval $[0,\infty)$. The* ***Laplace transform*** *of $f(t)$ is the* *function (of s)*

$$F(s) = \mathcal{L}\left\{f(t)\right\} = \int_0^\infty e^{-st} f(t)\, dt, \tag{6.26}$$

provided the improper integral exists. $f(t)$ is the ***inverse Laplace transform*** *of $F(s)$ means that $\mathcal{L}\left\{f(t)\right\} = F(s)$ and we write $\mathcal{L}^{-1}\left\{F(s)\right\} = f(t)$.*

Use the commands `laplace` and `invlaplace`, which are contained in the `inttrans` package, to compute Laplace transforms and inverse Laplace transforms.

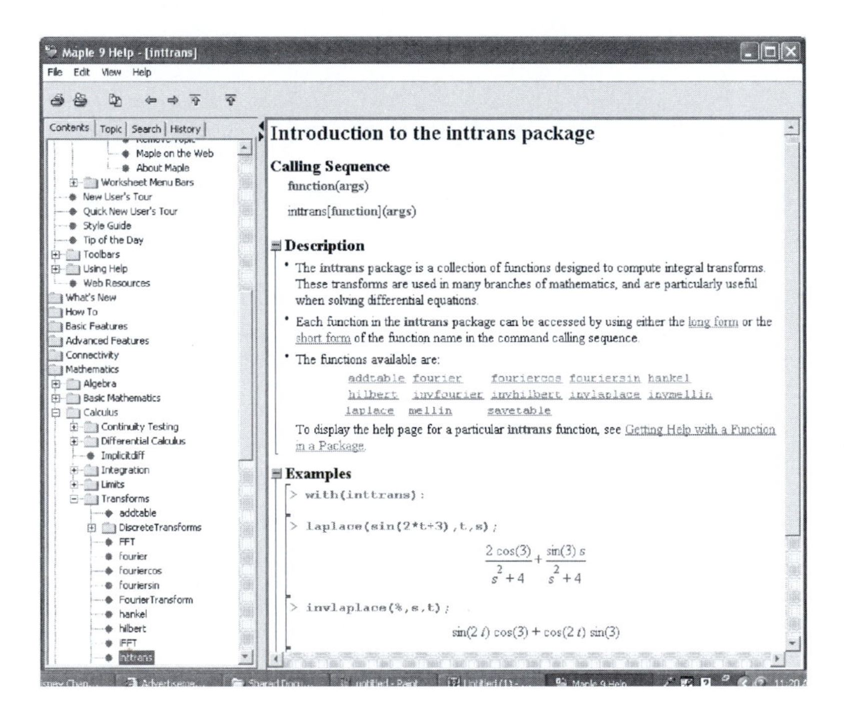

1. `laplace(f(t),t,s)` computes $\mathcal{L}\left\{f(t)\right\} = F(s)$.
2. `invlaplace(F(s),s,t)` computes $\mathcal{L}^{-1}\left\{F(s)\right\} = f(t)$.
3. `Heaviside(t)` returns $\mathcal{U}(t) = \begin{cases} 0, & t < 0 \\ 1, & t \geq 0 \end{cases}$.

Typically, when we use Laplace transforms to solve a differential equation for a function $y(t)$, we will compute the Laplace transform of each term of the differential equation, solve the resulting algebraic equation for the Laplace transform of $y(t)$, $\mathcal{L}\left\{y(t)\right\} = Y(s)$, and finally determine $y(t)$ by computing the inverse Laplace transform of $Y(s)$, $\mathcal{L}^{-1}\left\{Y(s)\right\} = y(t)$.

EXAMPLE 6.3.7: Let $y = f(t)$ be defined recursively by $f(t) = \begin{cases} 1, & 0 \le t < 1 \\ -1, & 1 \le t < 2 \end{cases}$ and $f(t) = f(t-2)$ if $t \ge 2$. Solve $y'' + 4y' + 20y = f(t)$.

SOLUTION: We begin by defining and graphing $y = f(t)$ for $0 \le t \le 5$ in Figure 6-22. Note that `elif` is used to avoid repeated `if...fi` statements.

```
> f:='f':y:='y':
> f:=proc(t) option remember;
> if t<1 and t>=0 then 1 elif
> t<2 and t>=1 then -1 else
> f(t-2) fi end:
> plot('f(t)','t'=0..6,numpoints=200);
```

We then define `LHS_Eq` to be the left-hand side of the equation $y'' + 4y' + 20y = f(t)$,

```
> with(inttrans):
> LHS_Eq:=diff(y(t),t$2)+4*diff(y(t),t)+20*y(t):
```

and compute the Laplace transform of `LHS_Eq` with `laplace`, naming the result `stepone`.

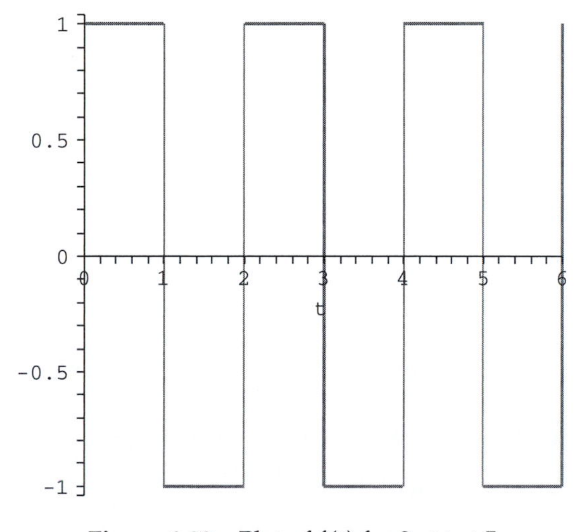

Figure 6-22 Plot of $f(t)$ for $0 \le t \le 5$

```
> stepone:=laplace(LHS_Eq,t,s);
```

$$stepone := s^2 laplace\left(y\left(t\right),t,s\right) - \mathrm{D}(y)\left(0\right) - sy\left(0\right) + 4\,slaplace\left(y\left(t\right),t,s\right) - 4\,y\left(0\right) + 20\,laplace\left(y\left(t\right),t,s\right)$$

Let `lr` denote the Laplace transform of the right-hand side of the equation, $f(t)$. We now solve the equation 20`ly`$+4s$`ly`$+s^2$`ly`$-4y(0)-sy(0)-y'(0 = lr$ for `ly` and name the resulting output `steptwo`.

```
> steptwo:=solve(stepone=lr,laplace(y(t),t,s));
```

$$steptwo := \frac{\mathrm{D}\left(y\right)\left(0\right) + sy\left(0\right) + 4\,y\left(0\right) + lr}{s^2 + 4\,s + 20}$$

```
> stepthree:=expand(steptwo);
```

$$stepthree := \frac{\mathrm{D}\left(y\right)\left(0\right)}{s^2+4s+20} + \frac{sy\left(0\right)}{s^2+4s+20} + 4\frac{y\left(0\right)}{s^2+4s+20} + \frac{lr}{s^2+4s+20}$$

To find $y(t)$, we must compute the inverse Laplace transform of $\mathcal{L}\left\{y(t)\right\}$; the formula for which is explicitly obtained from `steptwo` with `op(4,stepthree)`. First, we rewrite : $\mathcal{L}\left\{y(t)\right\}$. Then,

$$\begin{aligned} y(t) &= \mathcal{L}^{-1}\left\{\frac{\mathcal{L}\left\{y(t)\right\}}{s^2+4s+20} + \frac{4y(0)+sy(0)+y'(0)}{s^2+4s+20}\right\} \\ &= \mathcal{L}^{-1}\left\{\frac{\mathcal{L}\left\{y(t)\right\}}{s^2+4s+20}\right\} + \mathcal{L}^{-1}\left\{\frac{4y(0)+sy(0)+y'(0)}{s^2+4s+20}\right\}. \end{aligned}$$

Completing the square yields $s^2 + 4s + 20 = (s+2)^2 + 16$. Because

$$\mathcal{L}^{-1}\left\{\frac{b}{(s-a)^2+b^2}\right\} = e^{at}\sin bt \quad \text{and} \quad \mathcal{L}^{-1}\left\{\frac{s-a}{(s-a)^2+b^2}\right\} = e^{at}\cos bt,$$

the inverse Laplace transform of

$$\frac{4y(0)+sy(0)+y'(0)}{s^2+4s+20} = y(0)\frac{s+2}{(s+2)^2+4^2} + \frac{y'(0)+2y(0)}{4}\frac{4}{(s+2)^2+4^2}$$

is

$$y(0)e^{-2t}\cos 4t + \frac{y'(0)+2y(0)}{4}e^{-2t}\sin 4t,$$

which is defined as $y_1(t)$. We perform these steps with Maple by first using `invlaplace` to calculate $\mathcal{L}^{-1}\left\{\frac{4y(0)+sy(0)+y'(0)}{s^2+4s+20}\right\}$, naming the result `stepfour`.

```
> op(4,stepthree);
```

$$\frac{lr}{s^2+4s+20}$$

```
> stepfour:=stepthree-op(4,stepthree);
```

$$stepfour := \frac{\mathrm{D}(y)(0)}{s^2+4s+20} + \frac{sy(0)}{s^2+4s+20} + 4\frac{y(0)}{s^2+4s+20}$$

```
> y1:=simplify(invlaplace(stepfour,s,t));
```

$$y1 := 1/4 e^{-2t}\left(\sin(4t)\mathrm{D}(y)(0) + 2\sin(4t)y(0) + 4y(0)\cos(4t)\right)$$

```
> y1:=simplify(convert(y1,trig));
```

$$\begin{aligned} y1 := & -1/2\cos(2t)\sin(2t)\sinh(2t)\mathrm{D}(y)(0) \\ & - \cos(2t)\sin(2t)y(0)\sinh(2t) \\ & + 1/2\cos(2t)\sin(2t)\cosh(2t)\mathrm{D}(y)(0) \\ & + \cos(2t)\sin(2t)\cosh(2t)y(0) \\ & - 2(\cos(2t))^2 y(0)\sinh(2t) + 2(\cos(2t))^2\cosh(2t)y(0) \\ & + y(0)\sinh(2t) - \cosh(2t)y(0) \end{aligned}$$

```
> y1:=simplify(convert(y1,expsincos));
```

$$\begin{aligned} y1 := & 1/2\left(\cos(2t)\sin(2t)\mathrm{D}(y)(0) + 2\cos(2t)\sin(2t)y(0)\right. \\ & \left. + 4(\cos(2t))^2 y(0) - 2y(0)\right)e^{-2t} \end{aligned}$$

To compute the inverse Laplace transform of $\frac{\mathcal{L}\{f(t)\}}{s^2+4s+20}$, we begin by computing `lr` $= \mathcal{L}\{f(t)\}$. Let $\mathcal{U}_a(t) = \begin{cases} 1, & t \geq a \\ 0, & t < a \end{cases}$. Then,

$$\mathcal{U}_a(t) = \mathcal{U}(t-a) = \texttt{Heaviside(t - a)}.$$

The periodic function $f(t) = \begin{cases} 1, & 0 \leq t < 1 \\ -1, & 1 \leq t < 2 \end{cases}$ and $f(t) = f(t-2)$ if $t \geq 2$ can be written in terms of step functions as

$$\begin{aligned} f(t) &= \mathcal{U}_0(t) - 2\mathcal{U}_1(t) + 2\mathcal{U}_2(t) - 2\mathcal{U}_3(t) + 2\mathcal{U}_4(t) - \cdots \\ &= \mathcal{U}(t) - 2\mathcal{U}(t-1) + 2\mathcal{U}(t-2) - 2\mathcal{U}(t-3) + 2\mathcal{U}(t-4) - \cdots \\ &= \mathcal{U}(t) + 2\sum_{n=1}^{\infty}(-1)^n\mathcal{U}(t-n). \end{aligned}$$

The Laplace transform of $\mathcal{U}_a(t) = \mathcal{U}(t-a)$ is $\frac{1}{s}e^{-as}$ and the Laplace transform of $f(t)\mathcal{U}_a(t) = f(t)\mathcal{U}(t-a)$ is $e^{-as}F(s)$, where $F(s)$ is the Laplace transform of $f(t)$. Then,

$$\texttt{lr} = \frac{1}{s} - \frac{2}{s}e^{-s} + \frac{2}{s}e^{-2s} - \frac{2}{s}e^{-3s} + \cdots$$
$$= \frac{1}{s}\left(1 - 2e^{-s} + 2e^{-2s} - 2e^{-3s} + \cdots\right)$$

and

$$\frac{\texttt{lr}}{s^2+4s+20} = \frac{1}{s\left(s^2+4s+20\right)}\left(1 - 2e^{-s} + 2e^{-2s} - 2e^{-3s} + \cdots\right)$$
$$= \frac{1}{s\left(s^2+4s+20\right)} + 2\sum_{n=1}^{\infty}(-1)^n \frac{e^{-ns}}{s\left(s^2+4s+20\right)}.$$

Because $\dfrac{1}{s^2+4s+20} = \dfrac{1}{4}\dfrac{1}{(s+2)^2+4^2}$, $\mathcal{L}^{-1}\left\{\dfrac{1}{s\left(s^2+4s+20\right)}\right\} = \int_0^t \dfrac{1}{4}e^{-2\alpha}\sin 4\alpha\, d\alpha$, computed and defined to be the function $g(t)$.

```
> s:='s':
> stepfive:=simplify(op(4,stepthree)/(s*lr));
```

$$stepfive := \frac{1}{\left(s^2+4s+20\right)s}$$

```
> g:=convert(simplify(invlaplace(stepfive,s,t)),trig);
```

$$g := 1/20 - 1/20\left(\cosh(2t) - \sinh(2t)\right)\cos(4t)$$
$$- 1/40\left(\cosh(2t) - \sinh(2t)\right)\sin(4t)$$

```
> g:=simplify(convert(g,expsincos));
```

$$g := -1/40\left(-2e^{2t} + 2\cos(4t) + \sin(4t)\right)e^{-2t}$$

```
> array([seq([n,2*(-1)^n*subs(t=t-n,g)*
> Heaviside(t-n)],n=1..4)]);
```

$$\begin{bmatrix} 1 & 1/20\left(-2e^{2t-2} + 2\cos(4t-4) + \sin(4t-4)\right)e^{-2t+2}\mathit{Heaviside}(t-1) \\ 2 & -1/20\left(-2e^{2t-4} + 2\cos(4t-8) + \sin(4t-8)\right)e^{-2t+4}\mathit{Heaviside}(t-2) \\ 3 & 1/20\left(-2e^{2t-6} + 2\cos(4t-12) + \sin(4t-12)\right)e^{-2t+6}\mathit{Heaviside}(t-3) \\ 4 & -1/20\left(-2e^{2t-8} + 2\cos(4t-16) + \sin(4t-16)\right)e^{-2t+8}\mathit{Heaviside}(t-4) \end{bmatrix}$$

Then, $\mathcal{L}^{-1}\left\{2(-1)^n \frac{e^{-ns}}{s\left(s^2+4s+20\right)}\right\} = 2(-1)^n g(t-n)\mathcal{U}(t-n)$ and the inverse Laplace transform of

$$\frac{1}{s\left(s^2+4s+20\right)} + 2\sum_{n=1}^{\infty}(-1)^n \frac{e^{-ns}}{s\left(s^2+4s+20\right)}$$

is

$$y_2(t) = g(t) + 2\sum_{n=1}^{\infty}(-1)^n g(t-n)\mathcal{U}(t-n).$$

It then follows that

$$y(t) = y_1(t) + y_2(t) = y(0)e^{-2t}\cos 4t + \frac{y'(0)+2y(0)}{4}e^{-2t}\sin 4t$$

$$+ 2\sum_{n=1}^{\infty}(-1)^n g(t-n)\mathcal{U}(t-n),$$

where $g(t) = \frac{1}{20} - \frac{1}{20}e^{-2t}\cos 4t - \frac{1}{40}e^{-2t}\sin 4t$.

To graph the solution for various initial conditions on the interval $[0,5]$, we define $y_2(t) = g(t) + 2\sum_{n=1}^{5}(-1)^n g(t-n)\mathcal{U}(t-n)$, `sol`, and `inits`. (Note that we can graph the solution for various initial conditions on the interval $[0,m]$ by defining $y_2(t) = g(t) + 2\sum_{n=1}^{m}(-1)^n g(t-n)\mathcal{U}(t-n)$.)

```
> y2:='y2':
> y2:=g+2*sum('(-1)^n*subs(t=t-n,g)*
> Heaviside(t-n)','n'=1..4):
> sol:=y1+y2:
```

We then create a table of graphs of `sol` on the interval $[0,5]$ corresponding to replacing $y(0)$ and $y'(0)$ by the values $-1/2$, 0, and $1/2$ and then display the resulting graphics array in Figure 6-23.

```
> init_pos:=[-1/2,0,1/2]:
> init_vel:=[-1/2,1/2,1]:

> to_graph:=seq(seq(subs(y(0)=init_pos[i],
> D(y)(0)=init_vel[j],sol),i=1..3),j=1..3):

> to_show:=seq(plot(to_graph[i],t=0..5,
   color=BLACK),i=1..9):

> with(plots):
> graphics_array:=display(to_show,insequence=true):
> display(graphics_array);
```

■

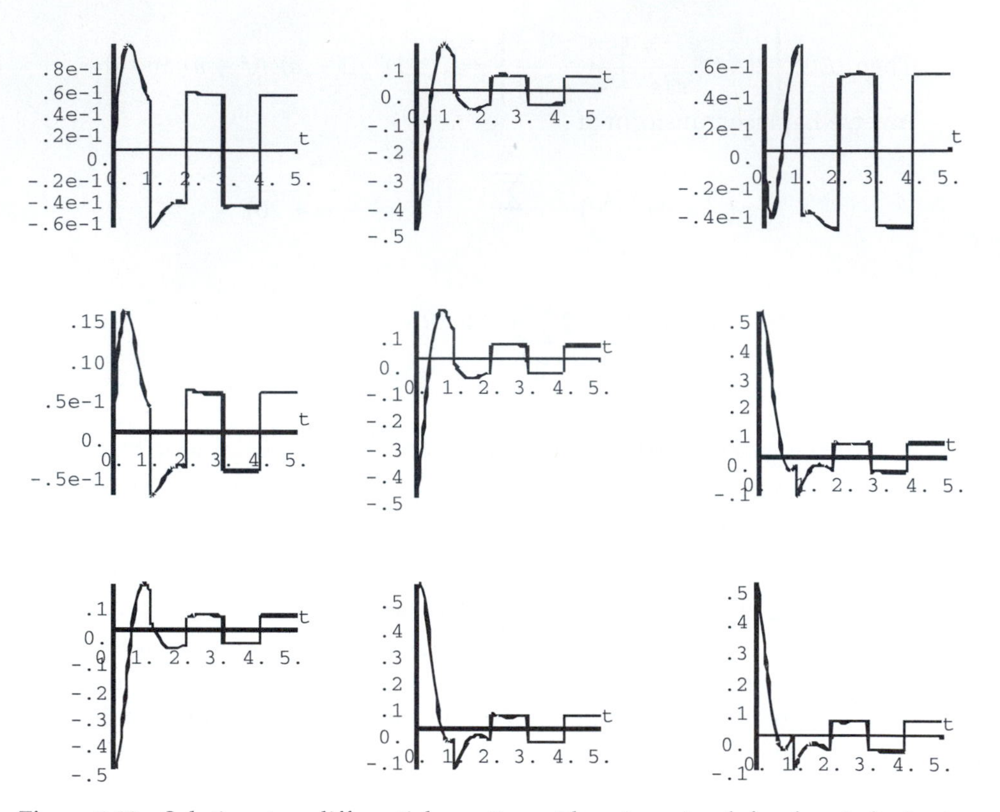

Figure 6-23 Solutions to a differential equation with a piecewise-defined periodic forcing function

Application: The Convolution Theorem

Sometimes we are required to determine the inverse Laplace transform of a product of two functions. Just as in differential and integral calculus when the derivative and integral of a product of two functions did not produce the product of the derivatives and integrals, respectively, neither does the inverse Laplace transform of the product yield the product of the inverse Laplace transforms. *The Convolution Theorem* tells us how to compute the inverse Laplace transform of a product of two functions.

Theorem 24 (The Convolution Theorem). *Suppose that $f(t)$ and $g(t)$ are piecewise continuous on $[0, \infty)$ and both are of exponential order. Further, suppose that the Laplace transform of $f(t)$ is $F(s)$ and that of $g(t)$ is $G(s)$. Then,*

$$\mathcal{L}^{-1}\left\{F(s)G(s)\right\} = \mathcal{L}^{-1}\left\{\mathcal{L}\left\{(f * g)(t)\right\}\right\} = \int_0^t f(t-v)g(v)\,dv. \tag{6.27}$$

Note that $(f * g)(t) = \int_0^t f(t-v)g(v)\,dv$ is called the **convolution integral**.

EXAMPLE 6.3.8 (L-R-C Circuits): The initial-value problem used to determine the charge $q(t)$ on the capacitor in an L-R-C circuit is

$$\begin{cases} L\frac{d^2Q}{dt^2} + R\frac{dQ}{dt} + \frac{1}{C}Q = f(t) \\ Q(0) = 0, \frac{dQ}{dt}(0) = 0 \end{cases},$$

where L denotes inductance, $dQ/dt = I$, $I(t)$ current, R resistance, C capacitance, and $E(t)$ voltage supply. Because $dQ/dt = I$, this differential equation can be represented as

$$L\frac{dI}{dt} + RI + \frac{1}{C}\int_0^t I(u)\,du = E(t).$$

Note also that the initial condition $Q(0) = 0$ is satisfied because $Q(0) = \frac{1}{C}\int_0^0 I(u)\,du = 0$. The condition $dQ/dt(0) = 0$ is replaced by $I(0) = 0$. (a) Solve this *integrodifferential equation*, an equation that involves a derivative as well as an integral of the unknown function, by using the Convolution theorem. (b) Consider this example with constant values $L = C = R = 1$, and $E(t) = \begin{cases} \sin t, 0 \le t < \pi/2 \\ 0, t \ge \pi/2 \end{cases}$. Determine $I(t)$ and graph the solution.

SOLUTION: We proceed as in the case of a differential equation by taking the Laplace transform of both sides of the equation. The Convolution theorem, (6.27), is used in determining the Laplace transform of the integral with

$$\mathcal{L}\left\{\int_0^t I(u)\,du\right\} = \mathcal{L}\left\{1 * I(t)\right\} = \mathcal{L}\left\{1\right\}\mathcal{L}\left\{I(t)\right\} = \frac{1}{s}\mathcal{L}\left\{I(t)\right\}.$$

Therefore, application of the Laplace transform yields

$$Ls\mathcal{L}\left\{I(t)\right\} - LI(0) + R\mathcal{L}\left\{I(t)\right\} + \frac{1}{C}\frac{1}{s}\mathcal{L}\left\{I(t)\right\} = \mathcal{L}\left\{E(t)\right\}.$$

Because $I(0) = 0$, we have $Ls\mathcal{L}\left\{I(t)\right\} + R\mathcal{L}\left\{I(t)\right\} + \frac{1}{C}\frac{1}{s}\mathcal{L}\left\{I(t)\right\} = \mathcal{L}\left\{E(t)\right\}$. Simplifying and solving for $\mathcal{L}\left\{I(t)\right\}$ results in $\mathcal{L}\left\{I(t)\right\} = \dfrac{Cs\mathcal{L}\left\{E(t)\right\}}{LCs^2 + RCs + 1}$.

```
> with(inttrans):
> laplace(int(i(u),u=0..t),t,s);
```

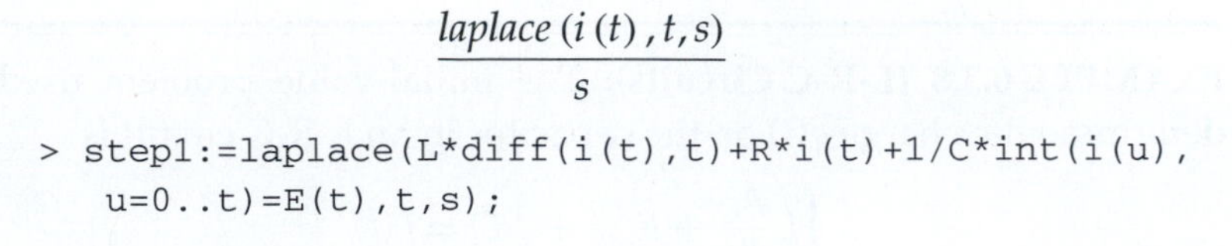

$$\frac{laplace\,(i\,(t)\,,t,s)}{s}$$

```
> step1:=laplace(L*diff(i(t),t)+R*i(t)+1/C*int(i(u),
  u=0..t)=E(t),t,s);
```

$$\begin{aligned} step1, := {} & L\left(slaplace\,(i\,(t)\,,t,s) - i\,(0)\right) + Rlaplace\,(i\,(t)\,,t,s) \\ & + \frac{laplace\,(i\,(t)\,,t,s)}{Cs} \\ & = laplace\,(E\,(t)\,,t,s) \end{aligned}$$

```
> step2:=solve(subs(i(0)=0,step1),laplace(i(t),t,s));
```

$$step2 := \frac{laplace\,(E\,(t)\,,t,s)\,Cs}{Ls^2C + RCs + 1}$$

so that $I(t) = \mathcal{L}^{-1}\left\{\frac{Cs\mathcal{L}\left\{E(t)\right\}}{LCs^2 + RCs + 1}\right\}$. For (b), we note that $E(t) = \begin{cases} \sin t,\ 0 \le t < \pi/2 \\ 0,\ t \ge \pi/2 \end{cases}$ can be written as $E(t) = \sin t\left(\mathcal{U}(t) - \mathcal{U}(t - \pi/2)\right)$. We define and plot the forcing function $E(t)$ on the interval $[0, \pi]$ in Figure 6-24.

We use lower-case letters to avoid any possible ambiguity with built-in Maple functions, like `I`.

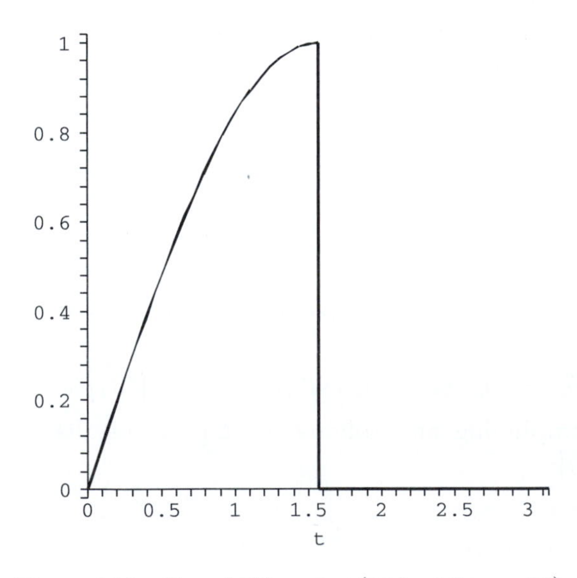

Figure 6-24 Plot of $E(t) = \sin t\left(\mathcal{U}(t) - \mathcal{U}(t - \pi/2)\right)$

Next, we compute the Laplace transform of $\mathcal{L}\left\{E(t)\right\}$ with `laplace`. We call this result `cape`.

```
> E:=t->sin(t)*Heaviside(Pi/2-t):
> Plot_E:=plot(E(t),t=0..Pi,color=BLACK):
> with(plots):
> display(Plot_E);
> cape:=laplace(E(t),t,s);
```

$$cape := -\frac{-1+se^{-1/2\pi s}}{s^2+1}$$

The Laplace transform of $I(t)$, called `capi`, is computed next.

```
> capi:=simplify(cape/(s^2+s+1));
```

$$capi := -\frac{-1+se^{-1/2\pi s}}{\left(s^2+1\right)\left(s^2+s+1\right)}$$

```
> capi_2:=expand(capi);
```

$$capi\_2 := \frac{1}{\left(s^2+1\right)\left(s^2+s+1\right)} - \frac{se^{-1/2\pi s}}{\left(s^2+1\right)\left(s^2+s+1\right)}$$

We determine $I(t)$ with `invlaplace`.

```
> i:=invlaplace(capi,s,t);
```

$$\begin{aligned} i := &\left(2/3\sqrt{3}e^{-1/2t+1/4\pi}\sin\left(1/4\sqrt{3}\left(-\pi+2t\right)\right)+\cos\left(t\right)\right) \\ &\times Heaviside\left(-1/2\pi+t\right)-\cos\left(t\right)+e^{-1/2t}\cos\left(1/2\sqrt{3}t\right) \\ &+1/3\sqrt{3}e^{-1/2t}\sin\left(1/2\sqrt{3}t\right) \end{aligned}$$

This solution is plotted in and displayed with the forcing function (in gray) in Figure 6-25. Notice the effect that the forcing function has on the solution to the differential equation.

```
> Plot_i:=plot(i,t=0..10):
> display(Plot_i,Plot_E);
```

In this case, we see that we can use `dsolve` together with the option `method=laplace` to solve the initial-value problem

$$\begin{cases} Q''+Q'+q=E(t) \\ Q(0)=0,\ Q'(0)=0 \end{cases}$$

as well.

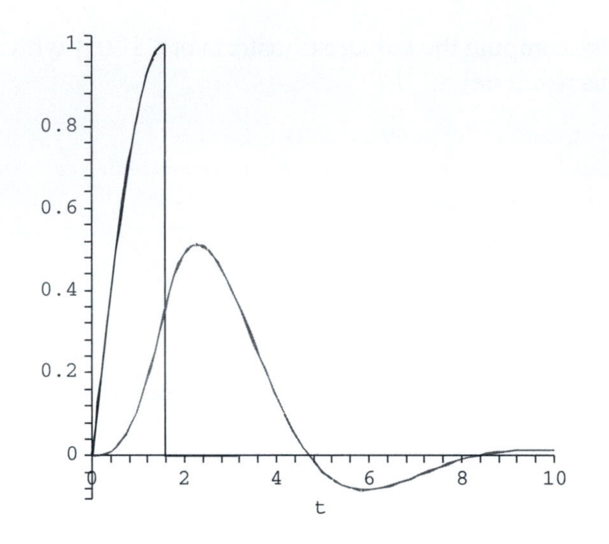

Figure 6-25 $I(t)$ (in black) and $E(t)$ (in gray)

```
> solb:=dsolve(diff(Q(t),t$2)+diff(Q(t),t)+Q(t)=E(t),
   Q(0)=0,D(Q)(0)=0,Q(t),method=laplace);
```

$$solb := Q(t) = \left(2/3\sqrt{3}e^{-1/2t+1/4\pi}\sin\left(1/4\sqrt{3}(-\pi+2t)\right)+\cos(t)\right)$$
$$\times Heaviside\left(-1/2\pi+t\right)-\cos(t)+e^{-1/2t}\cos\left(1/2\sqrt{3}t\right)$$
$$+1/3\sqrt{3}e^{-1/2t}\sin\left(1/2\sqrt{3}t\right)$$

■

Application: The Dirac Delta Function

Let $\delta(t-t_0)$ denote the (generalized) function with the two properties

1. $\delta(t-t_0)=0$ if $t\neq t_0$, and
2. $\int_{-\infty}^{\infty}\delta(t-t_0)\,dt=1$,

which is called the **Dirac delta function** and is quite useful in the definition of impulse forcing functions that arise in some differential equations. The Laplace transform of $\delta(t-t_0)$ is $\mathcal{L}\{\delta(t-t_0)\}=e^{-st_0}$. The Maple function `Dirac` represents the δ distribution.

EXAMPLE 6.3.9: Solve $\begin{cases} x''+x'+x=\delta(t)+\mathcal{U}(t-2\pi) \\ x(0)=0,\ x'(0)=0 \end{cases}$.

SOLUTION: We define Eq to be the equation $x''+x'+x = \delta(t)+\mathcal{U}(t-2\pi)$ and then use laplace to compute the Laplace transform of Eq, naming the resulting output lap_Eq. The symbol laplace(x(t),t,s) represents the Laplace transform of x(t). We then apply the initial conditions $x(0) = 0$ and $x'(0) = 0$ to lap_Eq and name the resulting output sub_conds.

```
> Eq:=diff(x(t),t$2)+diff(x(t),t)+x(t)=
> Dirac(t)+Heaviside(t-2*Pi);
```

$$Eq := \frac{d}{d\,`\$'\,(t,2)}x(t) + \frac{d}{dt}x(t) + x(t) = Dirac(t) + Heaviside(t - 2\pi)$$

```
> lap_Eq:=laplace(Eq,t,s);
```

$$lap\_Eq := s^2 laplace(x(t), t, s) - \mathrm{D}(x)(0) - sx(0)$$
$$+\, slaplace(x(t), t, s) - x(0) + laplace(x(t), t, s) = 1 + \frac{e^{-2\pi s}}{s}$$

```
> sub_conds:=subs(x(0)=0,D(x)(0)=0,lap_Eq);
```

$$sub\_conds := s^2 laplace(x(t), t, s) + slaplace(x(t), t, s) + laplace(x(t), t, s)$$
$$= 1 + \frac{e^{-2\pi s}}{s}$$

Next, we use solve to solve the equation sub_conds for the Laplace transform of $x(t)$.

```
> lap_x:=solve(sub_conds,laplace(x(t),t,s));
```

$$lap\_x := \frac{s + e^{-2\pi s}}{s\left(s^2 + s + 1\right)}$$

To find $x(t)$, we must compute the inverse Laplace transform of the Laplace transform of $\mathcal{L}\left\{x(t)\right\}$ obtained in lap_x. We use invlaplace to compute the inverse Laplace transform of lap_x.

```
> invlaplace(lap_x,s,t);
```

$$2/3\sqrt{3}e^{-1/2\,t}\sin\left(1/2\sqrt{3}t\right) -$$
$$\frac{2/3\,iHeaviside(t - 2\pi)\left(2i\sqrt{3} + \left(1 - i\sqrt{3}\right)e^{-1/2\left(1+i\sqrt{3}\right)(t-2\pi)} - \left(1 + i\sqrt{3}\right)e^{-1/2\left(1-i\sqrt{3}\right)(t-2\pi)}\right)\sqrt{3}}{\left(1 - i\sqrt{3}\right)\left(1 + i\sqrt{3}\right)}$$

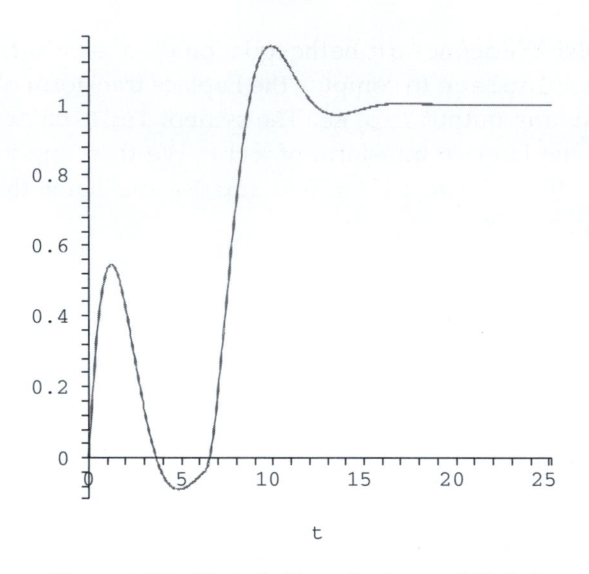

Figure 6-26 Plot of $x(t)$ on the interval $[0, 8\pi]$

We use `plot` to graph the solution on the interval $[0, 8\pi]$ in Figure 6-26.

```
> plot(rhs(sola),t=0..8*Pi,discont=true,numpoints=500);
```

Finally, we note that `dsolve` together with the `method=laplace` option is able to solve the initial-value problem directly as well.

```
> sola:=dsolve(Eq,x(0)=0,D(x)(0)=0,x(t),method=laplace);
```

$$sola := x(t) = 2/3\sqrt{3}e^{-1/2t}\sin\left(1/2\sqrt{3}t\right) -$$
$$\frac{2/3\, i\, Heaviside\,(t-2\pi)\left(2i\sqrt{3}+\left(1-i\sqrt{3}\right)e^{-1/2\left(1+i\sqrt{3}\right)(t-2\pi)}-\left(1+i\sqrt{3}\right)e^{-1/2\left(1-i\sqrt{3}\right)(t-2\pi)}\right)\sqrt{3}}{\left(1-i\sqrt{3}\right)\left(1+i\sqrt{3}\right)}$$

■

6.3.5 Nonlinear Higher-Order Equations

Generally, rigorous results regarding nonlinear equations are very difficult to obtain. In some cases, analysis is best carried out numerically and/or graphically. In other situations, rewriting the equation as a system can be of benefit, which is discussed in the next section (see Examples 6.4.5, 6.4.4, and 6.4.7).

6.4 Systems of Equations

6.4.1 Linear Systems

We now consider first-order linear systems of differential equations:

$$\mathbf{X}' = \mathbf{A}(t)\mathbf{X} + \mathbf{F}(t), \tag{6.28}$$

where

$$\mathbf{X}(t) = \begin{pmatrix} x_1(t) \\ x_2(t) \\ \vdots \\ x_n(t) \end{pmatrix}, \quad \mathbf{A}(t) = \begin{pmatrix} a_{11}(t) & a_{12}(t) & \ldots & a_{1n}(t) \\ a_{21}(t) & a_{22}(t) & \ldots & a_{2n}(t) \\ \vdots & \vdots & \ldots & \vdots \\ a_{n1}(t) & a_{n2}(t) & \ldots & a_{nn}(t) \end{pmatrix}, \quad \text{and} \quad \mathbf{F}(t) = \begin{pmatrix} f_1(t) \\ f_2(t) \\ \vdots \\ f_n(t) \end{pmatrix}.$$

Homogeneous Linear Systems

The corresponding homogeneous system of equation (6.28) is

$$\mathbf{X}' = \mathbf{A}\mathbf{X}. \tag{6.29}$$

In the same way as with the previously discussed linear equations, a **general solution** of (6.28) is $\mathbf{X} = \mathbf{X}_h + \mathbf{X}_p$ where $\mathbf{X}_h$ is a *general solution* of (6.29) and $\mathbf{X}_p$ is a *particular solution* of the nonhomogeneous system (6.28).

If $\Phi_1, \Phi_2, \ldots, \Phi_n$ are n linearly independent solutions of (6.29), a **general solution** of (6.29) is

$$\mathbf{X} = c_1\Phi_1 + c_2\Phi_2 + \cdots + c_n\Phi_n = \begin{pmatrix} \Phi_1 & \Phi_2 & \cdots & \Phi_n \end{pmatrix} \begin{pmatrix} c_1 \\ c_2 \\ \vdots \\ c_n \end{pmatrix} = \Phi\mathbf{C},$$

where

$$\Phi = \begin{pmatrix} \Phi_1 & \Phi_2 & \cdots & \Phi_n \end{pmatrix} \quad \text{and} \quad \mathbf{C} = \begin{pmatrix} c_1 \\ c_2 \\ \vdots \\ c_n \end{pmatrix}.$$

A **particular solution** to a system of ordinary differential equations is a set of functions that satisfy the system but do not contain any arbitrary constants. That is, a particular solution to a system is a set of specific functions, *containing no arbitrary constants*, that satisfy the system.

Φ is called a **fundamental matrix** for (6.29). If Φ is a fundamental matrix for (6.29), $\Phi' = \mathbf{A}\Phi$ or $\Phi' - \mathbf{A}\Phi = \mathbf{0}$.

After loading the `DEtools` package and defining **A**, the command

```
matrixDE(A,t)
```

attempts to find a fundamental matrix for (6.29).

A(t) constant

Suppose that $\mathbf{A}(t) = \mathbf{A}$ has constant real entries. Let λ be an eigenvalue of $\mathbf{A}$ with corresponding eigenvector $\mathbf{v}$. Then, $\mathbf{v}e^{\lambda t}$ is a solution of $\mathbf{X}' = \mathbf{AX}$.

If $\lambda = \alpha + \beta i$, $\beta \neq 0$, is an eigenvalue of $\mathbf{A}$ and has corresponding eigenvector $\mathbf{v} = \mathbf{a} + \mathbf{b}i$, two linearly independent solutions of $\mathbf{X}' = \mathbf{AX}$ are

$$e^{\alpha t}\left(\mathbf{a}\cos\beta t - \mathbf{b}\sin\beta t\right) \quad \text{and} \quad e^{\alpha t}\left(\mathbf{a}\sin\beta t + \mathbf{b}\cos\beta t\right). \tag{6.30}$$

EXAMPLE 6.4.1: Solve each of the following systems: (a) $\mathbf{X}' = \begin{pmatrix} -1/2 & -1/3 \\ -1/3 & -1/2 \end{pmatrix}\mathbf{X}$; (b) $\begin{cases} x' = \frac{1}{2}y \\ y' = -\frac{1}{8}x \end{cases}$; (c) $\begin{cases} dx/dt = -\frac{1}{4}x + 2y \\ dy/dt = -8x - \frac{1}{4}y \end{cases}$.

SOLUTION: (a) With `eigenvects`, which is contained in the `linalg` package, we see that the eigenvalues and eigenvectors of $\mathbf{A} = \begin{pmatrix} -1/2 & -1/3 \\ -1/3 & -1/2 \end{pmatrix}$ are $\lambda_1 = -1/6$ and $\lambda_2 = -5/6$ and $\mathbf{v}_1 = \begin{pmatrix} -1 \\ 1 \end{pmatrix}$ and $\mathbf{v}_2 = \begin{pmatrix} 1 \\ 1 \end{pmatrix}$, respectively.

```
> with(linalg):
> with(DEtools):
> A:=matrix(2,2,[-1/2,-1/3,-1/3,-1/2]);
```

$$A := matrix\left([[-1/2, -1/3], [-1/3, -1/2]]\right)$$

```
> eigenvects(A);
```

$$[-1/6, 1, \{vector\left([-1, 1]\right)\}], [-5/6, 1, \{vector\left([1, 1]\right)\}]$$

Then $\mathbf{X}_1 = \begin{pmatrix} -1 \\ 1 \end{pmatrix} e^{-t/6}$ and $\mathbf{X}_2 = \begin{pmatrix} 1 \\ 1 \end{pmatrix} e^{-5t/6}$ are two linearly independent solutions of the system so a general solution is $\mathbf{X} = \begin{pmatrix} -e^{-t/6} & e^{-5t/6} \\ e^{-t/6} & e^{-5t/6} \end{pmatrix}\begin{pmatrix} c_1 \\ c_2 \end{pmatrix}$; a fundamental matrix is $\Phi = \begin{pmatrix} -e^{-t/6} & e^{-5t/6} \\ e^{-t/6} & e^{-5t/6} \end{pmatrix}$, which we confirm using `matrixDE`.

```
> matrixDE(A,t);
```

$$\left[matrix\left([[e^{-1/6t}, e^{-5/6t}], [-e^{-1/6t}, e^{-5/6t}]]\right), vector\left([0, 0]\right)\right]$$

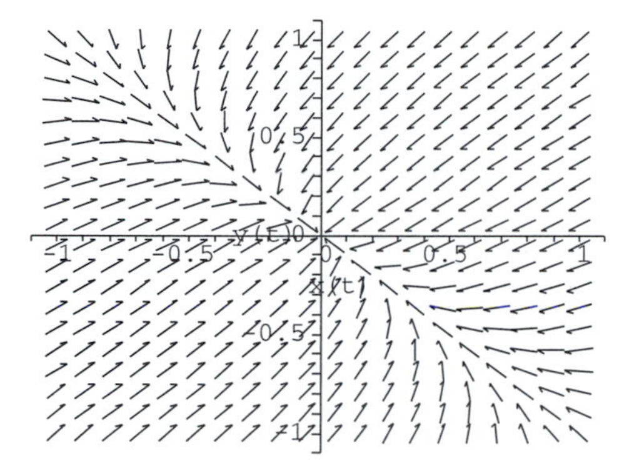

Figure 6-27 Direction field for $\mathbf{X}' = \mathbf{AX}$

We use `dsolve` to find a general solution of the system by entering

```
> gensol:=dsolve(diff(x(t),t)=-1/2*x(t)-1/3*y(t),
  diff(y(t),t)=-1/3*x(t)-1/2*y(t),x(t),y(t));
```

$$gensol := \left\{x(t) = \_C1\, e^{-5/6t} + \_C2\, e^{-1/6t}, y(t) = \_C1\, e^{-5/6t} - \_C2\, e^{-1/6t}\right\}$$

We graph the direction field with `DEplot`, which is contained in the `DEtools` package, in Figure 6-27.

Remark. After you have loaded the `DEplot` package,

```
DEplot([diff(x(t),t)=f(x(t),y(t)),diff(y(t),t)=g(x(t),
y(t))],x=a..b,y=c..d,scene=[x(t),y(t)])
```

generates a basic direction field for the system $\{x' = f(x,y), y' = g(x,y)\}$ for $a \leq x \leq b$ and $c \leq y \leq d$.

```
> DEplot([diff(x(t),t)=-1/2*x(t)-1/3*y(t),diff(y(t),t)
  =-1/3*x(t)-1/2*y(t)],
> [x(t),y(t)],t=-1..1,x=-1..1,y=-1..1,scene=[x(t),
> y(t)],scaling=CONSTRAINED,color=BLACK);
```

Several solutions are also graphed with `DEplot` and shown together with the direction field in Figure 6-28.

```
> ivals:=seq(-1+.25*i,i=0..8):
> i1:=seq([x(0)=1,y(0)=i],i=ivals):
> i2:=seq([x(0)=i,y(0)=1],i=ivals):
> i3:=seq([x(0)=-1,y(0)=i],i=ivals):
> i4:=seq([x(0)=i,y(0)=-1],i=ivals):
```

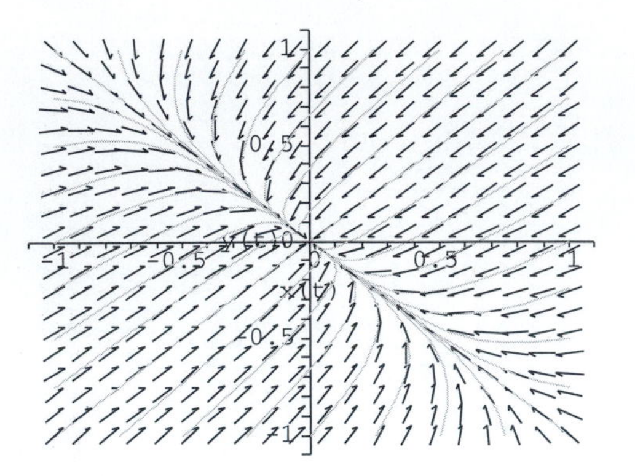

Figure 6-28 Direction field for $\mathbf{X}' = \mathbf{AX}$ along with various solution curves

```
> DEplot([diff(x(t),t)=-1/2*x(t)-1/3*y(t),
> diff(y(t),t)=-1/3*x(t)-1/2*y(t)],[x(t),y(t)],t=0..10,
> [i1,i2,i3,i4],x=-1..1,y=-1..1,scene=[x(t),y(t)],
    scaling=CONSTRAINED,
> color=BLACK,linecolor=GRAY,thickness=1);
```

(b) In matrix form the system is equivalent to the system $\mathbf{X}' = \begin{pmatrix} 0 & 1/2 \\ -1/8 & 0 \end{pmatrix}\mathbf{X}$. As in (a), we use `eigenvects` to see that the eigenvalues and eigenvectors of $\mathbf{A} = \begin{pmatrix} 0 & 1/2 \\ -1/8 & 0 \end{pmatrix}$ are $\lambda_{1,2} = 0 \pm \frac{1}{4}i$ and $\mathbf{v}_{1,2} = \begin{pmatrix} 1 \\ 0 \end{pmatrix} \pm \begin{pmatrix} 0 \\ 1/2 \end{pmatrix} i$.

```
> A:=matrix(2,2,[0,1/2,-1/8,0]);
```

$$A := matrix\left([[0,1/2],[-1/8,0]]\right)$$

```
> eigenvects(A);
```

$$[1/4\,i,1,\{vector\left([1,1/2\,i]\right)\}],\ [-1/4\,i,1,\{vector\left([1,-1/2\,i]\right)\}]$$

Two linearly independent solutions are then $\mathbf{X}_1 = \begin{pmatrix} 1 \\ 0 \end{pmatrix}\cos\frac{1}{4}t - \begin{pmatrix} 0 \\ 1/2 \end{pmatrix}\sin\frac{1}{4}t = \begin{pmatrix} \cos\frac{1}{4}t \\ -\frac{1}{2}\sin\frac{1}{4}t \end{pmatrix}$ and $\mathbf{X}_2 = \begin{pmatrix} 1 \\ 0 \end{pmatrix}\sin\frac{1}{4}t + \begin{pmatrix} 0 \\ 1/2 \end{pmatrix}\cos\frac{1}{4}t = \begin{pmatrix} \sin\frac{1}{4}t \\ \frac{1}{2}\cos\frac{1}{4}t \end{pmatrix}$ and a general solution is $\mathbf{X} = c_1\mathbf{X}_1 + c_2\mathbf{X}_2$

$$= \begin{pmatrix} \cos\frac{1}{4}t & \sin\frac{1}{4}t \\ -\frac{1}{2}\sin\frac{1}{4}t & \frac{1}{2}\cos\frac{1}{4}t \end{pmatrix} \begin{pmatrix} c_1 \\ c_2 \end{pmatrix} \text{ or } x = c_1\cos\frac{1}{4}t + c_2\sin\frac{1}{4}t \text{ and } y$$

$$= -c_1\frac{1}{2}\sin\frac{1}{4}t + \frac{1}{2}c_2\cos\frac{1}{4}t.$$

```
> matrixDE(A,t);
```

$$[matrix([[\cos(1/4\,t), \sin(1/4\,t)], [-1/2\,\sin(1/4\,t), 1/2\,\cos(1/4\,t)]]), vector([0,0])]$$

As before, we use `dsolve` to find a general solution.

```
> gensol:=dsolve(diff(x(t),t)=1/2*y(t),diff(y(t),t)
   =-1/8*x(t),x(t),y(t));
```

$$gensol := \{y(t) = 1/2\_C1\,\cos(1/4\,t) - 1/2\_C2\,\sin(1/4\,t), x(t) = \_C1\,\sin(1/4\,t) + \_C2\,\cos(1/4\,t)\}$$

Initial-value problems for systems are solved in the same way as for other equations. For example, entering

```
> partsol:=dsolve(diff(x(t),t)=1/2*y(t),diff(y(t),t)
   =-1/8*x(t),x(0)=1,y(0)=-1,x(t),y(t));
```

$$partsol := \{y(t) = -\cos(1/4\,t) - 1/2\,\sin(1/4\,t), x(t) = -2\,\sin(1/4\,t) + \cos(1/4\,t)\}$$

finds the solution that satisfies $x(0) = 1$ and $y(0) = -1$.

We graph $x(t)$ and $y(t)$ together as well as parametrically with `plot` in Figure 6-29.

```
> assign(partsol):
> plot([x(t),y(t)],t=0..8*Pi,color=[BLACK,GRAY]);
> plot([x(t),y(t),t=0..8*Pi],color=BLACK,scaling=
   CONSTRAINED);
```

We can also use `DEplot` to graph the direction field and/or various solutions as we do next in Figure 6-30.

```
> x:='x':y:='y':
> i1:=seq([x(0)=0,y(0)=-1+.25*i],i=0..8):
> DEplot([diff(x(t),t)=1/2*y(t),diff(y(t),t)
   =-1/8*x(t)],[x(t),y(t)],
> t=0..8*Pi,[i1],x=-1..1,y=-1..1,scene=[x(t),y(t)],
   scaling=CONSTRAINED,
> color=GRAY,linecolor=BLACK,thickness=1,arrows=LARGE);
```

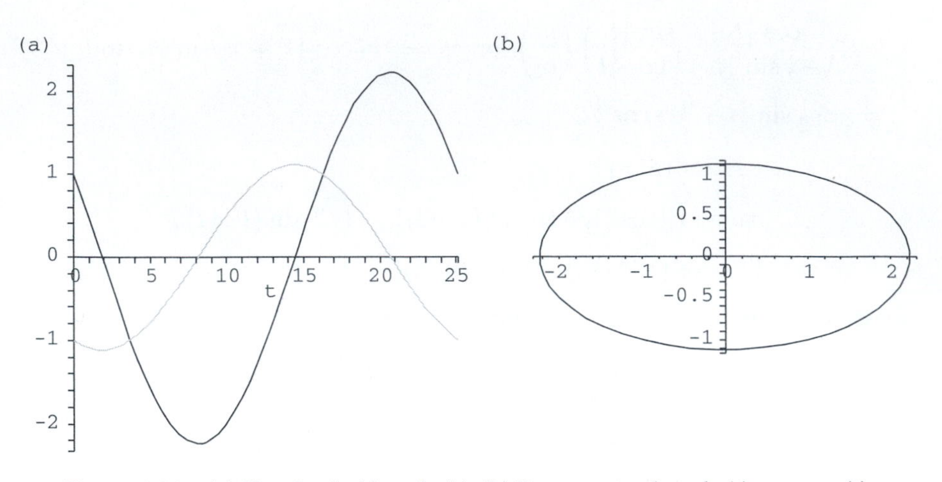

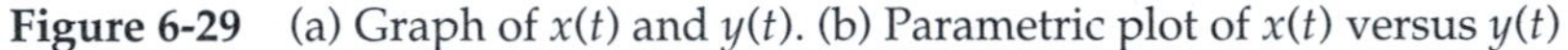

Figure 6-29 (a) Graph of $x(t)$ and $y(t)$. (b) Parametric plot of $x(t)$ versus $y(t)$

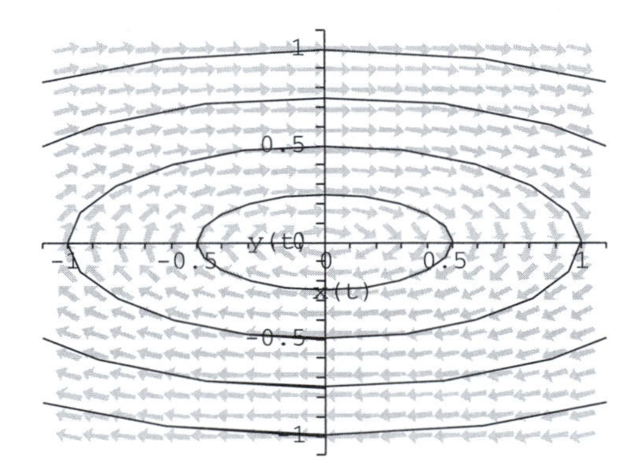

Figure 6-30 Notice that all nontrivial solutions are periodic

(c) In matrix form, the system is equivalent to the system $\mathbf{X}' = \begin{pmatrix} -\frac{1}{4} & 2 \\ -8 & -\frac{1}{4} \end{pmatrix} \mathbf{X}$. The eigenvalues and corresponding eigenvectors of $\mathbf{A} = \begin{pmatrix} -\frac{1}{4} & 2 \\ -8 & -\frac{1}{4} \end{pmatrix}$ are found to be $\lambda_{1,2} = -\frac{1}{4} \pm 4i$ and $\mathbf{v}_{1,2} = \begin{pmatrix} 0 \\ 2 \end{pmatrix} \pm \begin{pmatrix} 1 \\ 0 \end{pmatrix} i$ with `eigenvects`.

```
> A:=matrix(2,2,[-1/4,2,-8,-1/4]):
> eigenvects(A);
```

$$[-1/4 + 4\,i, 1, \{vector\,([1, 2\,i])\}], [-1/4 - 4\,i, 1, \{vector\,([1, -2\,i])\}]$$

A general solution is then

$$\mathbf{X} = c_1\mathbf{X}_1 + c_2\mathbf{X}_2$$
$$= c_1 e^{-t/4}\left(\begin{pmatrix}1\\0\end{pmatrix}\cos 4t - \begin{pmatrix}0\\2\end{pmatrix}\sin 4t\right) + c_2 e^{-t/4}\left(\begin{pmatrix}1\\0\end{pmatrix}\sin 4t + \begin{pmatrix}0\\2\end{pmatrix}\cos 4t\right)$$
$$= e^{-t/4}\left[c_1\begin{pmatrix}\cos 4t\\-2\sin 4t\end{pmatrix} + c_2\begin{pmatrix}\sin 4t\\2\cos 4t\end{pmatrix}\right] = e^{-t/4}\begin{pmatrix}\cos 4t & \sin 4t\\-2\sin 4t & 2\cos 4t\end{pmatrix}\begin{pmatrix}c_1\\c_2\end{pmatrix}$$

or $x = e^{-t/4}\left(c_1\cos 4t + c_2\sin 4t\right)$ and $y = e^{-t/4}\left(2c_2\cos 4t - 2c_1\sin 4t\right)$.

```
> matrixDE(A,t);
```

$$\left[matrix\left([[e^{-1/4\,t}\cos(4\,t), e^{-1/4\,t}\sin(4\,t)], [-2\,e^{-1/4\,t}\sin(4\,t), 2\,e^{-1/4\,t}\cos(4\,t)]]\right), vector([0,0])\right]$$

We confirm this result using `dsolve`.

```
> dsolve(diff(x(t),t)=-1/4*x(t)+2*y(t),diff(y(t),t)
   =-8*x(t)-1/4*y(t),
> x(t),y(t));
```

$$\left\{y(t) = -2\,e^{-1/4\,t}\left(-\_C1\,\cos(4\,t) + \_C2\,\sin(4\,t)\right),\right.$$
$$\left. x(t) = e^{-1/4\,t}\left(\_C1\,\sin(4\,t) + \_C2\,\cos(4\,t)\right)\right\}$$

We use `DEplot` to graph the direction field associated with the system along with various solutions in Figure 6-31.

```
> ivals:=seq(-1+.25*i,i=0..8):
> i1:=seq([x(0)=1,y(0)=i],i=ivals):
> i2:=seq([x(0)=i,y(0)=1],i=ivals):
> i3:=seq([x(0)=-1,y(0)=i],i=ivals):
> i4:=seq([x(0)=i,y(0)=-1],i=ivals):
> DEplot([diff(x(t),t)=-1/4*x(t)+2*y(t),diff(y(t),t)
   =-8*x(t)-1/4*y(t)],
> [x(t),y(t)],t=0..10,[i1,i2,i3,i4],x=-1..1,y=-1..1,
   scene=[x(t),y(t)],
> scaling=CONSTRAINED,color=GRAY,linecolor=BLACK,
   thickness=1,
> stepsize=.05);
```

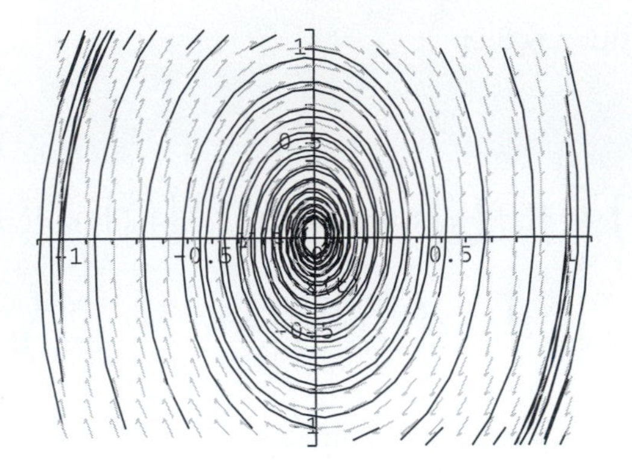

Figure 6-31 Various solutions and direction field associated with the system

Last, we illustrate how to solve an initial-value problem and graph the resulting solutions by finding the solution that satisfies the initial conditions $x(0) = 100$ and $y(0) = 10$ and then graphing the results with `plot` in Figure 6-32.

```
> x:='x':y:='y':
> partsol:=dsolve(diff(x(t),t)=-1/4*x(t)+2*y(t),
> diff(y(t),t)=-8*x(t)-1/4*y(t),x(0)=100,y(0)=10,x(t),
   y(t));
> assign(partsol):
> plot([x(t),y(t)],t=0..20,color=[BLACK,GRAY]);
> plot([x(t),y(t),t=0..20],color=BLACK);
```

$$partsol := \{y(t) = -2\,e^{-1/4\,t}\,(-5\cos(4\,t) + 100\sin(4\,t))\,, x(t) = e^{-1/4\,t}\,(5\sin(4\,t) + 100\cos(4\,t))\}$$

■

Application: The Double Pendulum

The motion of a double pendulum is modeled by the system of differential equations

$$\begin{cases} (m_1 + m_2)\,l_1{}^2 \dfrac{d^2\theta_1}{dt^2} + m_2 l_1 l_2 \dfrac{d^2\theta_2}{dt^2} + (m_1 + m_2)\,l_1 g\theta_1 = 0 \\ m_2 l_2{}^2 \dfrac{d^2\theta_2}{dt^2} + m_2 l_1 l_2 \dfrac{d^2\theta_1}{dt^2} + m_2 l_2 g\theta_2 = 0 \end{cases}$$

using the approximation $\sin\theta \approx \theta$ for small displacements. θ_1 represents the displacement of the upper pendulum and θ_2 that of the lower pendulum. Also, m_1 and

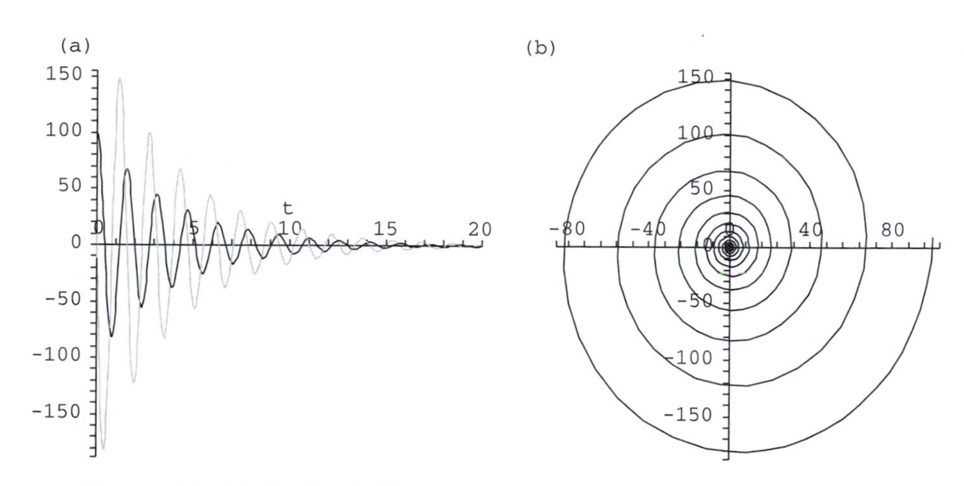

Figure 6-32 (a) Graph of $x(t)$ and $y(t)$. (b) Parametric plot of $x(t)$ versus $y(t)$

m_2 represent the mass attached to the upper and lower pendulums, respectively, while the length of each is given by l_1 and l_2.

EXAMPLE 6.4.2: Suppose that $m_1 = 3$, $m_2 = 1$, and each pendulum has length 16. If $\theta_1(0) = 1$, $\theta_1{}'(0) = 0$, $\theta_2(0) = -1$, and $\theta_2{}'(0) = 0$, solve the double pendulum problem using $g = 32$. Plot the solution.

SOLUTION: In this case, the system to be solved is

$$\begin{cases} 4 \cdot 16^2 \dfrac{d^2\theta_1}{dt^2} + 16^2 \dfrac{d^2\theta_2}{dt^2} + 4 \cdot 16 \cdot 32\theta_1 = 0 \\ 16^2 \dfrac{d^2\theta_2}{dt^2} + 16^2 \dfrac{d^2\theta_1}{dt^2} + 16 \cdot 32\theta_2 = 0 \end{cases},$$

which we simplify to obtain

$$\begin{cases} 4\dfrac{d^2\theta_1}{dt^2} + \dfrac{d^2\theta_2}{dt^2} + 8\theta_1 = 0 \\ \dfrac{d^2\theta_2}{dt^2} + \dfrac{d^2\theta_1}{dt^2} + 2\theta_2 = 0 \end{cases}.$$

First, we use `dsolve` to solve the initial value problem.

```
> Eq1:=4*diff(theta[1](t),t$2)+diff(theta[2](t),t$2)+
> 8*theta[1](t)=0:
```

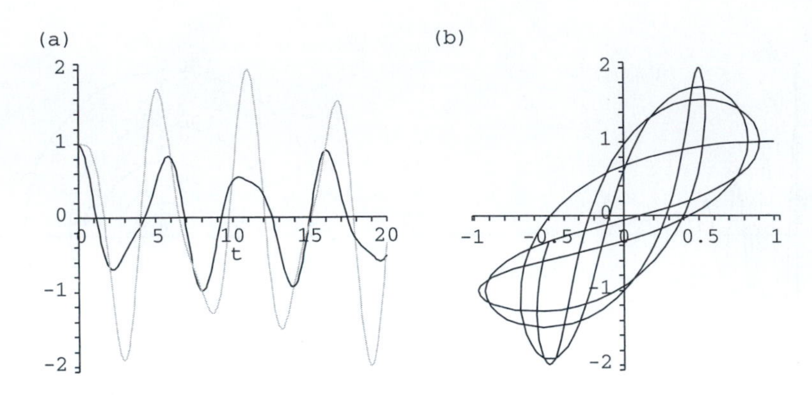

Figure 6-33 (a) $\theta_1(t)$ (in black) and $\theta_2(t)$ (in gray) as functions of t. (b) Parametric plot of $\theta_1(t)$ versus $\theta_2(t)$

```
> Eq2:=diff(theta[1](t),t$2)+diff(theta[2](t),t$2)+
> 2*theta[2](t)=0:
```

The **Laplace transform** of $y = f(t)$ is $F(s)=\mathcal{L}\{f(t)\} = \int_0^\infty e^{-st} f(t)\,dt$.

To solve the initial-value problem using traditional methods, we use the *method of Laplace transforms*. To do so, we define sys to be the system of equations and then use dsolve together with the option method=laplace to compute the Laplace transform of each equation.

```
> sola:=dsolve(Eq1,Eq2,theta[1](0)=1,D(theta[1])(0)=0,
  theta[2](0)=1,
> D(theta[2])(0)=0,theta[1](t),theta[2](t),
  method=laplace);
```

$$sola := \left\{\theta_1(t) = 1/4\cos(2t) + 3/4\cos\left(2/3\sqrt{3}t\right), \theta_2(t) = 3/2\cos\left(2/3\sqrt{3}t\right) - 1/2\cos(2t)\right\}$$

These two functions are graphed together in Figure 6-33(a) and parametrically in Figure 6-33(b).

```
> assign(sola):
> plot([theta[1](t),theta[2](t)],t=0..20,
  color=[BLACK,GRAY]);

> plot([theta[1](t),theta[2](t),t=0..20],color=BLACK);
```

We can illustrate the motion of the pendulum as follows. First, we define the function pen2.

```
> pen2:=proc(t0,len1,len2)
> local pt1,pt2,xt0,yt0;
```

```
> xt0:=evalf(subs(t=t0,theta[1](t)));
> yt0:=evalf(subs(t=t0,theta[2](t)));
> pt1:=[len1*cos(3*Pi/2+xt0),len1*sin(3*Pi/2+xt0)];
> pt2:=[len1*cos(3*Pi/2+xt0)+len2*cos(3*Pi/2+yt0),
> len1*sin(3*Pi/2+xt0)+len2*sin(3*Pi/2+yt0)];
> plot([[0,0],pt1,pt2],xtickmarks=2,ytickmarks=2,
> view=[-32..32,-32..0]);
> end:
```

Next, we define `ivals` to be a list of 16 evenly spaced numbers between 0 and 10. `seq` is then used to apply `pen2` to the list of numbers in `ivals`. The resulting set of graphics is displayed as an array using `display` with the option `insequence=true` in Figure 6-34.

```
> with(plots):
```

```
> ivals:=[seq(10*i/15,i=0..15)]:
```

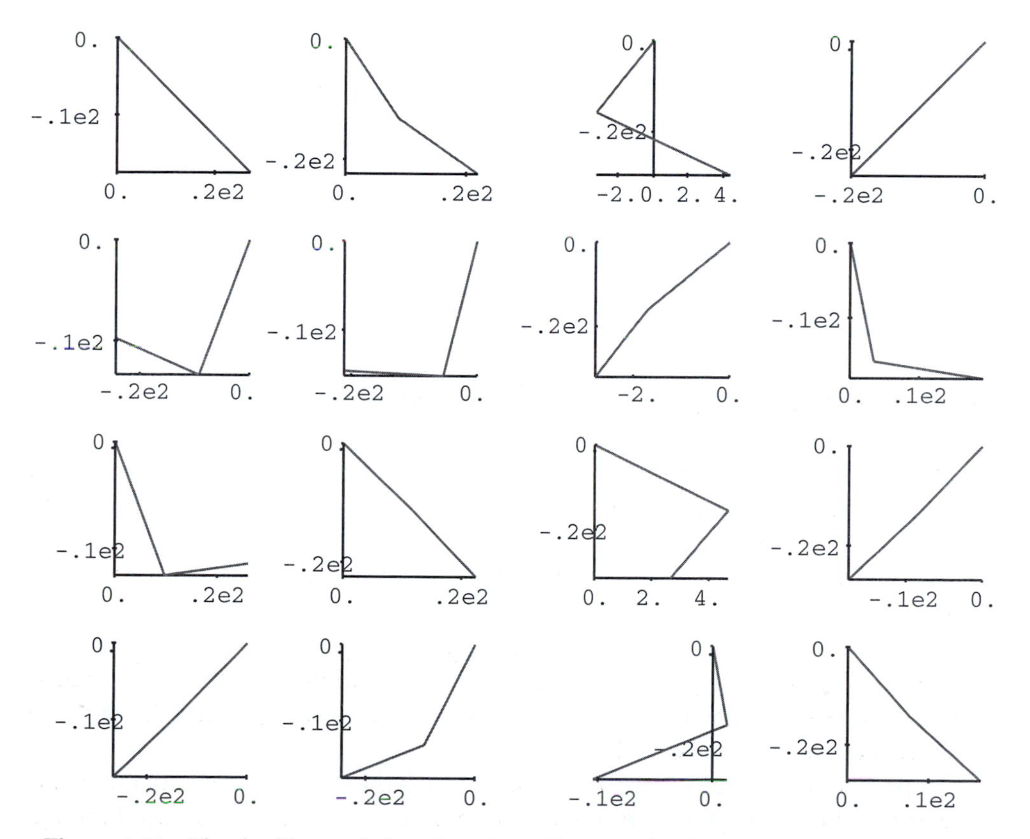

Figure 6-34 The double pendulum for 16 equally spaced values of t between 0 and 10

```
> toshow:=[seq(pen2(i,16,16),i=ivals)]:
> nops(toshow);
```

$$16$$

```
> anarray:=display(toshow,insequence=true):
> display(anarray);
```

We can also use `display` to generate an animation. We show one frame from the animation that results from the following command.

```
> display(toshow,insequence=true);
```

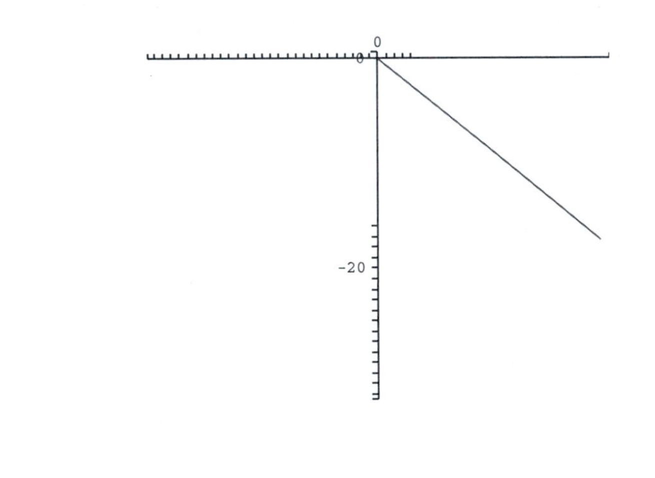

■

6.4.2 Nonhomogeneous Linear Systems

Generally, the method of undetermined coefficients is difficult to implement for nonhomogeneous linear systems as the choice for the particular solution must be very carefully made.

Variation of parameters is implemented in much the same way as for first-order linear equations.

Let $\mathbf{X}_h$ be a general solution to the corresponding homogeneous system of (6.28), $\mathbf{X}$ a general solution of (6.28), and $\mathbf{X}_p$ a particular solution of (6.28). It then follows that $\mathbf{X} - \mathbf{X}_p$ is a solution to the corresponding homogeneous system so $\mathbf{X} - \mathbf{X}_p = \mathbf{X}_h$ and, consequently, $\mathbf{X} = \mathbf{X}_h + \mathbf{X}_p$.

A particular solution of (6.28) is found in much the same way as with first-order linear equations. Let Φ be a fundamental matrix for the corresponding homogeneous system. We assume that a particular solution has the form $\mathbf{X}_p = \Phi\mathbf{U}(t)$.

Differentiating $\mathbf{X}_p$ gives us

$$\mathbf{X}_p{}' = \Phi'\mathbf{U} + \Phi\mathbf{U}'.$$

Substituting into (6.28) results in

$$\begin{aligned}\Phi'\mathbf{U} + \Phi\mathbf{U}' &= \mathbf{A}\Phi\mathbf{U} + \mathbf{F}\\ \Phi\mathbf{U}' &= \mathbf{F}\\ \mathbf{U}' &= \Phi^{-1}\mathbf{F}\\ \mathbf{U} &= \int \Phi^{-1}\mathbf{F}\,dt,\end{aligned}$$

where we have used the fact that $\Phi'\mathbf{U} - \mathbf{A}\Phi\mathbf{U} = \left(\Phi' - \mathbf{A}\Phi\right)\mathbf{U} = \mathbf{0}$. It follows that

$$\mathbf{X}_p = \Phi \int \Phi^{-1}\mathbf{F}\,dt. \tag{6.31}$$

A general solution is then

$$\begin{aligned}\mathbf{X} &= \mathbf{X}_h + \mathbf{X}_p\\ &= \Phi\mathbf{C} + \Phi \int \Phi^{-1}\mathbf{F}\,dt\\ &= \Phi\left(\mathbf{C} + \int \Phi^{-1}\mathbf{F}\,dt\right) = \Phi \int \Phi^{-1}\mathbf{F}\,dt,\end{aligned}$$

where we have incorporated the constant vector $\mathbf{C}$ into the indefinite integral $\int \Phi^{-1}\mathbf{F}\,dt$.

EXAMPLE 6.4.3: Solve the initial-value problem

$$\mathbf{X}' = \begin{pmatrix} 1 & -1 \\ 10 & -1 \end{pmatrix}\mathbf{X} - \begin{pmatrix} t\cos 3t \\ t\sin t + t\cos 3t \end{pmatrix}, \quad \mathbf{X}(0) = \begin{pmatrix} 1 \\ -1 \end{pmatrix}.$$

Remark. In traditional form, the system is equivalent to

$$\begin{cases} x' &= x - y - t\cos 3t \\ y' &= 10x - y - t\sin t - t\cos 3t \end{cases}, \quad x(0) = 1,\ y(0) = -1.$$

SOLUTION: The corresponding homogeneous system is $\mathbf{X}_h' = \begin{pmatrix} 1 & -1 \\ 10 & -1 \end{pmatrix} \mathbf{X}_h$. The eigenvalues and corresponding eigenvectors of $\mathbf{A} = \begin{pmatrix} 1 & -1 \\ 10 & -1 \end{pmatrix}$ are $\lambda_{1,2} = \pm 3i$ and $\mathbf{v}_{1,2} = \begin{pmatrix} 1 \\ 10 \end{pmatrix} \pm \begin{pmatrix} -3 \\ 0 \end{pmatrix} i$, respectively.

```
> with(linalg):
> with(DEtools):
> A:=matrix(2,2,[1,-1,10,-1]):
> eigenvects(A);
```

$$[3\,i, 1, \{vector\,([1, 1 - 3\,i])\}], [-3\,i, 1, \{vector\,([1, 1 + 3\,i])\}]$$

A fundamental matrix is $\Phi = \begin{pmatrix} \sin 3t & \cos 3t \\ \sin 3t - 3\cos 3t & \cos 3t + 3\sin 3t \end{pmatrix}$

```
> fm:=matrixDE(A,t);
```

$$fm := [matrix\,([[\sin(3\,t), \cos(3\,t)], [\sin(3\,t) - 3\,\cos(3\,t), \cos(3\,t)\,, +3\,\sin(3\,t)]])\ vector\,([0,0])]$$

```
> fm[1];
```

$$matrix\,([[\sin(3\,t), \cos(3\,t)], [\sin(3\,t) - 3\,\cos(3\,t), \cos(3\,t) + 3\,\sin(3\,t)]])$$

```
> fminv:=simplify(inverse(fm[1]));
```

$$\begin{bmatrix} 1/3\,\cos(3\,t) + \sin(3\,t) & -1/3\,\cos(3\,t) \\ -1/3\,\sin(3\,t) + \cos(3\,t) & 1/3\,\sin(3\,t) \end{bmatrix}$$

We now compute $\Phi^{-1}\mathbf{F}(t)$

```
> ft:=matrix(2,1,[-t*cos(3*t),-t*sin(t)-t*cos(3*t)]);
```

$$\begin{bmatrix} -t\cos(3\,t) \\ -t\sin(t) - t\cos(3\,t) \end{bmatrix}$$

```
> step1:=evalm(fminv &* ft);
```

$$\begin{bmatrix} -(1/3\cos(3t)+\sin(3t))t\cos(3t)-1/3\cos(3t)\,(-t\sin(t)-t\cos(3t)) \\ -(-1/3\sin(3t)+\cos(3t))t\cos(3t)+1/3\sin(3t)\,(-t\sin(t)-t\cos(3t)) \end{bmatrix}$$

For length considerations, we display only the final results. To see each result as it is generated, replace the colons with semi-colons.

and $\int \Phi^{-1}\mathbf{F}(t)\,dt$.

```
> step2:=map(int,step1,t):
```

A general solution of the nonhomogeneous system is then $\Phi\left(\int \Phi^{-1}\mathbf{F}(t)\,dt + \mathbf{C}\right)$.

```
> simplify(evalm(fm[1] &* step2)):
```

It is easiest to use `matrixDE`

```
> check1:=matrixDE(A,ft,t):
```

```
> check1[1];
```

$$\begin{bmatrix} \cos(3t) & \sin(3t) \\ \cos(3t) + 3\sin(3t) & \sin(3t) - 3\cos(3t) \end{bmatrix}$$

```
> check1[2];
```

$$\Big[-\frac{1}{72}\cos(3t) - 1/32\cos(t) - 1/12\sin(3t)\,t - 1/4\cos(3t)\,t^2 + 1/8t\sin(t),$$
$$-\frac{1}{72}\cos(3t) - 1/32\cos(t) - 1/12\sin(3t)\,t - 1/4\cos(3t)\,t^2 + 1/8t\sin(t) + 1/24\sin(3t)$$
$$-\frac{5}{32}\sin(t) + 3/4t\cos(3t) - 3/4\sin(3t)\,t^2 + \frac{23}{8}t\cos(t) - 4t\,(\cos(t))^3\Big]$$

or `dsolve` to solve the initial-value problem directly.

```
> check2:=dsolve(diff(x(t),t)=x(t)-y(t)-t*cos(3*t),
> diff(y(t),t)=10*x(t)-y(t)-t*sin(t)-t*cos(3*t),x(0)=1,y(0)=-1,
> x(t),y(t));
```

$$check2 := \Big\{y(t) = -\frac{31}{32}\cos(3t) + \frac{123}{32}\sin(3t) - \frac{5}{32}\sin(t) - 1/4\,t\cos(3t) - 1/8\,t\cos(t)$$
$$-3/4\sin(3t)\,t^2 - 1/12\sin(3t)\,t + 1/8\,t\sin(t) - 1/32\cos(t) - 1/4\cos(3t)\,t^2,$$
$$x(t) = 2/3\sin(3t) + \frac{33}{32}\cos(3t) - 1/12\sin(3t)\,t + 1/8\,t\sin(t)$$
$$-1/32\cos(t) - 1/4\cos\left(3t\right)t^2\Big\}$$

The solutions are graphed with `plot` in Figure 6-35.

```
> assign(check2):
> plot([x(t),y(t)],t=0..8*Pi,color=[BLACK,GRAY]);
> plot([x(t),y(t),t=0..8*Pi],color=BLACK,scaling=CONSTRAINED);
```

■

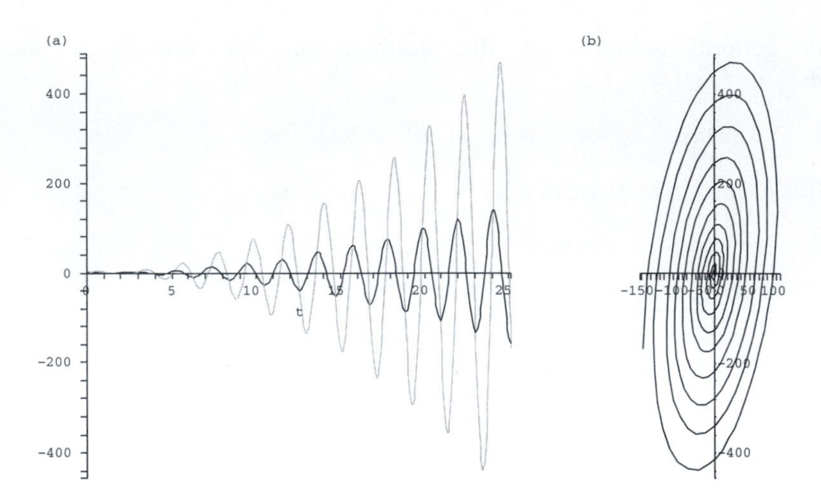

Figure 6-35 (a) Graph of $x(t)$ (in black) and $y(t)$ (in gray). (b) Parametric plot of $x(t)$ versus $y(t)$

6.4.3 Nonlinear Systems

Nonlinear systems of differential equations arise in numerous situations. Rigorous analysis of the behavior of solutions to nonlinear systems is usually very difficult, if not impossible.

To generate numerical solutions of equations, use `dsolve` with the `numeric` option.

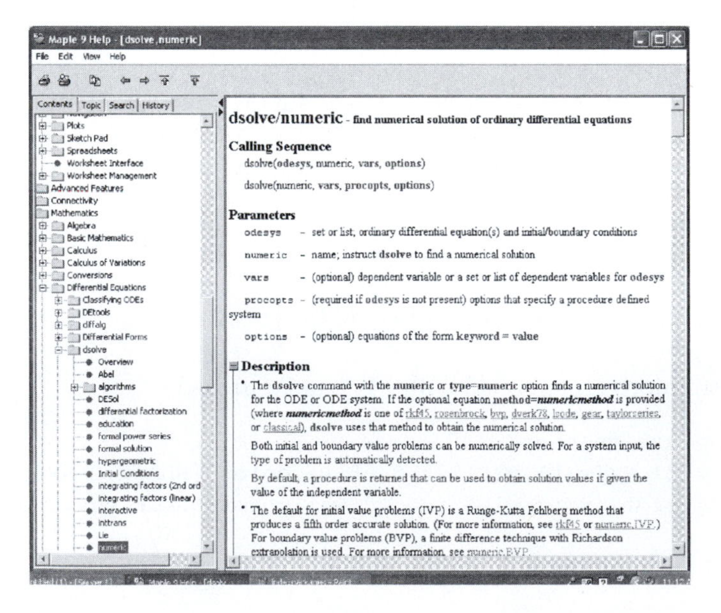

Also see Example 6.4.7.

EXAMPLE 6.4.4 (Van-der-Pol's equation): Van-der-Pol's equation $x''+\mu\left(x^2-1\right)x'+x=0$ can be written as the system

$$\begin{aligned} x' &= y \\ y' &= -x-\mu\left(x^2-1\right)y. \end{aligned} \qquad (6.32)$$

If $\mu = 2/3$, $x(0) = 1$, and $y(0) = 0$, (a) find $x(1)$ and $y(1)$. (b) Graph the solution that satisfies these initial conditions.

SOLUTION: We use `dsolve` with the `numeric` option to solve (6.32) with $\mu = 2/3$ subject to $x(0) = 1$ and $y(0) = 0$. We name the resulting numerical solution `numsol`.

```
> with(plots):
> numsol:=dsolve([diff(x(t),t)=y(t),diff(y(t),t)=-x(t)
  -2/3*(x(t)^2-1)*y(t),
> x(0)=1,y(0)=0],[x(t),y(t)],numeric);
```

```
Warning, the name changecoords has been redefined
```

$$numsol := proc(x_rkf45)...endproc$$

We evaluate `numsol` if $t = 1$ to see that $x(1) \approx .5128$ and $y(1) \approx -.9692$.

```
> numsol(1);
```

$$[t = 1.0, x(t) = 0.512847902997304538, y(t) = -0.969203620640395002]$$

`odeplot` is used to graph $x(t)$ and $y(t)$ together in Figure 6-36(a); a three-dimensional plot, $(t, x(t), y(t))$, is shown in Figure 6-36(b); a parametric plot is shown in Figure 6-36(c); and the limit cycle is shown more clearly in Figure 6-36(d) by graphing the solution for $20 \le t \le 30$.

```
> odeplot(numsol,[[t,x(t)],[t,y(t)]],0..15,numpoints
  =200);

> odeplot(numsol,[t,x(t),y(t)],0..15,axes=BOXED,
  numpoints=200);

> odeplot(numsol,[x(t),y(t)],0..15,numpoints=200);

> odeplot(numsol,[x(t),y(t)],20..30,numpoints=200);
```

■

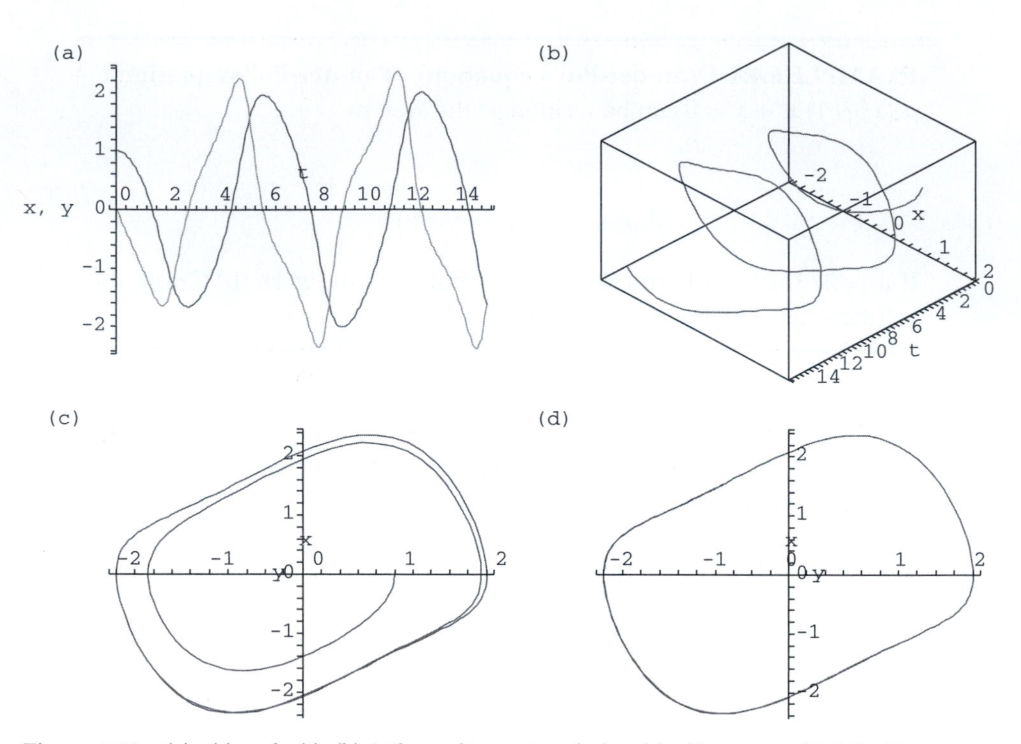

Figure 6-36 (a) $x(t)$ and $y(t)$. (b) A three-dimensional plot. (c) $x(t)$ versus $y(t)$. (d) $x(t)$ versus $y(t)$ for $20 \leq t \leq 30$

Linearization

An **autonomous system** does not explicitly depend on the independent variable, t. That is, if you write the system omitting all arguments, the independent variable (typically t) does not appear.

Consider the autonomous system of the form

$$\begin{aligned} x_1{}' &= f_1\left(x_1, x_2, \ldots, x_n\right) \\ x_2{}' &= f_2\left(x_1, x_2, \ldots, x_n\right) \\ &\vdots \\ x_n{}' &= f_n\left(x_1, x_2, \ldots, x_n\right). \end{aligned} \tag{6.33}$$

An **equilibrium** (or **rest**) **point**, $E = (x_1{}^*, x_2{}^*, \ldots, x_n{}^*)$, of (6.33) is a solution of the system

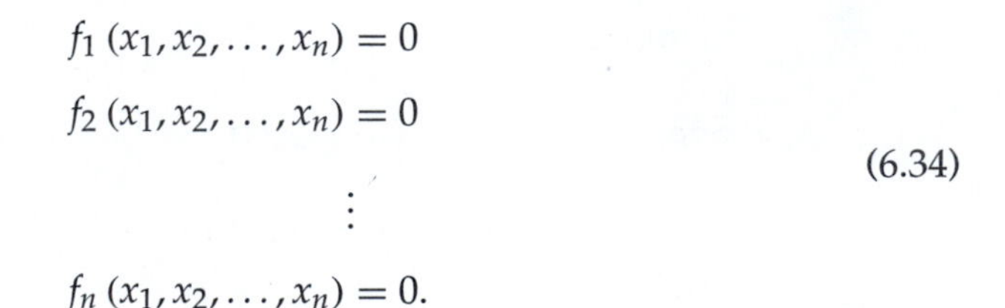

$$\begin{aligned} f_1\left(x_1, x_2, \ldots, x_n\right) &= 0 \\ f_2\left(x_1, x_2, \ldots, x_n\right) &= 0 \\ &\vdots \\ f_n\left(x_1, x_2, \ldots, x_n\right) &= 0. \end{aligned} \tag{6.34}$$

The **Jacobian** of (6.33) is

$$\mathrm{J}(x_1, x_2, \ldots, x_n) = \begin{pmatrix} \frac{\partial f_1}{\partial x_1} & \frac{\partial f_1}{\partial x_2} & \cdots & \frac{\partial f_1}{\partial x_n} \\ \frac{\partial f_2}{\partial x_1} & \frac{\partial f_2}{\partial x_2} & \cdots & \frac{\partial f_2}{\partial x_n} \\ \vdots & \vdots & \cdots & \vdots \\ \frac{\partial f_n}{\partial x_1} & \frac{\partial f_n}{\partial x_2} & \cdots & \frac{\partial f_n}{\partial x_n} \end{pmatrix}.$$

Use the `jacobian` function, which is contained in the `linalg` package, to compute the Jacobian matrix for a set of functions.

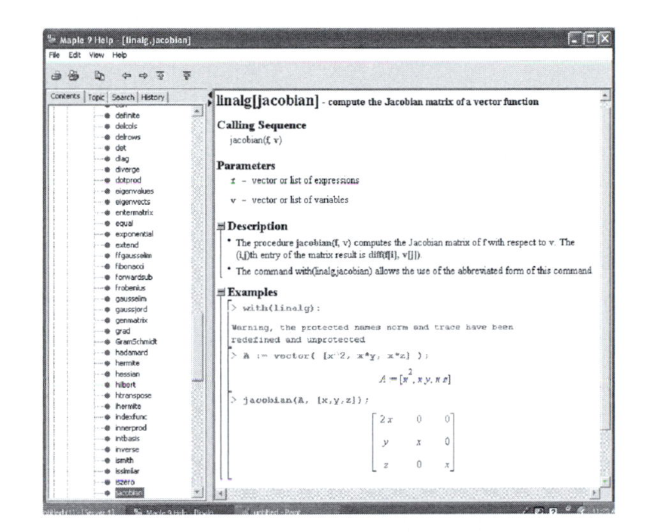

The rest point, E, is **locally stable** if and only if all the eigenvalues of $\mathbf{J}(E)$ have negative real part. If E is not locally stable, E is **unstable**.

EXAMPLE 6.4.5 (Duffing's Equation): Consider the forced **pendulum equation** with damping,

$$x'' + kx' + \omega \sin x = F(t). \tag{6.35}$$

Recall the Maclaurin series for $\sin x$: $\sin x = x - \frac{1}{3!}x^3 + \frac{1}{5!}x^5 - \frac{1}{7!}x^7 + \cdots$. Using $\sin x \approx x$, (6.35) reduces to the linear equation $x'' + kx' + \omega x = F(t)$.

On the other hand, using the approximation $\sin x \approx x - \frac{1}{6}x^3$, we obtain $x'' + kx' + \omega\left(x - \frac{1}{6}x^3\right) = F(t)$. Adjusting the coefficients of x and

x^3 and assuming that $F(t) = F\cos\omega t$ gives us **Duffing's equation**:

$$x'' + kx' + cx + \epsilon x^3 = F\cos\omega t, \tag{6.36}$$

where k and c are positive constants.

Let $y = x'$. Then, $y' = x'' = F\cos\omega t - kx' - cx - \epsilon x^3 = F\cos\omega t - ky - cx - \epsilon x^3$ and we can write (6.36) as the system

$$\begin{aligned} x' &= y \\ y' &= F\cos\omega t - ky - cx - \epsilon x^3. \end{aligned} \tag{6.37}$$

Assuming that $F = 0$ results in the autonomous system

$$\begin{aligned} x' &= y \\ y' &= -cx - \epsilon x^3 - ky. \end{aligned} \tag{6.38}$$

The rest points of system (6.38) are found by solving

$$\begin{aligned} x' &= y \\ y' &= -cx - \epsilon x^3 - ky, \end{aligned}$$

resulting in $E_0 = (0, 0)$.

```
> with(DEtools):
> with(linalg):
```

```
> solve(y=0,-c*x-epsilon*x^3-k*y=0,x,y);
```

$$\{y = 0, x = 0\}, \left\{y = 0, x = RootOf\left(c + \epsilon\,\_Z^2, label = \_L1\right)\right\}$$

We find the Jacobian of (6.38) in `s1`, evaluate the Jacobian at E_0,

```
> s1:=jacobian([y,-c*x-epsilon*x^3-k*y],[x,y]);
```

$$\begin{bmatrix} 0 & 1 \\ -c - 3\,\epsilon\,x^2 & -k \end{bmatrix}$$

```
> s2:=subs(x=0,eval(s1));
```

$$\begin{bmatrix} 0 & 1 \\ -c & -k \end{bmatrix}$$

and then compute the eigenvalues with `eigenvalues`.

```
> s3:=eigenvalues(s2);
```

$$s3 := -1/2\,k + 1/2\,\sqrt{k^2 - 4\,c},\ -1/2\,k - 1/2\,\sqrt{k^2 - 4\,c}$$

Because k and c are positive, $k^2 - 4c < k^2$ so the real part of each eigenvalue is always negative if $k^2 - 4c \neq 0$. Thus, E_0 is locally stable.

For the autonomous system

$$\begin{aligned} x' &= f(x,y) \\ y' &= g(x,y), \end{aligned} \tag{6.39}$$

Bendixson's theorem states that if $f_x(x,y)+g_y(x,y)$ is a continuous function that is either always positive or always negative in a particular region R of the plane, then system (6.39) has no limit cycles in R. For (6.38) we have

$$\frac{d}{dx}(y) + \frac{d}{dy}\left(-cx - \epsilon x^3 - ky\right) = -k,$$

which is always negative. Hence, (6.38) has no limit cycles and it follows that E_0 is globally, asymptotically stable.

```
> diff(y,x)+diff(-c*x-epsilon*x^3-k*y,y);
```

$$-k$$

We use `DEplot` to illustrate two situations that occur. In Figure 6-37(a), we use $c = 1$, $\epsilon = 1/2$, and $k = 3$. In this case, E_0 is a *stable node*. On the other hand, in Figure 6-37(b), we use $c = 10$, $\epsilon = 1/2$, and $k = 3$. In this case, E_0 is a *stable spiral*.

```
> ivals:=seq(-2.5+.5*i,i=0..10):
> i1:=seq([x(0)=2.5,y(0)=i],i=ivals):
> i2:=seq([x(0)=i,y(0)=2.5],i=ivals):
> i3:=seq([x(0)=-2.5,y(0)=i],i=ivals):
> i4:=seq([x(0)=i,y(0)=-2.5],i=ivals):
> DEplot([diff(x(t),t)=y(t),diff(y(t),t)
   =-1*x(t)-1/2*x(t)^3-3*y(t)],
> [x(t),y(t)],t=0..10,[i1,i2,i3,i4],x=-2.5..2.5,
   y=-2.5..2.5,
> scene=[x(t),y(t)],scaling=CONSTRAINED,color=GRAY,
   linecolor=BLACK,
> thickness=1,stepsize=.05);

> ivals:=seq(-1+.25*i,i=0..8): i1:=seq([x(0)=1,y(0)=i],
   i=ivals):
> i2:=seq([x(0)=i,y(0)=1],i=ivals):
> i3:=seq([x(0)=-1,y(0)=i],i=ivals):
```

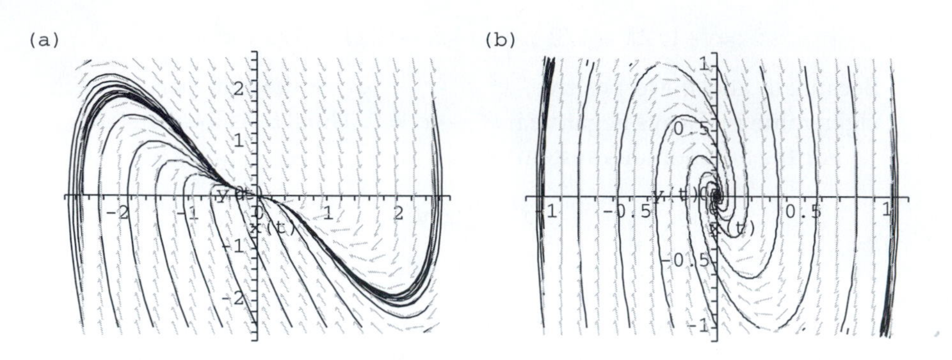

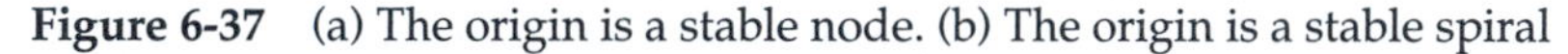

Figure 6-37 (a) The origin is a stable node. (b) The origin is a stable spiral

```
> i4:=seq([x(0)=i,y(0)=-1],i=ivals):
> DEplot([diff(x(t),t)=y(t),diff(y(t),t)
   =-10*x(t)-1/2*x(t)^3-3*y(t)],
> [x(t),y(t)],t=0..10,[i1,i2,i3,i4],x=-1..1,y=-1..1,
> scene=[x(t),y(t)],scaling=CONSTRAINED,color=GRAY,
   linecolor=BLACK,
> thickness=1,stepsize=.01);
```

EXAMPLE 6.4.6 (Predator – Prey): The **predator–prey** equations take the form

$$\frac{dx}{dt} = ax - bxy$$
$$\frac{dy}{dt} = dxy - cy$$

where a, b, c, and d are positive constants. x represents the size of the prey population at time t while y represents the size of the predator population at time t. We use `solve` to calculate the rest points. In this case, there is one boundary rest point, $E_0 = (0,0)$, and one interior rest point, $E_1 = (c/d, a/b)$.

```
> with(linalg):
> with(DEtools):

> rps:=solve(a*x-b*x*y=0,d*x*y-c*y=0,x,y);
```

$$rps := \{x = 0, y = 0\}, \left\{x = \frac{c}{d}, y = \frac{a}{b}\right\}$$

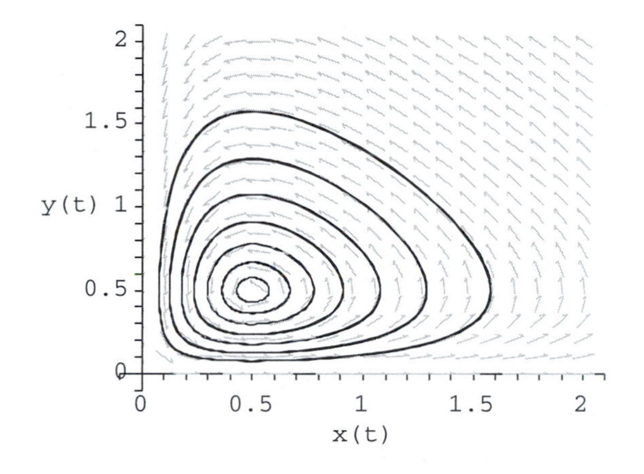

Figure 6-38 Multiple limit cycles about the interior rest point

The Jacobian is then found using `jacobian`.

```
> jac:=jacobian([a*x-b*x*y,d*x*y-c*y],[x,y]);
```

$$jac := matrix\left([[a - by, -xb], [yd, dx - c]]\right)$$

E_0 is unstable because one eigenvalue of $\mathbf{J}(E_0)$ is positive. For the linearized system, E_1 is a center because the eigenvalues of $\mathbf{J}(E_1)$ are complex conjugates.

```
> eigenvalues(subs(rps[1],eval(jac)));
```

$$a, -c$$

```
> eigenvalues(subs(rps[2],eval(jac)));
```

$$\sqrt{-ca}, -\sqrt{-ca}$$

In fact, E_1 is a center for the nonlinear system as illustrated in Figure 6-38, where we have used $a = 1$, $b = 2$, $c = 2$, and $d = 1$. Notice that there are multiple limit cycles around $E_1 = (1/2, 1/2)$.

```
> inits:=seq([x(0)=i/20,y(0)=i/20],i=3..10):

> DEplot([diff(x(t),t)=x(t)-2*x(t)*y(t),
> diff(y(t),t)=2*x(t)*y(t)-y(t)],[x(t),y(t)],
   t=0..50,[inits],
> x=0..2,y=0..2,scene=[x(t),y(t)],scaling=CONSTRAINED,
   color=GRAY,
> linecolor=BLACK,thickness=1,stepsize=.1);
```

In this model, a stable interior rest state is not possible.

The complexity of the behavior of solutions to the system increases based on the assumptions made. Typical assumptions include adding satiation terms for the predator (y) and/or limiting the growth of the prey (x). The **standard predator – prey equations of Kolmogorov type,**

$$\begin{aligned} x' &= \alpha x\left(1-\frac{1}{K}x\right)-\frac{mxy}{a+x} \\ y' &= y\left(\frac{mx}{a+x}-s\right), \end{aligned} \tag{6.40}$$

incorporate both of these assumptions.

We use `solve` to find the three rest points of system (6.40). Let $E_0 = (0,0)$ and $E_1 = (k,0)$ denote the two boundary rest points, and let E_2 represent the interior rest point.

```
> with(linalg):
> with(DEtools):
> rps:=solve(alpha*x*(1-1/k*x)-m*x*y/(a+x)=0,
  y*(m*x/(a+x)-s)=0,x,y);
```

$$rps := \{x=0, y=0\}, \{y=0, x=k\},$$
$$\left\{y = -\frac{\alpha\, a\,(-mk+ks+sa)}{k\,(m^2-2\,ms+s^2)}, x = -\frac{sa}{-m+s}\right\}$$

The Jacobian, $\mathbf{J}$, is calculated next in `s1` with `jacobian`.

```
> s1:=jacobian([alpha*x*(1-1/k*x)-m*x*y/(a+x),
  y*(m*x/(a+x)-s)],[x,y]);
```

$$\begin{bmatrix} 1-2\,x-\frac{y}{1/10+x}+\frac{xy}{(1/10+x)^2} & -\frac{x}{1/10+x} \\ y\left((1/10+x)^{-1}-\frac{x}{(1/10+x)^2}\right) & \frac{x}{1/10+x}-s \end{bmatrix}$$

Because $\mathbf{J}(E_0)$ has a positive eigenvalue, E_0 is unstable.

```
> e0:=subs(rps[1],eval(s1));
```

$$\begin{bmatrix} 1 & 0 \\ 0 & -s \end{bmatrix}$$

```
> eigenvalues(e0);
```

$$\alpha, -s$$

The stability of E_1 is determined by the sign of $m - s - am/(a+k)$.

```
> e1:=subs(rps[2],eval(s1));
```

$$\begin{bmatrix} -1 & -\frac{10}{11} \\ 0 & \frac{10}{11} - s \end{bmatrix}$$

```
> eigs1:=eigenvalues(e1);
```

$$eigs1 := -\alpha, -\frac{-mk + ks + sa}{a + k}$$

The eigenvalues of $J(E_2)$ are quite complicated.

```
> e2:=subs(rps[3],eval(s1)):
> eigenvalues(e2);
```

$$1/2 \frac{s^2\alpha a + s^2\alpha k - s\alpha mk + \alpha sam + \sqrt{\begin{array}{l} s^4\alpha^2a^2 + 2s^4\alpha^2ak + 2s^3\alpha^2a^2m + s^4\alpha^2k^2 - 2s^3\alpha^2k^2m + s^2\alpha^2m^2k^2 \\ -2s^2\alpha^2m^2ka + \alpha^2s^2a^2m^2 + 4mks^4\alpha a + 4mk^2s^4\alpha - 12m^2k^2s^3\alpha \\ -8m^2ks^3\alpha a + 12m^3k^2s^2\alpha + 4m^3ks^2\alpha a - 4m^4k^2s\alpha \end{array}}}{mks - m^2k},$$

$$1/2 \frac{s^2\alpha a + s^2\alpha k - s\alpha mk + \alpha sam - \sqrt{\begin{array}{l} s^4\alpha^2a^2 + 2s^4\alpha^2ak + 2s^3\alpha^2a^2m + s^4\alpha^2k^2 - 2s^3\alpha^2k^2m + s^2\alpha^2m^2k^2 \\ -2s^2\alpha^2m^2ka + \alpha^2s^2a^2m^2 + 4mks^4\alpha a + 4mk^2s^4\alpha - 12m^2k^2s^3\alpha \\ -8m^2ks^3\alpha a + 12m^3k^2s^2\alpha + 4m^3ks^2\alpha a - 4m^4k^2s\alpha \end{array}}}{mks - m^2k}$$

Instead of using the eigenvalues to classify E_2, we compute the characteristic polynomial of $J(E_2)$, $p(\lambda) = c_2\lambda^2 + c_1\lambda + c_0$, and examine the coefficients. Notice that c_2 is always positive.

```
> cpe2:=charpoly(e2,lambda);
```

$$cpe2 := -\frac{s^3\alpha k + s^3\alpha a + s^2\lambda\alpha a + s^2\lambda\alpha k - 2s^2\alpha mk - s^2\alpha am - s\lambda^2km - s\lambda\alpha mk + s\alpha m^2k + s\lambda\alpha ma + \lambda^2km^2}{m(-m+s)k}$$

```
> c0:=simplify(subs(lambda=0,eval(cpe2)));
```

$$c0 := -\frac{(-mk + ks + sa)\,s\alpha}{mk}$$

```
> c1:=simplify(coeff(cpe2,lambda));
```

$$c1 := -\frac{s\alpha\;(sa + ks - mk + am)}{m\,(-m + s)\,k}$$

```
> c2:=simplify(coeff(cpe2,lambda^2));
```

$$c2 := 1$$

On the other hand, c_0 and $m - s - am/(a + k)$ have the same sign because

```
> simplify(c0/eigs1);
```

$$\frac{(-mk+ks+sa)\,s}{mk}$$

is always positive. In particular, if $m - s - am/(a+k) < 0$, E_1 is stable. Because c_0 is negative, by Descartes' rule of signs, it follows that $p(\lambda)$ will have one positive root and hence E_2 will be unstable.

On the other hand, if $m - s - am/(a+k) > 0$ so that E_1 is unstable, E_2 may be either stable or unstable. To illustrate these two possibilities let $\alpha = K = m = 1$ and $a = 1/10$. We recalculate.

```
> alpha:=1:k:=1:m:=1:a:=1/10:
> rps:=solve(alpha*x*(1-1/k*x)-m*x*y/(a+x)=0,
  y*(m*x/(a+x)-s)=0,x,y);
```

$$rps := \{x=0, y=0\}, \{x=1, y=0\},$$
$$\left\{x = -1/10\,\frac{s}{-1+s}, y = -\frac{1}{100}\,\frac{-10+11\,s}{(-1+s)^2}\right\}$$

```
> s1:=jacobian([alpha*x*(1-1/k*x)-m*x*y/(a+x),
  y*(m*x/(a+x)-s)],[x,y]);
```

$$\begin{bmatrix} 1-2\,x-\frac{y}{1/10+x}+\frac{xy}{(1/10+x)^2} & -\frac{x}{1/10+x} \\ y\left((1/10+x)^{-1}-\frac{x}{(1/10+x)^2}\right) & \frac{x}{1/10+x}-s \end{bmatrix}$$

```
> e2:=subs(rps[3],eval(s1)):
> cpe2:=charpoly(e2,lambda);
```

$$cpe2 := -1/10\,\frac{10\,\lambda^2-10\,\lambda^2 s-9\,\lambda\,s+11\,\lambda\,s^2+10\,s-21\,s^2+11\,s^3}{-1+s}$$

```
> c0:=simplify(subs(lambda=0,cpe2));
```

$$c0 := -1/10\,s\,(-10+11\,s)$$

```
> c1:=simplify(coeff(cpe2,lambda));
```

$$c1 := -1/10\,\frac{s\,(-9+11\,s)}{-1+s}$$

```
> c2:=simplify(coeff(cpe2,lambda^2));
```

$$c2 := 1$$

Using `solve`, we see that

1. c_0, c_1, and c_2 are positive if $9/11 < s < 10/11$, and
2. c_0 and c_2 are positive and c_1 is negative if $0 < s < 9/11$.

```
> solve(c0>0 and c1>0,s);
```

$$\left\{\frac{9}{11} < s, s < \frac{10}{11}\right\}$$

```
> solve(c0>0 and c1 <0,s);
```

$$\left\{0 < s, s < \frac{9}{11}\right\}$$

In the first situation, E_2 is stable; in the second, E_2 is unstable.

Using $s = 19/22$, we graph the direction field associated with the system as well as various solutions in Figure 6-39. In the plot, notice that all nontrivial solutions approach $E_2 \approx (.63, .27)$; E_2 is stable – a situation that cannot occur with the standard predator–prey equations.

```
> subs(s=19/22,rps[3]);
```

$$\left\{x = \frac{19}{30}, y = \frac{121}{450}\right\}$$

```
> ivals:=seq(i/14,i=0..14):
> i1:=seq([x(0)=1,y(0)=i],i=ivals):
> i2:=seq([x(0)=i,y(0)=1],i=ivals):
> DEplot([diff(x(t),t)=alpha*x(t)*(1-1/k*x(t))
   -m*x(t)*y(t)/(a+x(t)),
> diff(y(t),t)=y(t)*(m*x(t)/(a+x(t))-19/22)],
> [x(t),y(t)],t=0..25,[i1,i2],x=0..1,y=0..1,
   scene=[x(t),y(t)],
> scaling=CONSTRAINED,color=GRAY,linecolor=BLACK,
> thickness=1,stepsize=.075);
```

On the other hand, using $s = 8/11$ (so that E_2 is unstable) in Figure 6-40 we see that all nontrivial solutions appear to approach a limit cycle.

```
> DEplot([diff(x(t),t)=alpha*x(t)*(1-1/k*x(t))
   -m*x(t)*y(t)/(a+x(t)),
> diff(y(t),t)=y(t)*(m*x(t)/(a+x(t))-8/11)],
> [x(t),y(t)],t=0..50,[i1,i2],x=0..1,y=0..1,
   scene=[x(t),y(t)], > scaling=CONSTRAINED,color=GRAY,
   linecolor=BLACK,thickness=1,stepsize=.075);
```

The limit cycle is shown more clearly in Figure 6-41.

```
> DEplot([diff(x(t),t)=alpha*x(t)*(1-1/k*x(t))
   -m*x(t)*y(t)/(a+x(t)),
> diff(y(t),t)=y(t)*(m*x(t)/(a+x(t))-8/11)],
> [x(t),y(t)],t=0..50,[[x(0)=.759,y(0)=.262]],x=0..1,y=0..1,
   scene=[x(t),y(t)],
```

```
> scaling=CONSTRAINED,color=GRAY,linecolor=BLACK,
   thickness=1,
> arrows=NONE,stepsize=.075);
```

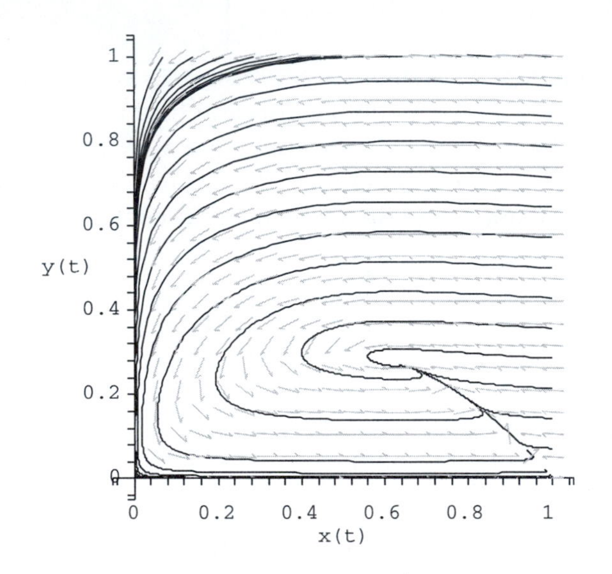

Figure 6-39 $s = 19/22$

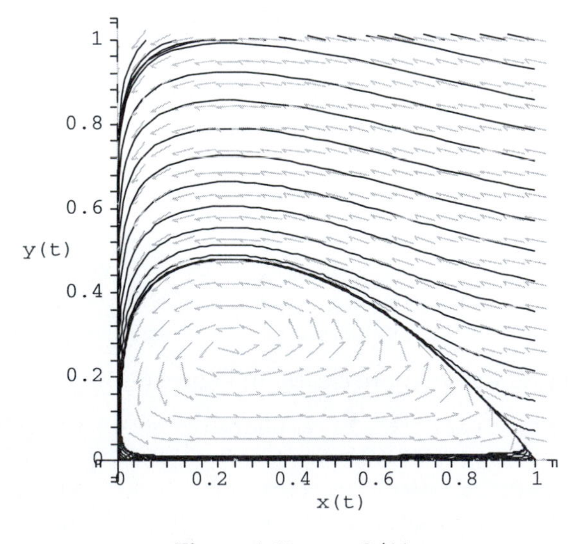

Figure 6-40 $s = 8/11$

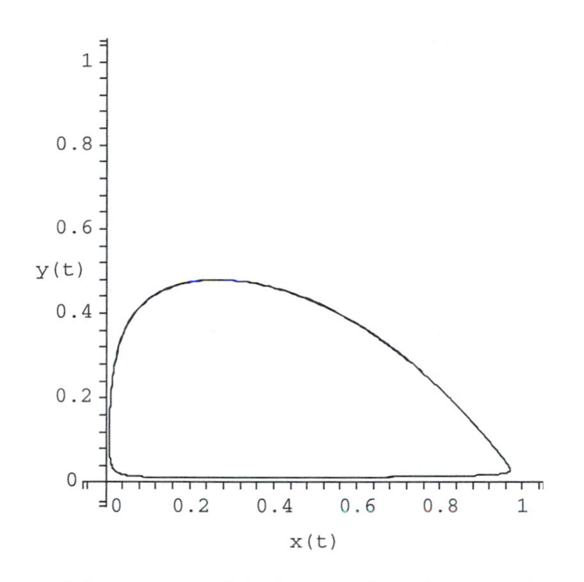

Figure 6-41 A better view of the limit cycle without the direction field

Also see Example 6.4.4.

EXAMPLE 6.4.7 (Van-der-Pol's equation): In Example 6.4.4 we saw that **Van-der-Pol's equation** $x'' + \mu\left(x^2 - 1\right)x' + x = 0$ is equivalent to the system

$$\begin{cases} x' = y \\ y' = \mu\left(1 - x^2\right)y - x \end{cases}.$$

Classify the equilibrium points, use `dsolve` with the `numeric` option, to approximate the solutions to this nonlinear system, and plot the phase plane.

SOLUTION: We find the equilibrium points by solving $\begin{cases} y = 0 \\ \mu\left(1 - x^2\right)y - x = 0 \end{cases}$. From the first equation, we see that $y = 0$. Then, substitution of $y = 0$ into the second equation yields $x = 0$. Therefore, the only equilibrium point is $(0, 0)$. The Jacobian matrix for this system is

$$\mathrm{J}(x, y) = \begin{pmatrix} 0 & 1 \\ -1 - 2\mu xy & -\mu\left(x^2 - 1\right) \end{pmatrix}.$$

The eigenvalues of $\mathbf{J}(0,0)$ are $\lambda_{1,2} = \frac{1}{2}\left(\mu \pm \sqrt{\mu^2 - 4}\right)$.

```
> with(DEtools):
> with(linalg):
> f:=(x,y)->y:
> g:=(x,y)->-x-mu*(x^2-1)*y:
> jac:=jacobian([f(x,y),g(x,y)],[x,y]);
```

$$\begin{bmatrix} 0 & 1 \\ -1 - 2\,\mu\,xy & -\mu\,\left(x^2 - 1\right) \end{bmatrix}$$

```
> eigenvalues(subs([x=0,y=0],jac));
```

$$-1/2\,\mu\,x^2 + 1/2\,\mu + 1/2\,\sqrt{\mu^2 x^4 - 2\,\mu^2 x^2 + \mu^2 - 4 - 8\,\mu\,xy},$$
$$-1/2\,\mu\,x^2 + 1/2\,\mu - 1/2\,\sqrt{\mu^2 x^4 - 2\,\mu^2 x^2 + \mu^2 - 4 - 8\,\mu\,xy}$$

Alternatively, the sequence of commands

```
> lin_mat:=array([[0,1],[-1,mu]]):
> with(linalg):
> eigs:=eigenvals(lin_mat);
```

$$eigs := 1/2\,\mu + 1/2\,\sqrt{\mu^2 - 4},\, 1/2\,\mu - 1/2\,\sqrt{\mu^2 - 4}$$

gives us the same result.

Notice that if $\mu > 2$, then both eigenvalues are positive and real. Hence, we classify $(0,0)$ as an **unstable node**. On the other hand, if $0 < \mu < 2$, then the eigenvalues are a complex conjugate pair with a positive real part. Hence, $(0,0)$ is an **unstable spiral**. (We omit the case $\mu = 2$ because the eigenvalues are repeated.)

```
> sys:=mu->[diff(x(t),t)=y(t),diff(y(t),t)
   =mu*(1-x(t)^2)*y(t)-x(t)];
```

$$sys := \mu \mapsto \left[\frac{d}{dt}x(t) = y(t), \frac{d}{dt}y(t) = \mu\,\left(1 - (x(t))^2\right) y(t) - x(t)\right]$$

We now show several curves in the phase plane that begin at various points for various values of μ by using `seq` to generate a list of ordered pairs that will correspond to the initial conditions in the initial-value problem.

```
> inits1:=seq([x(0)=0.1*cos(2*Pi*i/4),
   y(0)=0.1*sin(2*Pi/4)],i=0..4);
```

$$inits1 := \{[x(0) = 0.1, y(0) = 0.1], [x(0) = 0.0, y(0) = 0.1], [x(0) = -0.1, y(0) = 0.1]\}$$

```
> inits2:=seq([x(0)=-5,y(0)=-5+10*i/9],i=0..9);
```

$$inits2 := \left\{\left[x(0) = -5, y(0) = \frac{35}{9}\right], [x(0) = -5, y(0) = 5], [x(0) = -5, y(0) = -5], \left[x(0) = -5, y(0) = -\frac{35}{9}\right], \left[x(0) = -5, y(0) = -\frac{25}{9}\right], [x(0) = -5, y(0) = -5/3], [x(0) = -5, y(0) = -5/9], [x(0) = -5, y(0) = 5/9], [x(0) = -5, y(0) = 5/3], \left[x(0) = -5, y(0) = \frac{25}{9}\right]\right\}$$

```
> inits3:=seq([x(0)=5,y(0)=-5+10*i/9],i=0..9):
> inits4:=seq([x(0)=-5+10*i/9,y(0)=-5],i=0..9):
> inits5:=seq([x(0)=-5+10*i/9,y(0)=5],i=0..9):
> initconds:=`union`(inits1,inits2,inits3,inits4,inits5):
> nops(initconds);
```

$$39$$

We then use `phaseportrait` in the same way as we use `DEplot` to graph various solutions.

```
> phaseportrait(sys(1/2),[x(t),y(t)],t=0..20,initconds,
   x=-5..5,y=-5..5,
> arrows=NONE,linecolor=BLACK,stepsize=0.05);
> phaseportrait(sys(1),[x(t),y(t)],t=0..20,initconds,
   x=-5..5,y=-5..5,
> arrows=NONE,linecolor=BLACK,stepsize=0.05);
> phaseportrait(sys(3/2),[x(t),y(t)],t=0..20,initconds,
   x=-5..5,y=-5..5,
> arrows=NONE,linecolor=BLACK,stepsize=0.05);
> phaseportrait(sys(3),[x(t),y(t)],t=0..20,initconds,
   x=-5..5,y=-5..5,
> arrows=NONE,linecolor=BLACK,stepsize=0.05);
```

We show all four graphs together in Figure 6-42. In each figure, we see that all of the curves approach a curve called a *limit cycle*. Physically, the fact that the system has a limit cycle indicates that for all oscillations,

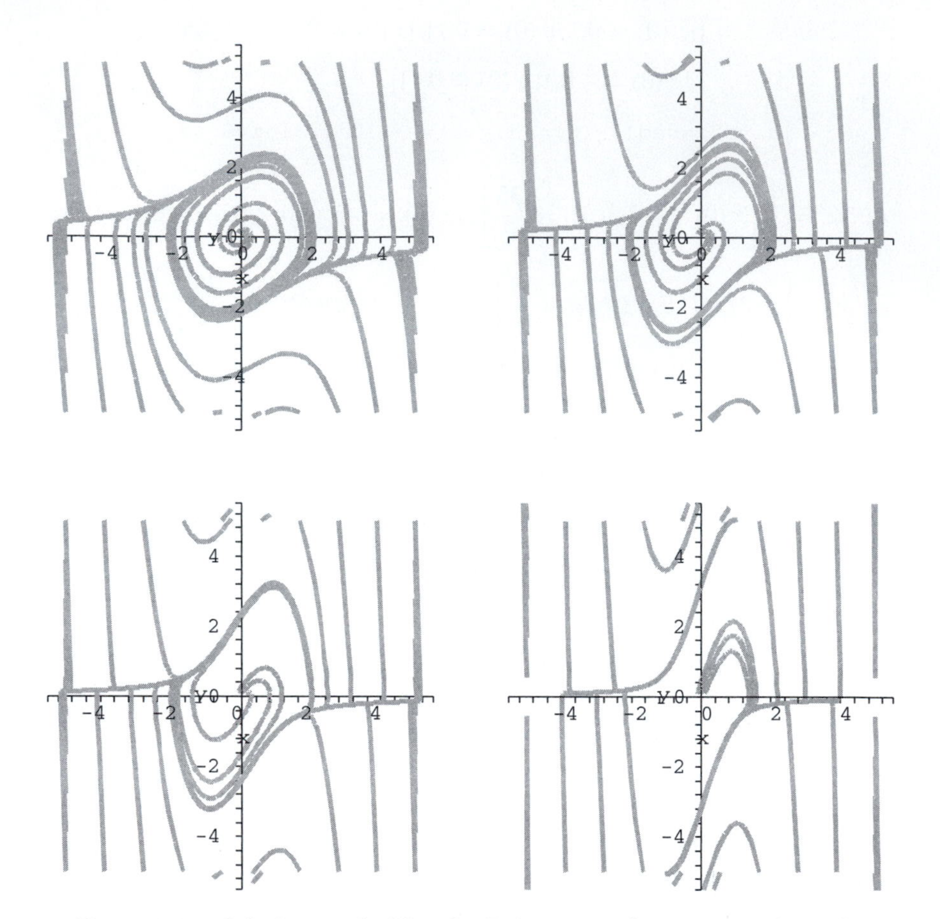

Figure 6-42 Solutions to the Van-der-Pol equation for various values of μ

the motion eventually becomes periodic, which is represented by a closed curve in the phase plane.

■

6.5 Some Partial Differential Equations

We now turn our attention to several partial differential equations. Several examples in this section will take advantage of commands contained in the `PDEtools` package. Information regarding the functions contained in the `PDEtools` package is obtained with `?PDEtools`.

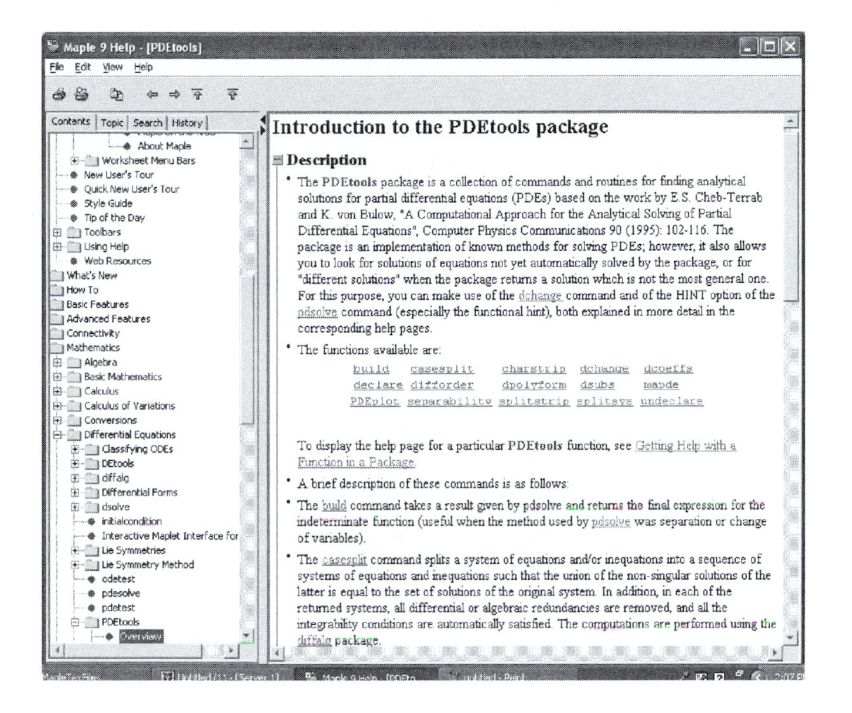

6.5.1 The One-Dimensional Wave Equation

Suppose that we pluck a string (like a guitar or violin string) of length p and constant mass density that is fixed at each end. A question that we might ask is: What is the position of the string at a particular instance of time? We answer this question by modeling the physical situation with a partial differential equation, namely the wave equation in one spatial variable:

$$c^2 \frac{\partial^2 u}{\partial x^2} = \frac{\partial^2 u}{\partial t^2} \qquad \text{or} \qquad c^2 u_{xx} = u_{tt}. \tag{6.41}$$

In (6.41), $c^2 = T/\rho$, where T is the tension of the string and ρ is the constant mass of the string per unit length. The solution $u(x,t)$ represents the displacement of the string from the x-axis at time t. To determine u we must describe the boundary and initial conditions that model the physical situation. At the ends of the string, the displacement from the x-axis is fixed at zero, so we use the homogeneous boundary conditions $u(0,t) = u(p,t) = 0$ for $t > 0$. The motion of the string also depends on the displacement and the velocity at each point of the string at $t = 0$. If the initial displacement is given by $f(x)$ and the initial velocity by $g(x)$, we have the initial conditions $u(x,0) = f(x)$ and $u_t(x,0) = g(x)$ for $0 \leq x \leq p$. Therefore, we determine the displacement of the string with the initial-boundary-value problem

$$\begin{cases} c^2 \dfrac{\partial^2 u}{\partial x^2} = \dfrac{\partial^2 u}{\partial t^2}, \ 0 < x < p, \ t > 0 \\ u(0,t) = u(p,t) = 0, \ t > 0 \\ u(x,0) = f(x), \ u_t(x,0) = g(x), \ 0 < x < p \end{cases} . \tag{6.42}$$

This problem is solved through separation of variables by assuming that $u(x,t) = X(x)T(t)$. Substitution into (6.41) yields

λ is a constant.

$$c^2 X''T = XT'' \qquad \text{or} \qquad \frac{X''}{X} = \frac{T''}{c^2 T} = -\lambda$$

so we obtain the two second-order ordinary differential equations $X'' + \lambda X = 0$ and $T'' + c^2 \lambda T = 0$. At this point, we solve the equation that involves the homogeneous boundary conditions. The boundary conditions in terms of $u(x,t) = X(x)T(t)$ are $u(0,t) = X(0)T(t) = 0$ and $u(p,t) = X(p)T(t) = 0$, so we have $X(0) = 0$ and $X(p) = 0$. Therefore, we determine $X(x)$ by solving the *eigenvalue problem*

$$\begin{cases} X'' + \lambda X = 0, \ 0 < x < p \\ X(0) = X(p) = 0 \end{cases} .$$

The eigenvalues of this problem are $\lambda_n = (n\pi/p)^2$, $n = 1, 2, 3, \ldots$ with corresponding eigenfunctions $X_n(x) = \sin(n\pi x/p)^2$, $n = 1, 2, 3, \ldots$. Next, we solve the equation $T'' + c^2 \lambda_n T = 0$. A general solution is

$$T_n(t) = a_n \cos\left(c\sqrt{\lambda_n} t\right) + b_n \sin\left(c\sqrt{\lambda_n} t\right) = a_n \cos\frac{cn\pi t}{p} + b_n \sin\frac{cn\pi t}{p},$$

where the coefficients a_b and b_n must be determined. Putting this information together, we obtain

$$u_n(x,t) = \left(a_n \cos\frac{cn\pi t}{p} + b_n \sin\frac{cn\pi t}{p}\right) \sin\frac{n\pi x}{p},$$

so by the Principle of Superposition, we have

$$u(x,t) = \sum_{n=1}^{\infty} \left(a_n \cos\frac{cn\pi t}{p} + b_n \sin\frac{cn\pi t}{p}\right) \sin\frac{n\pi x}{p}.$$

Applying the initial displacement $u(x,0) = f(x)$ yields

$$u(x,0) = \sum_{n=1}^{\infty} a_n \sin\frac{n\pi x}{p} = f(x),$$

so a_n is the *Fourier sine series coefficient* for $f(x)$, which is given by

$$a_n = \frac{2}{p}\int_0^p f(x)\sin\frac{n\pi x}{p}\,dx, \quad n = 1, 2 \ldots$$

In order to determine b_n, we must use the initial velocity. Therefore, we compute

$$\frac{\partial u}{\partial t}(x,t) = \sum_{n=1}^{\infty}\left(-a_n\frac{cn\pi}{p}\sin\frac{cn\pi t}{p} + b_n\frac{cn\pi}{p}\cos\frac{cn\pi t}{p}\right)\sin\frac{n\pi x}{p}.$$

Then,

$$\frac{\partial u}{\partial t}(x,0) = \sum_{n=1}^{\infty} b_n\frac{cn\pi}{p}\sin\frac{n\pi x}{p} = g(x)$$

so $b_n\frac{cn\pi}{p}$ represents the Fourier sine series coefficient for $g(x)$ which means that

$$b_n = \frac{p}{cn\pi}\int_0^p g(x)\sin\frac{n\pi x}{p}\,dx, \quad n = 1, 2 \ldots$$

EXAMPLE 6.5.1: Solve $\begin{cases} u_{xx} = u_{tt},\ 0 < x < 1,\ t > 0 \\ u(0,t) = u(1,t) = 0,\ t > 0 \\ u(x,0) = \sin\pi x,\ u_t(x,0) = 3x + 1,\ 0 < x < 1 \end{cases}$.

SOLUTION: The initial displacement and velocity functions are defined first.

```
> f:=x->sin(Pi*x):
> g:=x->3*x+1:
```

Next, the functions to determine the coefficients a_n and b_n in the series approximation of the solution $u(x,t)$ are defined. Here, $p = c = 1$.

```
> a[1]:=2*int(f(x)*sin(Pi*x),x=0..1);
```

$$a_1 := 1$$

```
> n:='n':
> a[n]:=2*int(f(x)*sin(n*Pi*x),x=0..2);
```

$$a_n := 4\,\frac{\sin(n\pi)\cos(n\pi)}{\pi\left(-1+n^2\right)}$$

Because n represents an integer, these results indicate that $a_n = 0$ for all $n \geq 2$. In fact, when we instruct Maple to assume that n is an integer with `assume`, Maple determines that $a_n = 0$.

```
> assume(n,integer):
> a[n]:=2*int(f(x)*sin(n*Pi*x),x=0..2);
```

$$a_n := 0$$

Similarly, assuming that n is an integer, we see that $b_n = \frac{2\left(1+(-4)^{n+1}\right)}{\pi^2 n^2}$.

```
> b[n]:=2/(n*Pi)*int(g(x)*sin(n*Pi*x),x=0..1);
```

$$b_n := -2\,\frac{-1+4\,(-1)^n}{n^2\pi^2}$$

The function u defined next computes the nth term in the series expansion. Thus, uapprox determines the approximation of order k by summing the first k terms of the expansion, as illustrated with approx(10).

Notice that we define uapprox(n) so that Maple "remembers" the terms uapprox that are computed. That is, Maple need not recompute uapprox(n-1) to compute uapprox(n) provided that uapprox(n-1) has already been computed.

```
> u:=n->-2*(4*(-1)^n-1)/(n^2*Pi^2)*sin(n*Pi*t)*sin(n*Pi*x):
```

```
> uapprox:=proc(n) option remember;
> uapprox(n-1)+u(n);
> end:
> uapprox(0):=cos(Pi*t)*sin(Pi*x):
```

```
> uapprox(10);
```

$$\cos(\pi\,t)\sin(\pi\,x)+10\,\frac{\sin(\pi\,t)\sin(\pi\,x)}{\pi^2}-3/2\,\frac{\sin(2\pi\,t)\sin(2\pi\,x)}{\pi^2}$$
$$+\frac{10}{9}\,\frac{\sin(3\pi\,t)\sin(3\pi\,x)}{\pi^2}-3/8\,\frac{\sin(4\pi\,t)\sin(4\pi\,x)}{\pi^2}+2/5\,\frac{\sin(5\pi\,t)\sin(5\pi\,x)}{\pi^2}$$
$$-1/6\,\frac{\sin(6\pi\,t)\sin(6\pi\,x)}{\pi^2}+\frac{10}{49}\,\frac{\sin(7\pi\,t)\sin(7\pi\,x)}{\pi^2}-\frac{3}{32}\,\frac{\sin(8\pi\,t)\sin(8\pi\,x)}{\pi^2}$$
$$+\frac{10}{81}\,\frac{\sin(9\pi\,t)\sin(9\pi\,x)}{\pi^2}-\frac{3}{50}\,\frac{\sin(10\pi\,t)\sin(10\pi\,x)}{\pi^2}$$

To illustrate the motion of the string, we graph uapprox(10), the tenth partial sum of the series, on the interval [0, 1] for 16 equally spaced values of t between 0 and 2. One frame from the resulting animation is shown in Figure 6-43.

```
> with(plots):
> animate(uapprox(10),x=0..1,t=0..2,frames=16);
```

On the other hand, entering

```
> anarray:=animate(uapprox(10),x=0..1,t=0..2,frames=16,
  color=BLACK):
> display(anarray);
```

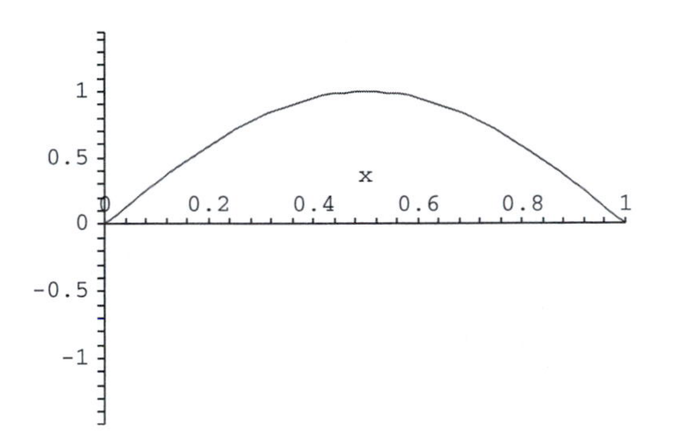

Figure 6-43 The motion of the spring for 16 equally spaced values of t between 0 and 2

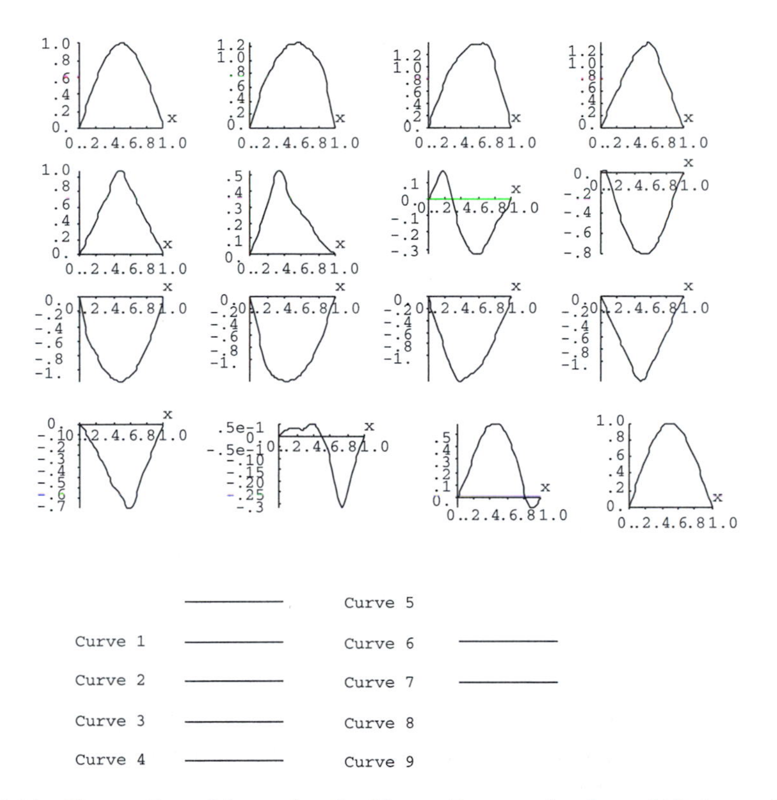

Figure 6-44 The motion of the spring for 16 equally spaced values of t between 0 and 2

graphs `uapprox(10)`, for 16 equally spaced values of t between 0 and 2 with `animate` and then displays the resulting graphs as the array shown in Figure 6-44.

`pdsolve` attempts to solve partial differential equations in the same way that `dsolve` attempts to solve ordinary differential equations.

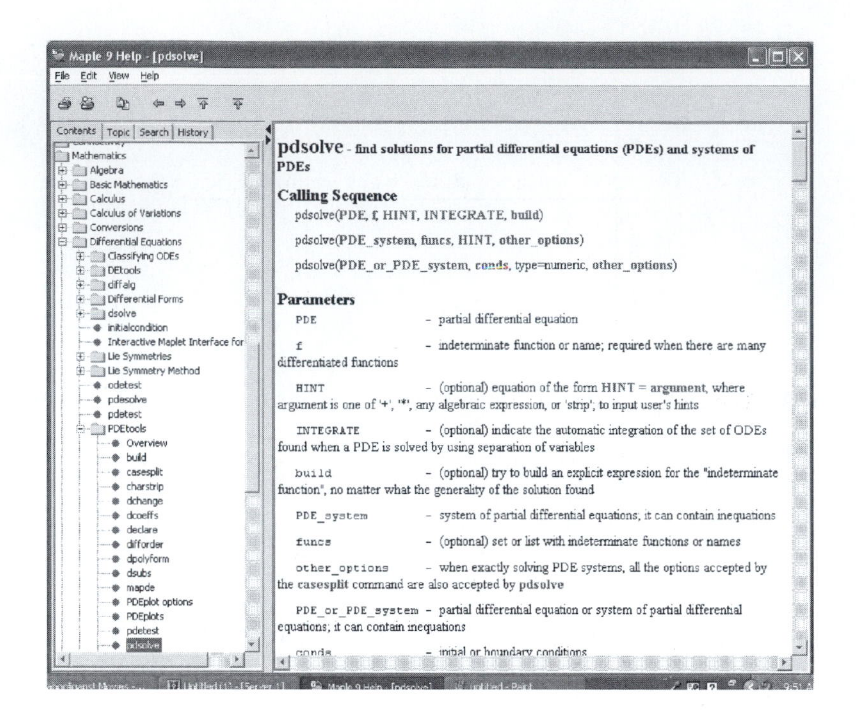

Using `pdsolve`, we obtain D'Alembert's solution.

```
> pdsolve(pde);
```

$$u(x,t) = \_F1(t+x) + \_F2(t-x)$$

■

6.5.2 The Two-Dimensional Wave Equation

One of the more interesting problems involving two spatial dimensions (x and y) is the wave equation. The two-dimensional wave equation in a circular region which is radially symmetric (not dependent on θ) with boundary and initial conditions is expressed in polar coordinates as

$$\begin{cases} c^2\left(\dfrac{\partial^2 u}{\partial r^2} + \dfrac{1}{r}\dfrac{\partial u}{\partial r}\right) = \dfrac{\partial^2 u}{\partial t^2},\ 0 < r < \rho,\ t > 0 \\ u(\rho,t) = 0,\ |u(0,t)| < \infty,\ t > 0 \\ u(r,0) = f(r),\ \dfrac{\partial u}{\partial t}(r,0) = g(r),\ 0 < r < \rho \end{cases}.$$

Notice that the boundary condition $u(\rho, t) = 0$ indicates that u is fixed at zero around the boundary; the condition $|u(0, t)| < \infty$ indicates that the solution is bounded at the center of the circular region. Like the wave equation discussed previously, this problem is typically solved through separation of variables by assuming a solution of the form $u(r, t) = F(r)G(t)$. Applying separation of variables yields the solution

$$u(r,t) = \sum_{n=1}^{\infty} \left(A_n \cos ck_n t + B_n \sin ck_n t\right) J_0\left(k_n r\right),$$

where $\lambda_n = c\alpha_n / \rho$, and the coefficients A_n and B_n are found through application of the initial displacement and velocity functions. With

α_n represents the nth zero of the Bessel function of the first kind of order zero.

$$u(r,0) = \sum_{n=1}^{\infty} A_n J_0\left(k_n r\right) = f(r)$$

and the orthogonality conditions of the Bessel functions, we find that

$$A_n = \frac{\int_0^\rho r\, f(r) J_0\left(k_n r\right) dr}{\int_0^\rho r \left[J_0\left(k_n r\right)\right]^2 dr} = \frac{2}{\left[J_1\left(\alpha_n\right)\right]^2} \int_0^\rho r\, f(r) J_0\left(k_n r\right) dr,\ n = 1, 2, \ldots$$

Similarly, because

$$\frac{\partial u}{\partial t}(r,0) = \sum_{n=1}^{\infty} \left(-ck_n A_n \sin ck_n t + ck_n B_n \cos ck_n t\right) J_0\left(k_n r\right)$$

we have

$$u_t(r,0) = \sum_{n=1}^{\infty} ck_n B_n J_0\left(k_n r\right) = g(r).$$

Therefore,

$$B_n = \frac{\int_0^\rho r\, g(r) J_0\left(k_n r\right) dr}{ck_n \int_0^\rho r \left[J_0\left(k_n r\right)\right]^2 dr} = \frac{2}{ck_n \left[J_1\left(\alpha_n\right)\right]^2} \int_0^\rho r\, g(r) J_0\left(k_n r\right) dr,\ n = 1, 2, \ldots$$

As a practical matter, in nearly all cases, these formulas are difficult to evaluate.

EXAMPLE 6.5.2: Solve $\begin{cases} \dfrac{\partial^2 u}{\partial r^2} + \dfrac{1}{r}\dfrac{\partial u}{\partial r} = \dfrac{\partial^2 u}{\partial t^2},\ 0 < r < 1,\ t > 0 \\ u(1,t) = 0,\ |u(0,t)| < \infty,\ t > 0 \\ u(r,0) = r(r-1),\ \dfrac{\partial u}{\partial t}(r,0) = \sin \pi r,\ 0 < r < 1 \end{cases}.$

SOLUTION: In this case, $\rho = 1, f(r) = r(r-1)$, and $g(r) = \sin \pi r$. To calculate the coefficients, we will need to have approximations of the zeros of the Bessel functions, so we re-enter the table of zeros that were found earlier in Chapter 4. Then, for $1 \le n \le 8$, α_n is the nth zero of J_0.

```
> alpha:=array([2.4048, 5.5201, 8.6537, 11.792,
> 14.931, 18.071, 21.212, 24.352]):
```

```
> alpha[1];
```

$$2.4048$$

Next, we define the constants ρ and c and the functions $f(r) = r(r-1)$, $g(r) = \sin \pi r$, and $k_n = \alpha_n/\rho$.

```
> c:=1:
> rho:=1:
> f:=r->r*(r-1):
> g:=r->sin(Pi*r):
> k:=n->alpha[n]/rho:
```

The formulas for the coefficients A_n and B_n are then defined so that an approximate solution may be determined. (We use lower-case letters to avoid any possible ambiguity with built-in Maple functions.) Note that we use `evalf` and `Int` to approximate the coefficients and avoid the difficulties in integration associated with the presence of the Bessel function of order zero.

```
> a:=proc(n) option remember;
> 2/BesselJ(1,alpha[n])^2*evalf(Int(r*f(r)*BesselJ
   (0,k(n)*r),r=0..rho))
> end:
```

```
> b:=proc(n) option remember;
> 2/(c*k(n)*BesselJ(1,alpha[n])^2)*evalf(Int(r*g(r)
   *BesselJ(0,k(n)*r),r=0..rho))
> end:
```

We now compute the first eight values of A_n and B_n. Because `a` and `b` are defined using `proc` with the `remember` option, Maple remembers these values for later use.

```
> array([seq([n,a(n),b(n)],n=1..8)]);
```

$$\begin{bmatrix} 1 & -0.3235010276 & 0.5211819702 \\ 2 & 0.2084692034 & -0.1457773395 \\ 3 & 0.007640292446 & -0.01342290349 \\ 4 & 0.03838004574 & -0.008330225220 \\ 5 & 0.005341000922 & -0.002504216150 \\ 6 & 0.01503575901 & -0.002082788164 \\ 7 & 0.003340078858 & -0.0008805687934 \\ 8 & 0.007857367112 & -0.0008134612340 \end{bmatrix}$$

The nth term of the series solution is defined in u. Then, an approximate solution is obtained in uapprox by summing the first eight terms of u.

```
> u:='u':n:='n':
> u:=(n,r,t)->(a(n)*cos(c*k(n)*t)+b(n)*sin(c*k(n)*t))
   *BesselJ(0,k(n)*r):
> uapprox:=sum('u(n,r,t)','n'=1..8);
```

$$\begin{aligned} uapprox := & (-0.3235010276 \cos(2.4048\, t) + 0.5211819702 \sin(2.4048\, t))\, BesselJ(0, 2.4048\, r) \\ & + (0.2084692034 \cos(5.5201\, t) - 0.1457773395 \sin(5.5201\, t))\, BesselJ(0, 5.5201\, r) \\ & + (0.007640292446 \cos(8.6537\, t) - 0.01342290349 \sin(8.6537\, t))\, BesselJ(0, 8.6537\, r) \\ & + (0.03838004574 \cos(11.792\, t) - 0.008330225220 \sin(11.792\, t))\, BesselJ(0, 11.792\, r) \\ & + (0.005341000922 \cos(14.931\, t) - 0.002504216150 \sin(14.931\, t))\, BesselJ(0, 14.931\, r) \\ & + (0.01503575901 \cos(18.071\, t) - 0.002082788164 \sin(18.071\, t))\, BesselJ(0, 18.071\, r) \\ & + (0.003340078858 \cos(21.212\, t) - 0.0008805687934 \sin(21.212\, t))\, BesselJ(0, 21.212\, r) \\ & + (0.007857367112 \cos(24.352\, t) - 0.0008134612340 \sin(24.352\, t))\, BesselJ(0, 24.352\, r) \end{aligned}$$

We graph uapprox for several values of t in Figure 6-45.

```
> with(plots):
> drumhead:=animate3d([r*cos(theta),r*sin(theta),uapprox],
   r=0..1,theta=-Pi..Pi,t=0..1.5,
> frames=9):

> display(drumhead);
```

In order to actually watch the drumhead move, we can use animate loop to generate an animation. Be aware, however, that generating many three-dimensional graphics and then animating the results uses a great deal of memory and can take considerable time, even on a relatively powerful computer. We show one frame from the animation that results from the following animation command in Figure 6-46.

```
> ?animate3d
```

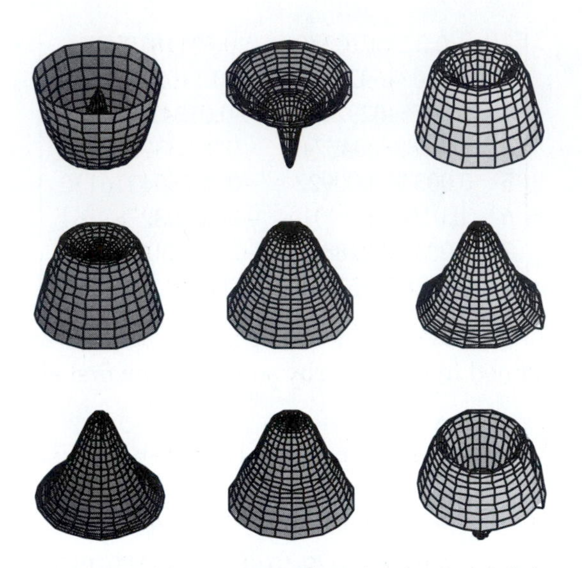

Figure 6-45 The drumhead for nine equally spaced values of t between 0 and 1.5

Figure 6-46 A drumhead

```
> tvals:=[seq(1.5*i/8,i=0..9)]:
> animate3d([r*cos(theta),r*sin(theta),uapprox],r=0..1,
> theta=-Pi..Pi,t=0..1.5,frames=9);
```

■

If the displacement of the drumhead is not radially symmetric, the problem that describes the displacement of a circular membrane in its general case is

$$\begin{cases} c^2\left(\dfrac{\partial^2 u}{\partial r^2}+\dfrac{1}{r}\dfrac{\partial u}{\partial r}+\dfrac{1}{r^2}\dfrac{\partial^2 u}{\partial \theta^2}\right)=\dfrac{\partial^2 u}{\partial t^2},\ 0<r<\rho,\ -\pi<\theta<\pi,\ t>0 \\ u(\rho,\theta,t)=0,\ |u(0,\theta,t)|<\infty,\ -\pi\le\theta\le\pi,\ t>0 \\ u(r,\pi,t)=u(r,-\pi,t),\ \dfrac{\partial u}{\partial \theta}(r,\pi,t)=\dfrac{\partial u}{\partial \theta}(r,-\pi,t),\ 0<r<\rho,\ t>0 \\ u(r,\theta,0)=f(r,\theta),\ \dfrac{\partial u}{\partial t}(r,\pi,0)=g(r,\theta),\ 0<r<\rho,\ -\pi<\theta<\pi \end{cases}. \tag{6.43}$$

Using separation of variables and assuming that $u(r,\theta,t) = R(t)H(\theta)T(t)$, we obtain that a general solution is given by

$$u(r,\theta,t) = \sum_n a_{0n} J_0 (\lambda_{0n} r) \cos (\lambda_{0n} ct) + \sum_{m,n} a_{mn} J_m (\lambda_{mn} r) \cos (m\theta) \cos (\lambda_{mn} ct)$$
$$+ \sum_{m,n} b_{mn} J_m (\lambda_{mn} r) \sin (m\theta) \cos (\lambda_{mn} ct) + \sum_n A_{0n} J_0 (\lambda_{0n} r) \sin (\lambda_{0n} ct)$$
$$+ \sum_{m,n} A_{mn} J_m (\lambda_{mn} r) \cos (m\theta) \sin (\lambda_{mn} ct)$$
$$+ \sum_{m,n} B_{mn} J_m (\lambda_{mn} r) \sin (m\theta) \sin (\lambda_{mn} ct),$$

where J_m represents the mth Bessel function of the first kind, α_{mn} denotes the nth zero of the Bessel function $y = J_m(x)$, and $\lambda_{mn} = \alpha_{mn}/\rho$. The coefficients are given by the following formulas.

$$a_{0n} = \frac{\int_0^{2\pi} \int_0^{\rho} f(r,\theta) J_0 (\lambda_{0n} r)\, r\, dr\, d\theta}{2\pi \int_0^{\rho} [J_0 (\lambda_{0n} r)]^2\, r\, dr} \qquad a_{mn} = \frac{\int_0^{2\pi} \int_0^{\rho} f(r,\theta) J_m (\lambda_{mn} r) \cos (m\theta)\, r\, dr\, d\theta}{\pi \int_0^{\rho} [J_m (\lambda_{mn} r)]^2\, r\, dr}$$

$$b_{mn} = \frac{\int_0^{2\pi} \int_0^{\rho} f(r,\theta) J_m (\lambda_{mn} r) \sin (m\theta)\, r\, dr\, d\theta}{\pi \int_0^{\rho} [J_m (\lambda_{mn} r)]^2\, r\, dr} \qquad A_{0n} = \frac{\int_0^{2\pi} \int_0^{\rho} g(r,\theta) J_0 (\lambda_{0n} r)\, r\, dr\, d\theta}{2\pi \lambda_{0n} c\pi \int_0^{\rho} [J_0 (\lambda_{0n} r)]^2\, r\, dr}$$

$$A_{mn} = \frac{\int_0^{2\pi} \int_0^{\rho} g(r,\theta) J_m (\lambda_{mn} r) \cos (m\theta)\, r\, dr\, d\theta}{\pi \lambda_{mn} c \int_0^{\rho} [J_m (\lambda_{mn} r)]^2\, r\, dr} \qquad B_{mn} = \frac{\int_0^{2\pi} \int_0^{\rho} g(r,\theta) J_m (\lambda_{mn} r) \sin (m\theta)\, r\, dr\, d\theta}{\pi \lambda_{mn} c \int_0^{\rho} [J_m (\lambda_{mn} r)]^2\, r\, dr}$$

EXAMPLE 6.5.3: Solve

$$\begin{cases} 10^2 \left(\dfrac{\partial^2 u}{\partial r^2} + \dfrac{1}{r} \dfrac{\partial u}{\partial r} + \dfrac{1}{r^2} \dfrac{\partial^2 u}{\partial \theta^2} \right) = \dfrac{\partial^2 u}{\partial t^2}, \ 0 < r < 1, \\ \quad -\pi < \theta < \pi, \ t > 0 \\ u(1,\theta,t) = 0, \ |u(0,\theta,t)| < \infty, \ -\pi \le \theta \le \pi, \ t > 0 \\ u(r,\pi,t) = u(r,-\pi,t), \ \dfrac{\partial u}{\partial \theta}(r,\pi,t) = \dfrac{\partial u}{\partial \theta}(r,-\pi,t), \ \ldots, \\ \quad 0 < r < 1, \ t > 0 \\ u(r,\theta,0) = \cos(\pi r/2) \sin \theta, \\ \quad \dfrac{\partial u}{\partial t}(r,\pi,0) = (r-1) \cos(\pi\theta/2), \ 0 < r < 1, \\ \quad -\pi < \theta < \pi \end{cases}$$

SOLUTION: To calculate the coefficients, we will need to have approximations of the zeros of the Bessel functions, so we re-enter the table of zeros that were found earlier in Chapter 4. A function alpha is then defined so that these zeros of the Bessel functions can more easily be obtained from the list.

```
> ALPHA:=array([
> [2.4048, 5.5201, 8.6537, 11.792,
> 14.931, 18.071, 21.212, 24.352],
> [3.8317, 7.0156, 10.173, 13.324,
> 16.471, 19.616, 22.760, 25.904],
> [5.1356, 8.4172, 11.620, 14.796,
> 17.960, 21.117, 24.270, 27.421],
> [6.3802, 9.7610, 13.015, 16.223,
> 19.409, 22.583, 25.748, 28.908],
> [7.5883, 11.065, 14.373, 17.616,
> 20.827, 24.019, 27.199, 30.371],
> [8.7715, 12.339, 15.700, 18.980,
> 22.218, 25.430, 28.627, 31.812],
> [9.9361, 13.589, 17.004, 20.321,
> 23.586, 26.820, 30.034, 33.233]]):
```

```
> alpha:=table():
> for i from 0 to 6 do
> for j from 1 to 8 do
> alpha[i,j]:=ALPHA[i+1,j] od od:
```

The appropriate parameter values as well as the initial condition functions are defined as follows. Notice that the functions describing the initial displacement and velocity are defined as the product of functions. This enables the subsequent calculations to be carried out using evalf and Int.

```
> c:=10:
> rho:=1:
```

```
> f:='f':
> f1:=r->cos(Pi*r/2):
> f2:=theta->sin(theta):
> f:=proc(r,theta) option remember;
> f1(r)*f2(theta)
> end:
```

```
> g1:=r->r-1:
> g2:=theta->cos(Pi*theta/2):
> g:=proc(r,theta) option remember;
> g1(r)*g2(theta)
> end:
```

The coefficients a_{0n} are determined with the function a0.

```
> a0:=proc(n) option remember;
> evalf(Int(f1(r)*BesselJ(0,alpha[0,n]*r)*r,
  r=0..rho)*Int(f2(t),
  t=0..2*Pi)/(2*Pi*Int(r*BesselJ(0,alpha[0,n]*r)^2,
  r=0..rho)))
> end:
```

We use seq to calculate the first five values of a_{0n}.

```
> seq(a0(n),n=1..5):
```

Because the denominator of each integral formula used to find a_{mn} and b_{mn} is the same, the function bjmn which computes this value is defined next. A table of nine values of this coefficient is then determined.

```
> bjmn:=proc(m,n) option remember;
> evalf(Int(r*BesselJ(m,alpha[m,n]*r)^2,r=0..rho))
> end:
> seq(seq(bjmn(m,n),m=1..3),n=1..3):
```

We also note that in evaluating the numerators of a_{mn} and b_{mn} we must compute $\int_0^\rho rf_1(r)J_m(\alpha_{mn}r)\,dr$. This integral is defined in fbjmn and the corresponding values are found for $n = 1, 2, 3$ and $m = 1, 2, 3$.

```
> fbjmn:=proc(m,n) option remember;
> evalf(Int(f1(r)*BesselJ(m,alpha[m,n]*r)*r,r=0..rho))
> end:
> seq(seq(fbjmn(m,n),m=1..3),n=1..3):
```

The formula to compute a_{mn} is then defined and uses the information calculated in fbjmn and bjmn. As in the previous calculation, the coefficient values for $n = 1, 2, 3$ and $m = 1, 2, 3$ are determined.

```
> a:=proc(m,n) option remember;
> evalf(fbjmn(m,n)*Int(f2(t)*cos(m*t),
  t=0..2*Pi)/(Pi*bjmn(m,n)))
> end:
> seq(seq(a(m,n),m=1..3),n=1..3):
```

A similar formula is then defined for the computation of b_{mn}.

```
> b:=proc(m,n) option remember;
> evalf(fbjmn(m,n)*Int(f2(t)*sin(m*t),
  t=0..2*Pi)/(Pi*bjmn(m,n)))
```

```
> end:
> seq(seq(b(m,n),m=1..3),n=1..3):
```

Note that we define the coefficients using proc with the remember option so that Maple "remembers" previously computed values, reducing computation time. The values of A_{0n} are found similar to those of a_{0n}. After defining the function capa to calculate these coefficients, a table of values is then found.

```
> capa0:=proc(n) option remember;
> evalf(Int(g1(r)*BesselJ(0,alpha[0,n]*r)*r,
   r=0..rho)*Int(g2(t),t=0..2*Pi)/(2*Pi*c*alpha[0,n]
   *Int(r*BesselJ(0,alpha[0,n]*r)^2,r=0..rho)))
> end:

> seq(capa0(n),n=1..6):
```

The value of the integral of the component of g, g1, which depends on r and the appropriate Bessel functions, is defined as gbjmn.

```
> gbjmn:=proc(m,n) option remember;
> evalf(Int(g1(r)*BesselJ(m,alpha[m,n]*r)*r,r=0..rho))
> end:

> seq(seq(gbjmn(m,n),m=1..3),n=1..3):
```

Then, A_{mn} is found by taking the product of integrals, gbjmn depending on r and one depending on θ. A table of coefficient values is generated in this case as well.

```
> capa:='capa':

> capa:=proc(m,n) option remember;
> evalf(gbjmn(m,n)*Int(g2(t)*cos(m*t),t=0..2*Pi)
   /(Pi*alpha[m,n]*c*bjmn(m,n)))
> end:

> seq(seq(capa(m,n),m=1..3),n=1..3):
```

Similarly, the B_{mn} are determined.

```
> capb:=proc(m,n) option remember;
> evalf(gbjmn(m,n)*Int(g2(t)*sin(m*t),t=0..2*Pi)
   /(Pi*alpha[m,n]*c*bjmn(m,n)))
> end:

> seq(seq(capb(m,n),m=1..3),n=1..3):
```

Now that the necessary coefficients have been found, we construct an approximate solution to the wave equation by using our results. In the following, `term1` represents those terms of the expansion involving a_{0n}, `term2` those terms involving a_{mn}, `term3` those involving b_{mn}, `term4` those involving A_{0n}, `term5` those involving A_{mn}, and `term6` those involving B_{mn}.

```
> term1:=sum('a0(n)*BesselJ(0,alpha[0,n]*r)
  *cos(alpha[0,n]*c*t)',n=1..5);

> n:='n':m:='m':
> term2:=sum('sum('a(m,n)*BesselJ(m,alpha[m,n]*r)
  *cos(m*theta)*cos(alpha[m,n]*c*t)',
> n=1..3)',m=1..3):

> n:='n':m:='m':
> term3:=sum('sum('b(m,n)*BesselJ(m,alpha[m,n]*r)
  *sin(m*theta)*cos(alpha[m,n]*c*t)',
> n=1..3)',m=1..3):

> n:='n':
> term4:=sum('capa0(n)*BesselJ(0,alpha[0,n]*r)
  *sin(alpha[0,n]*c*t)',n=1..5):

> n:='n':m:='m':

> term5:=sum('sum('capa(m,n)*BesselJ(m,alpha[m,n]*r)
  *cos(m*theta)*sin(alpha[m,n]
  *c*t)',
> n=1..3)',m=1..3):

> n:='n':m:='m':
> term6:=sum('sum('capb(m,n)*BesselJ(m,alpha[m,n]*r)
  *sin(m*theta)*sin(alpha[m,n]
  *c*t)',
> n=1..3)',m=1..3):
```

Therefore, our approximate solution is given as the sum of these terms as computed in `u`.

```
> u:=term3+term4+term5+term6:
```

A table of nine plots for nine equally spaced values of t from $t = 0$ to $t = 1$ using increments of $1/8$ is then generated with `animate`. This table of graphs is displayed as a graphics array in Figure 6-47.

```
> with(plots):
```

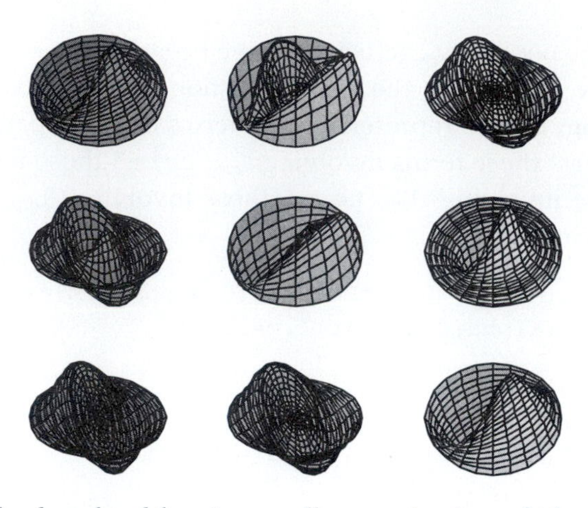

Figure 6-47 The drumhead for nine equally spaced values of t from $t = 0$ to $t = 1$

```
> somegraphs:=animate3d([r*cos(theta),r*sin(theta),u],
   r=0..1,
> theta=-Pi..Pi,t=0..1,frames=9):
> display(somegraphs);
```

Of course, we can generate many graphs with a Do loop and animate the result as in the previous example. Be aware, however, that generating many three-dimensional graphics and then animating the results uses a great deal of memory and can take considerable time, even on a relatively powerful computer.

■

6.5.3 Other Partial Differential Equations

A partial differential equation of the form

$$a(x,y,u)\frac{\partial u}{\partial x} + b(x,y,u)\frac{\partial u}{\partial y} = c(x,y,u) \tag{6.44}$$

is called a **first-order, quasi-linear partial differential equation**. In the case when $c(x,y,u) = 0$, (6.44) is **homogeneous**; if a and b are independent of u, (6.44) is **almost linear**; and when $c(x,y,u)$ can be written in the form $c(x,y,u) = d(x,y)u + s(x,y)$, (6.44) is **linear**. Quasi-linear partial differential equations can frequently be solved using the *method of characteristics*.

EXAMPLE 6.5.4: Use the *method of characteristics* to solve the initial-value problem $\begin{cases} -3xtu_x + u_t = xt \\ u(x,o) = x \end{cases}$.

SOLUTION: Note that `pdsolve` can find a general solution to the equation, but not solve the initial-value problem.

```
> with(PDEtools):
> u:='u':x:='x':t:='t':
> pdsolve(-3*x*t*diff(u(x,t),x)+diff(u(x,t),t)=x*t);
```

$$u(x,t) = -1/3\,x + \_F1\left(2/3\ln(x) + t^2\right)$$

For this problem, the *characteristic system* is

$$\partial x/\partial r = -3xt, \qquad x(0,s) = s$$
$$\partial t/\partial r = 1, \qquad t(0,s) = 0$$
$$\partial u/\partial r = xt, \qquad u(0,s) = s$$

We begin by using `dsolve` to solve $\partial t/\partial r = 1$, $t(0,s) = 0$

```
> d1:=dsolve(diff(t(r),r)=1,t(0)=0,t(r));
```

$$d1 := t(r) = r$$

and obtain $t = r$. Thus, $\partial x/\partial r = -3xr$, $x(0,s) = s$, which we solve next

```
> d2:=dsolve(diff(x(r),r)=-3*x(r)*r,x(0)=s,x(r));
```

$$d2 := x(r) = se^{-3/2\,r^2}$$

and obtain $x = se^{-3r^2/2}$. Substituting $r = t$ and $x = se^{-3r^2/2}$ into $\partial u/\partial r = xt$, $u(0,s) = s$ and using `dsolve` to solve the resulting equation yields the following result, named `d3`.

```
> u:='u':
> d3:=dsolve(diff(u(r),r)=exp(-3/2*r^2)*s*r,
> u(0)=s,u(r));
```

$$d3 := u(r) = -1/3\,se^{-3/2\,r^2} + 4/3\,s$$

To find $u(x,t)$, we must solve the system of equations

$$\begin{cases} t = r \\ x = se^{-3r^2/2} \end{cases}$$

for r and s. Substituting $r = t$ into $x = se^{-3r^2/2}$ and solving for s yields $s = xe^{3t^2}/2$.

```
> vals:=solve(x=exp(-3/2*r^2)*s,t=r,r,s);
```

$$vals := \left\{ r = t, s = \frac{x}{e^{-3/2\,t^2}} \right\}$$

Thus, the solution is given by replacing the values obtained above in the solution obtained in `d3`. We do this below by using `assign` to assign r and s the values in `vals` and `assign` to assign $u(r)$ the value obtained in `d3`. We then evaluate $u(r)$. The resulting output represents the solution to the initial-value problem.

```
> assign(vals):
> assign(d3):
> u(r);
```

$$-1/3\,x + 4/3\,\frac{x}{e^{-3/2\,t^2}}$$

Finally, we verify that this result is the solution to the problem.

```
> simplify(-3*x*t*diff(u(r),x)+diff(u(r),t));
```

$$xt$$

The initial condition $u(x, 0) = x$ has parametrization $\begin{cases} x = s \\ t = 0 \\ u = s \end{cases}$. We use `PDEplot`, which is contained in the `PDEtools` package, to graph the solution for $0 \le s \le 15$ in Figure 6-48. With the first command, we graph

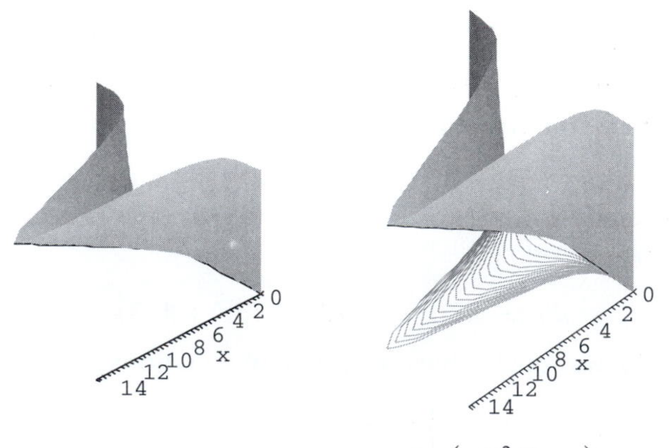

Figure 6-48 Plot of $u(x, t) = \frac{1}{3}x\left(4e^{3t^2/2} - 1\right)$

the solution drawing 20 characteristics (the default is 10) and in the second command, we draw 20 characteristics in addition to including the option `basechar=true` so that the base characteristics are also displayed in the resulting graph.

```
> u:='u':
> s:='s':
> PDEplot(-3*x*t*diff(u(x,t),x)+diff(u(x,t),t)
   =x*t,[s,0,s],s=0..15,
> numchar=20);
> PDEplot(-3*x*t*diff(u(x,t),x)+diff(u(x,t),t)
   =x*t,[s,0,s],s=0..15,
> numchar=20,basechar=true);
```

Bibliography

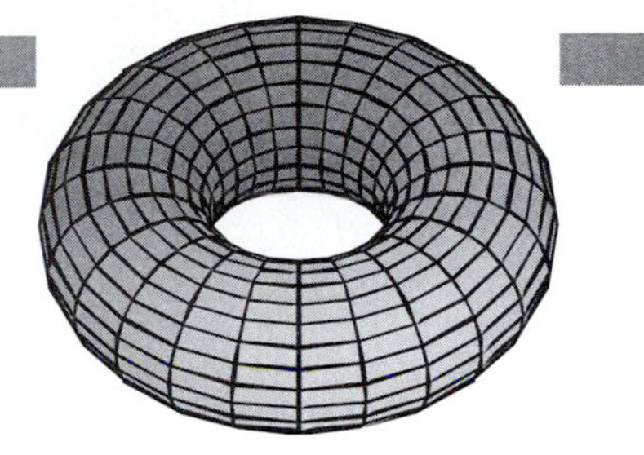

[1] Abell, Martha and Braselton, James, *Differential Equations with Maple*, Third Edition, Academic Press, 2004.

[2] Abell, Martha and Braselton, James, *Modern Differential Equations*, Second Edition, Harcourt, 2001.

[3] Abell, Martha L., Braselton, James P., and Rafter, John A., *Statistics with Maple*, Academic Press, 1999.

[4] Barnsley, Michael, *Fractals Everywhere*, Second Edition, Morgan Kaufmann, 2000.

[5] Braselton, James P., Abell, Martha L., and Braselton, Lorraine M., "When is a surface *not* orientable?", *International Journal of Mathematical Education in Science and Technology*, Volume 33, Number 4, 2002, pp. 529–541.

[6] Devaney, Robert L. and Keen, Linda (eds.), *Chaos and Fractals: The Mathematics Behind the Computer Graphics*, Proceedings of Symposia in Applied Mathematics, Volume 39, American Mathematical Society, 1989.

[7] Edwards, C. Henry and Penney, David E., *Calculus with Analytic Geometry*, Fifth Edition, Prentice Hall, 1998.

[8] Edwards, C. Henry and Penney, David E., *Differential Equations and Boundary Value Problems: Computing and Modeling*, Third Edition, Pearson/Prentice Hall, 2004.

[9] Gaylord, Richard J., Kamin, Samuel N., and Wellin, Paul R., *Introduction to Programming with Maple*, Second Edition, TELOS/Springer-Verlag, 1996.

[10] Graff, Karl F., *Wave Motion in Elastic Solids*, Oxford University Press/Dover, 1975/1991.

[11] Gray, Alfred, *Modern Differential Geometry of Curves and Surfaces*, Second Edition, CRC Press, 1997.

[12] Kyreszig, Erwin, *Advanced Engineering Mathematics*, Seventh Edition, John Wiley & Sons, 1993.

[13] Larson, Roland E., Hostetler, Robert P., and Edwards, Bruce H., *Calculus with Analytic Geometry*, Sixth Edition, Houghton Mifflin, 1998.

[14] Robinson, Clark, *Dynamical Systems: Stability, Symbolic Dynamics, and Chaos*, Second Edition, CRC Press, 1999.

[15] Smith, Hal L. and Waltman, P., *The Theory of the Chemostat: Dynamics of Microbial Competition*, Cambridge University Press, 1995.

[16] Stewart, James, *Calculus: Concepts and Contexts*, Second Edition, Brooks/Cole, 2001.

[17] Weisstein, Eric W., *CRC Concise Encyclopedia of Mathematics*, CRC Press, 1999.

[18] Wolfram, Stephen, *A New Kind of Science*, Wolfram Media, 2002.

[19] Zwillinger, Daniel, *Handbook of Differential Equations*, Second Edition, Academic Press, 1992.

Subject Index

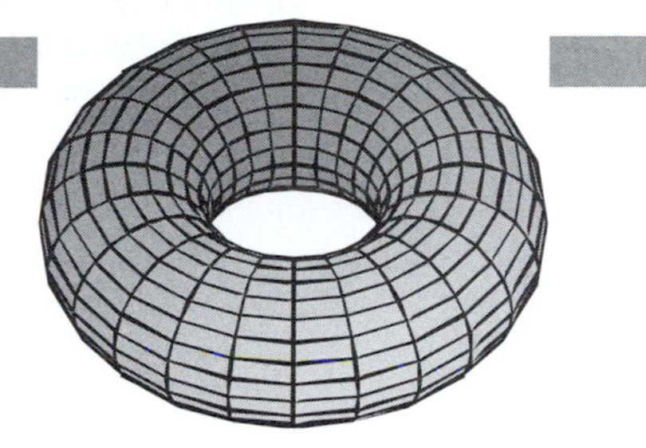